IMPLEMENTING THE ENVIRONMENTAL PROTECTION REGIME FOR THE ANTARCTIC

ENVIRONMENT & POLICY
VOLUME 28

The titles published in this series are listed at the end of this volume.

Implementing the Environmental Protection Regime for the Antarctic

Edited by
Davor Vidas

*The Fridtjof Nansen Institute,
Lysaker, Norway*

KLUWER ACADEMIC PUBLISHERS
DORDRECHT / BOSTON / LONDON

A C.I.P. Catalogue record for this book is available from the Library of Congress.

ISBN 0-7923-6609-3 (HB)

Published by Kluwer Academic Publishers,
P.O. Box 17, 3300 AA Dordrecht, The Netherlands.

Sold and distributed in North, Central and South America
by Kluwer Academic Publishers,
101 Philip Drive, Norwell, MA 02061, U.S.A.

In all other countries, sold and distributed
by Kluwer Academic Publishers,
P.O. Box 322, 3300 AH Dordrecht, The Netherlands.

Cover design is based on a painting *Mi tierra, se cae* (My Earth falls), created specially for the cover of this book by the Peruvian artist Walter Giraldo. The editor wishes to thank his good friend Walter for permission to use this beautiful painting to illustrate the theme of the book.

© Walter Giraldo (cover illustration)

Printed on acid-free paper

All Rights Reserved
© 2000 Kluwer Academic Publishers
No part of the material protected by this copyright notice may be reproduced or
utilized in any form or by any means, electronic or mechanical,
including photocopying, recording or by any information storage and
retrieval system, without written permission from the copyright owner.

Printed in the Netherlands.

MOJEM DRAGOM MALOM FEDORU ZA PRVI ROĐENDAN,

TATA

Contents

Notes on Contributors xi

Preface and Acknowledgements xv

List of Abbreviations xix

Introductory Overview

1 Entry into Force of the Environmental Protocol and Implementation Issues: An Overview 1
   Davor Vidas

Part I

ISSUES OF JURISDICTION, CONTROL AND ENFORCEMENT

2 Means and Methods of Implementation of Antarctic Environmental Regimes and National Environmental Instruments: An Exercise in Comparison 21
   William Bush

3 Port State Jurisdiction in Antarctica: A New Approach to Inspection, Control and Enforcement 45
   Francisco Orrego Vicuña

4 Regulating Tourism in the Antarctic: Issues of Environment and Jurisdiction 71
   Mike G. Richardson

Part II

INSTITUTIONAL SUPPORT TO THE IMPLEMENTATION OF THE PROTOCOL

5   Institutional Issues for the Antarctic Treaty System with the
    Protocol in Force: An Overview 93
    Jorge Berguño

6   The Committee for Environmental Protection: Its Establishment,
    Operation and Role within the Antarctic Treaty System 107
    Olav Orheim

7   Establishment of an Antarctic Treaty Secretariat: Pending Legal
    Issues 125
    Francesco Francioni

8   The ATS on the Web: Introducing Modern Information Technology
    in Antarctic Affairs 141
    Davor Vidas and *Birgit Njåstad*

Part III

NORMATIVE SUPPORT TO THE IMPLEMENTATION OF THE PROTOCOL: AN ANTARCTIC LIABILITY REGIME

9   Liability Annex or Annexes to the Environmental Protocol:
    A Review of the Process within the Antarctic Treaty System 163
    Mari Skåre

10 The Legal Need for an Antarctic Environmental Liability Regime 181
    René Lefeber

11 The Prospects for an Antarctic Environmental Liability Regime 199
    René Lefeber

Part IV

RELATIONSHIP WITH OTHER INTERNATIONAL INSTRUMENTS AND ARRANGEMENTS

12 Relationship between the Environmental Protocol and UNEP Instruments 221
    Donald R. Rothwell

13 Towards Guidelines for Antarctic Shipping: A Basis for Cooperation between the Antarctic Treaty Consultative Parties and the IMO... 243
    Tullio Scovazzi

14 The Antarctic Continental Shelf Beyond 200 Miles: A Juridical Rubik's Cube 261
    Davor Vidas

15 CCAMLR and the Environmental Protocol: Relationships and Interactions 273
    Richard A. Herr

Part V

IMPLEMENTING THE PROTOCOL DOMESTICALLY: THE CONSULTATIVE PARTIES' LEGISLATION AND PRACTICE
(selected case studies)

16 Implementing the Environmental Protocol Domestically: An Overview 287
    Kees Bastmeijer

17 Australian Implementation of the Environmental Protocol............................ 309
    William Bush

18 Implementation of the Antarctic Environmental Protocol by Chile: History, Legislation and Practice.. 337
    María Luisa Carvallo and *Paulina Julio*

19 A Self-Executing Treaty? Italian Legislation and Practice in Implementing the Environmental Protocol.. 355
    Laura Pineschi

20 Norway: Implementing the Protocol on Environmental Protection.................. 379
   Birgit Njåstad

21 South Africa: Implementing the Protocol on Environmental
   Protection.. 399
   Klaus Dodds

22 The United States: Legislation and Practice in Implementing the
   Protocol 417
   Christopher C. Joyner

   Index 439

Notes on Contributors

KEES BASTMEIJER is Senior Lecturer and Researcher on Environmental Law at the Law Faculty of the Tilburg University, the Netherlands, where he is writing a doctoral thesis on the legal implementation of the Antarctic Environmental Protocol. He has been a member of the Dutch delegation to Antarctic Treaty meetings since 1992 and he has worked on the legal incorporation of the Protocol into Dutch law.

JORGE BERGUÑO is Deputy Director of the Chilean Antarctic Institute. He has represented Chile at various international negotiations on the matters concerning the Antarctic, the oceans, environment and disarmament. He has served as Ambassador to Australia, Canada, UNESCO, GATT, and other international organisations. He has chaired the CCAMLR Commission and is currently Vice-Chairman of the Committee for Environmental Protection.

WILLIAM BUSH is an international lawyer and the former Director of the Treaties and the Antarctic Sections of the Department of Foreign Affairs and Trade in Canberra, Australia. He is the editor of *Antarctica and International Law: A Collection of Inter-State and National Documents*, published by Oceana since 1982.

MARIA LUISA CARVALLO is the Legal Adviser of the Chilean Antarctic Institute. She has been a member of the Chilean delegation to many Antarctic Treaty Consultative Meetings. She has published several articles about various aspects of the Antarctic Treaty System. She was awarded the 'Hamilton S. Amerasinghe' fellowship by the United Nations in 1988.

KLAUS DODDS is Lecturer in Geography at Royal Holloway, University of London. He is the author of *Geopolitics in Antarctica: Views from the Southern Oceanic Rim* (Wiley, 1997) and *Geopolitics in a Changing World* (Longman, 2000) and co-editor of *Geopolitical Traditions: A Century of Geopolitical Thought* (Routledge, 2000).

FRANCESCO FRANCIONI is Professor of International Law and Vice-Rector, University of Siena, Italy, and also Visiting Professor at the University of Oxford and at the School of Law of the University of Texas. He was President of the

UNESCO World Heritage Committee, 1997–98, and is the legal adviser of the Italian Ministry of Foreign Affairs on Antarctic Treaty. He is the author of many publications on Antarctic law and international environmental law.

RICHARD A. HERR is Reader in Political Science, University of Tasmania, Hobart. He is the author or editor of numerous publications on South Pacific regional affairs, Antarctic politics, marine resource policy, parliamentary democracy and electoral analysis. He has also served as a consultant to South Pacific governments on regional affairs and development.

CHRISTOPHER JOYNER is Professor of Government and International Law at Georgetown University, Washington, DC. He is the author or editor of numerous publications. His recent books on Antarctic affairs include *Governing the Frozen Commons: The Antarctic Regime and Environmental Protection* (Univ. of South Carolina Press, 1998) and *Eagle Over the Ice: US Foreign Policy and Antarctica* (with E. Theis, Univ. Press of New England, 1997).

PAULINA JULIO is a lawyer and the Head of Antarctic Department, Environmental Division of the Chilean Ministry of Foreign Affairs. She has been a member of the Chilean delegation in many Antarctic Treaty Consultative Meetings and has participated in the meetings of the Commission for the Conservation of Antarctic Marine Living Resources.

RENÉ LEFEBER is Legal Officer in the Netherlands Ministry of the Environment and a part-time lecturer on the Faculty of Law of the University of Utrecht. He has written *Transboundary Environmental Interference and the Origin of State Liability* (Kluwer Law International, 1996). He regularly contributes to the *Yearbook of International Environmental Law* on liability issues.

BIRGIT L. NJÅSTAD is Environmental Manager at the Norwegian Polar Institute. She is extensively involved in the implementation of the Antarctic environmental management regime in Norway, and for several years has also been a member of the Norwegian delegation to the Antarctic Treaty Consultative Meetings and the Committee for Environmental Protection.

OLAV ORHEIM is Managing Director of the Norwegian Polar Institute in Tromsø, and Adjunct Professor at the University of Bergen. His publications deal mainly with glaciers, the climate and other related environmental issues. He has participated in 14 research expeditions to Antarctica, and has been a member of the Norwegian delegation to the Antarctic Treaty Consultative Meetings for more than 20 years. He is Chairman of the Committee for Environmental Protection.

FRANCISCO ORREGO VICUÑA is Professor of International Law at the University of Chile, Member of the *Institut de Droit International*, Judge and Vice-President of the World Bank Administrative Tribunal, and President of the Panel of the UN Compensation Commission, among other functions. He is the author of, among other numerous publications, *Antarctic Resources Policy* (1983), and *Antarctic Mineral Exploitation* (1988), both published by Cambridge University Press.

LAURA PINESCHI is Associate Professor of the International Organisation at the Faculty of Law of the University of Parma. She is the author of various studies on the protection of the Antarctic environment, including a monograph *La protezione dell ámbiente in Antartide* (Protection of the Antarctic Environment), CEDAM, 1993.

MIKE G. RICHARDSON is Head of the Polar Regions Section of the United Kingdom Foreign and Commonwealth Office. He has represented the United Kingdom at Antarctic Treaty Consultative Meetings and CCAMLR since 1992 and was actively involved in the negotiations of the Environmental Protocol. His prior posts have included the UK's Nature Conservancy Council and the British Antarctic Survey.

DONALD R. ROTHWELL is Associate Professor at the Faculty of Law, University of Sydney. His major research interests include international environmental law, law of the sea, law of the polar regions, and the interaction of municipal and international law. His latest publication is *Navigational Rights and Freedoms and the New Law of the Sea* (Kluwer Law International, 2000), co-edited with Sam Bateman.

TULLIO SCOVAZZI is Professor of International Law at the University of Milan–Bicocca, Italy, and a legal expert for the Government of Italy to various international negotiations on the law of the sea. He is the author or editor of works on various aspects of international law, including *Lines in the Sea* (co-edited with G. Francalanci, Martinus Nijhoff, 1994) and *International Law for Antarctica* (co-edited with F. Francioni, Kluwer Law International, 1996).

MARI SKÅRE is a lawyer and Head of Division in the Norwegian Ministry of Foreign Affairs. She has served on the Norwegian delegation to several Antarctic Treaty Consultative Meetings and also served as a consultant on legal, policy and administrative issues.

DAVOR VIDAS is Director of the Polar Programme at the Fridtjof Nansen Institute. He has been a member of and adviser to the Norwegian delegation to Antarctic Treaty meetings since 1992. His books address Antarctic and Arctic affairs, the law of the sea and environmental law. His recent publications are *Protecting the Polar Marine Environment* (Cambridge University Press, 2000) and *Order for the Oceans at the Turn of the Century* (W. Østreng, co-editor, Kluwer Law International, 1999).

Preface and Acknowledgements

On 14 January 1998, the day on which the Protocol on Environmental Protection to the Antarctic Treaty entered into force, the Fridtjof Nansen Institute launched a two-year international research project entitled 'Implementing the Environmental Protection Regime for the Antarctic'. The project sought to identify and analyse the outstanding implementation issues for the Environmental Protocol. This book is the final outcome of that project, which was directed by the editor of this book and in which the contributors to the book took part.

The project was a continuation of Fridtjof Nansen Institute's long-standing involvement in research on Antarctic affairs. The milestones in the Institute's activities in this respect have often corresponded with those of the Antarctic Treaty System itself. In 1973, when the issue of Antarctic minerals arose, an informal meeting of experts from the Antarctic Treaty countries was held under the auspices of the Institute to discuss problems related to possible exploration and exploitation of mineral resources in the Antarctic; this meeting provided one of the first occasions for the exchange of views on these matters.[1] In May 1990, soon after the decision of the Antarctic Treaty parties to convene a special Consultative Meeting to negotiate a regime for the comprehensive protection of the Antarctic environment, the Institute arranged a major international conference in Oslo on 'The Antarctic Treaty System in World Politics'. The conference, which was attended by some 150 participants from 28 countries from all the (inhabited) continents, was one of the first open and informal fora discussing the emerging environmental protection regime for the Antarctic.[2] Formal negotiations for the Environmental Protocol commenced several months later and resulted in the adoption of the Protocol in October 1991. Already on 1 January 1992, the Institute followed up the adoption of the Protocol by launching 'The International Antarctic Regime Project' (IARP), a three-year international research project involving experts from four continents. The IARP analysed the effectiveness and legitimacy of the Antarctic Treaty System as

---

[1] Outcomes of the meeting are reported in *Antarctic Resources: Report from the Informal Meeting of Experts at the Fridtjof Nansen Institute, 30 May – 9 June 1973* (Lysaker: Fridtjof Nansen Institute, 1974).

[2] For the conference proceedings see A. Jørgensen-Dahl and W. Østreng (eds.), *The Antarctic Treaty System in World Politics* (London: Macmillan, 1991).

supplemented by the Protocol, and it thus addressed the interplay of political and legal challenges and adaptations in Antarctic affairs.[3] An important feature of the IARP was the continuous *interaction* between the project research team and a large number of decision-makers in Antarctic affairs.

In the project on 'Implementing the Environmental Protection Regime for the Antarctic', we decided to change the model somewhat. In addition to interacting with Antarctic decision-makers, the research team itself was *composed* of both researchers and decision-makers. It should be stated here that the views expressed in this book of officials who participated in the project are views expressed in their own personal capacity and should not be attributed to any institution or governmental body. This also applies to all other publications that have resulted from this project.

In the course of the project, the team met once a year for project workshops, where various drafts for the chapters of this book were discussed. Special acknowledgement is given here to participants in the project who have made contributions to this book and also to the institutions that provided venues for the workshops: in November 1998, the Inter-University Centre in Dubrovnik, Croatia; and in September 1999, the Fridtjof Nansen Institute at Polhøgda near Oslo, Norway. In conjunction with these arrangements, I especially wish to extend my gratitude to several persons who provided organisational help: for the 1998 workshop in Dubrovnik, *hvala* to Berta Dragičević, Srećko Kržić and Dubravka Kapetanić; and for the 1999 workshop at Polhøgda, *takk* to Erling Hagen, Jan Magne Markussen, Maryanne Rygg and Hans Håkon Skjønsberg.

Special appreciation is also due to the publisher, in particular Henny Hoogervorst, Environmental Sciences publisher at the Kluwer Academic Publishers, for her professional assistance in various phases that led toward the production of this book – and not least for her patience and understanding in the final phase of manuscript completion.

My particular thanks go to Mary Hustad for careful and thorough language editing of the manuscript.

I wish to acknowledge the invaluable support received from my colleagues at the Fridtjof Nansen Institute, in particular Maryanne Rygg for her conscientious copy-editing and formatting of the manuscript, and for her highly professional assistance in various phases of the manuscript preparation and completion; Kari Lorentzen and Øystein B. Thommessen for bibliographical and documentary assistance; and Erling Hagen for various technical assistance.

The Peruvian artist, Walter Giraldo, receives my gratitude and applaud for creating specially for the cover of this book a painting entitled *Mi tierra, se cae* (My Earth falls) – an artistic vision of a cataclysm that could threaten all of us and the

---

[3] Final results of the project are published in O.S. Stokke and D. Vidas (eds.), *Governing the Antarctic: The Effectiveness and Legitimacy of the Antarctic Treaty System* (Cambridge University Press, 1996).

planet we live on if we do not behave in a responsible way towards the natural environment. Let us hope that through this book we have contributed in some small way toward preventing Walter's vision, as shown on the cover, from becoming a reality.

Funding for the project and for the editing of this book was provided by the Tinker Foundation, New York, the Royal Norwegian Ministry of Foreign Affairs, the Royal Norwegian Ministry of Justice, the Research Council of Norway and the Fridtjof Nansen Institute. The support of all of the aforesaid is gratefully acknowledged. I especially wish to acknowledge the engagement of the President of the Tinker Foundation, Renate Rennie, who took part in the final workshop of the project and provided invaluable support and encouragement in various phases of the project.

Finally, let me add that none of the above-mentioned individuals or organisations is responsible for any shortcomings of this book. Although the responsibility for the content of each chapter lies with the individual author, the editor alone is answerable for the book as such. If the reader should find satisfaction and enrichment in the pages that follow, then of course all those concerned deserve a measure of the credit.

Unless a more recent date has been specified at any particular place in the book, the book is up to date as of 29 December 1999.

Davor Vidas
Oslo, 22 May 2000

List of Abbreviations

AAT	Australian Antarctic Territory
ACA	(United States) Antarctic Conservation Act
AEON	Antarctic Environmental Officers Network
AER	(Norwegian) Regulations Relating to Protection of the Environment in Antarctica
AFMA	Australian Fisheries Management Authority
AMC	(South African) Antarctic Management Committee
ANARE	Australian National Antarctic Research Expedition
ANC	African National Congress
APPS	(United States) Act to Prevent Pollution from Ships
ASMAs	Antarctic Specially Managed Areas
ASOC	Antarctic and Southern Ocean Coalition
ASTCA	(United States) Antarctic Science, Tourism and Conservation Act
ASTIs	Areas of Special Tourist Interest
ATCM	Antarctic Treaty Consultative Meeting
ATEP Act	(Australian) Antarctic Treaty (Environment Protection) Act
ATS	Antarctic Treaty System
Basel Convention	Convention on the Control of Transboundary Movements of Hazardous Wastes and Their Disposal
CBD	Convention on Biological Diversity
CCAMLR	Convention on the Conservation of Antarctic Marine Living Resources

CEE	comprehensive environmental evaluation
CEMP	CCAMLR Ecosystem Monitoring Programme
CEP	Committee for Environmental Protection
CEQ	(United States) Council on Environmental Quality
CFCs	chlorofluorocarbons
CFR	(United States) Code of Federal Regulations
CIMAR	(Chilean) Maritime Training Centre
CITES	Convention on International Trade in Endangered Species
CNIA	Chilean SCAR Committee
COMNAP	Council of Managers of National Antarctic Programmes
CONAEIA	(Chilean) National Committee for Environmental Impact Assessments
CONAMA	(Chilean) National Environmental Commission
CPA	(Chilean) Council for Antarctic Policy
CRAMRA	Convention on the Regulation of Antarctic Mineral Resource Activities
DEA&T	(South African) Department of Environmental Affairs and Tourism
DIRECTEMAR	(Chilean) Maritime Territory and Merchant Marine Directorate
EC	European Community
EEZ	exclusive economic zone
EHSMS	Environmental, Health and Safety Management System
EIA	environmental impact assessment
ENEA	(Italian) National Agency for Atomic Energy
EO	Eighth Offering
EPA	(United States) Environmental Protection Agency
FAO	Food and Agriculture Organisation
FCCC	United Nations Framework Convention on Climate Change
FNI	Fridtjof Nansen Institute

GATT	General Agreement on Tariffs and Trade
GOSEAC	Group of Specialists on Environmental Affairs and Conservation
GU	*Gazzetta Ufficiale della Repubblica Italiana*
IAATO	International Association of Antarctic Tour Operators
IACS	International Association of Classification Societies
IAEA	International Atomic Energy Agency
ICAIR	International Centre for Antarctic Information and Research
ICAO	International Civil Aviation Organisation
ICJ	International Court of Justice
IDAT	International Database on Antarctic Tourism
IEE	initial environmental evaluation
IEM	integrated environmental management
IGY	International Geophysical Year
IHO	International Hydrographic Organisation
ILM	*International Legal Materials*
IMO	International Maritime Organisation
INACH	Instituto Antartico Chileno (Chilean Antarctic Institute)
INSROP	International Northern Sea Route Programme
ISBA	International Seabed Authority
IUCN	World Conservation Union
IUU fishing	illegal, unregulated and unreported fishing
LOS Convention	United Nations Convention on the Law of the Sea
MARPOL 73/78	International Convention for the Prevention of Pollution from Ships (1973), as modified by the Protocol of 1978
MEPC	Marine Environment Protection Committee (of the IMO)
MPAs	Multiple-use Planning Areas
MSC	Maritime Safety Committee (of the IMO)
NARE	Norwegian Antarctic Research Expedition

NEPA	(United States) National Environmental Policy Act
NGO	non-governmental organisation
NOAA	(United States) National Oceans and Atmospheric Administration
NP	Norsk Polarinstitutt (Norwegian Polar Institute)
NPWD	(South African) National Public Works Department
NSF	(United States) National Science Foundation
Ot.prp.	Odelstingsproposisjon (Norwegian Government Proposition to the Odelsting)
P&I Club	Protection and Indemnity Club
PA	preliminary assessment
PEIMC	Prince Edward Islands Management Committee
PEIMP	Prince Edward Islands Management Plan
PEMD	Polar Environmental Management Department (of the Norwegian Polar Institute)
PERM	Preliminary Environmental Review Memorandum
PNRA	Italian National Antarctic Research Programme
Polar Code	(Draft) International Code of Safety for Ships in Polar Waters
Protocol	Protocol on Environmental Protection to the Antarctic Treaty
PS(PPS) Act	(Australian) Protection of the Sea (Prevention of Pollution from Ships) Act
RDP	(South African) Reconstruction and Development Programme
SACAR	South African Committee for Antarctic Research
SANAE	South African National Antarctic Expedition
SANAP	South African National Antarctic Programme
SATO	South Atlantic Treaty Organisation
SCAR	Scientific Committee on Antarctic Research
Seals Convention	Convention for the Conservation of Antarctic Seals

SOLAS	International Convention for the Safety of Life at Sea
STCW	International Convention on Standards of Training, Certification, and Watchkeeping for Seafarers
TEWG	Transitional Environmental Working Group
UK	United Kingdom
UN	United Nations
UNCED	United Nations Conference on Environment and Development
UNCLOS III	Third United Nations Conference on the Law of the Sea
UNEP	United Nations Environment Programme
UNESCO	United Nations Educational, Scientific and Cultural Organisation
UNTS	*United Nations Treaty Series*
US	United States
USAP	United States Antarctic Program
USC	United States Code
USD	United States dollars
Vienna Convention	Convention on the Law of Treaties
WMO	World Meteorological Organisation
WTO	World Trade Organisation
WWW	World Wide Web

1

Entry into Force of the Environmental Protocol and Implementation Issues: An Overview

Davor Vidas

When the Protocol on Environmental Protection to the Antarctic Treaty[1] entered into force on 14 January 1998, a new phase commenced for the Antarctic Treaty System (ATS): the phase of *implementation* of a complex international environmental protection regime. The parties to the Protocol – which in practice predominantly coincide with the group of the Antarctic Treaty Consultative Parties[2] are today confronting questions of quite a different nature than those that were discussed when the Protocol was under negotiation, or even after it was adopted.

With the entry into force of the Protocol, *pending issues* related to the implementation of the Protocol, both in the international and domestic contexts were immediately opened. Consequently, a sense of urgency has developed in addressing these implementation issues more thoroughly and more systematically, with a demand for investing a serious policy input. What are the main implementation issues that need to be solved in order to enable and enhance the implementation of the Protocol? Which premises can be offered for their solution? What could be the consequences for the Consultative Parties of a possible failure in resolving the pending implementation issues? Which are, then, the options to chose from? These questions have thus far not been addressed in a systematic manner in the existing

---

[1] Text reprinted in ILM, Vol. 30, 1991, pp. 1,416ff.
[2] As of 29 December 1999, there were 28 parties to the Protocol, of which: 26 Consultative Parties which were original signatories of the Protocol, one additional Consultative Party (Bulgaria) whose consultative status was acknowledged at the Twenty-second Consultative Meeting in 1998, and only one among 17 non-Consultative Parties (Greece). This status of participation remained unchanged as of 2 April 2000. At the Twenty-third Consultative Meeting (Lima, 1999), Resolution 6 (1999) was adopted, urging those non-Consultative Parties which have not yet become parties to the Protocol (particularly those with Antarctic tourist activities organised in their territory) to adhere to the Protocol as soon as possible; see *Final Report from the Twenty-third Antarctic Treaty Consultative Meeting, Lima, Peru, 24 May – 4 June 1999* (Lima: Peruvian Ministry of Foreign Affairs, 1999), Annex C.

literature on Antarctic affairs, and the aim of this book is to provide such a systematic overview of implementation issues for the Environmental Protocol.

THE PROTOCOL IN A NUTSHELL: A POLITICAL PERSPECTIVE

In order to understand why the Protocol was adopted at all, one must be aware of the political context which set the stage for the negotiation of the Protocol. Also, the heritage of that context will be manifested in the current phase of the implementation of the Protocol. In providing this background, it is therefore necessary to begin with at least a sketch of the political circumstances under which the Protocol was negotiated and review how, in turn, the adoption of the Protocol has impacted that context.

The entire decade of the 1980s, which eventually led to the adoption of the Protocol in October 1991, was for the Consultative Parties a phase of *negotiation*. Concurrently, various types of 'external pressure' were being exerted on the Antarctic Treaty System, be these from the debate in the UN General Assembly on the 'Question of Antarctica', from environmental organisations or even from the domestic public in various countries.

The long-standing record of issue-specific approaches to Antarctic environmental protection, introduced to the ATS through recommendations and then through the 1972 Convention for the Conservation of Antarctic Seals[3] (Seals Convention) and the 1980 Convention on the Conservation of Antarctic Marine Living Resources[4] (CCAMLR), culminated in the adoption in June 1988 of the Convention on the Regulation of Antarctic Mineral Resource Activities[5] (CRAMRA), which was opened for signature in November the same year. Soon afterward, the 'CRAMRA crisis' shook the ATS: in the course of the spring of 1989, Australia and France announced that they would not sign CRAMRA, and an attitude of unwillingness to either sign or ratify[6] CRAMRA soon was adopted by several other Consultative Parties. Instead of proceeding with signing or ratifying CRAMRA, these countries proposed a new instrument that would ban all mineral

---

[3] Published in UNTS, Vol. 1080, pp. 175ff; text reprinted in ILM, Vol. 11, 1972, pp. 251ff.

[4] Text reprinted in ILM, Vol. 19, 1980, pp. 837ff.

[5] Text reprinted in ILM, Vol. 27, 1988, pp. 868ff. For an overview, see C.C. Joyner, 'The Effectiveness of CRAMRA', in O.S. Stokke and D. Vidas (eds.), *Governing the Antarctic: The Effectiveness and Legitimacy of the Antarctic Treaty System* (Cambridge University Press, 1996), pp. 152–162. See also F. Orrego Vicuña, *Antarctic Mineral Exploitation: The Emerging Legal Framework* (Cambridge University Press, 1988); and R. Wolfrum, *The Convention on the Regulation of Antarctic Mineral Resource Activities: An Attempt to Break New Ground* (Berlin: Springer–Verlag, 1991).

[6] When also New Zealand, who had already signed CRAMRA, followed with an announcement that it did not intend to ratify the Convention, it became clear that CRAMRA had no prospects of entry into force because CRAMRA in effect required ratification of *all* the claimant countries as a prerequisite for the entry into force; see Art. 62(1) in conjunction with provisions on the establishment of and membership in the institutions under CRAMRA.

activity (except of scientific research) in the Antarctic and that would introduce a comprehensive environmental protection system. Following a decision of the Fifteenth Antarctic Treaty Consultative Meeting, held in Paris in September 1989,[7] a Special Consultative Meeting was convened in order to negotiate a new environmental protection instrument. The negotiations were conducted in haste, and, in the record space of less than a year of these negotiations, a new legal instrument – the Protocol on Environmental Protection to the Antarctic Treaty – was adopted.[8]

The Consultative Parties, who initially may have appeared to be urgently negotiating and adopting the Protocol in preventing, or responding to, the threats to the Antarctic environment, were in fact reacting to two acute *political* problems. The first was the challenge to the Consultative Parties' legitimacy in governing the Antarctic from subjects external to the ATS. The second, and equally important problem was the struggle to maintain internal cohesion and balance within the System, especially regarding the sovereignty issue.

The Consultative Parties had thus had substantial incentives – in themselves not always directly or exclusively related to environmental protection – that prompted the Parties to agree expeditiously on issues relating to human activities and environmental protection in the Antarctic. The reason that the Consultative Parties made a new start after the abandonment of CRAMRA in 1989 was *not* because CRAMRA had contained insufficient environmental safeguards – these were in fact stringent[9] – but because of a complex combination of economic and political factors. Aside from the awareness that, for the foreseeable future, any mineral activities in the Antarctic would lack commercial significance, the major factors involved were 1) fears that CRAMRA would disturb the sensitive balance of sovereignty in the Antarctic, 2) political-ideological critique of the Consultative Parties, from a group of developing countries in the UN General Assembly, 3) pressures from environmental NGOs, and 4) domestic policy considerations which related to the above factors.

The very fact that the rules for environmental protection were adopted so rapidly in the most comprehensive ATS instrument in this field may be seen as

---

[7] See Recommendation XV–1 (1989); text reprinted in J.A. Heap (ed.), *Handbook of the Antarctic Treaty System*, 8th edition (Washington, DC: United States Department of State, 1994), pp. 2,005–2,007.

[8] The time-span used for negotiating the Protocol extended from opening of the Eleventh Special Consultative Meeting in November 1990 to the final outcome of that meeting being adopted in October 1991; see paras. 2, 3, 8 and 9 of the Final Report of the Eleventh Special Antarctic Treaty Consultative Meeting; text reprinted in Heap (ed.), *Handbook of the Antarctic Treaty System*, pp. 2,015–2,016.

[9] See W.M. Bush, 'The 1988 Wellington Convention: How Much Environmental Protection?', in J. Verhoeven, P. Sands and M. Bruce (eds.), *The Antarctic Environment and International Law* (London: Graham & Trotman, 1992), pp. 69–83; F. Orrego Vicuña, 'The Effectiveness of the Protocol on Environmental Protection to the Antarctic Treaty', in Stokke and Vidas (eds.), *Governing the Antarctic*, pp. 197–198; Sir Arthur Watts, *International Law and the Antarctic Treaty System* (Cambridge: Grotius Publications, 1992), p. 276; and Wolfrum, *The Convention on the Regulation of Antarctic Mineral Resource Activities*.

largely the outcome of the efforts made by the Parties to find an urgent solution to the 'CRAMRA crisis' within the System. The Protocol addressed environmental protection through two essentially different approaches: the blanket *prohibition* of mining – the one activity regulated under CRAMRA, and the detailed *regulation* of all other activities in the Antarctic (except activities already regulated under CCAMLR, the Seals Convention and the International Convention for the Regulation of Whaling).[10]

Article 7 of the Protocol states unambiguously that '[a]ny activity relating to mineral resources, other than scientific research, shall be prohibited'. This single provision is basically a response to the many criticisms voiced against CRAMRA. Firstly, the Article rendered the sovereignty issue redundant, insofar as a 'delimitation' in relation to mineral rights was no longer required. Secondly, Article 7 neutralised the criticism from a group of developing countries, which, since 1989, had been demanding in the UN General Assembly that a ban on mineral activities be introduced in the Antarctic. Thirdly, the provision allowed the Consultative Parties to present themselves as environmentally highly conscious, more so than anywhere else on the globe, thereby satisfying many of the demands for which environmental NGOs had campaigned. This latter point was instrumental for several of the Consultative Parties in dealing with domestic policy concerns.

In this assessment of the political context that led to the adoption of the Protocol, we see that the Antarctic Treaty Consultative Parties have adopted a legally binding instrument for the protection of the Antarctic environment, but that the main incentives for their doing this so quickly were primarily of a political rather than an environmental protection nature. The Environmental Protocol was, in a political sense, effective immediately upon adoption and as such has continued to be a success. Indeed, the Protocol has in many ways significantly contributed to both the strengthening of international cooperation within the ATS and to a change of perception about the ATS in the broader international community.[11] Soon after the adoption of the Protocol, the external criticism of the System was substantially weakened and eventually almost vanished. Currently, however, with the Protocol in force and the changed political context relating to Antarctic affairs, the real test for the Protocol will remain to be its implementation as an *environmental* protection and management tool.

THE PROTOCOL IN A NUTSHELL: A LEGAL PERSPECTIVE

Although the Environmental Protocol was negotiated in haste, the content of its provisions, by and large, did not provide a new writing. To a large extent, the Protocol and its annexes evolved from a 'cut and paste' operation. A majority of the

---

[10] See para. 7 of the Final Act of the Eleventh Special Antarctic Treaty Consultative Meeting; text reprinted in Heap (ed.), *Handbook of the Antarctic Treaty System*, pp. 2,016–2,018.

[11] See D. Vidas, 'The Antarctic Treaty System in the International Community: An Overview', in Stokke and Vidas (eds.), *Governing the Antarctic*, pp. 50–58.

provisions were taken over from recommendations adopted earlier.[12] Even some of the Protocol's basic environmental principles come from CRAMRA – the very instrument that the Protocol has superseded.[13] It is not the purpose of this introductory chapter to enter into any extensive analysis or description of the provisions of the Protocol; these are available elsewhere.[14] What will be provided here is a brief outline that aims to place the Protocol in the context of the wider ATS legal framework and to point out some aspects of the Protocol that, while perhaps of less interest at the time of its adoption, need to be more carefully considered in the current phase of implementation.

While the Protocol did not bring too much fresh regulation into the ATS, it did introduce several new elements. Firstly, the Protocol approached the protection of the Antarctic environment in a *comprehensive* manner; secondly, the Protocol 'codified' the existing recommendations into a *legally binding* instrument; and thirdly, the Protocol provided for the establishment of a *new institution* within the ATS, the Committee for Environmental Protection (CEP), that became operative in 1998.

The legal position of the Protocol in the overall ATS has, in itself, also been an innovation. The Protocol *supplements* the Antarctic Treaty and neither modifies nor amends the Treaty (Article 4(1)). Moreover, consistency of the Protocol with other components of the ATS is the subject of a special provision, Article 5. As to the annexes, Article 9(1) states that 'Annexes to this Protocol shall form an integral part thereof'. Annexes I–IV, which were adopted in the 'Protocol package' in Madrid, 4 October 1991, became effective simultaneously with the entry into force of the Protocol. Annex V, however, was embodied in Recommendation XVI–10, adopted at the Sixteenth Consultative Meeting in Bonn, only some weeks after the adoption of the Protocol. This annex has as yet still not become effective; meanwhile, it has prompted a discussion among the Consultative Parties as to what exactly the requirements are for this annex to become effective, i.e., whether the depositing of an instrument of 'acceptance' of Annex V would suffice or whether it is necessary to 'approve' Recommendation XVI–10.[15] While entering into a more detailed discussion of this matter might have been an interesting exercise for demonstrating the legal labyrinth of the ATS, the fact remains that today, almost a decade after adoption, Annex V on 'Area Protection and Area Management' is still not legally binding. This situation prompted the Consultative Parties to adopt Resolution 1

---

[12] See Orrego Vicuña, 'The Effectiveness of the Protocol', pp. 190–202.

[13] See C.C. Joyner, 'The Legitimacy of CRAMRA', in Stokke and Vidas (eds.), *Governing the Antarctic*, pp. 255–267.

[14] Among many analyses of the Protocol available to date, see especially: Orrego Vicuña, 'The Effectiveness of the Protocol', pp. 174–202; F. Orrego Vicuña, 'The Legitimacy of the Protocol on Environmental Protection to the Antarctic Treaty', in Stokke and Vidas (eds.), *Governing the Antarctic*, pp. 268–293; and F. Francioni, 'The Madrid Protocol on the Protection of the Antarctic Environment', *Texas International Law Journal*, Vol. 28, 1993, pp. 47–72.

[15] See M.G. Richardson, 'The Protocol on Environmental Protection Enters Into Force', *Polar Record*,

(1998), recommending that those Parties 'which have yet to approve Recommendation XVI–10 under the procedures of Article IX(4), take steps to do so as soon as possible'.[16]

The legal status of the Protocol as a supplement to the Antarctic Treaty, on the one hand, and the Protocol's proclaimed role as an environmental protection instrument, on the other, are the source of some inherent contradictions, which will have to be dealt with while the Protocol is being implemented.

It can be roughly said that the Protocol consists of two main units, which are determined by the type of activity in the Antarctic. Mineral activities gave rise to one of the Protocol's main units, which is contained in Articles 7 and 25: a *prohibition* of any such activities (except scientific research).[17] This approach can be regarded as entirely new, the direct opposite of the approach taken under CRAMRA. However, one can also argue that no substantial difference was introduced by the Protocol's mining ban. Under CRAMRA, a consensus of Parties was required to start a mining operation – the same is sufficient to revise the Protocol's mining ban.

Under the Protocol, the remaining human activities are *regulated*, and the other main unit of the Protocol is concerned with these activities. That unit comprises all the remaining provisions of the Protocol and its annexes, which create an environmental protection regime for the Antarctic. In this regard, Article 3(1) of the Protocol formulates environmental principles and, *inter alia*, states:

> The protection of the Antarctic environment and dependent and associated ecosystems and the intrinsic value of Antarctica, including its wilderness and aesthetic values and its value as an area for the conduct of scientific research, in particular research essential to understanding the global environment, shall be fundamental considerations in the planning and conduct of all activities in the Antarctic Treaty area.

For activities *not* prohibited in the Antarctic, except for those undertaken pursuant to CCAMLR or the Seals Convention,[18] the Protocol requires an environmental impact assessment at the planning stage; this is required if the activity is determined to have either a 'minor or transitory impact' or more than a 'minor or transitory impact' on the Antarctic environment or on dependent or associated ecosystems.[19] Some major contradictions of the Protocol stem from these crucial provisions.

Firstly, an unambiguous determination of the area of application of the Protocol, as an environmental protection instrument, is made difficult by the fact that the

---

Vol. 34, 1998, pp. 147–148.

[16] See Resolution 1 (1998), 'Annex V. Protected Areas', in *Final Report of the Twenty-second Antarctic Treaty Consultative Meeting, Tromsø, Norway, 25 May – 5 June 1998* (Oslo: Royal Norwegian Ministry of Foreign Affairs, 1998), Annex C.

[17] The Protocol does contain certain other prohibitions, such as the prohibition on introducing dogs onto land or ice shelves in Antarctica (Annex II, Art. 4(2)), but it is the mining ban that has been the outstanding prohibitory feature of the Protocol.

[18] See para. 8 of the Final Act of the Eleventh Antarctic Treaty Special Consultative Meeting.

[19] Art. 8 and Annex I to the Protocol.

Protocol lacks a specific provision as to its territorial scope. On the one hand, this apparent oversight would seem to be explained by the fact that the Protocol is meant to supplement the Antarctic Treaty. Thus, in the absence of any provision to the contrary, its area of application should be understood as identical to that of the Antarctic Treaty, i.e., south of 60° south latitude.[20] Moreover, the essence of the Protocol lies in Article 3, which comprises 'all activities in the *Antarctic Treaty area*'[21]. The Protocol uses the formulation 'Antarctic Treaty area' throughout the text of its provisions. Indeed, since the adoption of the Protocol, the Consultative Parties have declared at several of their gatherings (both formal and informal) that they agree that the area of application of the Protocol is same as that of the Antarctic Treaty.

On the other hand, confining the Protocol to such a geographic limit that seems inadequate in the context of the Protocol's environmental protection provisions appears contrary to the main (proclaimed) purpose of the instrument.[22] Article 3 of the Protocol demonstrates the contradiction of the Protocol being limited to 'activities in the Antarctic Treaty area' and at the same time introduces the concept of the 'protection of the Antarctic environment and dependent and associated ecosystems'. The latter are linked to the *natural*, not the political, boundaries of the Antarctic. In implementing the environmental principles of the Protocol,

> monitoring shall take place to facilitate early detection of the possible unforeseen effects of activities carried on both within and outside the Antarctic Treaty area on the Antarctic environment and dependent and associated ecosystems.[23]

This would indicate that the Antarctic Convergence should be considered the appropriate boundary for the seaward extent of the area of application of the Environmental Protocol. We find a precedent within the ATS itself, as this natural boundary of the Antarctic region was considered to be the relevant boundary in determining the area of application of CCAMLR.[24]

Yet, it has been questioned whether all the provisions of the Protocol should be understood to apply to the *entire* area south of 60° south latitude. For instance, does the mining ban contained in Article 7 of the Protocol apply to the portion of the 'seabed beyond the limits of national jurisdiction', which under the letter of the UN Convention on the Law of the Sea has the status of the international seabed area and

---

[20] Art. 4 of the Protocol, in conjunction with Art. VI of the Antarctic Treaty. See comment by W.M. Bush, *Antarctica and International Law: A Collection of Inter-State and National Documents* (Dobbs Ferry, NY: Oceana Publications, 1992–), Booklet AT91C, p. 2; at another place Bush comments, 'the area south of 60 degrees south latitude ... is the same as the area of operation of the protocol'; *ibid.*, Booklet AT91D, p. 11. Also, Orrego's view seems to be in line with such comment, though with an additional measure of precaution; see Orrego Vicuña, 'The Effectiveness of the Protocol', p. 182.

[21] Art. 3(1) of the Protocol (emphasis added). See also comment by Bush, *Antarctica and International Law* (1992–), Booklet AT91C, p. 2.

[22] Similarly, Bush, *Antarctica and International Law* (1992–), Booklet AT91C, pp. 2–3.

[23] Protocol, Art. 3(2)(e).

[24] CCAMLR, Art. I(1) and (4).

as such would fall under the competence of the International Seabed Authority?[25] Domestic legislation of different Consultative Parties provides different responses to this question.[26] Does the Protocol apply to the continental shelf off Antarctica but south of 60° south latitude; or does the Protocol apply to the Antarctic continental shelf extending even north of 60° south latitude?[27]

The other major issue to consider is the Protocol's reliance on the impact of activities measured by the standard of 'minor or transitory' impact. The contents of this standard would be, it has been said, developed through practice. However, devoid of a *common frame of reference*, the practice of various parties will inevitably vary. In this context, another contradiction of the Protocol becomes apparent. It would be assumed that standards such as 'minor or transitory impact' would be tested against the available syntheses of the current state of scientific knowledge on the Antarctic environment. There is, however, no such comprehensive synthesis of the state of the Antarctic environment hitherto available. In the bi-polar comparison, this aspect may appear ever more contradictory. Several comprehensive studies on the state of the Arctic environment have been compiled in recent years through international cooperation. Most important among these are the two state-of-the-art reports issued by the Arctic Monitoring and Assessment Programme in 1997 and 1998.[28] It is on the basis of these reports that the need for any new Arctic regional environmental protection legal instruments is considered by the Arctic countries. Regarding the Antarctic, it is only recently, in discussions between the Consultative Parties at the 1996 Consultative Meeting, that the possible need for a 'State of the Antarctic Environment' report was indicated.[29] Pursuant to the establishment of the CEP in 1998, the matter has been an item on CEP's agenda.[30] In the course of the past few years, both SCAR and an 'inter-sessional open-ended contact group' established by the CEP and chaired by Sweden have been involved in discussing how best to structure such an assessment'.[31] However, actual work on the

---

[25] For a detailed examination of this issue, see D. Vidas, 'Southern Ocean Seabed: Arena for Conflicting Regimes?', in D. Vidas and W. Østreng (eds.), *Order for the Oceans at the Turn of the Century* (The Hague: Kluwer Law International, 1999), pp. 291–314.

[26] See further Bastmeijer, Chapter 16, and Dodds, Chapter 21, in this book.

[27] For a Chilean interpretative declaration in this respect, made on the occasion of signing, as well as in connection with ratifying the Protocol, see Carvallo and Julio, Chapter 18 in this book.

[28] See Arctic Monitoring and Assessment Programme (AMAP), *AMAP Assessment Report: Arctic Pollution Issues* (Oslo: AMAP, 1998); and AMAP, *Arctic Pollution Issues: A State of the Arctic Environment Report* (Oslo: AMAP, 1997).

[29] See para. 163 of the *Final Report of the Twentieth Antarctic Treaty Consultative Meeting, Utrecht, 29 April – 10 May 1996* (The Hague: Netherlands Ministry for Foreign Affairs, 1997).

[30] See Orheim, Chapter 6 in this book.

[31] See especially Sweden, 'Report on the Work of the Intersessional Contact Group on SAER', doc. XXIII ATCM/WP 5, 1999; and Scientific Committee on Antarctic Research, 'Reporting on the State of the Antarctic Environment: The SCAR View', doc. XXIII ATCM/WP 6, 1999. For initial views see New Zealand, 'On the Need for a State of the Antarctic Environment Report', doc. XXI ATCM/WP 32, 1997; and Scientific Committee on Antarctic Research, 'State of the Antarctic Environment Report', doc. XXI ATCM/WP 19, 1997.

preparation of a 'State of the Antarctic Environment' report has not yet commenced; and a recent estimate on the possible completion of such a potential report is 'around year 2003–2004'.[32] The Consultative Parties have spent considerable time discussing which approach should be used in compiling such an assessment and have aired concerns about the costs and time involved in the work.[33] Incidentally, the time-range used only for discussing how to structure a future 'State of the Antarctic Environment' report has well exceeded the time used to negotiate and adopt the Protocol itself. While the approach employed in the Arctic would appear reasonable, albeit somewhat cautious and slow, the approach used in the Antarctic, however, appears peculiar: first adopting one of the most stringent international treaties for environmental protection, and then, on the eve of its entry into force, inquiring as to the actual state of the environment that the regulation is intended to protect. This approach lacks a common frame of reference against which an impact on the Antarctic environment (or on its associated and dependent systems) can be measured as being less or more than, or equal to, 'minor or transitory'.

To conclude this brief outline of some key features of the Protocol, it first must be stated that the Protocol has strengthened the *legal* regime for the protection of the Antarctic environment with several new elements: 1) a comprehensive instead of an issue-specific approach; 2) a legally binding nature of the instrument; and 3) a new institution with an advisory role in the implementation of the Protocol, which is also the first overall ATS institution. The practical implementation of the Protocol is to a large extent left to the domain of states parties and their national legislation and practice. While the Consultative Parties have negotiated the Protocol in a specific political context, the challenge now is to implement the resulting legal instrument. The Protocol has some inherent contradictions and it has also left certain issues unresolved. In the current phase of implementation, the main task of the parties to the Protocol will be to develop adequate *practical* responses to the pending issues; the CEP can in addition play an important role in helping to resolve the inherent contradictions of the Protocol.

IN THE AFTERMATH OF ADOPTION

It can be argued that the period of relaxation for the Consultative Parties that followed the adoption of the Protocol was all too premature. The time-span following the adoption of the Protocol in 1991 and preceding its entry into force in 1998 may perhaps be called a phase of *anticipation*. Certainly, various *initiatives* were given and some hefty debates occurred during this period. For example, there

---

[32] Sweden, 'Report on the Work of the Intersessional Contact Group on SAER', para. 47.

[33] See para. 141 of the *Final Report of the Twenty-first Antarctic Treaty Consultative Meeting, Christchurch, New Zealand, 19–30 May 1997* (Wellington: New Zealand Ministry of Foreign Affairs and Trade, 1997); and para. 58 of the Report of the Committee for Environmental Protection, in *Final Report of the XXII ATCM*, Annex E.

was the issue of Antarctic tourism[34] at the Venice Consultative Meeting in 1992, the question of an Antarctic Treaty Secretariat,[35] and the work of the Group of legal experts on liability for environmental damage in the Antarctic,[36] to mention the most prominent. However, no substantial *solution* was adopted for any of these issues: tourism, with its potential to reflect in the jurisdictional sphere, was soon toned down and embraced in Recommendation XVIII–1 adopted at the Kyoto Consultative Meeting in 1994; the establishment of a treaty secretariat has been stalled in a stalemate over its location; and the Group of legal experts on liability, after many years of meetings and drafting, was dissolved after submitting a final report on the main open issues to be solved for an Antarctic liability regime. Clearly, either there was no significant policy involvement on these issues during this 'phase of anticipation', or, if such involvement was present, it led to stalemates rather than to resolutions of the issues.[37]

Although it may be argued that the Consultative Parties, in the period from the adoption to the entry into force of the Protocol, did not succeed in addressing more seriously the up-coming challenges of implementation of the instrument they agreed upon, it would be both unjust and inaccurate to characterise this period as lacking implementation action. On the domestic scene, however, in the aftermath of adoption of the Protocol, the Consultative Parties have been involved in elaborating and adopting implementing legislation for the Protocol. Already at the adoption of the Protocol, the Parties agreed that, pending entry into force of the Protocol, it was desirable to apply Annexes I–IV, 'in accordance with their legal systems and to the extent practicable'.[38] This willingness of the parties to voluntarily implement aspects of the Protocol even ahead of its entry into force has been followed up through an information exchange at a series of Consultative Meetings that started at the Venice Meeting in 1992.

After the Protocol entered into force in 1998, the Consultative Parties (all of which parties to the Protocol) became mainly preoccupied with a complex set of primarily *practical* issues concerning the implementation of this legal instrument. However, the basic problem of the specific regional situation in the Antarctic remains the unresolved question of sovereignty and jurisdiction there, and thus also of control and enforcement when it comes to implementation of legal instruments for environmental protection in the Antarctic.

---

[34] See Richardson, Chapter 4 in this book.
[35] See Francioni, Chapter 7 in this book.
[36] See Skåre, Chapter 9 in this book.
[37] Indeed, one should note practical initiatives, such as creation of the Transitional Environmental Working Group at the 1994 Meeting and the introduction at the 1995 Seoul Meeting of three categories of measures under Art. IX of the Treaty instead of the previous single category of 'recommendations'. These were, still, only moderate highlights in the work of a row of, otherwise fairly passive, Consultative Meetings.
[38] Para. 14 of the Final Act from the Eleventh Special Antarctic Treaty Consultative Meeting. For an extensive discussion of various aspects of domestic implementing legislation for the Protocol, see Bastmeijer, Chapter 16 in this book.

LEVELS OF IMPLEMENTATION OF THE ENVIRONMENTAL PROTECTION REGIME FOR THE ANTARCTIC: THE STRUCTURE OF THIS BOOK

The Antarctic regulatory picture is unique in its emphasis on *regionally centralised* regulation. Domestic legislation of Antarctic Treaty parties that applies to activities of their nationals in the Antarctic of course exists, but this is regularly guided by the principle of personal, not territorial jurisdiction. Moreover, the pattern of this legislation is to implement the instruments agreed by the Consultative Parties through their regional ATS cooperation. No real sub-regional level in the protection of the environment exists either. As to the global instruments, these do apply to the Antarctic and its environment. However, the centralised role of the regional ATS level modifies this application as well. Through the decision-making impact of its members in global fora and in negotiations, the ATS has always managed to serve as a 'filter' for the application of global instruments, especially when these tended to become polar-specific.[39]

Whereas the adoption of the Protocol in 1991 provided a successful response to a number of largely political questions at that time, its entry into force has opened a different set of questions for the Consultative Parties. In the Antarctic context, implementation of the Environmental Protocol will occur at three major levels, to a large extent interrelated, and all of which demand a more systematic scrutiny:

- implementation through Antarctic Treaty System cooperation;
- implementation through interrelations with other international environmental arrangements and organisations; and
- implementation at the national level of parties to the Protocol.

Implementation Issues for the Antarctic Treaty System

There are three major sets of issue-areas on which the Consultative Parties will have to focus more closely in order to enable successful implementation of the Environmental Protocol. Firstly, there are the unresolved problems of jurisdiction and the consequences upon ensuring control and enforcement in implementation of the environmental regime for the Antarctic. Secondly, there are the ATS institutional issues, related to the establishment, role and relationship of institutions to other components of the System in implementation of the Protocol. And thirdly, there is the unfinished normative agenda of the Environmental Protocol – in particular, the liability regime for environmental damage in the Antarctic.

---

[39] Thus it came about that the Antarctic was included as a Special Area under several annexes of MARPOL 73/78. More recently, this has determined the destiny of the Polar Code, once intended to become a bi-polar instrument, in the IMO; see further Scovazzi, Chapter 13 in this book.

Issues of jurisdiction, control and enforcement. Ensuring the implementation of the Protocol requires the introduction of innovative mechanisms for control and enforcement in the Antarctic – all of which must be related to the need to establish an effective *jurisdiction* over activities in the Antarctic Treaty area. The Antarctic Treaty regulates jurisdiction in quite a limited manner,[40] failing to resolve the question of jurisdiction over nationals of Treaty parties who are *not* observers or exchanged scientists; nor does the Treaty address the question of jurisdiction over nationals of third states. The problem of the lack of a comprehensive jurisdictional regime was not of particular concern in the decades following the adoption of the Antarctic Treaty.

The question of jurisdiction will have to be revisited, particularly in the light of the growth in Antarctic tourism (and, in the CCAMLR context, illegal, unregulated and unreported fishing for Patagonian toothfish in the Southern Ocean). Otherwise, it will be increasingly difficult to ensure compliance with the Protocol in a situation where some 50 per cent of the vessels visiting Antarctica on tourist cruises fly flags of third states, often various 'flags of convenience'.[41]

The problem arises when an offence, committed in the Antarctic Treaty area by nationals (or vessels) of third states, is not a crime but a breach of legislation to implement environmental regulations and conservation measures, such as those under the Environmental Protocol and CCAMLR. In this respect, when flag state enforcement fails (as it often does), the need arises for a complementary means.

At their annual Meetings in 1996 and 1997, the Antarctic Treaty Consultative Parties initiated discussion on the possible need for introducing such complementary means in the Antarctic context.[42] Since all the regularly used gateway ports to the Antarctic are subject to the jurisdiction of the parties to the Protocol (Argentina, Australia, Chile, New Zealand and South Africa), a concept such as 'departure state jurisdiction' was proposed by the United Kingdom.[43] This would in practice mean the solving of jurisdiction *in* the Antarctic waters by dealing with it *outside* of the Antarctic Treaty area. However, although there is a general understanding among the Consultative Parties that there is a need for mechanisms to ensure the effective implementation of the Protocol, several parties have been opposed to far-reaching proposals such as 'departure state jurisdiction'. In the context of CCAMLR, however, an analogous jurisdictional concept (though on 'arrival' rather than 'departure') was successfully introduced when, in November 1997, CCAMLR adopted a 'Scheme to Promote Compliance by non-Contracting Party Vessels with

---

[40] Art. VIII of the Antarctic Treaty.

[41] Likewise when about half of some 100 vessels observed to be involved in unregulated fishing for Patagonian toothfish in the CCAMLR area in the past few seasons flew various 'flags of convenience', mainly flags of Belize, Namibia and Panama.

[42] See especially United Kingdom, 'Enhancing Compliance with the Protocol: Departure State Jurisdiction', doc. XXI ATCM/WP 22, 1997.

[43] *Ibid.*

CCAMLR Conservation Measures'.[44] In accordance with this Scheme, a non-party flagged vessel that has been sighted engaging in fishing activity in the CCAMLR regulatory area is *presumed* to be undermining the CCAMLR Conservation Measures. If the sighted vessel enters the port of any state party to the CCAMLR Convention, it must be inspected.

How can issues of jurisdiction be adequately solved and thus enable implementation of the environmental protection regime in the Antarctic, but not disturb the balance on sovereignty positions as preserved in Article IV of the Antarctic Treaty? This is probably the major issue for the Consultative Parties in the current phase of implementation of the Protocol. Part I of this book deals with this major issue.

Institutional support to the implementation of the Protocol. Permanent international institutions, of a type characteristic for most international cooperative processes, have for some time been absent from the ATS, and even from discussions among the Consultative Parties.[45] It is the Antarctic Treaty Consultative Meeting, a periodical gathering of the Treaty parties, that has been the main policy-making forum of the Antarctic cooperation ever since the Meeting's inception in 1961. Although once held biennially, these Meetings became annual following a decision of the XVI Consultative Meeting (Bonn, 1991), which was inspired by the adoption of the Protocol.[46] Be these Meetings annual or biennial, this policy forum is by no means sufficient to follow up on various *operative* demands associated with the implementation of the Protocol.

The Protocol established the Committee for Environmental Protection (CEP) as an advisory body to the parties in the implementation of the Protocol.[47] However, some major questions of the position of the CEP in the wider context of Antarctic cooperation remained to be solved in practice. These include a determination of the role the CEP will have in the wider context of science-related components in the ATS, in particular vis-à-vis the Council of Managers of National Antarctic

---

[44] Conservation Measure 118/XVI, in: CCAMLR, *Report of the Sixteenth Meeting of the Commission, Hobart, Australia, 27 October – 7 November 1997* (Hobart: CCAMLR, 1998), p. 39. The possible benefit from the experience of NAFO was mentioned in the Commission's deliberations that year; see *ibid.*, para. 5.10 (intervention by Norway). Indeed, in September the same year NAFO adopted an almost identical Scheme.

[45] For a review see P. Gautier, 'Institutional Developments in the Antarctic Treaty System', in F. Francioni and T. Scovazzi (eds.), *International Law for Antarctica*, 2nd edition (The Hague: Kluwer Law International, 1996), pp. 31–47.

[46] Whether this was a premature decision or not may be open to debate. At any rate, because of the pace of the Meetings since 1992, the organisational efforts invested and the results achieved, many Parties began advocating return to biennial Consultative Meetings. Perhaps it would have been better to have biennial meetings until the entry into force of the Protocol, and then intensify the frequency rather than the other way around? On some of the problems involved see Berguño, Chapter 5, Francioni, Chapter 7, and Vidas and Njåstad, Chapter 8, in this book.

[47] See Art. 12(1) of the Protocol.

Programmes (COMNAP) and the Scientific Committee on Antarctic Research (SCAR).

The resolution of many of the above-mentioned and related issues is, however, still pending. As to an ATS secretariat in particular, with the still unresolved matter of its location, the availability of creative policy options, as well as organisational and legal options, is becoming urgently needed. Recent developments at the Lima Consultative Meeting in 1999 seem to indicate that this matter might progress towards a solution acceptable to all the Parties.[48] The analysis undertaken in Part II of this book points firstly to controversial issues in the pending organisational ATS matters that impact the implementation of the Protocol and, secondly, traces the manner in which the Consultative Parties address these questions.

The unfinished normative agenda: elaborating a liability regime for environmental damage in the Antarctic. In Article 16 of the Protocol, the parties undertook 'to elaborate rules and procedures relating to liability for damage arising from activities taking place in the Antarctic Treaty area and covered by this Protocol. Those rules and procedures shall be included in one or more Annexes to be adopted in accordance with Article 9(2)'.

In the aftermath of the adoption of the Protocol, the Consultative Parties established at the XVII Consultative Meeting in Venice, 1992, the Group of legal experts on liability. Convened for its first working session in 1993, the Group initially showed steady progress. However, this has been increasingly slowed down in the later meetings, and signs of approaching a stalemate on several crucial elements have gradually become obvious. At the XXI Consultative Meeting in Christchurch, in 1997, the Consultative Parties required the Group of experts to present its written report to the next, XXII Consultative Meeting. When the report, which listed key pending issues for an Antarctic liability regime, was presented to the Parties in 1998, the Group was actually dissolved. In other words, as stated in official documents of the Consultative Meetings, the Consultative Parties decided that the 'Group of Legal Experts on Liability, by submitting its report, has fulfilled its task and its work in now completed; [and that] the further negotiation of an annex or annexes on liability be undertaken in Working Group I of the ATCM'.[49] Deliberations over liability, now in a policy rather than legal forum, continued at the XXIII Consultative Meeting in Lima, 1999. A renewed listing of key issues has been made, but a consensus of the Parties on how to solve these issues remains distant.[50] These matters will be further discussed at a Special Consultative Meeting in the Hague, September 2000.

---

[48] See especially para. 27 of the *Final Report of the XXIII ATCM*.
[49] See paras. 1 and 2 of Decision 3 (1998), 'Liability'; and paras. 61–84 of *Final Report of the XXII ATCM*.
[50] See paras. 75–98 of *Final Report of the XXIII ATCM*.

The sense of urgency, and the main change in the course that was agreed upon at the XXII Consultative Meeting in Tromsø, Norway, were prompted by the entry into force of the Protocol a few months prior to that Meeting. When viewed retrospectively, the task of the Group of legal experts resembled a 'mission impossible' – equipped with no real policy guidance, based on no risk assessment of actual activities in the Antarctic,[51] mainly devoid of natural science and technical expertise, this Group was left to discuss various legal options in a vacuum.

Nevertheless, major policy dilemmas, including the choice between a piecemeal or an overall approach in creating a liability regime under the Protocol, still persist. As to the legal aspects, the liability, in addition to being one of the most controversial problems of international law in general, is in the Antarctic context further complicated by the sovereignty balances and jurisdictional uncertainties, as well as by the special features of the Antarctic environment.

The chapters in Part III of this book contain, first, a review of the *process* toward an Antarctic liability regime thus far, and then an examination of the legal *need* and *prospects* for an Antarctic environmental liability regime. These chapters also contain a critical legal assessment of the proposals made to date as well as indicate some possible legal options open to the Consultative Parties for to proceed in this matter.

Relationship to Other Environmental Agreements and Organisations

The elevation of environmental values in the ATS, along with other changes within the System – especially regarding participation and information flow – has served to considerably weaken the basis for criticism against the ATS in the UN General Assembly. A consensus resolution was adopted at the 1994 General Assembly session that expressly acknowledged the merits of the ATS in the governance of Antarctic affairs.[52] This acknowledgement was reiterated and strengthened by the UN General Assembly resolution on the 'Question of Antarctica' of January 1997,[53] which also postponed any further discussion on that item until the 1999 General Assembly session[54]. While the emphasis on the UN General Assembly-based critique has been significantly toned down, if not subdued completely, the implementation of the Environmental Protocol diverts the focus to the relationship of the ATS to several other international organisations, also of the 'UN family', including the International Maritime Organisation, the UN Environmental Programme, and not least the institutions established under the 1982 UN Convention

---

[51] The first such risk assessment appeared only in 1999: COMNAP, 'An Assessment of Environmental Emergencies Arising from Activities in Antarctica', doc. XXIII ATCM/WP16, 1999.
[52] UNGA resolution 49/80.
[53] UNGA resolution 51/56.
[54] At the 1999 UN General Assembly session, the Assembly again adopted a resolution by consensus; see UNGA resolution 54/45.

on the Law of the Sea (the International Seabed Authority and the Commission on the Limits of the Continental Shelf).

When it comes to the normative level of these external ATS relationships, they are to a large extent related to the implications that the interplay between global and regional regimes may have in the Antarctic maritime area. An instrument under negotiation during the past few years, the draft Polar Code of Navigation, deserves special mention here. The Code will obviously not become the first global international instrument specifically crafted to equally apply in both polar oceans.[55] IMO's Maritime Safety Committee recently decided to exclude the Antarctic from the application of the Code unless Antarctic Treaty Parties decide otherwise. The Consultative Parties subsequently agreed to develop guidelines for Antarctic shipping within the ambit of the ATS and then to seek adoption of these by the IMO.[56] Hence the peculiarities of the Antarctic situation will remain preserved, and possibly even extended, through the IMO, to third parties.

As to the relationship of the Antarctic environmental protection regime to those established under the LOS Convention and its implementing agreements, several intricate law of the sea issues emerge for states cooperating in the management of the Antarctic region. What is to be done with the requirement contained in the LOS Convention relating to the submission of information on the outer limit of the continental shelf beyond 200 nautical miles to the Continental Shelf Commission? Further, who is competent to regulate, and accordingly to ban, mineral activities in the Southern Ocean seabed – the International Seabed Authority as the global body, or the Antarctic Treaty Consultative Parties through their regional regulation?

Finally, it is not only the relationship of the ATS in general, and the Protocol in particular, to the global agreements that needs to be scrutinised. The 'internal' relationship to other ATS components, and in particular the relationship with CCAMLR, requires careful analysis. This is so not least for the possible environmental impacts of fishing activities undertaken within the CCAMLR area. However, the Final Act of the Eleventh Special Consultative Meeting, which adopted the Protocol, stated that with respect to the activities referred to in Article 8 of the Protocol (environmental impact assessment), activities undertaken in the Antarctic Treaty area pursuant to CCAMLR and the Seals Convention were not intended to be included;[57] neither should anything in the Protocol derogate from the rights and obligations of the parties under CCAMLR, the Seals Convention and the International Convention for the Regulation of Whaling.[58]

---

[55] See discussion by Scovazzi, Chapter 13 in this book. See also L. Brigham, 'The Emerging International Polar Navigation Code: Bi-polar Relevance?', in D. Vidas (ed.), *Protecting the Polar Marine Environment: Law and Policy for Pollution Prevention* (Cambridge University Press, 2000), pp. 244–262.

[56] See Decision 2 (1999), 'Guidelines for Antarctic Shipping and Related Activities', in the *Final Report of the XXIII ATCM*, Annex B.

[57] See para. 8 of the Final Act.

[58] *Ibid.*, para. 7.

It is Part IV of this book that is devoted to the exploration of various international regime relationships over implementation issues.

Implementing the Environmental Protocol Domestically

The final part of the book contains domestic case-studies that analyse, in separate chapters, the implementation of the Protocol in several countries, all Consultative Parties to the Antarctic Treaty: Australia, Chile, Italy, Norway, South Africa, and the United States. While any such selection of a limited number of countries among the current 27 Consultative Parties (or 28 parties to the Protocol) must remain arbitrary to some extent, we have been led by several criteria when including in our scrutiny only the aforesaid countries. The cases involved include countries from different geographical regions. In addition to including both countries distant from, and neighbouring with, the Antarctic, the selected countries also represent different legal systems; the requirements under their domestic laws span from a direct incorporation of international instruments in the domestic legislative system ('self-executing treaties') to a detailed elaboration of domestic implementing legislation as a prerequisite for ratification of an international treaty. Moreover, the selected case-studies include both territorial claimant and non-claimant states in the Antarctic, both original parties to the Antarctic Treaty and those that acceded to the Treaty at later stages, and finally both developing and developed countries.

To enable a meaningful comparability of such a diversified group of countries, the main focus of the chapters was set on three aspects of domestic implementation of the Protocol.[59] Firstly, an examination of substantive issues, which includes discussion of how general obligations under the Protocol are implemented through domestic legislation, the type of activities covered by domestic legislation (and, especially highlighting the way in which activity by commercial, non-governmental interest is regulated), articulation of ATS-pending issues through domestic legislation (liability, emergency response action, etc.), and modes of regulation of jurisdiction, control and enforcement. Secondly, the analysis in the case-studies considers procedural and institutional issues. After providing a brief overview of the national legislative process and institutions involved, the focus here is on both the procedural aspects of the Antarctic (Protocol) domestic legislation in the overall domestic legal system and on the domestic institutions that are taking part in implementing the legislation. And finally, a third element of domestic case-studies contains an assessment of the effectiveness of domestic legislation and institutions in implementing the Protocol and of the contribution made to Antarctic law and policy by the national legal systems.

---

[59] In addition, there is an overview chapter included in Part V, which addresses various issues for domestic implementation of the Protocol; see Bastmeijer, Chapter 16 in this book.

Part I

ISSUES OF JURISDICTION, CONTROL AND ENFORCEMENT

2

Means and Methods of Implementation of Antarctic Environmental Regimes and National Environmental Instruments: An Exercise in Comparison

William Bush

Dr. Johnson remarked of a dog walking on its hind legs, 'It is not done well; but you are surprised to find it done at all'. Wonderment about the environmental regimes of the Antarctic Treaty System (ATS) depends on what one compares them to. The practice has been to compare them with other international regimes of, say, a regional environmental nature – and wonder. There is good reason to do so. No other regional environmental regime occupies a position akin to that of ATS with regard to the vastness of the region governed or the breadth and detail of the region's environmental regulation – regulation established under the Antarctic Treaty[1] and its Environmental Protocol[2], the Seals Convention[3] and CCAMLR[4], as well as conservation measures adopted by the Commission established under CCAMLR. Pride in that achievement can slip into complacency. As an antidote to that possibility, this chapter suggests that an instructive comparison would be what a state might be expected to do within its own undisputed sphere of competence. Justified wonderment that environmental management by a collection of states has worked at all should not blind one to the standards expected of national environmental management.

The decisions and processes at a national level that are necessary for maintaining an environmental regime have their parallels at the international level.

---

[1] UNTS, Vol. 402, pp. 71ff.
[2] ILM, Vol. 30, 1991, pp. 1,461ff.
[3] UNTS, Vol. 1080, pp. 175ff; and ILM, Vol. 11, 1972, pp. 251ff.
[4] ILM, Vol. 19, 1980, pp. 837ff.

Planning and consultation should precede rule-making. The assembled parties drawing up a new treaty, for example the Antarctic Treaty Consultative Meetings and the CCAMLR Commission, might be likened to national legislatures, namely the national executive empowered to pass subsidiary legislation. Legislation is implemented by the organs of a state. Here the analogy with an international regime becomes more complex because the regime often lacks direct organs of implementation; instead, implementation will generally be effected by individual member states. Thus, whereas national authorities have sole competence in the implementation of a national regime, under an international regime competence is shared. An international regime will typically incorporate procedural devices such as the exchange of information and obligations to consult in order to promote international coordination of individual, national action.

Comparing the facets of the Antarctic environmental regimes with their counterparts in a solely national regime can produce valuable insights into the ATS and where it should be heading.[5]

MEASURES OF ENVIRONMENTAL GOVERNANCE

The comparison of Antarctic environmental regimes with national regimes will be made using five criteria by which environmental governance can be measured. These criteria are based on the extent to which the regime

– is likely to produce informed decisions,
– successfully balances interests,
– ensures uniformity of application,
– is efficient, and
– is responsive to need.

Thus, procedures should be apparent by which decisions large and small are likely to be environmentally informed. This may involve the tapping of the requisite expertise to ensure sound planning as well as require, in the event of gaps in knowledge, the erring of decisions on the side of caution. An environmental regime should also ensure that relevant interests within a context of environmentally responsible decision-making, be they scientific research, tourism or exploitation of marine resources, are balanced, for example, in the process of environmental impact assessment (EIA) and the administration of a permit system. Uniformity of application could be promoted by the clarity of rules of the regime, uniformity of their interpretation and equality in application to all actors. Efficiency would have to do with the existing resources necessary for achieving a given environmental

---

[5] For the purpose of this chapter, the Environmental Protocol, the Seals Convention and CCAMLR will be regarded as having established individual environmental regimes. The collective whole will be referred to as the ATS.

outcome. In an Antarctic context, the individual national resources required may be able to be minimised through cooperation, for example, in carrying out monitoring and inspections. Finally, the efficiency of an environmental regime is determined by its responsiveness to need, i.e., the ability of the regime to respond promptly in order to optimise environmentally satisfactory outcomes.

Two facets common to both Antarctic environmental regimes and national regimes will be examined in light of the five criteria listed above for evaluating environmental governance. These facets are rule-making and implementation. Only two aspects of implementation will be examined in this way, namely EIA and permits. By way of contrast, a third, enforcement, which is not amenable to that form of analysis, will be examined. With such a limited analysis, it will not be possible to do more than illustrate the potentials of analysing the ATS as a national environmental regime. An interesting facet beyond the scope of this analysis is the interface of the Antarctic System with other international ('external') regimes;[6] the international status and number of the Antarctic environmental regimes make external cooperation more complex than is the case for a state in a similar situation.

RULE-MAKING

An assessment of rule-making should regard both the principal and subsidiary rules of an environmental regime. In a state, rules are typically represented by statutes which are enacted by a parliamentary legislature; subsidiary legislation is made by the executive. In an Antarctic context, the principal rules will for the most part be those set out in an instrument of treaty status under international law such as the Environmental Protocol. The instruments adopted by the CCAMLR Commission and Consultative Meetings in the form of 'conservation measures', 'measures' and the like should be assessed as the equivalent of national subsidiary legislation. These measures are sometimes in the form of instruments of general application. On other occasions, area-specific regimes are formulated under a protected area scheme. These too have national equivalents.

Whether Rule-making is Likely to be Informed

It is difficult to assess whether the negotiation of the principal environmental instruments of the ATS has been as environmentally enlightening as equivalent major national legislation. In both forums, non-environmental issues will intrude. Moreover, in international negotiations where majority decision-making does not occur, pressures to compromise are likely to be even stronger than in many parliamentary situations.

---

[6] See Rothwell, Chapter 12, Scovazzi, Chapter 13, and Vidas, Chapter 14, in this book.

The main conduit for expert advice in international negotiations is through national delegations. Expertise is also introduced through consultation with bodies such as the Scientific Committee on Antarctic Research (SCAR) or the convening of meetings of experts. In the case of the rushed negotiations of the Environmental Protocol, there was limited opportunity for this to occur. The openness of negotiations might be regarded as another indication of how informed the final instrument is likely to be. Apart from the attendance of some bodies such as the World Conservation Union (IUCN) and the Food and Agriculture Organisation (FAO) in the final diplomatic conference that drew up CCAMLR, the negotiation was conducted in secret.[7] In contrast, a similar number of organisations, including the Antarctic and Southern Ocean Coalition (ASOC), 'were invited to take part in' all sessions of the special Consultative Meeting that drew up the Environmental Protocol.[8]

It is obvious that a major effort has been made to ensure that the bodies formulating subsidiary measures are guided by expert advice. From the start, the Consultative Parties drew upon the expertise of SCAR, and SCAR was mentioned in the Seals Convention as the advisory source. The Council of Managers of National Antarctic Programmes (COMNAP), established in 1988, now plays an important role as well. Expertise is sought from time to time from specialised agencies of the United Nations such as FAO and the International Maritime Organisation (IMO). Moreover, a pattern has developed of accepting advice from non-governmental environmental organisations such as ASOC.[9] Expertise is also assured by the make up and role of the Scientific Committee in advising the CCAMLR Commission and the Committee for Environmental Protection (CEP) under the Environmental Protocol.[10] It is probably fair to say that the Antarctic environmental regimes are better advised than many national authorities that make subsidiary legislation.

The extent that an environmental regime, on the basis of the best expert advice, errs in situations of uncertainty on the side of caution in the regulation of activities is

---

[7] Final Act of the Conference on the Conservation of Antarctic Marine Living Resources; text reprinted in W.M. Bush (ed.), *Antarctica and International Law: A Collection of Inter-State and National Documents* (London: Oceana Publications, 1982–88), Vol. I, p. 394. See also J.N. Barnes, 'The Emerging Convention on the Conservation of Antarctic Marine Living Resources: An Attempt to Meet the New Realities of Resource Exploitation in the Southern Ocean', in J.I. Charney (ed.), *The New Nationalism and the Use of Common Spaces* (Totowa, NJ: Allanheld, Osmun Publishers, 1982), p. 249.

[8] Para. 6 of *Final Report of the Eleventh Antarctic Treaty Special Consultative Meeting*; text reprinted in W.M. Bush (ed.), *Antarctica and International Law: A Collection of Inter-State and National Documents* (Dobbs Ferry, NY: Oceana Publications, 1992–), Binder II, Booklet AT91D, pp. 5–7.

[9] ASOC is appointed an observer to the CEP in accordance with Rule 4(c) of the Rules of Procedure for the Committee for Environmental Protection (Decision 2 (1998)); text in *Final Report of the Twenty-second Antarctic Treaty Consultative Meeting, Tromsø, Norway, 25 May – 5 June 1998* (Oslo: Royal Norwegian Ministry of Foreign Affairs, 1998), Annex B, pp. 62–67. On the CEP's Rules of Procedure, see discussion by Orheim, Chapter 6 in this book.

[10] CCAMLR, Art. XIV and Protocol, Art. 11. On the role of the CEP, see Orheim, Chapter 6 in this book.

a measure of its rigour. In this respect the record of the environmental regimes within the ATS is patchy but probably comparable or better than what takes place generally at a national level. For all the merit of the 'principles of conservation' spelt out in Article II of CCAMLR, their application was frustrated for most of a decade by differences within the CCAMLR Scientific Committee. The principles requiring that parties adopt an ecosystem approach in making decisions to harvest marine living resources resulted in dissention on important issues such as identifying conservative precautionary catch levels in the absence of reliable surveys of the size of the target species.[11] Of course, now that the Commission, on the advice of the Scientific Committee, is adopting informed measures, it is ironic that illegal, unregulated and unreported (IUU) fishing by third party vessels threatens the sustainability of Southern Ocean fish stocks. Measures taken to reduce the serious decline in albatross numbers from incidental mortality in fishing operations constitute another example of the failure to apply precautionary principles.

The Environmental Protocol goes further than CCAMLR in addressing, in Article 3(2)(c), the need for information on which to base management decisions:

> activities in the Antarctic Treaty area shall be planned and conducted on the basis of information sufficient to allow prior assessments of, and informed judgements about, their possible impacts on the Antarctic environment and dependent and associated ecosystems and on the value of Antarctica for the conduct of scientific research.

Despite this clear directive for information, the Protocol falls short of specifically prohibiting actions in the absence of information being available that would satisfy the provision.[12] Because the mandatory EIA procedures of the Protocol focus principal attention on proposed activities likely to have a more significant impact ('more than a minor or transitory impact'),[13] the cumulative impact of less significant activities is likely to be the main loophole by which activities can proceed without adequate caution. A specific example is the reluctance of tour operators to accept constraints on visiting areas where the impact of such visits is unassessed. At the Twentieth Consultative Meeting, 'ASOC urged tour operators to apply a precautionary approach and not to add new sites until in-depth studies have been carried out regarding the impact of these visits'.[14] For reasons such as these, the environmental regime of the Protocol is far from being a model of the application of the precautionary principle. One procedure commonly found in domestic legislation

---

[11] See, e.g., S. Nicol, 'CCAMLR and Its Approaches to Management of Krill Fishery', *Polar Record*, Vol. 27, 1991, pp. 229–236; and F. Orrego Vicuña, 'The Effectiveness of the Decision-Making Machinery of CCAMLR: An Assessment', in A. Jørgensen-Dahl and W. Østreng (eds.), *The Antarctic Treaty System in World Politics* (London: Macmillan, 1991), pp. 30–32.

[12] Compare Protocol, Art. 3(2)(c) with CRAMRA, Art. 4(4).

[13] Protocol, Art. 8 and Annex I.

[14] *Final Report of the Twentieth Antarctic Treaty Consultative Meeting, Utrecht, 29 April – 10 May 1996* (The Hague: Netherlands Ministry of Foreign Affairs, 1997), para. 78. See also Richardson, Chapter 4 in this book.

that might be introduced is a system of interim protection of, say, a particular area threatened with activity pending a more detailed assessment. This would be similar to the establishment by the CCAMLR Commission of precautionary catch limits and other measures on new and developing fisheries.[15]

Balancing Interests in Rule-making

The extent to which interests are successfully balanced in making rules has a significant bearing on the viability of an environmental regime. The fewer the interests, the easier it is to reach an acceptable balance. In this sense the task faced by rule-makers of the Antarctic system has been much simpler than that faced by their national equivalents. Until the rapid growth in tourism in the late 1980s, the main preoccupation of environmental regulation at regular Consultative Meetings concerned scientific research and associated official expeditions. This circumscribed focus had been maintained by allocating the regulation of marine living resources and minerals to special Consultative Meetings and the respective separate regimes. Tour operators, through the International Association of Antarctic Tour Operators (IAATO), are now, together with environmental organisations, routinely represented at Consultative Meetings. Since 1988, ASOC has been an observer at meetings of the CCAMLR Commission. Industry groups are not similarly represented. Their interests are filtered through national delegations.

Because of the possibility of actors not being represented, there is no assurance that Antarctic forums will be able to balance the interests of all actors involved in the Antarctic. For one thing, it is always possible for a government like Pakistan, which is not a party to the Antarctic Treaty, to send expeditions to the Antarctic. For another, commercial interests which have no strong links to Antarctic Treaty parties may become involved in activities as diverse as fishing, tourism, iceberg harvesting or harvesting biological resources for commercial medical research.

Uniformity of Application of Rules

Because of the international character of the regimes and the special juridical conditions of the Antarctic, there is no assurance that rules of international Antarctic environmental regimes will have uniform application to all actors as they would be expected to have under a national regime. If regarded purely as treaty based regimes, there is the difficulty that these environmental regimes will not bind states that are not a party to them, nor will they bind private actors beyond the jurisdictional scope of states parties. In the Antarctic context, this basic consideration of treaty law is

---

[15] See, for example, Conservation Measures 31/X (in CCAMLR, *Report of the Tenth Meeting of the Commission, Hobart, Australia, 21 October – 1 November 1991* (Hobart: CCAMLR, 1992), pp. 27–28) and 65/XII (in CCAMLR, *Report of the Twelfth Meeting of the Commission, Hobart, Australia, 25 October – 5 November 1993* (Hobart: CCAMLR, 1994), pp. 38–41).

complicated by the reluctance from some quarters to formulate even clear, let alone comprehensive, jurisdictional rules for the various substantive obligations.[16] This reluctance is generated by divergent positions on sovereignty and sensitivities about extra-territorial application. The task of clarifying the rules is left to national authorities when international regimes are implemented through domestic law, with the likely result of imperfect jurisdictional coverage of Antarctic actors.

A further lack of uniformity is the exclusion from the mandatory disciplines of Annex IV of the Protocol of most ships supporting national expeditions on the ground of sovereign immunity. The parallel in the domestic legal system is the extent of immunity from process of the state; in national legal systems, government agencies are generally bound by statutes and subject to suit. Annex IV is not to apply to 'any warship, naval auxiliary or other ship owned or operated by a State and used, for the time being, only on government non-commercial service' and only 'so far as is reasonable and practicable'.[17]

In contrast to the lack of uniformity caused by the exclusion of ships, arguments of uniformity can be used to resist discriminatory controls designed to promote environmental ends. The Environmental Protocol recognises the value of Antarctica for scientific research and suggests that this activity should have some priority (particularly Article 3(3)). Placing priority on scientific research can be justified on the grounds of the importance of science in the history of human presence in the region, the insight it provides into global processes, and the intrinsic value of scientific research itself. However, an entrenched priority for one activity can impede the development of a desirable environmental regime if such a justified priority is used politically to oppose the introduction of more stringent regulation of other activities justifiable on environmental grounds. One such example of impeding desirable environmental development would seem to be the rejection, led by the United States and supported by ASOC, of the development of an annex or other rules applicable to tourism and non-governmental activity.[18]

Efficiency of Rule-making and Its Responsiveness to Environmental Need

Treaties, the traditional form for the embodiment of non-customary international legal rules, are notoriously time-consuming to create. Once in place, treaties are also difficult to amend. In contrast, national legislation can be put in place much more quickly. A majority of some sort, rather than a consensus, is all that is required in a

---

[16] See discussion by Orrego Vicuña, Chapter 3 in this book.
[17] Protocol, Annex IV, Art. 11. See Orrego Vicuña, Chapter 3 in this book. See also C.C. Joyner, 'Protection of the Antarctic Environment Against Marine Pollution under the 1991 Protocol', in D. Vidas (ed.), *Protecting the Polar Marine Environment: Law and Policy for Pollution Prevention* (Cambridge University Press, 2000), pp. 117–119.
[18] See discussion by Richardson, Chapter 4 in this book; also doc. XVI ATCM/INFO 77, 1991, p. 1; and Bush (ed.), *Antarctica and International Law* (1992–), Binder II, Booklet AT92C, pp. 29–31.

parliamentary system. Even so, just under two years were required to finalise the Antarctic Environmental Protocol, while the enactment of national implementing legislation and the introduction of other measures of compliance straggled along for a further seven years or so. Indeed, the common need for treaty approval or implementation by the legislature means that treaties tend to involve the worst of both worlds: slow to draw up and slow to approve. What is more, the slowest party to give the required consent will determine the pace at which a treaty will enter into force.

Both CCAMLR and, to a certain degree, the Environmental Protocol have put in place procedures for the adoption of subsidiary instruments. This, in the terminology of national law-making, would generally be termed subsidiary legislation and may be in the form of regulations implementing acts of parliament, 'decrees' implementing 'laws' or the like. National subsidiary legislation can be made with a great deal more ease than legislation enacted by the principal legislature. The decision on subsidiary legislation is generally exclusively in the hands of the executive arm of government. The legislation will stand unless it is successfully challenged on administrative legal grounds or any disallowance procedures are instituted. In contrast, the international procedures for adoption within Antarctic forums of subsidiary instruments are only marginally less onerous than those for the adoption of treaty instruments. In Antarctic Treaty forums, unanimity or consensus is required in both cases.[19]

Considering these circumstances, the more recent experience of the CCAMLR Commission has shown that the procedure for the adoption of subsidiary instruments is reasonably efficient. Typically, the Commission meeting in October and November each year adopts a series of conservation measures on catch limits and other issues applicable to the forthcoming season. Even though those conservation measures are not binding on adoption, they will become binding for each party unless it objects within 90 days (Article IX(6)).

There is no similar procedure for the standard run of subsidiary instruments adopted by the Antarctic Treaty Consultative Meeting in its role as the decision-making body under the Environmental Protocol. On the other hand, the Protocol envisages that 'any Annex may itself make provision for amendments and modifications to become effective on an accelerated basis' (Article 9(3)). In the form of a one-year opting-out provision, this clause is provided for in each of the annexes so far adopted. Moreover, Annex V on protected areas applies a 90-day opting-out procedure for the approval of management plans. This procedure applies to an important form of subsidiary instrument (Articles 6(1) and 8(1)). Although no opting-out procedure has been introduced for normal subsidiary instruments adopted

---

[19] Rule 24 of the Revised Rules of Procedure (1997), attached to Decision 1 (1997), in *Final Report of the Twenty-first Antarctic Treaty Consultative Meeting, Christchurch, New Zealand, 19–30 May 1997* (Wellington: New Zealand Ministry of Foreign Affairs and Trade, 1997), Annex B, pp. 121–127; and Art. XII(1) of CCAMLR.

by Consultative Meetings, the reforms introduced in Decision 1 (1995)[20] have streamlined the time at which some instruments enter into effect. If a matter can be characterised as 'an internal organisational matter' it should be embodied in an instrument referred to as a 'Decision' which will become 'operative at adoption or at such other time as may be specified'. 'Hortatory texts' are to be embodied in 'Resolutions'. Most substantive rule-making instruments on environmental protection under the Protocol are likely to fall within the category of those 'intended to be legally binding once it has been approved by all the Antarctic Treaty Consultative Parties'. These instruments 'will be expressed as a Measure recommended for approval in accordance with paragraph 4 of Article IX of the Antarctic Treaty'.

Subsidiary instrument-making in Antarctic forums is less responsive to need than is national subsidiary legislation in that such instruments must be adopted at meetings which at present are scheduled once a year rather than promptly when the need arises. At present, no procedure seems to exist for the intersessional adoption by Consultative Meetings of a subsidiary instrument even on an interim basis. Unforeseen circumstances may require prompter action. 'With the agreement of the ATCM', the CEP may meet and take decisions between annual meetings.[21] The urgent need to develop additional measures to combat IUU fishing prompted the CCAMLR Commission to hold its first intersessional meeting in April 1999; however, rather than make a final decision then, the meeting agreed that a draft scheme it had devised should be submitted for approval at their next regular Commission meeting, 25 October – 5 November 1999.[22] As unscheduled meetings of all parties are difficult to arrange and costly to run, it should be possible to establish a practice of intersessional decision-making using electronic communications.[23]

In summary, rule-making at an international level, particularly where unanimity or consensus is required, has a reputation for being cumbersome. While in Antarctic forums this may be justified for the development of new treaty instruments, the inefficiency and inflexibility of international rule-making is mitigated in a number of ways. Subsidiary instruments can be adopted under procedures by which the consent of parties is assumed in the absence of objection. Moreover, less tangible factors such as the specialisation of Antarctic Treaty forums and the ready availability within them of expert advice also promote efficiency, thus making rule creation responsive to need. Perhaps the most significant inflexible element is the

---

[20] See Decision 1 (1995), 'Measures, Decisions and Resolutions', in *Final Report of the Nineteenth Antarctic Treaty Consultative Meeting, Seoul, 8–19 May 1995* (Seoul: Ministry of Foreign Affairs of the Republic of Korea, 1995), Annex B, pp. 89–90.

[21] Rule 9 of Rules of Procedure for the CEP. See further Orheim, Chapter 6 in this book.

[22] Media release of the European Commission, in <http://europa.eu.int/comm/dg14/info/info55> dated 3 May 1999.

[23] On the various aspects of the introduction of electronic communication in the ATS, see Vidas and Njåstad, Chapter 8 in this book.

fact that decision-making is shackled to periodic meetings, presently only held annually.[24] The development of procedures for intersessional rule-making would introduce a flexibility that exists at the national level, at least for subsidiary legislation.

IMPLEMENTATION

Only three aspects of implementation of the Protocol will be touched on here: EIA, the issue of permits, and enforcement. Limited space prevents drawing national comparisons with procedures such as monitoring, suspension or cancellation of an activity, emergency response action and inspection except in so far as these may fall within the ambit of any of the three examined.

Environmental Impact Assessment

The EIA provisions of the Protocol have been criticised and defended.[25] It is undeniable, though, that the Protocol gives the provisions impressive prominence in the environmental principles in its Articles 3 and 8, and in Annex I.

EIA and informed decision-making. EIA procedures should be the quintessence of informed decision-making. They involve individual examination of the environmental implications of a proposed activity to serve as a basis for a decision on whether or not the activity should proceed. Whether prescribed under an international or national regime, the value of the assessment procedures will depend on the quality of the assessment and the relationship of the assessment to the ultimate decision.

The Protocol gives some assurance of the quality of assessments by specifying what draft assessments should address. The Protocol requires public scrutiny of some assessments and permits the CEP and Consultative Meeting to review them. Understandably, the Protocol in Annex I is most prescriptive of *comprehensive* environmental evaluations (CEEs) of activities likely to have more serious impacts, namely those involving proposed activities which are likely to have more than a minor or transitory impact.[26] So far only a handful of such evaluations have been

---

[24] Even the annual interval is not always possible to maintain, as seen from the fact that no regular ATCM is currently scheduled to be held in 2000 due to no host state being available; see also paras. 35–37 and 170, in: *Final Report of the Twenty-third Antarctic Treaty Consultative Meeting Lima, Peru, 24 May – 4 June 1999* (Lima: Peruvian Ministry of Foreign Affairs, 1999), Annex C; also available at website <www.rree.gob.pe/conaan/meeting1.htm>.

[25] See, for example, discussion in F. Orrego Vicuña, 'The Effectiveness of the Protocol on Environmental Protection to the Antarctic Treaty', in O.S. Stokke and D. Vidas (eds.), *Governing the Antarctic: The Effectiveness and Legitimacy of the Antarctic Treaty System* (Cambridge University Press, 1996), pp. 190–193.

[26] Protocol, Annex I, Arts. 3(2) (contents), 3(3), (4) and (6) (scrutiny), 3(6) (final assessment to

undertaken. Annex I says little about the form of *initial* environmental evaluations[27] and nothing about *preliminary* evaluations which it states are to be 'considered in accordance with appropriate national procedures'.[28] The environmental principles in Article 3 of the Protocol which apply to 'activities in the Antarctic Treaty area' state that these 'shall be planned and conducted on the basis of information sufficient to allow prior assessments of, and informed judgements about, their possible impacts' (Article 3(3)(c)). The Consultative Parties are filling gaps by resorting to means such as the non-mandatory 'Guidelines for environmental impact assessment in Antarctica' set out in Resolution 1 (1999).[29]

The Protocol links but does not bind to the assessment process the decision it places in national hands to proceed with an activity. This is done by implication in Article 3(2)(c) in stating what should be taken into account in such decisions and specifically in Annex I. For example, after completion of a CEE, the decision-maker still has the discretion to approve an activity 'based on' the evaluation 'as well as other relevant considerations' (Protocol, Annex I, Article 4). Only if the decision-maker decides that the expected impact exceeds a standard set out elsewhere in the Protocol (probably in Article 3(2)(b)) would the decision-maker be bound to disapprove the activity.

By these means the Protocol, now supplemented by the 1999 Guidelines and other instruments, offers a reasonable assurance that a wide range of decisions will be taken on an informed basis. This compares favourably with many national procedures.

EIA and balancing of interests. Environmental impact assessment procedures that are transparent and informed are also an effective means of balancing interests. For example, the procedures provide an opportunity for interested parties to learn of a proposed activity and make comments. One of the environmental principles of the Protocol is that in planning an activity 'full account' should be taken of 'whether the activity will detrimentally affect any other activity in the Antarctic Treaty area' (Article 3(2)(c)(iii)). CEEs under the Protocol should include 'consideration of the effects of the proposed activity on the conduct of scientific research and on other existing uses and values' (Annex I, Article 3(2)(i)). The obligation to take account of cumulative impacts also provides a basis for the balancing of interest. An evaluation should address the 'cumulative impacts of the proposed activity in the light of existing activities and other known planned activities' (Annex I, Article 3(2)(f)).[30]

---

address comments), as well as 3(4) and (5) (opportunity of the CEP and Consultative Meeting to review).

[27] Protocol, Annex I, Art. 2(1) (contents).

[28] *Ibid.*

[29] Note particularly reference to the importance of the 'initial scoping process' in Resolution 1 (1999), para. 3(1)(1); text in *Final Report of the XXIII ATCM*, Annex C.

[30] See also Protocol, Annex 1, Art. 2(1)(b) and, more generally, Protocol, Art. 3(2)(c)(ii).

These considerations suggest that the Environmental Protocol has mechanisms that compensate significantly for the added difficulty that an international (as opposed to a national) regime would have in balancing interests. This compensation should not be overstated because the less advertised and structured EIAs are, the less effective they are likely to be as a mechanism for balancing interests and taking account of cumulative impacts. This observation particularly affects taking account of activities such as tourist visits that are of relatively low impact but which may be repeated frequently by different operators. Under the Protocol the overwhelming proportion of all activities – tourism or otherwise – need not be advertised in advance; only the small minority subject to CEE are. This fact limits the value of EIA under the Protocol as a balancing procedure. As discussed below, the administration of a permit system provides more scope to balancing interests (and taking account of the cumulative effects) of a series of similar low impact activities.

EIA and uniformity of application. The Protocol itself provides little assurance of uniformity regarding the application of EIAs. This is mainly because decisions based on EIAs are in national hands and the procedures themselves are largely not prescribed, or, if prescribed, are left to the discretion of national authorities to follow. EIA obligations do not, of course, apply generally to activities of states (or their nationals) that are not a party to the Protocol; they do not even apply to all the activities of the parties and their nationals.

The general rule is that proposed activities should 'be considered in accordance with appropriate national procedures' (Protocol, Annex I, Article 1(1)). In the case of preliminary assessments, the Protocol does not lay down any international procedures for national authorities which apply beyond the general obligations in Article 3, referred to above, that all activities be subject to prior assessment. Some procedures are laid down for the following stage of an IEE (Protocol, Annex I, Article 2) and these evaluations are to be advertised and examinable after the event (Protocol, Annex I, Article 6). Only in the case of CEEs are procedures prescribed in some detail and an opportunity provided for comment on the draft.

With the relative paucity of direction given by the Protocol of the assessment process, the widely different national practices on EIA (such assessments are unknown in some countries) and the absence of expertise in some countries, it may be natural to expect wide variation in how each party will fulfil its EIA obligations.[31] On the other hand, the tradition of Antarctic Treaty cooperation is seeking to reduce this likelihood by, for example, the work of COMNAP and the development of guidelines, such as the recent Resolution 1 (1999). However, much more needs to be done in this area. Divergences of national practices are likely to be revealed by inspection reports and required notification of draft or completed evaluations. This should facilitate the refinement of the guidelines which, in time, may be reformulated as mandatory measures.

---

[31] See, for instance, chapters in Part V of this book.

The inapplicability of EIA obligations to some Antarctic actors is a serious defect. Activities of Pakistan and some tourist ventures aside, nothing to date has challenged the efficacy of the Protocol that can be compared with the scope and seriousness of the third party IUU fishing activity that is currently undermining CCAMLR. This sort of jurisdictional black hole is virtually unknown in a domestic legal context for activities carried on within a state's borders.

Uniformity of application of EIA is also affected by other limitations; for example, continuing activities that are characterised as 'pre-existing' need not be evaluated.[32] There is also the mysterious limitation of Article 8 and Annex I to 'any activities undertaken in the Antarctic Treaty area pursuant to scientific research programmes, tourism and all other governmental and non-governmental activities in the Antarctic Treaty area for which advance notice is required under Article VII(5) of the Antarctic Treaty, including associated logistic support activities'.[33] The arguably hortatory obligations included among the 'environmental principles' in Article 3(2)(c) of the Protocol apply without the qualification to 'activities in the Antarctic Treaty area'.

In summary, the Protocol does not provide assurance of uniform application of EIA of activities in Antarctica because: 1) it is inapplicable across all classes of activity, 2) it does not apply to all potential actors in the Antarctic, and 3) it has the potential for lack of coordination and lack of uniform practice by different national implementing authorities. In the tradition of Antarctic Treaty cooperation it is likely that a number of the gaps and divergences arising from these factors will be reduced over time.

On the basis of most of the factors examined, the EIA procedures of the Protocol would not compare favourably with the uniformity achievable in national best practice. This is not altogether a fair comparison in light of the fact that few, if any, countries require EIA over such a wide range of activities – minor as well as major – as does the Protocol. While it is likely that there are examples of domestic procedures more rigorous than those provided in the Protocol, the criterion for selection of activities to which domestic procedures are to apply are likely to be narrow or predicated upon political decisions often unrelated to the anticipated environmental impacts.

EIA: efficiency and responsiveness to need. By the very nature of its procedures EIA is particularly adaptable to different circumstances. By subjecting a proposed activity to individual attention, the assessment process should be able to take into account matters unique to that activity. On the other hand, this individual focus is

---

[32] 'The Madrid Protocol does not envisage . . . environmental assessment of pre-existing activities'; Australian Antarctic Division, 'Quarrying in Antarctica: An Environmental Policy Statement' (May 1996), para. 2(7) (text reprinted in Bush (ed.), *Antarctica and International Law* (1992–), Binder III, Booklet AU/REV 1995–96, p. 3).

[33] Protocol, Art. 8(2). Limitations that may be implied by this wording are discussed in Bush (ed.), *Antarctica and International Law* (1992–), Binder I, Booklet AT91C, pp. 27–29.

inherently inefficient. The procedures of preparing an evaluation and commenting on and assessing an activity make demands on expertise. Furthermore, effort is duplicated when similar activities are assessed. Tension between responsiveness to need and efficiency can be reflected in the ambitious role that an environmental management regime accords to EIA and the relative sparseness of procedures enjoined.

As mentioned above, the Protocol regime exceeds most national environmental management regimes in its breadth of coverage. It counterbalances this breadth by limiting the most onerous assessment procedures to activities likely to have the heaviest or most serious impacts. The threefold categorisation of less than, equal to and greater than a standard of 'minor or transitory impact' (Protocol, Article 8(1)) is associated with different levels of rigour in assessment. This can be seen as part of a triage process to ensure that a proposed activity receives 'treatment' appropriate to its needs.

An inherent difficulty, though, is that the less focus there is on the individual circumstances of a proposed activity, the less likely the assessment process will be able to inform adequately a decision based on the assessment. Sharing expertise, developing guidelines, establishing baseline measurements and developing markers of environmental impact, whether for particular areas or types of activity, are all steps that can improve both the quality and efficiency of the assessments. Even so, there is always likely to be a gap between the excellence of assessment and available resources of expertise, time and money. Considerations such as these squarely raise the question of whether, for most activities, better environmental outcomes are more likely from an informed, coordinated permit system than from arrangements under which decisions to proceed are linked to individual EIAs.

Permit System

Permits have long been an element of environmental management within the ATS. The 1964 Agreed Measures for the Conservation of Fauna and Flora required a permit to take native mammals, to control of activities within specially protected areas, and to regulate the introduction of non-indigenous species.[34] A parallel system of permits has been incorporated in Protocol annexes that replace the Agreed Measures: Annex II on conservation of Antarctic fauna and flora and Annex V on area protection and management. Furthermore, it is hard to envisage how, consistent with Article 13 of the Protocol, parties can give effect to the requirement to submit activities to EIA without linking this requirement to a permit system even though neither Article 8 of the Protocol nor Annex I make mention of permits. Permits are at the heart of the environmental management regimes under the Seals Convention and CCAMLR.

---

[34] Arts. VI, VIII and IX of the Agreed Measures for the Conservation of Antarctic Fauna and Flora.

Informed decision-making and a permit system. A permit system can produce environmentally appropriate decisions only if permits are issued within clear limits, either established on the basis of an informed judgement or, in the absence of such limits, issued by an authority that is adequately informed. In the latter case, where no limits apply, the governing instrument may leave the decision fairly open or may constrain it by criteria. National decisions under the Protocol on whether a higher impact activity that has been subject to a CEE may proceed are subject to relatively few substantive, as opposed to procedural, constraints. In this case the decision-maker stands to be informed by the evaluation 'as well as other relevant considerations' (Annex I, Article 4). A permit to take a specially protected species provides an example of an Antarctic environmental regime prescribing strict criteria for the issue of permits. One criterion is that the permit, if issued, 'will not jeopardize the survival or recovery of that species or local population' (Annex II, Article 3(5)). Expertise and information is required to render a decision informed on the basis of such a criterion. A national decision-maker may well depend on other parties to provide relevant information. Some information might be forthcoming in the course of regular exchanges of information among the parties but some may not be. For example, relevant to any decision to issue permits regarding Antarctic fauna and flora would be other permits affecting the same species in the same locality, yet Annex II does not require this information to be provided until the end of the current season (Article 6).

The practice of issuing fisheries permits (or 'licenses' as they are generally called) under CCAMLR provides the leading example within the ATS of permits issued within clear limits that have been set down on the basis of an informed judgement. National authorities issue fisheries licences within limits specified by conservation measures that the CCAMLR Commission has adopted 'on the basis of the best scientific evidence available' (CCAMLR, Article IX(1)(f)).

The expert deliberations of the Scientific Committee led to the CCAMLR Commission's adopting conservation measures which set, for example, the total allowable catch for a season in a defined ocean area. On the basis of this measure, the parties to the Convention may issue licences for taking that catch. In doing so there is no need for the national authorities to duplicate the expert assessment process undertaken by the Scientific Committee and Commission.

There is no permit system under the Environmental Protocol (or the Agreed Measures) associated with the coordination of information, expert advice and unified decision-making equivalent to that found under CCAMLR.

Balancing of interests and a permit system. A major weakness of permit systems within the ATS is the difficulty of balancing interests in the issue of permits. Fisheries licensing in accordance with conservation measures under CCAMLR leave to the national licensing authorities the allocation of licenses to operators. The authorities may license operators to take fish up to the global limit for the area set in a conservation measure. The operator can fish until the Commission gives notice that the catch limit taken by all operators in the area has been reached. In other words, the Commission does not allocate the global limit among the parties: the first and highest volume operators can scoop the pool.[35]

If fisheries licensing arrangements under CCAMLR seriously lack procedures to balance interests, then permit systems under the Environmental Protocol and Agreed Measures lack them even more. There is no provision for the Consultative Meetings to determine even sustainable global limits to which national authorities are subject for the granting of permits.

Exchange of information and consultation processes directly among national permit issuing authorities and through forums such as the CEP can probably secure the desirable degree of coordination for the national issue of permits for resources for which the demand is clearly sustainable. However, the inevitable lag involved in fully informing different national decision-makers – not least about the cumulative impact of their individual decisions – is likely to lead to unsatisfactory decisions which include demands for resources which approach unsustainable levels. The 'resource' may be the usefulness of an area for research[36] or its attraction for a commercial interest like tourism.

Practice might mitigate such difficulties. For example, work is being undertaken on standardisation and publication of advance notice of itineraries of tourist operations.[37] In addition, Annex V on protected areas envisages the possibility of management plans including 'provisions relating to the circumstances in which Parties should seek to exchange information in advance on activities which they propose to conduct' (Article 5(3)(k)). This invitation has not been generally accepted,[38] but some management plans, purportedly reflecting Annex V, show signs of moving in this direction.[39] Opposition has prevented the acceptance of proposals whereby authorities of one or more parties who have particular interests in a protected area should be given a coordinating role for that area.[40]

Consultative processes through SCAR and COMNAP give more assurance of the coordination of research activities than do other activities such as tourism. Not

---

[35] For criticism to this effect, see Australian Fisheries Management Authority, 'Heard Island and McDonald Island Exploratory Fishery (HIMIEF): Draft Interim Management Policy Covering the Period 2 November 1996 to 31 August 1997', reprinted in Bush (ed.), *Antarctica and International Law* (1992–), Binder III, Booklet AU96, p. 102.

[36] Recommendations XV–15 and XV–17; reprinted in Bush (ed.), *Antarctica and International Law* (1992–), Binder I, Booklet AT89D, pp. 126–127 and 131–132.

[37] See, for example, *Final Report of the XXI ATCM*, paras. 89–90 and Annex J.

[38] See 'Guidelines for Environmental Impact Assessment in Antarctica', appended to Resolution 1 (1999), in *Final Report of the XXIII ATCM*, Annex C.

[39] Para. 7, Specially Protected Area No. 25, annexed to Measure 1 (1997), in *Final Report of the XXI ATCM*, Annex A; para. 7(vi), Revised Management Plan for Site of Special Scientific Interest No. 23, Svarthamaren, Measure 1 (1999), in *Final Report of the XXIII ATCM*, Annex A.

[40] For a discussion of this see Bush (ed.), *Antarctica and International Law* (1992–), Binder II, Booklet AT91G, pp. 84–85.

all tour operators are members of IAATO, an organisation that might otherwise fulfil a consultative role.[41] Furthermore, tour operators have shown reluctance to commit themselves to limiting the number or intensity of visits to particular sites. In some measure this reluctance springs from the uncertainties of weather and ice conditions that on any particular voyage could rule out particular landing sites. Insistence on untrammeled rights of freedom of movement whether for tourism, scientific research or other activity is inconsistent with even the introduction of the imperfect practice of setting global limits that operate under CCAMLR. In other words, the Antarctic Treaty Consultative Meeting is a long way from limiting the number of visits to particular localities and allocating that limited number to the 'market'. There was no general support for the annex on tourism that might have provided a basis of such disciplines particular to tourism.[42]

The introduction of a permit system that could balance competing demands for access is further hindered by Annex V of the Protocol which stipulates that 'Entry into an Antarctic Specially Managed Area shall not require a permit' (Article 4(3)).[43] The category of Antarctic Specially Managed Areas (ASMAs) referred to in Annex V is the category of protected area that seemed most relevant to the regulation of non-scientific activities such as tourism.

It is possible to speculate on procedures to apportion defined global limits. One such procedure could involve operators bidding for units of a resource fixed in accordance with expert advice. This might be a share of a total allowable catch for fishing or a right to visit a tourist site. Proceeds could finance the body concerned as well as pay dividends to governments. Other procedures applied by national authorities to allocate limited resources might be used as a model. The introduction of any scheme such as this would require a greater degree of international coordination than has yet been evident in the ATS.

Enforcement

A credible set of enforcement procedures linked to sanctions is the norm for a national legal system whether it be as part of an environmental protection regime or otherwise. The political will has not existed to replicate this in an Antarctic context.

---

[41] See, for example, IAATO report in *Final Report of the XXII ATCM*, Annex G, pp. 281–292, and *ibid.*, para. 101; also *Final Report of the XXIII ATCM*, para. 115.

[42] *Final Report of the Seventeenth Antarctic Treaty Consultative Meeting, Venice, 11–20 November 1992* (Rome: Italian Ministry of Foreign Affairs, 1993), paras. 108–114; and discussion in Bush (ed.), *Antarctica and International Law* (1992–), Binder II, Booklet AT92C, pp. 30–31, and *ibid.*, AT92B, pp. 12–13. See discussion by Richardson, Chapter 4 in this book; also D. Vidas, 'Antarctic Tourism: A Challenge to the Legitimacy of the Antarctic Treaty System?', *German Yearbook of International Law*, Vol. 36, 1993, pp. 214–220.

[43] For British views on the hortatory nature of management plans for Antarctic specially managed areas, see United Kingdom, 'The Status of Protected Area Designations under Annex V to the Environmental Protocol', doc. XX ATCM/WP 16, 1996. The British argument was not generally accepted; see *Final Report of the XX ATCM*, paras. 154–156.

This is as much to do with the sense that law should have no place among those in the freemasonry engaged in heroic and scientific endeavour as with the singular international juridical status of the Antarctic. Although the explosion of illegal fishing in the last few years and the growth of legitimate private commercial activity have had the effect of shifting this perception, it still remains an influential attitude in any debate on the 'legalisation' of the ATS.

Even on these attitudinal questions the distance between the Antarctic Treaty and national legal systems is not as much as first might be imagined. In most national environmental regimes, it can be expected that enforcement occurs rarely. Most breaches are likely to go unnoticed. When breaches are noticed, there is a disinclination to take enforcement action because this is a resource intensive activity which requires both the gathering of evidence and the taking of legal action; a breach when observed will most likely elicit no more than a warning. Thus, even in a national legal system the threat of legal sanction generally remains in the background. Its contribution to conforming behaviour is still important, but in an effective environmental regime the threat of legal sanction will be a buttress rather than substitution for community support for what the environmental regime is trying to achieve.

Developing *attitudes* among a community of governments, operators and others visiting the region for work purposes or out of curiosity is something that the ATS has done well; it is of primary importance that this work be continued. The development of legal enforcement should complement, not replace that endeavour.

There are two means of enforcement, direct and indirect. Direct enforcement of an environmental regime would involve utilisation of criminal or quasi-criminal procedures to take actions to sanction those who have engaged in non-conforming behaviour. Indirect enforcement includes legal measures such as utilising civil claims for breach of a civil law or expropriating or otherwise preventing someone from benefiting from the proceeds of an illegal activity.

Direct enforcement. An effective system of direct enforcement of criminal or quasi-criminal law needs

- jurisdictional competence of the authorities taking the action,
- workable substantive laws formulated on the basis of that jurisdictional competence,
- a procedure for detecting breaches and obtaining evidence,
- procedures for prosecuting the breach which provide assurance that those accused of breach remain within jurisdiction,
- a penalty of sufficient severity to deter breaches,
- means of enforcing the penalty whether or not this involves a gaol term or, more likely, procedures to seize property whether inside or outside the jurisdiction in order to satisfy a monetary penalty.

What could be envisaged here is an elaborate international enforcement structure – an international corps of inspectors, backed up with adequate resources, entitled to prosecute breaches before an international tribunal – but enforcement is likely to remain in national hands. Thus, feasible improvement in enforcement is likely to involve *better coordination* of national action rather than the erection of new international structures. A major facet of coordination would probably involve rights of apprehension and the development of private international law links such as the taking of evidence and recognition of judgements. At least the possibility already exists for some degree of coordination of national substantive laws to implement the Protocol through the exchange of information on implementation.

Something as basic as national legislation giving effect to an international instrument cannot always be assumed. In 1997 the CCAMLR Commission found it necessary to adopt a conservation measure which required each party to prohibit its flag vessels from fishing except pursuant to the issue of a licence subject to conditions consistent with conservation measures (Conservation Measure 119/XVI). In accordance with Article 27 of the 1969 Vienna Convention on the Law of Treaties,[44] one might have thought that a party that had not already adopted such a law would risk breaching the parent convention on the grounds that the parties would be obliged to regulate their flag vessels: 'A party may not invoke the provisions of its internal law as justification for its failure to perform a treaty'.

Jurisdictional competence is sadly inadequate as an instrument for applying the Protocol to the range of actors in the Antarctic. This cannot be attributed to the lack of enough jurisdictional competence on the part of the domestic legal system of each party any more than to CCAMLR's lack of jurisdictional competence to regulate high seas fishing. A non-claimant under the Protocol has a range of jurisdictional competences that it can assert, including the most important ground of nationality. Parties to CCAMLR are in something of a similar position. To implement the Protocol a claimant could be expected to rely on additional jurisdictional grounds emanating from its territorial claim.

The main alleged obstacle which is hindering jurisdictional arrangements in implementing the Protocol is the continuing disagreement over sovereignty claims in the Antarctic. The Antarctic Treaty, the Protocol and the subsidiary instruments are almost silent on the issue. Until the parties reach an understanding on jurisdiction, it is not possible to develop the cooperative enforcement procedures essential for effecting a viable legal regime in the remote and hostile physical environment of the Antarctic. Under the marine living resources regime, the agreement on flag state jurisdiction means that the principal gap in jurisdictional competence relates to the activities of non-parties. Under the environmental regime of the Protocol, the third state problem is just another unresolved jurisdictional issue.

Members of official expeditions have long been subject to the control of home authorities, either by virtue of civil service, military and other disciplinary control or

---

[44] Published in UNTS, Vol. 1155, pp. 331ff; text reproduced in ILM, Vol. 8, 1969, pp. 654ff.

by virtue of the environmental legislation of the state of nationality of the expedition. Only with the rise of substantial unofficial activity has jurisdiction become a real issue. Under the environmental regime of the Protocol, the Consultative Parties showed some willingness to agree about the exercise of jurisdiction on 'all expeditions to and within Antarctica, on the part of its ships or nationals, and all expeditions to Antarctica organized in or proceeding from its territory', as mentioned in the Antarctic Treaty, Article VII(5)(a). The relevance of a jurisdictional net of this description will vary from one Protocol party to another, depending on the amount of activity linked to a party's territory or the strength of national links to that party. Thus, a party like the United States, which provides a big market for Antarctic tourism and in which major tourist operators are based, could exploit those jurisdictional links and have a real influence in bringing about tourist behaviour in the Antarctic that conforms with the Protocol. By contrast, a country from which only a few tourists originate and in which none of the major tour operators organising seaborne visits are based has a much smaller potential to exercise jurisdiction to implement the Protocol. Tour operators could simply avoid that country if it were seen to enforce more rigorous environmental regulation than other countries.

In referring to 'expeditions ... proceeding from its territory', Article VII(5)(a) of the Antarctic Treaty also suggests the acceptability of port and departure state jurisdiction.[45] In 1996, the Twentieth Consultative Meeting discussed 'the inspection of ships by port states in order to ensure compliance with the requirements of MARPOL 73/78 and Annex IV to the Protocol, on the prevention of marine pollution'.[46] The following year the United Kingdom proposed this as a basis of jurisdiction to enforce environmental regulations.[47] Some states, including Argentina and Germany, doubted 'the legality of asserting jurisdiction over future acts of foreign expeditions outside territorial waters'.[48] Practical issues also influence the attractiveness of asserting port or departure jurisdiction to police the Protocol. Enforcement is an onerous activity. Assuming operators did not rearrange their operations to avoid the ports concerned, the burden would fall on those Protocol parties controlling gateway ports.[49]

None of the grounds of jurisdiction suggested in the Antarctic Treaty provides anything like a comprehensive basis for a party to an Antarctic Treaty instrument to exercise jurisdiction over all private actors in the Antarctic including third party

---

[45] For a discussion of the possibilities of these two forms of jurisdiction, see Orrego Vicuña, Chapter 3 in this book, where also the distinction between port and departure state jurisdiction is described. For a general discussion of port state jurisdiction, see D. Anderson, 'Port States and Environmental Protection', in A.E. Boyle and D. Freestone (eds.) *International Law and Sustainable Development: Past Achievements and Future Challenges* (Oxford: Clarendon Press, 1999), pp. 325–344.

[46] *Final Report of the XX ATCM*, paras. 93–95.

[47] United Kingdom, 'Enhancing Compliance with the Protocol: Departure State Jurisdiction', doc. XXI ATCM/WP 22, 1997.

[48] *Final Report of the XXI ATCM*, para. 15; see Richardson, Chapter 4 in this book. Other discussion is at 'Liability – Report of the Group of Legal Experts', doc. XXII ATCM/WP 1, 1998, para. 23.

[49] See further discussion by Orrego Vicuña, Chapter 3 in this book.

nationals. A viable regime of this sort could exist only under some territorial basis of jurisdiction recognised by all parties. Proposals to develop such a basis from a condominium or some sort of trusteeship administered by parties have long fallen foul of divergent views about claims. Even so, the flexibility provided by Article IV(2) of the Antarctic Treaty to protect divergent positions on claims should not be underestimated:

> No acts or activities taking place while the present Treaty is in force shall constitute a basis for asserting, supporting or denying a claim to territorial sovereignty in Antarctica or create any rights of sovereignty in Antarctica.

Territorial jurisdiction could be based on regarding the ATS as an *objective regime*.[50] While there are a number of juridical concerns hindering such a regime from being generally acknowledged, the principal obstacle is the ambivalent attitude of the Antarctic Treaty parties themselves. This is in spite of the occasions that Consultative Parties have asserted special competences that suggest such a regime. The Consultative Parties have always balked at the hurdle when faced with a clear opportunity of asserting that the Antarctic Treaty and related instruments have established an objective regime.[51]

Acknowledgement of an objective regime need not be inconsistent with an assertion by a claimant that it is exercising jurisdiction on the basis of its claim rather than on the existence of an objective regime. Referring to Article IV(2) of the Antarctic Treaty and bifocalism, the claimant could assert that its exercise of jurisdiction was on the grounds of its claim; the other parties could understand the exercise of jurisdiction to have been on the basis of an objective regime.[52]

Acknowledgement of an objective regime may also assist environmental enforcement under CCAMLR. An objective regime could form the basis for the collective assertion of control over maritime zones off the Antarctic and permit cooperative enforcement arrangements similar to those that exist elsewhere.[53] On the other hand, such a possible innovation may be overtaken by general developments in the law of the sea through instruments such as the 1995 UN Fish Stocks Agreement that, in certain circumstances, permits enforcement against third party flag vessels.[54]

---

[50] The United Kingdom proposed this in its 1997 document 'Enhancing Compliance with the Protocol: Departure State Jurisdiction'.

[51] This argument is developed further in Bush (ed.), *Antarctica and International Law* (1992–), Binder II, Booklet AT/Rev. 1958–59, pp. 36–38.

[52] The argument is developed further in Bush (ed.), *Antarctica and International Law* (1992–), Binder I, Booklet AT91C, pp. 47–48.

[53] For example, arrangements within the Organisation of Eastern Caribbean States, mentioned in W. Edeson, 'Towards Long-Term, Sustainable Use: Some Recent Developments in the Legal Regime of Fisheries', in Boyle and Freestone (eds.), *International Law and Sustainable Development*, p. 183.

[54] Agreement for the Implementation of the Provisions of the United Nations Convention on the Law of the Sea of 10 December 1982 Relating to the Conservation and Management of Straddling Fish Stocks and Highly Migratory Fish Stocks; text reproduced in ILM, Vol. 34, 1995, pp. 1,547ff. See Orrego Vicuña, Chapter 3 in this book. For further discussion see B. Vukas and D. Vidas, 'Flags of

Indirect enforcement. Indirect means of enforcement hold promise; however, just as for direct means of enforcement, international coordination is necessary. Regard for trade and other international obligations is also necessary. An illustration of the coordination necessary is provided by moves to develop within the CCAMLR Commission a measure under which contracting parties would 'deny access to their markets of imports of *Dissostichus* spp. unless it was demonstrated that the *Dissostichus* spp. were caught in the Convention Area in accordance with CCAMLR conservation measures or were caught legitimately outside the Convention Area'.[55] Agreement to bar from ports tourists vessels that have not complied with Antarctic environmental regulations would be a step of similar complexity that parties to the Protocol might contemplate. ASOC and IUCN state that:

> the purpose of a liability regime is not only to pay for the costs of recovery and remediation ('polluter pays'). It is part of a preventive system that aims to stimulate greater care on the part of operators to avoid environmental harm and adverse impact on the other values recognized as important by the ATS, including its value for scientific research, by indicating the duties of operators and the consequences to them of environmental harm resulting from their operations.[56]

Lack of agreement about jurisdictional underpinnings of a liability regime seems almost as great an obstacle to the negotiation of a liability scheme to satisfy Article 16 of the Protocol as the substantive terms of the regime.

Liability could conceivably have a role in re-enforcing the environmental regime under CCAMLR as well as under the Protocol. The importance of the role of regional fisheries organisations such as the CCAMLR Commission in managing high seas resources, including straddling fish stocks, is now well recognised. It is not beyond the bounds of possibility that a domestic court may be persuaded to award damages in tort against a maverick fishing company that has imperilled the commercial sustainability of a stock by fishing in disregard of conservation measures imposed by the regional organisation.

Convenience and High Seas Fishing: The Emergence of a Legal Framework', in O.S. Stokke (ed.), *Governing High Seas Fisheries: The Interplay of Global and Regional Regimes* (Oxford University Press, forthcoming 2001).

[55] CCAMLR, *Report of the Seventeenth Meeting of the Commission*, para. 5(18). For the adoption of the catch documentation scheme, see media release dated 5 November 1999 by the Australian Minister for the Environment and Heritage and Minister for Agriculture, Fisheries and Forestry, at <www.antdiv.gov.au/news/>.

[56] ASOC and IUCN, 'Environmental Liability', doc. ATCM XXIII/INF 91, 1999. For further discussion on liability in the Antarctic context see especially Lefeber, Chapter 10 in this book; and, more generally, R. Lefeber, *Transboundary Environmental Interference and the Origin of State Liability* (The Hague: Kluwer Law International, 1996).

CONCLUSION

Comparison of the international environmental regimes in the Antarctic with what one would expect from national regimes reveals that there is much that stands up well. This is particularly true in the extent to which measures and other decisions of the international regimes are well informed. However, in other respects the ATS environmental regimes lag well behind what one would expect of a national system. An example is the establishment of only global catch limits for fisheries licensed under CCAMLR rather than setting national or operator quotas. A procedure does not even exist for setting global limits for permits under the Protocol. IUU fishing in the Southern Ocean has revealed just how much CCAMLR lacks effective enforcement procedures. Although agreement on jurisdictional competence covering private actors in the region seems essential, such agreement still seems distant.

As mass tourism cannot be many years away, the ATS risks destabilisation unless innovative approaches are adopted to cope with future, if not present, levels of human activity. At the very least, centrifugal regional tendencies will develop. The absence of effective cooperative procedures to enforce environmental regulations would undermine environmental management. This in turn could affect cooperation under the ATS generally and threaten its defence and other non-environmental values. This can hardly be in the interests of any party.

3

Port State Jurisdiction in Antarctica: A New Approach to Inspection, Control and Enforcement

Francisco Orrego Vicuña

EARLY JURISDICTIONAL BALANCES IN THE ANTARCTIC TREATY SYSTEM

The jurisdictional approach followed by the 1959 Antarctic Treaty[1] was the outcome of a difficult compromise between countries favouring the principle of nationality and those supporting the principle of territoriality.[2] Contrary to what might have been expected, the former principle was supported not only by countries that pertained to the category of non-claimants but also by important claimants such as the United Kingdom and Norway. It was in fact Britain that introduced the proposal for a jurisdictional system based on nationality,[3] while Norway maintained 'that any person in the Antarctic should be subject solely to the penal jurisdiction of the country of which he is a national'.[4] France, Chile and Argentina led the position relying on the territorial principle and its close association with the exercise of sovereignty claimed by these countries.[5]

---

[1] UNTS, Vol. 402, pp. 71ff.
[2] On jurisdiction in the Antarctic Treaty System, see generally F. Orrego Vicuña, *Antarctic Mineral Exploitation: The Emerging Legal Framework* (Cambridge University Press, 1988), pp. 90–117.
[3] United Kingdom, 'Statement in Relation to Art. VIII of the Antarctic Treaty at the Plenary Committee of the Conference on Antarctica, 30 November 1959', as reproduced in Chile, *Memoria del Ministerio de Relaciones Exteriores* (Santiago: Chilean Ministry of Foreign Affairs, 1959), pp. 698–704; as translated to English in W.M. Bush (ed.), *Antarctica and International Law: A Collection of Inter-State and National Documents* (London: Oceana Publications, 1982–88), Vol. I, pp. 40–41.
[4] For statement in relation to Art. VIII made by Norway, see Bush (ed.), *Antarctica and International Law* (1982–88), Vol. I, p. 41.
[5] For statements by France, Argentina and Chile in relation to Art. VIII, see *ibid*.

Article VIII of the Antarctic Treaty envisaged a formula intended not to prejudge the substantive problem. The formula dealt in fact with jurisdiction based on nationality in respect of only three categories of persons: observers in charge of inspections, scientific personnel under exchange arrangements, and members of the staffs accompanying such persons.[6] In addition, expeditions and stations of course followed 'flag state' jurisdiction, although this was addressed by the Treaty as an activity requiring advanced notice rather than explicitly as a jurisdictional matter.[7] Three specific safeguards were also provided in relation to jurisdiction. Firstly, Article VIII was without prejudice to the position of the parties relating to jurisdiction over all other persons in Antarctica.[8] Secondly, consultations would be immediately undertaken to resolve any dispute concerning the exercise of jurisdiction in Antarctica.[9] And thirdly, Consultative Meetings were entrusted with the task of considering measures in respect of jurisdictional questions.[10]

As the Antarctic Treaty was at the time concerned only with scientific research and the need to avoid confrontations of a political and military nature, its jurisdictional approach described was adequate and sufficient to ensure these limited functions and activities. In fact, this approach has worked well for over three decades.

JURISDICTIONAL SHORTCOMINGS IN AN EVOLVING CONTEXT

In the context of an increasingly complex Antarctic Treaty System, however, the early jurisdictional model has evidenced a number of shortcomings, many of which were anticipated from the very outset.[11] This has been evidenced by the questions of maritime jurisdiction in respect of fishing, the prevention of marine pollution, the issues concerning mineral exploitation, and the comprehensive approach to the protection of the Antarctic environment embodied in the 1991 Protocol[12] and related instruments. These issues made it quite evident that Article VIII of the 1959 Treaty was no longer self-sufficient to ensure the orderly development of activities in the continent and surrounding maritime areas. Nor have other important questions, such as aerial jurisdiction and private property in Antarctica, been directly addressed.

Consequently, the Antarctic has come to be characterised by a serious jurisdictional vacuum, particularly noticeable in respect of nationals, vessels and

---

[6] Antarctic Treaty, Art. VIII(1).
[7] *Ibid.*, Art. VII(5).
[8] *Ibid.*, Art. VIII(1).
[9] *Ibid.*, Art. VIII(2).
[10] *Ibid.*, Art. IX(1)(e).
[11] See in particular the discussion held at the House of Lords in 1960 led by Lords Denning and Shackleton, as referred to by Orrego Vicuña, *Antarctic Mineral Exploitation*, pp. 95–97.
[12] ILM, Vol. 30, 1991, pp. 1,461ff.

aircraft of other parties and third states.[13] Judicial decisions and the national legislation of some Consultative Parties have contributed to clarification of some aspects of the exercise of jurisdiction in Antarctica, but not sufficiently to alter the substantive provisions of the Antarctic Treaty.[14] In view of the different positions of Consultative Parties in respect of jurisdiction, it cannot be expected that a single or general formula is likely to meet with acceptance by all the Parties. Following a fatal accident involving Russian and Uruguayan personnel in Antarctica, Uruguay proposed in 1992 to complete the jurisdictional arrangements under the Treaty,[15] but the Consultative Parties, while recognising the importance of this question 'which was left deliberately open in Article IX(1) of the Antarctic Treaty', opted not to pursue it in view of the 'delicate and sensitive problems' that it raised.[16] In other words, each regime or problem will need to be addressed on its own merits in order to find the appropriate jurisdictional answer, and this is in fact the approach that has been followed under the ATS.

It was precisely as a result of these early discrepancies that the 1964 Agreed Measures for the Conservation of Antarctic Fauna and Flora could only agree that permits be granted by the 'appropriate authority', without specifying whether this authority would be the state asserting territorial jurisdiction or the state of nationality, or even the flag state.[17] Subsequent recommendations, however, introduced with due caution the concept of jurisdiction over nationals, to some effect. In this respect, national legislation followed different approaches, some relying on territorial jurisdiction and others relying on the principle of nationality. In this matter, the survival of forms of concurrent jurisdiction side by side in the Antarctic was made possible through restraint and cooperation. However limited the jurisdictional approach of the Agreed Measures might have been, it allowed for the first time under the ATS the exercise of jurisdiction over limited, adjacent maritime areas.[18] It should also be noted that a special kind of 'scientific inspection' has been developed regarding protected areas in order to update information on the state of fauna and flora.[19]

---

[13] See comments by Bush (ed.), *Antarctica and International Law* (1982–88), Vol. I, pp. 78–82.
[14] For an examination of national legislation and judicial decisions concerning jurisdiction in Antarctica, see Orrego Vicuña, *Antarctic Mineral Exploitation*, pp. 97–101 and 104–107.
[15] Uruguay, 'Issues Relating to the Exercise of Jurisdiction in Antarctica', doc. XVII ATCM WP 17, 1992. See further doc. XVIII ATCM/WP 32, Rev. 1, 1994.
[16] *Final Report of the Eighteenth Antarctic Treaty Consultative Meeting, Kyoto, Japan, 11–22 April 1994* (Tokyo: Japanese Ministry of Foreign Affairs, 1994), para. 122.
[17] For a discussion of the jurisdictional regime of the Agreed Measures, see Orrego Vicuña, *Antarctic Mineral Exploitation*, pp. 110–112.
[18] *Ibid.*, pp. 137–141.
[19] United Kingdom, 'Making and Inspection Survey of Specially Protected Areas in the South Orkney Islands, Antarctica, 07/01/94–17/02/94', doc. XVIII ATCM/INFO 34, 1994.

NATIONALITY, FLAG STATE JURISDICTION AND THE EMERGING IMBALANCE

Jurisdiction based on nationality and on the flag of registry became paramount in the conventional regimes which dealt with high seas resources,[20] first in the 1972 Convention for the Conservation of Antarctic Seals[21] (Seals Convention) and than in the 1980 Convention on the Conservation of Antarctic Marine Living Resources[22] (CCAMLR). While the latter does not specifically advocate jurisdiction based on nationality and flag state registry, such type of jurisdiction can be clearly inferred from some other of its provisions and from national legislation and practice.[23] This departure from the cautious approach of the Antarctic Treaty and the earlier regimes can be explained by the fact that both the CCAMLR and the Seals Convention dealt essentially with maritime areas where the principle of nationality and flag state jurisdiction are the normal standards. However, that also meant the introduction of some jurisdictional imbalances in the ATS.

These imbalances relate to three aspects of Antarctic cooperation. Firstly, the balance between territorial jurisdiction and jurisdiction based on nationality has definitively been tilted in favour of the latter. Indeed, the principle of nationality is no longer the exception but the general rule within the System, a situation paralleling non-claimants having outnumbered claimants among Consultative Parties and Antarctic Treaty parties generally. Secondly, jurisdiction based on nationality is no longer applied mainly in respect of maritime jurisdiction but has expanded to cover the activities on the continent itself. The Convention on the Regulation of Antarctic Mineral Resource Activities[24] (CRAMRA) offers one example, despite important efforts undertaken in its negotiation to accommodate other interests, including those of claimant countries, in a highly imaginative institutional structure.[25] This is also the case, again with limited exceptions, of the Environmental Protocol.

Rather than innovating on the pre-existing jurisdictional situation, the Protocol has maintained the balances and imbalances as embodied in the ATS. An example of this is the conservation of Antarctic fauna an flora. While Annexes II and V to the Environmental Protocol have perfected the regime with regard to conservation and the system of protected areas and management, these Annexes have not altered the essence of the jurisdictional approach of the Agreed Measures and subsequent recommendations. In matters such as the prevention of marine pollution, waste disposal and management and environmental impact assessment, the principle of

---

[20] Orrego Vicuña, *Antarctic Mineral Exploitation*, pp. 112–113.
[21] UNTS, Vol. 1080, pp. 175ff; and ILM, Vol. 11, 1972, pp. 251ff.
[22] ILM, Vol. 19, 1980, pp. 837ff.
[23] Orrego Vicuña, *Antarctic Mineral Exploitation*, pp. 145–155.
[24] ILM, Vol. 27, 1988, pp. 868ff.
[25] Orrego Vicuña, *Antarctic Mineral Exploitation*, Chapter 7.

nationality or flag state jurisdiction prevails. Similarly, a potentially greater institutional role has been retained in respect of matters such as inspection since this was already established in earlier legal developments within the ATS. However, some innovations pertaining to jurisdictional issues and having a potential for their development have been introduced. These will be examined further below.

The third and most serious aspect of Antarctic cooperation affected by jurisdictional imbalance relates to the question of effectiveness within the ATS. As experience in the high seas has indicated, flag state jurisdiction or jurisdiction based solely on nationality has not been the most reliable approach for ensuring the implementation of agreed rules and standards. Inspection and prosecution carried out in such context have had a very limited effect. This situation has led to the emergence of alternative approaches that have the possibility of ensuring a more effective implementation of rules and regimes, first in terms of coastal state jurisdiction and more recently in terms of port state jurisdiction. The Antarctic cooperation has already begun to evidence some of the same trends.

THE CHANGING SCOPE OF THE INSPECTION SYSTEM UNDER THE ATS

In conjunction with the jurisdictional issues discussed above, the 1959 Antarctic Treaty addressed the question of the inspection system. Because the prevalent needs at the time were linked to the development of scientific research and the promotion of peaceful uses, the inspection arrangements devised under Article VII(1)–(4) of the Treaty were of a limited scope but sufficient for those purposes. These provisions and the practice developed in their implementation proved to be successful in the light of the early needs of the ATS.[26] However, as with jurisdiction, the period relating to the development of resources and environmental protection has raised new and different needs.

Because the area of application of the inspection system envisaged under the Antarctic Treaty referred only to the continent, the area had to be expanded for the purpose of the resource regimes. Although the Seals Convention had timidly hinted at the need for the inspection system to be applied essentially to the marine environment of sealing, such a system was never implemented. Some of the inspections undertaken under the Antarctic Treaty were also concerned with the conservation of living resources, but their scope was very limited and mostly symbolic.[27] It is the role of the CCAMLR to provide for a more comprehensive inspection system for the area of this Convention as a whole, thus properly extending the system to the high seas.

---

[26] Under Art. IX(1)(d) of the Antarctic Treaty, the Consultative Parties were entrusted with the adoption of measures regarding the 'facilitation of the exercise of the rights of inspection provided for in Article VII of the Treaty'.

[27] Orrego Vicuña, *Antarctic Mineral Exploitation*, pp. 250–252.

Another aspect of the inspection system envisaged by the 1959 Treaty concerns the inspection of ships and aircraft. Article VII(3) of the Antarctic Treaty provides that such inspections may only be made at points 'in Antarctica' where cargoes or personnel are discharged or embarked. Although vessels have regularly been inspected at such points,[28] this provision has been narrowly interpreted so to include only inspections of vessels flying the flag of a Treaty party. Vessels of third states are thus excluded, as are inspections on the high seas or of vessels underway, unless the consent of the master of the vessel has been obtained.[29]

While the reference of Article VII(3) clearly indicates the Antarctic continent, to the extent that installations or other facilities might be located beyond the continent, there is no reason why a similar rationale for inspection might not be applied. This is in fact the situation that arose under CRAMRA, which first provided for the inspection under Article VII of the Antarctic Treaty for the purposes of this Treaty and then for an inspection system geared to the particular needs of CRAMRA.[30] The latter system applied to the area of CRAMRA as a whole as well as to 'ships and aircraft supporting such activities at points of discharging or embarking cargoes or personnel anywhere in that area'. It follows that the CRAMRA inspection system is considerably larger than that of the Antarctic Treaty. CRAMRA allows for inspections at sea and on the continental shelf, or any other area covered by the Convention.[31]

Another area for needed improvement falls under the resource regimes and concerns the need to carry out the inspection in a manner compatible with the development of operations and, above all, with the protection of proprietary data and other justifiable requirements of confidentiality. This was particularly evident in the case of CRAMRA, but it is equally relevant under CCAMLR and other resource activities. It should be pointed out that CRAMRA envisaged two additional principles concerning inspection that are also relevant in other contexts. First, inspections carried out by national observers or by institutional observers should be compatible and reinforce each other but not impose an undue burden on the operation of stations, installations and equipment visited.[32] Second, no activity should be undertaken in a given area until effective provision has been made for inspection in that area.[33]

---

[28] For references to inspections of vessels from Liberia, Russia and Germany see, for example, *Final Report of the XVIII ATCM*, para. 76.

[29] United Kingdom, 'Inspections under the Antarctic Treaty. Inspection Checklists', doc. XIX ATCM/WP 2, 1995; in conjunction with *Final Report of the Nineteenth Antarctic Treaty Consultative Meeting, Seoul, 8–19 May 1995* (Seoul: Ministry of Foreign Affairs of the Republic of Korea, 1995), paras. 78 and 82.

[30] CRAMRA, Arts. 11 and 12.

[31] Orrego Vicuña, *Antarctic Mineral Exploitation*, p. 251.

[32] CRAMRA, Art. 12(6).

[33] *Ibid.*, Art. 12(8).

By applying practical experience, a number of useful improvements have been made to the inspection arrangements under the Antarctic Treaty. Joint inspections have been regularly undertaken[34] and reports are regularly distributed. However, there have been complaints that inspections concentrate on few stations and thereby affect the station's normal work.[35] This situation has probably arisen from the fact that conducting inspections has become more of a status symbol in the ATS than a mechanism directed to ensure compliance with the Treaty. But what is more important is that checklists have been approved for conducting inspections of stations and associated installations as well as of vessels, abandoned stations and dumping sites.[36]

THE NATURE OF THE INSPECTION SYSTEM UNDER THE ATS

The most serious problem affecting inspections under the 1959 Antarctic Treaty and related conventions concerned the nature of the inspection system. Because of the close links between inspection and jurisdiction, national appointments prevailed over institutional appointments. This was clearly the approach of Article VII of the Treaty in application of the principle of nationality, and it was also the approach of CCAMLR under the principle of flag state jurisdiction. Indeed, under the latter Convention, inspectors are designated by the members of the Commission: they remain subject to the jurisdiction of the contracting party of which they are nationals; they report to the member of the Commission by which they have been designated; and it is this member that reports to the Commission. Furthermore, flag state jurisdiction is also envisaged for the purposes of prosecution and sanctions.[37]

In spite of these restrictions, Article XXIV of CCAMLR hinted at some improvements in the inspection system. An important improvement was agreeing that 'a system of observation and inspection' would be established to take 'account of the existing international practice'.[38] A relevant case of international practice was that of the International Convention for the Regulation of Whaling.[39] Having originally established inspection based only on national appointments, the Convention later introduced amendments authorising observers under bilateral

---

[34] Joint inspections have been undertaken, for example, by New Zealand and the United Kingdom, New Zealand and the USSR, and France and Germany. See generally United Kingdom, 'Reports of Inspections under Art. VII of the Antarctic Treaty', doc. XVI ATCM/INFO 5, 1991.

[35] Argentina, 'Inspections under Art. VII of the Antarctic Treaty', doc. XVIII ATCM/WP 30, 1994. The need to rationalise Antarctic inspections has been noted in the *Final Report of the XIX ATCM*, para. 76.

[36] See United Kingdom, 'Inspections under the Antarctic Treaty'. See further SCAR, 'Antarctic Inspection Checklist', doc. XVIII ATCM/WP 22, 1994.

[37] CCAMLR, Art. XXIV.

[38] *Ibid.*, Art. XXIV(2)(a).

[39] UNTS, Vol. 161, pp. 74ff.

agreements and other arrangements, meaning a greater institutional role.[40] Moreover, the system envisaged under CCAMLR provided *inter alia* for the appointment of national observers and the flag state role discussed above, thus opening the way for other alternatives. Finally, observers and inspectors would operate under the terms and conditions established by the Commission, again meaning a greater institutional role. Most important was the practice of carrying out observation and inspection on board vessels engaged in scientific research or harvesting in the Convention area. This practice of potentially allowing for the inspection of vessels of any nationality has been confirmed under the CCAMLR, particularly under the terms of Conservation Measure 118/XVI of 1997.[41]

The CCAMLR arrangements were rightly criticised as weak, particularly considering that during the first five years of operation of the Convention, only voluntary guidelines for inspection based on bilateral agreements and reciprocity were agreed to.[42] However, this was to change as of 1988 when the Commission adopted the recommendation of a Standing Committee on Observation and Inspection, set up the previous year. That recommendation provided for a compulsory inspection and observation system.[43] In spite of these improvements, the practice has adhered to the approach of national inspections in respect of vessels flying the same flag. It has been reported, for example, that during 1990 the Soviet Union carried out 118 inspections of its own vessels, only one of which resulted in prosecution for violation of a Commission regulation on mesh size. In 1991 and 1992, 16 of 18 inspections carried out were conducted by inspectors on board vessels of their own countries.[44] It follows that nationality and the flag state are still very much the rule under this system with all the shortcomings this entails.

Under the CRAMRA system of inspection, the designated observers would only be subject to the jurisdiction of the party of which they were nationals, without prejudice to the positions relating to jurisdiction over all other persons in the area of the Convention. This meant enlarging of the geographical scope of the Antarctic Treaty. More important, however, was the fact that in addition to national observers designated by the members of the Commission, observers would also be designated by the Commission itself or by the relevant Regulatory Committees, thereby indicating a more centralised and institutionalised system of inspection.[45] Furthermore, reports would be transmitted to the Commission and to any Regulatory

---

[40] Orrego Vicuña, *Antarctic Mineral Exploitation*, p. 250.

[41] CCAMLR, Art. XXIV(2)(b). Conservation Measure 118/XVI was adopted at the XVI Meeting of the CCAMLR Commission, November 1997.

[42] L. Cordonnery, 'Environmental Protection in Antarctica: Drawing Lessons from the CCAMLR Model for the Implementation of the Madrid Protocol', *Ocean Development and International Law*, Vol. 29, 1998, pp. 135–136.

[43] F. Orrego Vicuña, 'The Regime of Antarctic Marine Living Resources', in F. Francioni and T. Scovazzi (eds.), *International Law for Antarctica* (The Hague: Kluwer Law International, 1996), pp. 143–145.

[44] Cordonnery, 'Environmental Protection in Antarctica', p. 136.

Committee which had competence in the areas where the inspections took place. Aerial inspections would also be carried out at any time over the CRAMRA area, again implying a broader role for the inspection system.

INSPECTION UNDER THE PROTOCOL: A STEP BACK

The inspection system envisaged under the Environmental Protocol is very narrow indeed, particularly considering the broad purposes of the Protocol to promote the protection of the Antarctic environment and dependent and associated ecosystems. True enough, attempts were made to develop a more comprehensive system of international inspection during the negotiations of the Protocol, but in the end the Protocol failed to innovate on the pre-existing jurisdictional rules and deliberately ignored the significant progress that CRAMRA had intended in this field.

A first significant limitation of the Protocol is that the inspection system is confined to the scope of Article VII of the 1959 Treaty, both in terms of jurisdiction and its geographical ambit.[46] This means that the Protocol has not availed itself of the evolving practice under the ATS, which among many other aspects had established a clear trend relating to maritime areas beyond the strict confines of the continent. Under the Protocol and the related understandings of the Eleventh Special Antarctic Treaty Consultative Meeting, nothing in the Protocol shall derogate from the rights and obligations of the parties under CCAMLR, the Seals Convention and the International Convention for the Regulation of Whaling; nor does environmental impact assessment apply to activities under either CCAMLR or the Seals Convention.[47] However, this does not mean at all that the marine environment is irrelevant for the purpose of inspection. On the contrary, the Protocol is concerned with the Antarctic Treaty area as a whole: the impact on dependent and associated ecosystems, specific questions of protection of the marine environment, and even the conservation principles of CCAMLR. Therefore, there would have been every reason to rely on the trend for broadening the scope of the Protocol's inspection system.

A particularly striking consequence of the above is that the system envisaged refers only to stations, installations, equipment, ships and aircraft open to inspection under Article VII(3) of the Antarctic Treaty, that is, the confinement of inspection to the continent and to ships and aircraft 'at points of discharging or embarking cargoes or personnel in Antarctica'.[48] This step back to square one is hardly conducive to creating an effective inspection system and is even less conducive to ensuring

---

[45] CRAMRA, Art. 12(1)(b).

[46] Protocol, Art. 14(1).

[47] *Ibid.*, Art. 4; also paras. 7 and 8 of the Final Act of the Eleventh Antarctic Treaty Special Consultative Meeting, text reproduced in J.A. Heap (ed.), *Handbook of the Antarctic Treaty System*, 8th edition (Washington, DC: US Department of State, 1994), pp. 2,016–2,018.

[48] Protocol, Art. 14(3).

compliance with the Protocol, as stated among the main purposes of the system under its Article 14(1).

In effect, there is little or no difference between the inspection system of the Antarctic Treaty and that of the Protocol. The inspections carried out under the Antarctic Treaty have increasingly been concerned with the protection of the environment, even before the Protocol entered into force.[49] The checklists referred to above have included specific items in respect of the protection of the environment.[50]

However, some aspects of the inspection system under the Protocol point toward an increased, potential institutionalisation and can therefore be considered more compatible with the evolving nature of such arrangements under the ATS. This is the case regarding the commitment of Consultative Parties to arrange inspections 'individually or collectively', and also the case of the 'dual track' which allows observers to be designated either by a Consultative Party of which they are nationals or by the Antarctic Treaty Consultative Meetings.[51] In the latter case, observers can in principle be of a nationality different from that of the inspected party and shall operate under the procedures established by such Consultative Meetings.[52] Non-governmental organisations have proposed to institutionalise an 'inspectorate' in the form of an Environmental Monitor Group.[53]

Two other features of the Protocol inspection system are a greater concern for the publicity of inspection reports and the inspection of records maintained by the inspected station. Inspection reports shall be sent to the inspected party, which will have the opportunity to comment; both the reports and the comments shall then be circulated to all the parties and to the Committee for Environmental Protection for consideration at the next Antarctic Treaty Consultative Meeting. Thereafter, they shall be made publicly available.[54] This procedure will allow for greater public scrutiny. The records maintained pursuant to the Protocol by the inspected station,

---

[49] See, for example, the emphasis placed on environmental protection in the inspections carried out by Australia, China, Germany and France, as well as by Norway, as reported in *Final Report of the Sixteenth Antarctic Treaty Consultative Meeting, Bonn, 7-18 October 1991* (Bonn: German Federal Ministry of Foreign Affairs, 1991), para. 59. See also, for example, the inspection reports submitted by Argentina (doc. XIX ATCM/INF 50, 1995) and Sweden (doc. XVIII/INFO 45, 1994). Most inspections include as a matter of routine the situation of conservation of flora and fauna and of protected areas in the vicinity of the station inspected; see, for example, Norway, 'Report of the Norwegian Antarctic Inspection under Art. VII of the Antarctic Treaty, December 1996', doc. XXI ATCM/IP 37, 1997; also New Zealand, 'Inspection Manual Adopted by New Zealand in Respect of the Implementation of the Antarctic (Environmental Protection) Act 1994', doc. XXI ATCM/WP 26, 1997.

[50] See United Kingdom, 'Inspections under the Antarctic Treaty'; also SCAR, 'Antarctic Inspection Checklist'. Para. 72 of the *Final Report of the XIX ATCM* also assigns special importance to inspection of waste disposal facilities.

[51] Protocol, Art. 14(1) and (2).

[52] Cordonnery, 'Environmental Protection in Antarctica', p. 136.

[53] ASOC, 'The Case for Strengthening the Mechanisms for Conducting Inspections under the Protocol', doc. XVIII ATCM/INFO 95, 1994.

[54] Protocol, Art. 14(4).

installation, ship or aircraft, shall also be made subject to inspection.[55] Since these records presumably do not refer to confidential or proprietary data, they are not subject to special restrictions in this context.

As these institutional developments are inserted within the limited scope of the inspection system referred to above, they are also of limited significance. Furthermore, as has been the case with CCAMLR, it is quite likely that practice will tend to favour national inspections over any independent mechanism.[56] The Protocol approach has been rightly criticised in terms of both having missed a unique opportunity to open new ground in the matter[57] and failing to extend to important aspects, such as environmental impact assessment procedures.[58]

COASTAL STATE JURISDICTION AS AN ALTERNATIVE APPROACH TO INSPECTION, CONTROL AND ENFORCEMENT

As noted above, the shortcomings of the principle of nationality and flag state jurisdiction and of the related systems of inspection, control and enforcement have led to a reaction first expressed in terms of asserting coastal state jurisdiction. While this has been a typical phenomenon of the high seas, with particular reference to questions of marine pollution and fisheries, it can be also found in the practice under the Antarctic Treaty and related developments.

The proclamation by United Kingdom of a 200-mile maritime zone for South Georgia and South Sandwich islands is one such case of asserting coastal jurisdiction.[59] While not taking the form of an exclusive economic zone (EEZ) claim, the proclamation poses serious jurisdictional problems for CCAMLR, not least in matters of inspection and enforcement. Indeed, although the intention of the jurisdiction is to supplement and reinforce, not to replace, the role of CCAMLR, the jurisdiction purports to control alleged violations of conservation measures while legislation to monitor and protect fish stocks in the zone has also been enacted.[60] The fact that the proclamation states that jurisdiction shall be exercised in accordance with the rules of international law does not add much clarification.

The legislation enacted pursuant to this proclamation excludes the area of application of the Antarctic Treaty but not that of CCAMLR. It also provides that fisheries officers called to intervene shall have regard for the provisions of the

---

[55] *Ibid.*, Art. 14(3).

[56] Cordonnery, 'Environmental Protection in Antarctica', p. 136.

[57] M. Bruce, 'Epilogue', in J. Verhoeven, P. Sands and M. Bruce (eds.), *The Antarctic Environment and International Law* (London: Graham & Trotman, 1992), p. 185.

[58] Cordonnery, 'Environmental Protection in Antarctica', p. 141.

[59] Proclamation (Maritime Zone) No. 1, 1993, *The South Georgia and the South Sandwich Islands Gazette*, No. 1, May 1993.

[60] The Fisheries (Conservation and Management) Ordinance 1993, 23 July 1993; The Fishing (Maritime Zone) Order 1993, 26 July 1993.

Convention, but the question of whether they have done so 'shall not be inquired into in any court of law'. This means in practice that national inspection and enforcement prevail over any arrangement under CCAMLR. Questions of renewed cooperation under CCAMLR to strengthen conservation and enforcement,[61] as well as the eventual application of CCAMLR dispute settlement procedures,[62] have also been raised in this context.

A similar situation arises under the EEZ proclamation made by Australia in 1994, which extends to waters partly under CCAMLR in conjunction with Australian External Territories and the Australian Antarctic Territory.[63] Thus far this legislation does not apply to foreign nationals and vessels, except for the Whale Protection Act of 1980, but the possibility of so doing is impending and as such poses a potential conflict with CCAMLR.

There are a number of dormant EEZ claims in the Antarctic which are made directly or indirectly yet not enforced under considerations of restraint and accommodation.[64] If national claims are asserted by some countries for the purposes of inspection, control and enforcement, this may activate all sorts of jurisdictional questions by other countries as well.

PORT STATE JURISDICTION:
A NEW APPROACH TO THE PROTECTION OF THE MARINE ENVIRONMENT AND FISHERIES CONSERVATION

The development of cooperation and the establishment of organisations for the purpose of protecting the environment and ensuring conservation and management of living resources would be rendered fruitless if effective compliance with the rules were not adequately ensured. Two lines of action have proven necessary to this effect: the first to strengthen the role of the flag state and thus increase its obligations and commitments in terms of compliance and enforcement in respect of vessels flying its flag; the second to supplement this role, which has proven historically to be insufficient, with other actions undertaken by means of international cooperation and other forms of participation involving international organisations and even coastal states or other states with a relevant interest in the matter.

---

[61] Joint Statement by Argentina and the United Kingdom of 7 May 1993, as cited by E. Meltzer, 'Global Overview of Straddling and Highly Migratory Fish Stocks: The Nonsustainable Nature of High Seas Fisheries', *Ocean Development and International Law*, Vol. 25, 1994, p. 278.

[62] Statement by Chile, in CCAMLR Commission, 'Preliminary Report of the Thirteenth Meeting, 1994', para. 7(4).

[63] Australia, Proclamation of the Outer Limit of the Exclusive Economic Zone, 26 July 1994, *Commonwealth of Australia Gazette*, Special, No. S 290, 29 July 1994, for which see S. Kaye and D.R. Rothwell, 'Australia's Antarctic Maritime Claims and Boundaries', *Ocean Development and International Law*, Vol. 26, 1995, pp. 195–226.

[64] For the continental shelf, see Vidas, Chapter 14 in this book.

The strengthening of the obligations of the flag state has already been the subject of important developments under international law, which is evidenced in general by the numerous conventions on international environmental law and in particular by the conventions of the International Maritime Organisation (IMO). This is also the case in respect of fisheries of the 1993 Agreement to Promote Compliance with International Conservation and Management Measures by Fishing Vessels on the High Seas,[65] of the 1995 Code of Conduct for Responsible Fisheries[66] and of the 1995 UN Fish Stocks Agreement.[67] All these agreements and conventions are relevant for the Antarctic. In addition this has also been the line pursued by most of the recent developments in the ATS itself.

It is precisely because exclusive flag state enforcement in the high seas as envisaged under traditional international law[68] has become a rather limited and many times unreliable mechanism for the adequate observance of the legal order that other alternatives began to emerge.[69] Boarding and inspection by non-flag state officials are provided for in bilateral agreements and in a number of treaties that establish regional fisheries organisations.[70] Seizure and arresting of vessels are in addition provided for in other cases.[71] Most of these treaties allow only for flag state prosecution and sanctioning, and no treaty allows for prosecution by a non-flag state, while all treaties are regional in nature. This trend, however, has also become global under the UN Fish Stocks Agreement of 1995.

The most significant development in this respect in contemporary international law relates to the concept of port state jurisdiction, which allows for a different kind of jurisdictional authority to intervene in matters concerning the marine environment

---

[65] Agreement to Promote Compliance with International Conservation and Management Measures by Fishing Vessels on the High Seas, ILM, Vol. 33, 1994, pp. 968ff (FAO Compliance Agreement).

[66] Code of Conduct for Responsible Fisheries, Resolutions 4/95 and 5/95 of the FAO Conference, 28th Session, 20–31 October 1995, doc. C 95/REP, Annex I (Code of Conduct).

[67] Agreement for the Implementation of the Provisions of the United Nations Convention on the Law of the Sea of 10 December 1982 relating to the Conservation and Management of Straddling Fish Stocks and Highly Migratory Fish Stocks, ILM, Vol. 34, 1995, pp. 1,542ff.

[68] For the discussion of the principles of international law governing jurisdiction over vessels on the high seas, see generally the *Lotus*, Permanent Court of International Justice, Series A, No. 10, p. 28.

[69] For an overview of this development in the 1990s, see B. Vukas and D. Vidas, 'Flags of Convenience and High Seas Fishing: The Emergence of a Legal Framework', in O.S. Stokke (ed.), *Governing High Seas Fisheries: The Interplay of Global and Regional Regimes* (Oxford University Press, forthcoming 2001).

[70] See in particular Art. 18 of the 1978 Convention on Future Multilateral Cooperation in the Northwest Atlantic Fisheries, UNTS, Vol. 1135, pp. 369ff; Art. 11(6) of the 1994 Convention on the Conservation and Management of Pollack Resources in the Central Bering Sea, ILM, Vol. 34, 1994, pp. 67ff; Art. 9(5) of the 1967 Convention on Conduct of Fishing Operations in the North Atlantic, ILM, Vol. 6, 1967, pp. 760ff; para. 3 of the 1965 Agreement between the Government of the United States and the Government of the U.S.S.R. relating to Fishing for King Crab, UNTS, Vol. 541, pp. 97ff.

[71] See, for example, Art. 10(1)(a)–(b) of the 1952 International Convention for the High Seas Fisheries of the North Pacific Ocean, UNTS, Vol. 205, pp. 77ff; and the 1992 Convention for the Conservation of Anadromous Stocks in the North Pacific Ocean, *Law of the Sea Bulletin*, No. 22, 1993, p. 21.

and fisheries. A limited role for port state intervention had been envisaged under the MARPOL 73/78 Convention regarding inspection of certificates and the reporting and prosecution of certain violations.[72] In order to allow for investigation and prosecution of violations which took place in the high seas and other areas, this approach was considerably enlarged in Article 218 of the UN Convention on the Law of the Sea (LOS Convention) in respect of marine pollution.[73]

The 1993 FAO Compliance Agreement expanded the concept of port state jurisdiction to the field of fisheries by requiring the port state to promptly notify the flag state when a fishing vessel is voluntarily in its port and there are reasonable grounds for believing that it has engaged in an activity undermining the effectiveness of international conservation and management measures.[74] However, beyond notification, special arrangements would have to be made for the port state to undertake investigations. The 1995 UN Fish Stocks Agreement provides that a port state has the right and duty to take measures, in accordance with international law, to promote the effectiveness of sub-regional, regional and global conservation and management measures, with the condition of not discriminating in form or in fact against the vessels of any state.[75]

In the exercise of these powers under the Fish Stocks Agreement the port state may, *inter alia*, inspect documents, fishing gear and catch on board the fishing vessels.[76] The '*inter alia*' clause indicates that other measures may be taken as well. Although detention, arrest or continued boarding are not referred to in the text of the Agreement, such situations may arise in the case of inaction of the flag state. It should also be noted that the port state may take action in its own right and it does

---

[72] MARPOL 73/78, Arts. 5(2)–(3) and 6 (2)–(5); see discussion by P.W. Birnie and A.E. Boyle, *International Law and the Environment* (Oxford: Clarendon Press, 1992), pp. 268–269. For an amalgamation of IMO resolutions on port state control see Resolution A. 787/19, 1995; for European Union action on port state control see European Council Directive 95/21/EC, and the corresponding amendments to the 1982 Paris Memorandum of Understanding on Port State Control. Various memoranda of understanding on port state control have since been developed for regions such as Asia-Pacific, Latin America, Caribbean, Mediterranean, the Gulf area, West and Central Africa, and Indian Ocean. See generally on this development M. Valenzuela, 'Enforcing Rules against Vessel-Source Degradation of the Marine Environment: Coastal, Flag and Port State Jurisdiction', in D. Vidas and W. Østreng (eds.), *Order for the Oceans at the Turn of the Century* (The Hague: Kluwer Law International, 1999), pp. 496–501.

[73] P.-M. Dupuy and M. Remond-Gouilloud, 'La préservation du milieu marin', in R.J. Dupuy and D. Vignes (eds.), *Traité du Nouveau Droit de la Mer* (Paris and Bruxelles: Economica and Brylant, 1985), pp. 979–1,045; D. Vignes, 'La juridiction de l'Etat du port et le navire en droit international', in Societé Française pour le Droit International, *Le navire en droit international* (Paris: Editions A. Pedone, 1992), pp. 127–150; G.C. Kasoulides, *Port State Control and Jurisdiction. Evolution of the Port State Regime* (Dordrecht: Martinus Nijhoff, 1993). See also generally Choung Il Chee, 'Jurisdiction of Port State over Private Foreign Vessel in International Law', *The Korean Journal of International Law*, Vol. 39, 1994, pp. 55–67.

[74] FAO Compliance Agreement, Art. V(2).

[75] Fish Stocks Agreement, Art. 23(1).

[76] *Ibid.*, Art. 23(2).

not need a request from another state to do so. The Agreement further requires that the vessel is voluntarily in the ports of the port state or at its offshore terminals.

Another aspect of particular significance is concerned with the access of fishing vessels to foreign ports. The Fish Stocks Agreement has rightly stated the governing principle of international law in terms that nothing in Article 23 of the Agreement 'affects the exercise by States of their sovereignty over ports in their territory in accordance with international law'.[77] In fact, it is a generally accepted principle of international law that states are under no legal obligation to grant merchant vessels access to their ports, although a presumption of being ports-opened may operate in certain circumstances.[78] The decision in the *Aramco* case to uphold a right of entry to foreign ports[79] has been criticised as not correctly stating the law.[80] States may of course grant rights of access by means of treaties and other agreements that create a legal obligation, as is particularly the case of Treaties of Friendship, Commerce and Navigation.[81] However, fishing vessels are usually excluded from such treaties.

The 1923 Statute of the International Regime of Maritime Ports provided for equality of treatment on the basis of reciprocity to vessels of the contracting parties, but the Statute expressly stated that it 'does not in any way apply to fishing vessels or to their catches'.[82] Even in the specific context of fishing agreements it is quite common that access to ports and other services will be restricted, particularly when violations of conservation and management measures have taken place.[83] Still, fishing agreements have more often established prohibitions of landing of catches, trans-shipment and other operations, again with particular emphasis on the question of violations.[84] There has been a great deal of national legislation requiring licenses for fishing vessels to enter a port and establishing restrictions or prohibitions on landings, trans-shipment, and other similar activities.[85] It is therefore that Article 62

---

[77] *Ibid.*, Art. 23(4).

[78] A.V. Lowe, 'The Right of Entry into Maritime Ports in International Law', *San Diego Law Review*, Vol. 14, 1977, p. 622. See also T. Treves, 'Navigation', in R.J. Dupuy and D. Vignes (eds.), *A Handbook on the New Law of the Sea*, Vol. 2 (Dordrecht: Martinus Nijhoff, 1991), pp. 940–942.

[79] 'Saudi Arabia v. Arabian American Oil Co.', *International Law Reports*, Vol. 27, 1958, p. 212.

[80] Lowe, 'The Right of Entry into Maritime Ports', p. 598.

[81] See *Military and Paramilitary Activities in and against Nicaragua, (Nicaragua v. United States)* (Merits), ICJ Reports 1986, p. 98, paras. 270–282, also reproduced in ILM, Vol. 25, 1986, pp. 1,023ff; see also *Case concerning Oil Platforms (Islamic Republic of Iran v. United States of America)*, International Court of Justice, Judgment on preliminary objection, *Communiqué*, No. 96/33, 12 December 1996.

[82] Art. 14 of the Convention and Statute of the International Regime of Maritime Ports, *League of Nations Treaty Series*, Vol. 58, pp. 285ff.

[83] See, for example, Art. 3(2)(d) of the 1989 Convention for the Prohibition of Fishing with Long Driftnets in the South Pacific, ILM, Vol. 29, 1990, pp. 1,449ff.

[84] See, for example, Art. VIII of the 1957 Interim Convention on Conservation of North Pacific Fur Seals, *U.S. Marine Mammal Commission Compendium of Selected Treaties*, Vol. II, pp. 1,581ff; and Art. 7 of the 1962 Agreement between Denmark, the Federal Republic of Germany and Sweden on the Protection of Salmon in the Baltic Sea.

[85] See for example United Kingdom, Fishery Limits Act 1976, section 3(6), *FAO Fishery and*

of the LOS Convention has authorised the coastal state to enact laws and regulations relating to 'the landing of all or any part of the catch by such vessels in the ports of the coastal State'.[86] Although the latter provision refers to conservation in the exclusive economic zone, when the principle of compatibility under the 1995 Fish Stocks Agreement is applied, the same standard will naturally apply to fishing activities beyond such zone.

The Fish Stocks Agreement allows states to adopt regulations that empower national authorities to prohibit landings and trans-shipments when it has been established that the catch has been taken in a manner that undermines the effectiveness of sub-regional, regional or global conservation and management measures on the high seas.[87] These measures can be taken by any state individually and they do not require collective action. Coordination of measures relating to port access by countries of a same region has also been suggested, with particular reference to the member countries of the Permanent Commission of the South Pacific.

The argument has been made that restrictions to the access of fishing vessels to foreign ports would be contrary to free trade and the provisions of GATT/WTO;[88] and it has been argued in particular that the free transit provisions of Article V would be compromised by such restrictions. However, it should be noted that nothing in the GATT/WTO derogates from the basic principle of state sovereignty over ports and no right of entry is established under these provisions. On the contrary, the authoritative interpretation of Article V relies on the terms of the 1923 Statute of the International Regime of Maritime Ports,[89] which as explained above expressly excludes fishing vessels from its provisions. Furthermore, the kind of transit envisaged in Article V bears no relationship to the high seas. Beyond the question of entry, measures restricting or prohibiting landings and trans-shipments, when associated with a violation of conservation and management measures, are a normal feature of environmental agreements, indeed those aiming at the conservation of natural resources.

Agricultural Legislation, Vol. 26, 1977, p. 89; Sri Lanka, Fisheries Act No. 59 of 1979, Regulation of Foreign Fishing Boats, *FAO Fishery and Agricultural Legislation*, Vol. 29, 1980, p. 89; Trinidad and Tobago, Archipelagic Waters and Exclusive Economic Zone Act 1986, section 32, *FAO Fishery and Agricultural Legislation*, Vol. 36, 1987, p. 107.

[86] LOS Convention, Art. 62(4)(h).
[87] Fish Stocks Agreement, Art. 23(3).
[88] D.H. Anderson, 'The Straddling Stocks Agreement of 1995. An Initial Assessment', *International and Comparative Law Quarterly*, Vol. 45, 1996, p. 472.
[89] WTO, *Guide to GATT Law and Practice*, Vol. 1, 1995, p. 214, note 1.

PORT STATE JURISDICTION IN THE ANTARCTIC TREATY SYSTEM

The ATS has long been linked to major global conventions,[90] which in turn have introduced the concept of port state jurisdiction. Consequently, this concept actually became, or potentially became, applicable to the Antarctic or activities there. This is particularly the case of the IMO conventions, MARPOL 73/78 and above all SOLAS, as well as of the 1982 LOS Convention. Under the latter Convention, port state jurisdiction for the purposes of marine pollution and preservation of the environment also became available to the Antarctic, of course limited by the extent to which the Antarctic Treaty Consultative Parties were also parties to these global instruments. More recently, this has also been the case of the 1993 FAO Compliance Agreement and the 1995 Fish Stocks Agreement. Under these agreements, port state jurisdiction in respect of fisheries has also become available to the Antarctic, though it is again subject to the participation of the Antarctic Treaty Parties in these instruments.

It follows that the general framework of international law is broad enough in respect of port state jurisdiction to allow major developments under the ATS, or under its principal conventions, such as CCAMLR. This conclusion may even be reached independently from conventional developments of the law, since this type of jurisdiction appears also increasingly present in terms of customary international law after it departed from the limited ruling of the Permanent Court of International Justice in the *Lotus* case. It is worth noting that this approach has been present in the historical practice under the Antarctic Treaty. In fact, when Italy, before acceding to the Treaty, organised a private expedition in 1976, the Argentine authorities denied the ship permission to sail from the port of Buenos Aires, a situation which prompted a discussion about the rules of international law, the Antarctic Treaty and the meaning of its Article X.[91]

It is also important to point out that some regional arrangements and conventions could be relevant for the exercise of jurisdiction and inspection in respect of Antarctic activities for parties to the Antarctic Treaty that also participate in such conventions. It has been suggested, for example, that the inspection system established in the 1982 Paris Memorandum of Understanding could allow the parties to the Treaty to inspect in European ports their ships which are bound for Antarctica.[92] The same would be true of other regional arrangements for port state inspection established under the auspices of IMO, already referred to above. Other

---

[90] See, for example, United Kingdom, 'The Relationship between the Protocol on Environmental Protection to the Antarctic Treaty and Other International Agreements of a Global or Regional Scope', doc. XX ATCM/WP 10, 1996; and on the same subject, Chile, 'Relation between the Protocol on Environmental Protection to the Antarctic Treaty and Other International Agreements of a Global and Regional Scope', doc. XIX ATCM/WP 20, 1995.

[91] See S. Zavatti, 'La spedizione antarctica italiana e il diritto internazionale', *Il Polo*, Vol. 32, 1976, pp. 44–46.

[92] Orrego Vicuña, *Antarctic Mineral Exploitation*, pp. 282–283.

conventions which allow coastal state intervention in the high seas could permit the exercise of jurisdiction and inspection in the Antarctic or the high seas in the event of, for example, pollution emergencies, as has been done by the Agreement for the Protection of the Marine Environment and Coastal Zones of the Southeast Pacific under the auspices of the Permanent Commission of the South Pacific.

Such a development could open the way for inspection not only 'in' the Antarctic or in the high seas areas which are covered by the various conventions under the ATS, but also and foremost in ports located *outside* areas in which Antarctic activities are conducted. In addition, these developments open up the possibility of exercising jurisdiction over ships and aircraft of third parties, either within or in connection with the area covered by the System. This perspective has also been a matter of concern in the discussions of the ATS in spite of the narrow interpretation mentioned above. The end result would be a more comprehensive system of inspection, control and enforcement.

THE CONTRIBUTION OF THE PROTOCOL TO PORT STATE JURISDICTION IN AND IN RELATION TO THE ANTARCTIC

Although the Protocol did not innovate on jurisdiction in the Antarctic but rather followed what was already built into the ATS, it indirectly, and perhaps inadvertently, made a major contribution to the development of both jurisdiction and inspection. This was firstly done in respect of compliance under Article 13 of the Protocol and later in respect of the prevention of marine pollution under its Annex IV.

To ensure compliance with the Protocol, Article 13 commits each party to take appropriate measures within that party's competence, that is within its jurisdiction, including the adoption of laws and regulations, administrative actions and enforcement measures. These measures can obviously include inspection. The fact noted above that inspection under the Protocol is restricted to Article VII of the Antarctic Treaty and its narrow geographical ambit, while limiting the scope of inspections, should not necessarily be taken to mean that the progress made under the ATS must be entirely ignored. The general safeguards relating to the relationship and consistency with the other components of the ATS, established in Articles 4 and 5 of the Protocol, are indeed helpful to this effect. It follows that to the extent that under the ATS conventions, recommendations and other instruments jurisdiction and inspection might have gone beyond Article VII of the 1959 Treaty, this might be lawfully retained for the purposes of the Protocol in the same matters. As discussed above, port state jurisdiction is one such potential development.

State practice developing under the Protocol confirms this perspective. While following in some respects jurisdiction based on nationality, the Australian Antarctic

Treaty (Environmental Protection) Act 1980,[93] as amended to include the obligations under the Protocol, applies in the Australian Antarctic Territory (AAT) to any persons and property, 'including foreign persons and property'. It must be noted that while inspectors as a general rule shall not search a foreign aircraft or a foreign vessel, this clause does not apply when foreign aircraft or vessels are in the AAT, except if they are vessels or aircraft of war.[94] The United Kingdom Antarctic Act 1994 and subsequent regulations require any person on a British expedition to hold a permit for entering and remaining in Antarctica;[95] if the place of final departure for Antarctica was in the United Kingdom, the operator of the vessel or aircraft and the master of that vessel or the commander of that aircraft shall be guilty of an offence in case of contravention of that obligation.[96] Here again is a kind of port state jurisdiction, further confirmed by the fact that, among other officials, a harbour master or an aerodrome manager are invested with the power of inspection.[97] Similarly, the New Zealand (Environmental Protection) Act 1994, which implements the Protocol in that country, applies *inter alia* to any person who is a member of an expedition to Antarctica, or any person responsible for organising such an expedition, which proceeds from New Zealand as the final point of departure for Antarctica.[98] No objections by third states are known to have been made in respect of this kind of port state development of jurisdiction and inspection.

It is particularly interesting to note that the negotiations concerning liability under the Protocol have considered contributions to be made to the Fund 'on the basis of taxes levied on the departure for Antarctica'. These 'contributions' would be collected by the state party in which the environmental assessment procedure has been undertaken.[99] They were later referred to as 'fees' levied on the 'departure of ships and aircraft'.[100] Whatever the implications of this approach for the financing of the Fund, the fact remains that the practice involves an exercise of a kind of port state jurisdiction for the purpose of collection. The question of the port of departure has also been raised in the context of the identification of the liable entity.

In accord with the practice under the ATS, port state jurisdiction might refer both to the jurisdiction that might be exercised 'in' the Antarctic and to that which might be exercised beyond the Antarctic but in connection with Antarctic activities and operations. The Twenty-first Antarctic Treaty Consultative Meeting made note

---

[93] Australia, Antarctic Treaty (Environmental Protection) Act 1980, section 4(1).
[94] *Ibid.*, sections 17(2)–(3).
[95] United Kingdom, Antarctic Act 1994, section 3.
[96] *Ibid.*, section 3(7).
[97] United Kingdom, Statutory Instruments, 1995, No. 490, The Antarctic Regulations 1995, section 9(5).
[98] New Zealand, Antarctica (Environmental Protection) Act 1994, as discussed by P.R. Dingwall, 'Legal, Policy, and Administrative Developments for Management of Tourism in the Ross Sea Region and New Zealand sub-Antarctic Islands', *Polar Record*, Vol. 34, 1998, pp. 143–145.
[99] Protocol Annex on Environmental Liability, Chairman's Third Offering, 1995, Art. 10(4).
[100] *Ibid.*, Fourth Offering, 1995, Art. 10(4).

of a statement by the United Kingdom to the effect that the United Kingdom, New Zealand and to some extent Finland have established port state jurisdiction in their domestic legislation in respect of Antarctica.[101]

PORT STATE JURISDICTION UNDER ANNEX IV

Annex IV to the Protocol, which is concerned with the prevention of marine pollution, has been generally considered a weaker instrument than MARPOL 73/78.[102] While under MARPOL 73/78 both the flag state and the port state exercise specific controls in terms of certificates, inspection and other measures, these mechanisms appear to be somewhat looser, or non-existent, under Annex IV. The broad sovereign immunity exclusion under Article 11 of this Annex has compounded the problem since most of the ships operating in the Antarctic are warships, naval auxiliary or other ships owned or operated by a state;[103] the fact that each party shall take into account the importance of protecting the Antarctic environment in this context is certainly not a sufficient guarantee.[104]

With respect to each party, Annex IV applies to ships flying the party's flag, or to other ships engaged in or supporting the party's Antarctic operations, while operating in the Antarctic Treaty area.[105] This jurisdictional approach is in accordance with the principle embodied in the ATS in respect of the high seas.

Notwithstanding the limitations of the Annex, two provisions are of particular interest for a potential development in the exercise of jurisdiction, control and inspection. The first is providing adequate facilities for the reception of sludge and other residues by each party 'at whose ports ships depart *en route* to or arrive from the Antarctic Treaty area'. This commitment, which entails some form of port state intervention, also relates to the ports of other parties, including those adjacent to the Antarctic Treaty area.[106]

The second and more important provision is that of safeguarding the rights and obligations of parties to the Protocol which are also parties to MARPOL 73/78. In accordance with Article 14 of Annex IV, nothing in that Article shall derogate from

---

[101] *Final Report of the Twenty-first Antarctic Treaty Consultative Meeting, Christchurch, New Zealand, 19–30 May 1997* (Wellington: New Zealand Ministry of Foreign Affairs and Trade, 1997), para. 15.

[102] T. Scovazzi, 'The Application of the Antarctic Treaty System to the Protection of the Antarctic Marine Environment', in F. Francioni (ed.), *International Environmental Law for Antarctica* (Milan: Giuffrè, 1992), pp. 113–134; L. Pineschi, *La protezione dell'ambiente in Antartide* (Padova: CEDAM, 1993), pp. 333–373.

[103] See discussion by C.C. Joyner, 'Protection of the Antarctic Environment against Marine Pollution under the 1991 Protocol', in D. Vidas (ed.), *Protecting the Polar Marine Environment: Law and Policy for Pollution Prevention* (Cambridge University Press, 2000), pp. 117–119.

[104] Protocol, Annex IV, Art. 11(2).

[105] *Ibid.*, Art. 2.

[106] *Ibid.*, Art. 9(2) and (3).

those rights and obligations. In respect of MARPOL 73/78 parties, this means that the more stringent provisions of this instrument shall prevail.[107] This provision of course includes the designation of the Antarctic as a special area under MARPOL 73/78 and its relevant annexes, providing for the prohibition of discharges from 'any' ship and other measures.[108] Although Article 14 refers only to MARPOL 73/78 parties, the implications of the MARPOL Convention are potentially more comprehensive since its principles have become generally agreed rules and standards under the 1982 LOS Convention, which can reach a larger number of states and eventually could also qualify as a rule of customary international law. Such standards could therefore eventually apply to all parties to the Protocol irrespective of their participation in MARPOL 73/78. Moreover, these same principles could also eventually apply to vessels of third states operating in the Antarctic and which are bound by the general rules of the LOS Convention.

The subject of port state jurisdiction has not been absent from the deliberations of the Antarctic Treaty Consultative Meetings. Following its inclusion as a separate agenda item,[109] a document submitted by the Netherlands referred to the need to supplement flag state jurisdiction with jurisdiction of the gateway ports.[110] However, a distinction was drawn between breaches of Annex IV and other breaches not as qualifying or situations in which the port state is not a party to MARPOL 73/78. Breaches of Annex IV are also breaches of MARPOL 73/78 and could be subject to measures by the port state which is a party to MARPOL 73/78. In the case of other breaches, the ship could be inspected but not prevented from departing. Possible alternatives that were suggested for port state jurisdiction beyond the strict confines of MARPOL-related situations were voluntary arrangements, notification of the captain of any defects found, and information to Antarctic observers and to the Consultative Meeting. Basing its argument on the existing acquiescence and regarding the ATS as having become an objective regime, the United Kingdom has encouraged the development of port state jurisdiction in respect of Antarctic Treaty parties, Protocol parties or third states.[111] The importance of this development is enhanced by the fact that almost 50 per cent of tourist vessels operating in the Antarctic are registered with non-Treaty parties, mainly flying flags of convenience.

However, these initiatives have been met with some reluctance, partly either because they could imply the exercise of jurisdiction by non-flag states without a

---

[107] Currently, however, there is little practical difference: after New Zealand's accession to MARPOL 73/78 in September 1998, all the current parties to the Protocol are also parties to MARPOL 73/78.

[108] Scovazzi, 'The Application of the Antarctic Treaty System', p. 127.

[109] *Final Report of the Twentieth Antarctic Treaty Consultative Meeting, Utrecht, 29 April – 10 May 1996* (The Hague: Netherlands Ministry of Foreign Affairs, 1997), agenda item 10.

[110] The Netherlands, 'Inspection of Ships in Gateway Ports to Antarctica, on the Basis of MARPOL 73/78, and in Antarctic Ports under the Environmental Protocol (Annex IV) to the Antarctic Treaty', doc. XX ATCM/WP 9, 1996.

[111] United Kingdom, 'Enhancing Compliance with the Protocol: Departure State Jurisdiction', doc. XXI ATCM/WP 22, 1997.

specific treaty commitment or because they could reach beyond the LOS Convention.[112] In particular Germany, which is the flag state for an important number of tourist vessels operating in the Antarctic, has stressed the need for maximum development of the controls under MARPOL 73/78, SOLAS and related conventions; but Germany has also advised against similar controls under Annex IV alone since there would be a lack of legal foundation for such a development.[113] Chile, on the other hand, has reported that it exercises port state jurisdiction not only under MARPOL 73/78, SOLAS and related national legislation, but also in respect of vessels from states not parties to MARPOL 73/78 or to the Antarctic Treaty 'so as to ensure non preferential treatment of these vessels'.[114]

It is of interest to note that in the checklist of inspections of vessels used by the United Kingdom, Italy and South Korea, there is an indication that inspectors should particularly concentrate on the observance of wildlife and Protected Areas regulation, waste disposal and prevention of marine pollution.[115] This has given rise to questions regarding awareness of oil prevention measures of the Protocol, the vessel being flagged with a signatory to MARPOL 73/78, awareness of the IMO Special Area provisions for Antarctica, and the existence of a formal contract with a port reception facility for oil residues outside the Antarctic Treaty area.[116] This all evidences the close relationship between inspections under the Antarctic Treaty and the Protocol and the broader framework of arrangements under international law.

Because Annex IV is restricted to ships operating in the Antarctic Treaty area, there could be some doubt as to whether port state jurisdiction could be exercised outside this area. However, in the light of the practice referred to above and of the fundamental purport of the Protocol and its annexes to protect the Antarctic environment, it can be concluded that port state jurisdiction can, to the extent that it becomes accepted, be lawfully exercised outside the Antarctic Treaty area in respect of activities that take place in this area, such as the operation of ships.

The doubt has also been raised about whether such jurisdiction can be properly exercised in the Antarctic considering that 'points of discharging or embarking cargoes or personnel' might not qualify as true ports or that it might be difficult to identify which state exercises jurisdiction over such points.[117] But this is more a technical problem rather than one of substance, and in any event there will always be a state operating the facilities at such points.

---

[112] *Final Report of the XXI ATCM*, para. 15; *Final Report of the XX ATCM*, para. 93.
[113] Germany, 'Inspection of Ships by Port States', doc. XXI ATCM/WP 16, 1997.
[114] Chile, 'Enforcement of International Maritime Conventions and Domestic Standards on the Inspection of Ships that Operate in the Antarctic', doc. XX ATCM/WP 17, 1996.
[115] United Kingdom, 'Report of a Joint Inspection under Art. VII of the Antarctic Treaty by United Kingdom, Italian and Korean Observers, January–February 1993' (with particular reference to the checklists attached), doc. XVIII ATCM/INFO 7, 1994, Annex C. 1.
[116] *Ibid.*, Annex C. 3 (c).
[117] Pineschi, *La protezione dell'ambiente*, p. 367.

The question of what actions might be undertaken by the port state should of course be agreed under the very mechanism that establishes such a system. In this respect, under the 1982 LOS Convention the limited actions under MARPOL 73/78 were considerably enlarged to include inspection, investigation and the instituting of some limited proceedings. This trend continued under the 1995 Fish Stocks Agreement.

CLOSING THE JURISDICTIONAL LOOPHOLE IN ANTARCTICA

It may be concluded from the above discussion that the introduction of port state jurisdiction in and related to the Antarctic would help close the jurisdictional loophole existing in the ATS, certainly in respect of inspection and control. At present, Article VII of the Antarctic Treaty and the practice developed thereunder may be considered adequate for the purposes of inspections concerning activities under the Treaty; this is also the case of the inspection system under the Protocol in conjunction with the same Article VII for the purpose of ensuring compliance with environmental protection. However, in respect of high seas areas, inspection has been developed, but with some shortcomings, only in relation to fishing activities under CCAMLR and not for general navigational purposes or the operation of ships. It is precisely on this point that the Protocol and Annex IV, eventually in conjunction with MARPOL 73/78, might provide the mechanisms for agreeing on a port state inspection system that would close a major loophole in the ATS. The same may be said of the Fish Stocks Agreement concerning the exercise of port state jurisdiction to ensure compliance with conservation arrangements under CCAMLR. In addition, a similar approach may be followed in respect of tourism and other activities organised in, or departing from, the territory of a party to the Protocol, for which the party has obligations under Article 3(4) and must conduct EIAs in accordance with under Article 8 of this instrument.

These developments would not only strengthen the jurisdictional arrangements in respect of parties to the Antarctic Treaty, its Protocol and special conventions, but they would above all also strengthen the position of third states which until now have had a rather elusive situation under the ATS. It is important to note that developments under CCAMLR have also pointed in the direction of asserting port state jurisdiction. In point of fact, under Conservation Measure 118/XVI, vessels of non-contracting parties sighted engaging in fishing activities in the Convention area are presumed to be undermining the effectiveness of CCAMLR Conservation Measures, a presumption that also applies to trans-shipment activities in or outside the Convention area. In such cases, besides the information that must be transmitted to the vessel and the CCAMLR Commission, it is expressly provided that a vessel entering a port of any contracting party shall be inspected and landings and trans-shipments of fish shall be prohibited in all contracting party ports if found in violation of conservation measures in force.[118] Similarly, a fishing vessel of a CCAMLR contracting party can be inspected in the ports of another contracting party to ensure that it has conducted its activities in accordance with the conservation measures in force. In case of a contravention, the flag state shall be informed and cooperative arrangements shall be made for investigation and application of sanctions under the flag state national legislation.[119]

These trends do not mean that flag state jurisdiction should be abolished, but as international developments strongly suggest, there is a need to supplement this jurisdiction with that of the port state, or another independent mechanism. The limitations affecting flag state jurisdiction would otherwise provide an incentive for coastal state intervention, as developments under international law and practice under the very ATS clearly indicate.

The introduction of port state jurisdiction in respect of the Antarctic would not only close the existing jurisdictional loophole in the ATS but it would also redress the jurisdictional balance that has been upset in the evolution of the port state regime taking place. Port state jurisdiction would be exercised in the ports of any party to the Protocol or the respective arrangements, but above all by those countries whose ports are closer to the Antarctic and used for departure or arrival, the so called 'gateway ports'. While gateway port countries, which are mainly located in the Southern Hemisphere, might incur financial commitments resulting from these obligations, or some form of liability for failure to comply with the inspection obligations, these situations might be solved through the various forms of assistance and cooperation established in the regimes established for this purpose.

Irrespective of their position as claimants or non-claimants, these countries would acquire a new role as port states. The fact that most gateway port countries are claimant countries does not in any way alter the added effectiveness that port state jurisdiction would bring into the ATS.

To this end it would be essential that the mechanism for enactment and implementation of port state jurisdiction be collectively defined within the ATS, since otherwise different standards and levels of stringency could be applied. This harmonisation of standards can be achieved within the framework of a joint regime. The distinction between the port of departure, mainly concerned with the inspection relating to personnel and expeditions, and port state jurisdiction, which relates to the inspection of the ship and, its activities, may be appropriately introduced in this context. A kind of Regional Memorandum of Understanding for port state jurisdiction in the Antarctic might also be an adequate approach to this effect, under which special certificates of Antarctic worthiness may also be issued. As with many developments under the ATS, it could suffice to enact the port state jurisdiction and

---

[118] CCAMLR Conservation Measure 118/XVI, paras. 4 and 5.

[119] See the Conservation Measure on 'Cooperation between Contracting Parties to Ensure Compliance with CCAMLR Conservation Measures with Regard to Their Vessels', CCAMLR, Seventeenth Meeting, Hobart, 26 October – 6 November 1998.

inspection regime by means of a resolution of an Antarctic Treaty Consultative Meeting.

Such an enactment would enhance compliance with the rules governing environmental protection within the System, and particularly those of the Protocol and its annexes.

4

Regulating Tourism in the Antarctic: Issues of Environment and Jurisdiction

Mike G. Richardson

During the early 1990s, debate on the adequacy of the regulation of tourism in Antarctica provided one of the more vigorous exchanges between Antarctic Treaty parties. Coming so soon after the adoption of the Environmental Protocol,[1] the debate may have fuelled doubts amongst some commentators as to whether the environmental protection regime foreseen by the Protocol was indeed as comprehensive as Treaty parties had intended. The debate came at a sensitive time. In the late 1980s and early 1990s, just when the Antarctic Treaty[2] was attempting to enhance its environmental credentials to a wider audience, including the United Nations, tourism, seemingly unregulated, was escalating at a considerable rate. With the increase in tourism came the concern that potential impacts might not only affect the Antarctic environment but also have adverse implications for other legitimate activities, notably scientific research and its associated logistics.

The history and trends of tourism in Antarctica have been documented adequately elsewhere[3] and need no particular elaboration here. Suffice it to say that the number of tourists visiting Antarctica yearly continues to increase. During the 1998-99 austral season, the number of ship-borne tourists exceeded 10,000 for the first time,[4] and the number of tourists now exceeds that of governmental scientists

---

[1] ILM, Vol. 30, 1991, pp. 1,461ff.
[2] UNTS, Vol. 402, pp. 71ff.
[3] R.K. Headland, 'Historical Development of Antarctic Tourism', *Annals of Tourism Research*, Vol. 21, 1994, pp. 269–280; D.J. Enzenbacher, 'Tourists in Antarctica: Numbers and Trends', *Polar Record*, Vol. 28, 1992, pp. 17–22; and D.J. Enzenbacher, 'The Management of Antarctic Tourism: Environmental Issues, the Adequacy of Current Regulations and Policy Options within the Antarctic Treaty System', Ph.D. diss., University of Cambridge, 1995, p. 300.
[4] International Association of Antarctic Tour Operators (IAATO), 'Projected Trends in Antarctic Tourism', data from IAATO and the US National Science Foundation presented to the 10th IAATO Meeting, Hamburg, Germany, 30 June 1999.

and support staff.[5] However, it has been calculated that the cumulative presence of tourists, i.e., the number of man-days per season, remains considerably less than the cumulative presence of the staff of governmental operators combined.[6]

Nevertheless, by the end of the 1990s, Treaty parties were confronted by a fully developed industry which was expanding considerably. Tourism was primarily confined to ship-borne cruises, mainly targeting the Antarctic Peninsula, though the Ross Sea area was fast becoming a secondary destination along with continental circumnavigation. Adventure tourism, though operating on a smaller scale, occupied a share of the market and was also on the increase, as was the number of private and charter yachts[7] visiting Antarctica. Large-scale aviation tourism, which resumed in the 1994–95 season following the Mount Erebus disaster of 1979, now runs some nine to ten flights yearly, bringing an annual total of around 3,500 passengers to Antarctica.[8] Overall, by all appearances it looks as though the trends of the past few years will continue or even escalate with the advent of a larger generation of tourist vessels due in Antarctic waters in the 1999–2000 season. Provisional estimates suggest a further 42 per cent upturn in ship-borne tourism in the coming two seasons 1999–2000 and 2000–2001, with numbers topping 14,000.[9]

This chapter will examine the history of the regulation of tourism within the Antarctic Treaty System. It will assess whether, in practice, the Environmental Protocol has proved to be an effective mechanism for regulating and managing tourism activities, and it will identify gaps in the existing regulatory regime, suggesting means to address those deficiencies.

THE MEANS OF REGULATION

The need for adequate regulation of tourism, and also for regulation of other non-governmental activities in Antarctica, has never been questioned by the Treaty parties. Rather, it was the means by which such regulation should be effected that generated the exchanges in the early 1990s.

There are various means by which tourism in Antarctica can be regulated. These include mandatory and non-mandatory provisions under the Antarctic Treaty System, self-regulation by the industry itself, and national regulations based on a range of jurisdictions. Through employing these means, some claimant states have achieved varying effectiveness in exercising control over flag-vessels, nationals, and territory in Antarctica, and more recently jurisdiction over third-party vessels and aircraft at 'gateway' points of departure for Antarctica.

---

[5] J.C.M. Beltramino, *The Structure and Dynamics of Antarctic Population* (New York: Vantage Press, 1993), p. 105.
[6] Headland, 'Historical Development of Antarctic Tourism', pp. 269–280.
[7] United Kingdom, 'Yacht Visits to Antarctica, 1970–98', doc. XXII ATCM/IP 1, 1998, p. 6.
[8] T.G. Bauer, 'Antarctic Tourism: Island Tourism with a Difference', *Islander Magazine*, 1999, p. 12.
[9] IAATO, 'Projected Trends in Antarctic Tourism'.

Under the procedures of Article IX of the Antarctic Treaty, both mandatory measures and non-mandatory guidelines have been adopted in the form of recommendations. In addition, a suite of parallel codes or guidelines, introduced by bodies such as SCAR and COMNAP, but most notably by IAATO, have augmented the latter.[10] Whilst the normal responsibilities of a state over its flag-vessels have been exercised within the Antarctic Treaty area, other national measures have been less frequently employed: for example, to ensure that either extra-territorial jurisdiction over nationals can be maintained, or to give full effect to Article VII of the Treaty whereby control can also be exercised over expeditions organised in, or departing to Antarctica from, the territory of a Consultative Party. The latter has particular relevance because of the references to tourism in Articles 3, 8 and 15 of the Environmental Protocol. These three articles draw together: 1) the references to expeditions under Article VII, 2) the references to tourism specifically addressed within the general environmental principles of the Protocol, and 3) the provisions on environmental impact assessment and emergency response action and contingency planning.

HISTORY OF REGULATION

Pre-Protocol

Although commercial tourism in Antarctica has been taking place for the past 40 years,[11] the numbers of tourists involved remained relatively modest until the late 1980s. The major escalation in the numbers of tourists coincided with the heightened awareness of Antarctic issues generated by the debate before and throughout the negotiations of the Protocol. Tourism in effect capitalised on the wide-ranging exposure given to Antarctica by main environmental NGOs such as WWF and Greenpeace. Unwittingly, their prominent campaigns for enhanced environmental protection fuelled a wider public desire to experience Antarctica first-hand. As a consequence, since 1989–90 the number of ship-borne tourists has increased around 105 per cent.

The recognition that tourism is a natural development in Antarctica dates back to at least the mid-1970s, when Recommendation VIII-9 was adopted. Although tourism was not perceived to be at odds with the 'peaceful purposes' set out in Article I of the Antarctic Treaty, it required regulation. Issuing a statement of 'Accepted Principles' and a 'Code of Guidance for visitors to the Antarctic' proved to be the most substantive of the earlier measures on tourism.

---

[10] Scientific Committee on Antarctic Research (SCAR), 'A Visitor's Guide to the Antarctic and its Environment', SCAR Sub-Committee on Conservation of the Working Group on Biology, 1984, p. 36; Council of Managers of National Antarctic Programmes (COMNAP), 'Visitor's Guide to the Antarctic', 1990, p. 1; International Association of Antarctic Tour Operators (IAATO), 'Guidelines for Tour Operators', *International Association of Antarctic Tour Operators*; Oceanites, 'Antarctic Traveller's Code for Visitors and Tour Operators', *Antarctic Century*, No. 7, 1991, pp. 6–7.

[11] Chile, 'Tourism and Other Activities in Antarctica', doc. XVI ATCM/WP 29, 1990, p. 11.

However, between 1966 and 1982 only six recommendations were adopted under the Treaty System which specifically addressed tourism and non-governmental activities. Although other measures, such as the introduction of protected area designations (e.g., Multiple-use Planning Areas (MPAs)), would have provided further opportunities to manage or regulate tourism activities indirectly, these were never capitalised on. Earlier recommendations on tourism targeted specific issues such as prior notification and reporting of tourist and NGO activities, the need to safeguard scientific research, self-sufficiency, and requirements for tour guides. But these recommendations also demonstrated a number of major shortcomings: The adoption of regulations was piecemeal and *ad hoc*. Notably, there was no overarching regulatory strategy or coherent, comprehensive overview. The regulations were largely hortatory, couched in 'soft law' terms and therefore not binding for the parties. Even when measures of a very specific nature were adopted, such as the introduction of Areas of Special Tourist Interest (ASTIs), no direct action resulted and no such areas were ever designated. ASTIs lapsed into obscurity; furthermore, the lengthy consensual approval mechanism under Article IX of the Antarctic Treaty almost ensured that the few regulations that were adopted failed to be approved subsequently and were therefore not implemented rapidly enough to address the pressing concerns of tourism.

Through all this, the Treaty parties acted in a somewhat uncharacteristic way; they failed to address tourism in the same manner that they had tackled other resource-based industries within the Treaty area. An important trait of the Consultative Parties had been their willingness to embrace a pro-active approach – negotiating regulations, often of an innovative form, well ahead of the actual need for control and therefore doing so in the absence of developmental pressures. Such pre-emptive far-sightedness had epitomised the handling of the Agreed Measures in 1964, the Convention for the Conservation of Seals[12] in 1972, the Convention on the Conservation of Antarctic Marine Living Resources[13] (CCAMLR) in 1980 and the Convention on the Regulation of Antarctic Mineral Resource Activities[14] (CRAMRA) in 1988. However, it was perhaps the direct result of the considerable attention focussed on these last two conventions throughout the 1970s and 1980s that resulted in correspondingly less political attention being devoted to tourism. The consequence was that at a time when significant expansion was taking place in the industry, no recommendations were adopted during the period 1982–1992.[15]

Hence, by the late 1980s the regulations relating to Antarctic tourism were inadequate and open to criticism.

---

[12] UNTS, Vol. 1080, pp. 175ff; and ILM, Vol. 11, 1972, pp. 251ff.
[13] ILM, Vol. 19, 1980, pp. 837ff.
[14] ILM, Vol. 27, 1988, pp. 868ff.
[15] D. Vidas, 'Antarctic Tourism: A Challenge to the Legitimacy of the Antarctic Treaty System?', *German Yearbook of International Law*, Vol. 36, 1993, pp. 187–224.

Tourism Regulation and the Environmental Protocol

With the escalating trend in tourist numbers, the mood of Treaty parties by 1989 was that something needed to be done; precisely what, in regulatory terms, could not collectively be agreed upon. Various options were proposed. The UK called for codification of the existing recommendations,[16] Germany for discussions, particularly on liability and insurance,[17] whilst Chile pressed both for greater control of nationals by official agencies and for the need to plan tourism and NGO activities during periods of the year when interference with scientific programmes would be minimal.[18] The parties did, however, agree on one thing – the need for a major review. Having just instigated work on comprehensive measures for the protection of the Antarctic environment (ultimately to become the Environmental Protocol), they also determined that such a review should be undertaken simultaneously through a Special Antarctic Treaty Consultative Meeting.[19]

At the same time, some parties also raised the perceived need for additional specific regulations targeting tourism. The aim was to ensure that the impact of tourism on the environment, as well as on scientific and related logistic activities, would be reduced.[20]

Considering the more pressing political objectives for a comprehensive environmental regime that existed at the time, it remains questionable whether adequate attention was paid to tourism during that review process, or indeed during the subsequent negotiations which led to the Protocol. Although the need for specific measures to regulate tourism was raised on more than one occasion during sessions of the Eleventh Special Consultative Meeting, particularly by Chile and France, there was no consensus on this approach. Instead, the prevailing view was that the Protocol and its Annexes should be designed to address generically all activities in Antarctica. No differentiation should be made regarding lawfully prosecuted activities, including tourism. The political imperative was to adopt the Protocol, preferably ahead of the 30th anniversary of the entry into force of the Antarctic Treaty. The scarceness of negotiating time available precluded specific examination of the complex issues of tourism. As adopted, the Protocol in many ways was indeed the product of 'considerable and commendable haste'.[21]

---

[16] United Kingdom, 'Effective Tourism and Non-Governmental Expeditions in the Antarctic Treaty Area', doc. XV ATCM/WP 33 and Annex (Ant/XIV/6), 1989, p. 13.

[17] Germany, doc. XV ATCM/WP 30, 1989, p. 3.

[18] Chile, 'Draft Recommendation submitted by the Delegation of Chile', doc. XV ATCM/WP 44, 1989, p. 1.

[19] United Kingdom. 'Effective Tourism and non-Governmental Expeditions in the Antarctic Treaty Area', p. 13. Recommendation XV–I, 'Comprehensive Measures for the Protection of the Antarctic Environment and Dependent and Associated Ecosystems', in *Final Report of the Fifteenth Antarctic Treaty Consultative Meeting, Paris, France, 9–20 October 1989* (Paris: French Ministry of Foreign Affairs, 1989), pp. 43–46.

[20] *Final Report of the XV ATCM*, paras. 155–158.

[21] Sir Arthur Watts, *International Law and the Antarctic Treaty System* (Cambridge: Grotius, 1992), p.

As an alternative means of demonstrating the comprehensiveness of the document, succinct references to tourism were incorporated into three key areas of the Protocol: Article 3 (on general environmental principles), Article 8 (on environmental impact assessment), and Article 15 (on emergency response action and contingency planning). However, because of the atypical nature of Antarctic tourism with its major involvement by third parties and the peculiarities of jurisdiction in Antarctica, the ability of the Protocol adequately to regulate tourism was not universally accepted by Treaty parties. Fundamental differences existed over the interpretation and application of the Protocol regarding the extent to which it addressed tourism and over the need for additional, tougher regulations in the form of a further mandatory Annex to the Protocol.

As was to be shown by subsequent discussions within the context of liability for environmental damage, ambiguity was evident even as to which activities in the Antarctic the Protocol addressed. Within key Article 3 on principles, three contrasting references to the scope of the Protocol can be found. On the one hand there is a reference to 'all activities in the Antarctic Treaty area'; on the other, more generally to just 'activities' and then more precisely to

> Activities undertaken in the Antarctic Treaty area pursuant to scientific research programmes, tourism and all other governmental and non-governmental activities in the Antarctic Treaty area for which advance notice is required in accordance with Article VII (5) of the Antarctic Treaty, including associated logistic support activities ...

Similar wording is reiterated in Articles 8 and 15 of the Protocol. Indeed, these three articles contain the only references to tourism found throughout the Protocol and its five Annexes.

If read literally, the above text from Article 3 is clearly more specific regarding both 'activities' and 'all activities' than is Article VII(5) of the Antarctic Treaty which only relates to activities (expeditions) organised in, or proceeding to Antarctica from, the territories of the parties. Arguably, many tourism-related activities might fall outside this definition.

That ambiguity prevails is apparent also from the various domestic legislation introduced so far to enact the Protocol. For example, some parties to the Protocol have incorporated the notion of port state control to provide for regulation of those expeditions (including tourist cruises) departing to Antarctica from their territory irrespective of whether the vessel is registered with that party or whether the expedition members are nationals of that party and thus subject to national jurisdiction.[22]

Following the adoption of the Protocol, major disagreement over its ability adequately to address tourism was immediately evident. Discussion at the 1991

496.
[22] United Kingdom, 'Antarctic Act, 1994' (UK Legislation enacting the Environmental Protocol), (London: Her Majesty's Stationery Office, 1994), p. 14. New Zealand, Antarctica (Environmental Protection) Act, 1994, No. 119 (New Zealand legislation enacting the Environmental Protocol).

Consultative Meeting in Bonn, within two weeks of the signing of the Protocol, reiterated earlier demands from some parties for a comprehensive review of tourism and the need for a further mandatory Annex. Consequently, Recommendation XVI–13 was adopted.[23] This Recommendation called for an informal meeting of Parties to make proposals to the Consultative Meeting on the comprehensive regulation of tourism, including the proposal for a further Annex.

The Fate of the Draft Annex

In 1992, at a two-day informal meeting on tourism in Venice to address Recommendation XVI–13, Treaty parties were strongly divided as to how best to proceed. They remained divided throughout the entire Venice Consultative Meeting following immediately after the informal meeting on tourism and at subsequent consultative meetings. However, despite strong and contrasting views, the debate converged mainly amongst only nine Consultative Parties, with some NGO participation. As a comprehensive description of events of the Venice tourism meeting is presented elsewhere,[24] only a synopsis is given here. On the one side, five countries (Chile, France, Germany, Italy and Spain) presented a draft sixth Annex to the Protocol on tourism.[25] These countries argued a legal, political and practical rationale for such an instrument. Legally, a comprehensive framework (i.e., Annex) would allow states to 1) enact a homogenous body of laws and regulations, and in doing so avoid inconsistencies, 2) provide a set of rules with indisputable legal binding force, and 3) enable all parties to the Protocol to be engaged in regulation, not just those involved actively in tourism. Presentationally, for political reasons, the 'group of five' argued that an Annex was the most appropriate form within which to incorporate such regulations, whilst practically it was important that tour operators and organisers of non-governmental expeditions should be made clearly aware of the rules pertaining to their activities.

The proponents of the draft Annex stressed that their sole motive was to make the provisions of the Protocol clearer. There was no desire to impose further regulatory constraints on tourist and non-governmental activities. Yet even a cursory glance at the principal articles of the draft Annex[26] indicated that in many key areas its provisions extended well beyond those set out in the Protocol, for example: re-introducing the concept of tourist areas; prohibiting access to the remainder of

---

[23] Recommendation XVI–13, 'Tourism and non-Governmental Activities in the Antarctic Treaty Area', in *Final Report of the XVI ATCM*, pp. 131–132.
[24] D. Vidas, 'Antarctic Tourism', pp. 187–224; and D. Vidas, 'The Legitimacy of the Antarctic Tourism Regime', in O.S. Stokke and D. Vidas (eds.), *Governing the Antarctic: The Effectiveness and Legitimacy of the Antarctic Treaty System* (Cambridge University Press, 1996), p. 464.
[25] Chile, France, Germany, Italy and Spain, 'Annex VI to the Protocol on Environmental Protection to the Antarctic Treaty. Regulations Concerning Tourism and non-Governmental Activity', doc. XVII ATCM/WP 1, 1992, p. 15.
[26] *Ibid.*

Antarctica unless a comprehensive environmental evaluation (CEE) was undertaken; extending the period of prior notification to a minimum of fifteen months; mandating the Committee for Environmental Protection (CEP) with executive powers; setting standards for non-governmental operators (e.g., for ships) well in excess of those expected from governmental operators; and requiring the holding of insurance.

Other parties, notably Australia, the UK and the USA,[27] argued that further mandatory provisions by means of an Annex were unnecessary: the Protocol explicitly addressed tourism. Its provisions were mandatory, yet sufficiently generic to encompass all governmental and non-governmental activities, and the proposals in favour of a further Annex should be discounted.

Furthermore, the parties argued that inconsistencies between parties in their national law-making, although an inherent problem in the domestic enactment of any international treaty, would not be enhanced by introducing yet another, more detailed instrument. Indeed, inconsistencies between the draft Annex and the Protocol might have resulted in even more contrasting interpretations. Moreover, the Protocol already set out indisputable, legally-binding rules, whilst the application of its rules and regulations to tourism could be made readily apparent to tour operators and organisers of non-governmental activities alike. A further Annex which would target just one sector of the Antarctic community (i.e., tourism) was, from a point of principle, not appropriate. It would create a divisive, twin-tracking of standards in some areas, for example on shipping. From an environmental protection standpoint, this was undesirable. Neither, it was contended, was the notion that an Annex would enable the parties to regulate tourism in a more timely manner credible. Any such Annex, to become effective, would first have to fulfil the approval procedures of Article IX of the Treaty. As experience with Annex V, on 'Area Protection and Management' has since shown,[28] such procedures may prove to be more protracted than ratification of the Protocol itself.

Australia concluded, in its review of Recommendation XVI–13,[29] on the adequacy of existing Treaty regulations and the implementation and application of the Protocol, that no additional regulations were needed at that stage, but that there was a need to provide guidance to tour operators and NGOs.

---

[27] United States, 'Antarctic Tourism and the Environmental Protocol', doc. XVII ATCM/WP 6, 1992, p. 10. United Kingdom, 'The Regulation of Tourism and non-Governmental Activities in the Antarctic Treaty Area', doc. XVII ATCM/WP 2, 1992, p. 11. United Kingdom, 'The Regulation of Tourism and non-Governmental Activities in the Antarctic Treaty Area (II)', doc. XVII ATCM/WP 6, 1992, p. 24. Australia, 'Tourism and non-Governmental Activities in the Antarctic Treaty Area', doc. XVII ATCM/WP 14, 1992, p. 7.

[28] See 'Report of the Depository Government of the Antarctic Treaty and its Protocol (USA) in Accordance with Recommendation XIII–2', in *Final Report of the Twenty-second Antarctic Treaty Consultative Meeting, Tromsø, Norway, 25 May – 5 June 1998* (Oslo: Royal Norwegian Ministry of Foreign Affairs, 1998), Annex F, pp. 223–237.

[29] Australia, 'Tourism and non-Governmental Activities in the Antarctic Treaty Area', p. 7.

In this debate, Australia held the middle-ground between the 'group of five' and the principal opponents of that group, i.e. the USA and the UK, who stressed that timely and effective implementation of the Protocol and its Annexes and the subsequent enforcement of their provisions was the most appropriate way to regulate tourism.

Curiously, throughout these exchanges, the Antarctic and Southern Ocean Coalition (ASOC), an environmental NGO which had usually been to the fore in urging for tighter environmental controls in Antarctica, expressed views which ran contrary to the proponents of an Annex.[30] The reason was perhaps that ASOC foresaw additional constraints on their own activities in Antarctica.

Despite the increasing calls for regulation ahead of the Venice discussions, the 1992 Consultative Meeting ended with no agreement amongst parties.[31] Even the suggestion from some, that further international discussions should be held, failed to attract overall support.[32] The hiatus had been reached, and although Chile and France would return the following year with a re-packaged Annex under the guise of 'Agreed Measures on Tourism',[33] the issue in terms of a major new regulatory instrument on tourism had, at least for the time being, been exhausted.

Subsequent Regulatory and Management Developments

Since the 1992 Consultative Meeting in Venice, Treaty parties have reverted to their former low-key approach to tourism management and regulation. Although the issues of tourism and non-governmental activity have continued to feature on the agenda of each Consultative Meeting since 1992, few initiatives have been taken to either optimise the effective implementation of the Protocol with regard to tourism with the introduction of more regulations, or indeed to even enhance the information base necessary for the sound management of tourism.

The most notable progress was the adoption in 1994 of Recommendation XVIII–1,[34] which provided non-mandatory guidelines for tourists and tour operators and significantly up-dated the Code of Guidance previously adopted back in 1975. Over a period of three years the Treaty parties also deliberated on the format of the data-reporting forms to be used both for advance notification and post-activity

---

[30] ASOC, 'Regulation of non-Governmental and Tourism Activities in Antarctica', doc. XVII ATCM/IP 52, 1992, p. 2.

[31] *Final Report of the Seventeenth Antarctic Treaty Consultative Meeting, Venice, Italy, 11–20 November 1992* (Rome: Italian Ministry of Foreign Affairs, 1993), paras. 108–114.

[32] *Ibid.*, para. 114.

[33] Chile and France, 'Agreed Measures on Tourism in Antarctica', doc. XVIII ATCM/WP 11, Rev. 1, 1994.

[34] Recommendation XVIII–1, 'Tourism and non-Governmental Activities', in *Final Report of the Eighteenth Antarctic Treaty Consultative Meeting, Kyoto, Japan, 11–22 April 1994* (Tokyo: Japanese Ministry of Foreign Affairs, 1994), pp. 35–45.

reports; they eventually adopted Resolution 3 (1997)[35] after two periods of trialing report forms. But even here the Treaty parties failed to retain effective control of their own information system. In the absence of any central repository within the Treaty System,[36] parties have shown a collective disinterest in collecting, collating and analysing data on tourism. Instead, they have relied mainly on IAATO or the US National Science Foundation, which has long maintained an important database on Antarctic tourism. To a lesser extent they have relied on external projects such as IDAT (International Database on Antarctic Tourism) which was funded through the International Centre for Antarctic Information and Research (ICAIR) by France, Italy, Netherlands, New Zealand and the UK. Yet data on the numbers of vessels and personnel are crucial not only for the management and regulation of tourism, but also for acquiring an understanding of how complex issues such as environmental impact assessment and cumulative impact might best be addressed.

The Treaty System has also been inadequate in other areas of data collection. Since the inception of the tourism debate, lack of information on the effects of tourism on the Antarctic environment has been wanting. Few data existed, and those that did were equivocal. With the information available, it was not possible to determine whether or not tourism had an impact on the Antarctic environment, and if it did, its magnitude. That remains much the situation today, though studies such as the Antarctic Site Inventory Project by Oceanites[37] have begun to gather data on sites most frequently visited by tourists; but again, such projects have received little corporate support.

The Dilemma of Antarctic Tourism: Optimum Regulation

In retrospect, it is pertinent to question whether the lack of consensus for an Annex, or for some other separate mandatory instrument, was desirable, or has the existing mechanism highlighted a major legal and political lacuna in the regulatory framework of the ATS? Can it be construed that the Consultative Parties have, in part, abdicated their responsibilities to regulate in favour of other mechanisms, including self-regulation by the industry itself? Were other regulatory options available, and have they been deployed to maximum effect?

The effectiveness of regulation needs to be judged against two parameters peculiar to Antarctica: on the one hand, the nature of the Antarctic tourist industry and in particular its international dimension; and on the other, the juridical status of Antarctica with its unique political and jurisdictional basis. Both parameters impose

---

[35] Resolution 3 (1997), 'Standard Form for Advance Notification and Post-visit Reporting on Tourism and non-Governmental Activities in Antarctica', in *Final Report of the Twenty-first Antarctic Treaty Consultative Meeting, Christchurch, New Zealand, 19–30 May 1997* (Wellington: New Zealand Ministry of Foreign Affairs and Trade, 1997), p. 135.
[36] See further discussion by Berguño, Chapter 5; and Francioni, Chapter 7 in this book.
[37] Oceanites, *Compendium of Antarctic Peninsula Visitor Sites* (Chevy Chase, MD: Oceanites, 1997),

major obstacles to achieving optimum regulation through comprehensive, implementable laws. This is largely because the normal mechanisms of control such as territorial jurisdiction cannot be exercised.

Parallels have been identified by some authors[38] between the regulation of tourism in the Arctic (Svalbard) or the sub-Antarctic (New Zealand islands)[39] and what might be achieved in the Antarctic. However, such comparisons, if drawn too closely, are unrealistic, because what is quite attainable in regulatory terms through a national system under territorial jurisdiction cannot be achieved within the complex international situation in the Antarctic.

Distinction also needs to be drawn between the legal basis for regulation and the feasibility of implementation *in practice*. It is clear that the mandatory provisions of the Protocol embrace tourism. But it is also apparent that despite interim implementation of the Protocol since 1991 and its entry into force in early 1998, the terms of the Protocol have yet to be used to full effect in the regulation of tourism.

To cite but four examples of ineffective regulation measures: Firstly, with the adoption of Decision 2 (1999),[40] Treaty parties are only now beginning to focus on standards relevant to Antarctic shipping[41]. These are regulatory mechanisms that will be particularly crucial with the advent of greatly increased tourist-related tonnage in Antarctic waters. Secondly, the use of Antarctic Specially Managed Areas (ASMAs) under Annex V is a regulatory mechanism particularly adaptable for tourist management; yet to date only one ASMA Management Plan (Admiralty Bay in the South Shetland Islands) has been put forward, though not yet designated formally, pending entry into force of Annex V to the Protocol.[42] Thirdly, EIA under Article 8 and Annex I of the Protocol has been so far inadequately utilised for tourist activities. Parties have demonstrated only partial appreciation for how EIA might best be used to mitigate any potential environmental impacts of tourism. In this area, parties have instead tended to adopt a more passive approach to EIA, relying to a degree on tour operators to take the initiative. Parties have yet to seize fully on the mechanism presented by Article 8(4) of the Protocol which provides for activities planned jointly by more than one party to coordinate EIA procedures under a lead party. This has just as much relevance for tourism as for other activities such as

---

p. 243.

[38] D.J. Enzenbacher, 'The Regulation of Antarctic Tourism', in C.M. Hall and M.E. Johnston (eds.), *Polar Tourism. Tourism in the Arctic and Antarctic Regions* (Chichester: Wiley, 1995), pp. 179–215. B. Stonehouse and K. Crosbie, 'Impacts and Management in the Antarctic Peninsula Area', in Hall and Johnston (eds.), *Polar Tourism*, pp. 217–233. M. Hall and M. Wouters, 'Managing Nature Tourism in the sub-Antarctic', *Annals of Tourism Research*, Vol. 21, 1994, pp. 255–274.

[39] G.R. Cessford and P.R. Dingwall, 'Tourism on New Zealand's sub-Antarctic Islands', *Annals of Tourism Research*, Vol. 21, 1994, pp. 318–332.

[40] Decision 2 (1999), 'Guidelines for Antarctic Shipping and related Activities', in *Final Report of the Twenty-third Antarctic Treaty Consultative Meeting, Lima, Peru, 24 May – 4 June 1999* (Lima: Peruvian Ministry of Foreign Affairs, 1999), Annex B.

[41] See Article 10 of Annex IV to the Protocol.

[42] *Final Report of the Twentieth Antarctic Treaty Consultative Meeting, Utrecht, Netherlands, 29 April – 10 May 1996* (The Hague: Netherlands Ministry of Foreign Affairs, 1997), para. 148.

scientific research in the Antarctic. For example, in 1997 the UK and the USA could have cooperated on the EIA undertaken for 'Orient Lines', the operator of the MV *Marco Polo,* considering that the registered office of the company was in the UK and most of the tour bookings were made in the US. Similarly, in the coming 2000-2001 season with the planned visit of the largest Antarctic tourist vessel MV *Rotterdam* to the Antarctic, EIA procedures[43] could have been jointly implemented between the US (where 'Holland-America Line' is registered) and the flag-state of the vessel, the Netherlands. Finally, the inspection provisions of Article 14 of the Protocol (or indeed of Article VII of the Antarctic Treaty) have been extended infrequently to tourist activities or to tourist vessels. Since 1961 only four such vessels have been inspected in the Treaty area, two in 1993 under the UK/Italy/Korea inspection programme,[44] and two in 1999 by UK/German inspections.[45] That vessels can only be inspected at points of embarkation or disembarkation of personnel or cargo is not, in practical terms, a significant restriction, and both the above inspection programmes included vessels flagged with non-Treaty parties, albeit the inspections were carried out with the permission of the ships' masters and indeed with active support from IAATO.

In negotiating comprehensive measures for the protection of the Antarctic environment the clear intent of the Antarctic Treaty parties had been that all who undertake activities in the Antarctic Treaty area should do so with full regard for the standards of the Protocol. However, realisation of this ideal is hampered by the considerable involvement of third-party states in Antarctic tourism.

A significant proportion (around 40 per cent in 1995–96)[46] of tourist vessels operating in the Antarctic are flagged with states which are either non-Treaty parties or non-Consultative Parties which have yet to accede to the Protocol. These vessels are therefore not bound by the terms of the Protocol because, as a general rule of international law, only states that are party to a treaty are held under its provisions.[47]

The complex multi-national nature of today's Antarctic tourist industry, which was eloquently predicted by Nicholson,[48] has indeed proved to be the norm, at least

---

[43] Holland-America Line, 'MS *Rotterdam.* Antarctic Cruise Expedition 2000. Initial Environmental Evaluation', p. 17 (plus annexes).

[44] United Kingdom, Italy and Korea, 'Report of a Joint Inspection under Article VII of the Antarctic Treaty by United Kingdom, Italian and Korean Observers, January-February, 1993' (London: Foreign and Commonwealth Office/Rome: Ministry of Foreign Affairs/Seoul: Ministry of Foreign Affairs, 1993), p. 37.

[45] United Kingdom and Germany, 'Report of a Joint Inspection under Article VII of the Antarctic Treaty by United Kingdom and German Observers' (London: Foreign and Commonwealth Office/Bonn: German Federal Ministry of Foreign Affairs, 1999), p. 132.

[46] United Kingdom, 'Recent Developments in Antarctic Tourism', doc. XX ATCM/IP 15, 1996, p. 20.

[47] P. Birnie, 'The Antarctic Regime and Third States', in R. Wolfrum (ed.), *Antarctic Challenge II. Conflicting Interests, Co-operation, Environmental Protection, Economic Development* (Berlin: Dunker & Humbolt, 1986), pp. 239–262.

[48] I.E. Nicholson, 'Antarctic Tourism – The Need for a Legal Regime?', in Wolfrum (ed.), *Antarctic Challenge II*, pp. 191–203. I.E. Nicholson, 'Antarctic Tourism – The Need for a Legal Regime', in A. Jorgensen-Dahl and W. Østreng (eds.), *The Antarctic Treaty System in World Politics* (London:

as far as ship-borne tourism to Antarctica is concerned. Tour companies may be registered in one country, operated from another, and marketed widely internationally whilst also sub-contracting to one or more third-party companies based elsewhere. The vessels they operate may be owned, chartered, or sub-chartered. They may be flagged with either Treaty or non-Treaty parties whilst the tourists themselves may comprise a wide spectrum of nationalities. Set against this international kaleidoscope it is difficult to envisage any wholly effective regulatory regime reliant only on the normal jurisdictions over flag-vessels, nationals or territory. Some parties may have sound regulatory controls in one area; an example of this is Sweden's domestic legislation[49] which enacts tough provisions and permitting requirements over all Swedish nationals in the Antarctic. However, the jurisdictional basis for a comprehensive regulatory regime amongst parties overall remains deficient.

Despite this deficiency, practical compliance by tour operators and their clients with the standards of tourism regulation has to date largely been achieved. The formula has been a mixture of mandatory and hortatory provisions through governmental control, coupled with self-regulation by the tourism industry itself. The latter has proved particularly important in ensuring adherence to standards set by Consultative Meetings. Crucial to this whole approach has been the attitude and influence of IAATO which represents the great majority of tour operators active in the Antarctic. Two factors have undoubtedly influenced IAATO policy: Firstly, the realisation that, faced with the possibilities of draconian measures on tourism (i.e., an Annex to the Protocol), it was preferable to introduce self-regulation than have regulations enforced on the industry. Secondly, there was the desire, prompted by an environmentally knowledgeable and critical tourism clientele, to maintain high environmental standards. Tourists now expect and demand compliance with the highest standards. Maintaining these standards through corporate guidelines drawn up, for example by IAATO,[50] is therefore of paramount importance to the industry. The value of self-regulation was also well recognised by the Consultative Parties at their 1996 Meeting[51] which urged IAATO to:

– ensure that their members conform fully with the provisions of the Protocol,
– disseminate Consultative Meeting recommendations and other texts relevant to tourism;
– produce other further guidelines and codes of conduct where appropriate,
– encourage all tour companies in Antarctica to become members of the Association.

Macmillan, 1991), pp. 415–427.

[49] Sweden, 'Antarctica Act. Government Bill, 1992–1993' (Legislation enacting the Environmental Protocol), 1992, p. 140.

[50] IAATO, 'Guidelines for Tour Operators'.

[51] *Final Report of the XX ATCM,* para. 84.

Against this background it is difficult to determine whether or not the imposition of a mandatory Annex would have enhanced compliance. Certainly, an Annex would have had presentational value and it would have reinforced the need to control tourism. However, a mandatory instrument which purported to regulate, yet only partially managed to do so because of the inherent problems of effective jurisdiction and internationalisation, could well have been less effective than the current blend of mandatory measures and hortatory guidelines. Furthermore, a more draconian approach, as advocated through an Annex, might even have seen a deterioration in the situation; and resulted in less adherence to environmental standards through tour operators 'flagging out' or relocating their operations in the territories of non-Treaty parties as a means of circumventing formal regulation.

Concerns over the impact of tourism on science have now in part dissipated. Complaints of recent years over the disruption to base programmes by overly-frequent tourist vessels have largely disappeared. Some operators, e.g., the UK, indirectly promote IAATO by refusing access to their facilities to non-IAATO vessels; others actively discourage tour operators from visiting their stations, and still others have introduced local management and regulation to minimise the impact of tourism on the operations of their stations.

On the Antarctic Peninsula, Esperanza Station (Argentina), Palmer Station (US) and Arctowski Station (Poland) have well-defined tourism management policies and strictly regulate tourism in the vicinity of their stations. Recently, a tourist information facility has been constructed at the Arctowski station specifically to divert tourists away from the scientific facilities of that station. Similar regulations are also being employed to manage Historic Sites and Monuments such as the historic huts in the Ross Sea area, or at Port Lockroy (HSM No 61), which is managed by the UK. The latter is now the most heavily visited site in the Antarctic. During the 1998–99 season 5,800 tourists from 72 tourist ship visits and 30 yachts visited Port Lockroy. This site demonstrates a sound working relationship between an operator and IAATO whilst it at the same time safeguards both the historic structures and the wider environment at the site.

But the broader concern of the scientific community remains – that the ever increasing demand of the tourist industry to explore hitherto pristine areas will unwittingly create impact to the detriment of science. The tourism industry must recognise and respond to this concern. If voluntary restraint on 'no-go areas' cannot be enacted by the industry (e.g., by curtailing helicopter flights into sensitive areas), then formal restrictions will inevitably need to be reconsidered.

But self-regulation is also threatened, as is the important pole-position that IAATO has maintained to date in the Antarctic tourist industry. The first signs of this were in 1996 when the 400 tourist-per-vessel limit set by IAATO bylaws was broached. This caused the operator in question 'Orient Lines' to be excluded from IAATO, even though the company continued to abide by the stringent guidelines of the Association. However, the 1999-2000 season will witness a new dimension in Antarctic ship-borne tourism, with a new generation of three larger tour ships commencing operation in the Treaty area.[52] The MV *Rotterdam*, operated by 'Holland-America Line' and flagged with the Netherlands, is for example anticipated to carry 1,000 passengers and 600 crew, but the ship will not be discharging passengers ashore.[53] This type of Antarctic cruising could be construed as a more 'environmentally friendly' approach because such a vessel will do no more than transit waters within the Antarctic Treaty area and thus leave no trace of its presence; however, larger ships do have the potential for considerable environmental impact or for search and rescue implications in the instance of a serious accident. Unlike smaller diesel powered vessels, such large ships carry significant quantities of heavy bunker fuel oil, and pollution emanating from these vessels during an accident could generate major environmental impact. In the words of one US environmental NGO, the deployment of the *Rotterdam* into the Antarctic would be the first, truly large-scale, mainstream intrusion that 'would be the most dramatic development in the private use of Antarctica since commercial tourism began in the late 1950s'.[54] The pros and cons of large versus small tourist vessels were expanded on in a 1999 Antarctic Treaty Inspection Report.[55] If IAATO is to maintain its well-deserved reputation as a responsible organisation and ensure that self-regulation remains effective, a mechanism will need to be found to bring the operators of these large vessels within the regulatory regime. This may pose the greatest challenge to IAATO in the immediate future.

An allied but separate issue which continues to challenge the regulatory capacity of the system is the operation of tourist vessels in the Antarctic by those non-Consultative Parties which have yet to become parties to the Protocol.[56] These countries are not bound by the legal provisions of the Protocol, nor have they necessarily enacted domestic legislation to control the activities of their nationals (including bodies corporate) or vessels in the Antarctic. The greatest cause for concern relates to the non-Consultative Parties that have a major tourist interest in the Antarctic. For example, the Canadian registered company 'Marine Expeditions' has a dominant share of the Antarctic ship-borne tourism industry.

In 1996, Treaty parties drew attention to their concerns in the report of the Twentieth Consultative Meeting and called upon: 'non-Consultative Parties with a particular interest in, or responsibility for, tourist companies operating in Antarctica to ratify the Protocol and its Annexes at the earliest opportunity and to introduce any necessary domestic enabling legislation to ensure compliance'.[57]

---

[52] *Final Report of the XXIII ATCM*, para. 113.
[53] Holland-America Line, 'MS *Rotterdam*. Antarctic Cruise Expedition 2000. Initial Environmental Evaluation', p. 17.
[54] 'Is Antarctic Tourism Changing Forever?', *The Antarctic Project*, Vol. 8, 1999, p. 4.
[55] United Kingdom and Germany, 'Report of a Joint Inspection', p. 132.
[56] As of 29 December 1999, only one of the 17 non-Consultative Parties had yet acceded to the Protocol. This status remained unchanged as of 8 June 2000.
[57] *Final Report of the XX ATCM*, reference to non-Consultative Parties, para. 85.

Similar texts were adopted in both 1997 and 1998, and the message was strengthened with the adoption of Resolution 6 in 1999.[58] In an effort to address this matter, other parties have taken on responsibilities for regulating Canadian operators in the Antarctic. For example, the US has required Canadian firms to undertake EIAs under US law, whilst the UK has granted permits to the Canadian-registered company 'Adventure Network International' on the grounds that the company is based in the UK.

Until parties such as Canada take action to ratify the Protocol, the ATS will continue to have within its own ranks a legal lacuna in respect of tourism regulations.

THE SHORTCOMINGS OF REGULATION

More general flaws in the current system are only too evident and have been commented on by others.[59] The existing tourist guidelines set out in Recommendation XVIII–1 are hortatory and thus not enforceable; not all tour operators are members of IAATO and thus bound by its by-laws and guidelines, and adequate means for imposing sanctions against infringements do not exist. Within the mandatory framework of the Protocol, consistent standards between parties have yet to be achieved. Particularly in relation to tourism, it has become increasingly apparent that there are problematic areas within the Protocol on which further attention should be focussed with regard to the terms of both regulation and implementation. Within this category are:

- how most appropriately to address EIA of tourism operations, especially the complex matter of cumulative impact,[60]
- how to determine baselines for, and monitoring of, tourist activities,
- how tourism data and other information for management purposes should be collected, collated, analysed and disseminated,
- how most effectively to address emergency response action and contingency planning relating to tourist activities.

---

[58] Resolution 6 (1999), 'Adherence to the Environmental Protocol by non-Consultative Parties', in *Final Report of the XXIII ATCM*, Annex C.

[59] B.A. Boczek, 'The Legal Status of Visitors, Including Tourists, and non-Governmental Expeditions in Antarctica', in R. Wolfrum (ed.), *Antarctic Challenge III: Conflicting Interests, Co-operation, Environmental Protection, Economic Development* (Berlin: Dunker & Humbolt, 1997). IUCN, 'Tourism in Antarctica', doc. XVII ATCM/IP 18, 1992, p. 7. M.E. Johnston and C.M. Hall, 'Visitor Management and the Future of Tourism in the Polar Regions', in Hall and Johnston (eds.), *Polar Tourism*, pp. 297–313.

[60] IUCN, 'Cumulative Environmental Impacts in Antarctica. Minimisation and Management', doc. XXII ATCM/IP 30, 1998, p. 17.

All these problem-areas illustrate the all too evident disparate nature of the Antarctic tourist industry and the particular difficulties that confront the Treaty parties – not least as to where responsibility (and possible liability) might lie. The somewhat complex proposals to allocate such state responsibility, based on a series of tiered criteria proposed by Australia,[61] found little favour: for example, should responsibility lie with the party in whose territory the expedition was organised or with the party from whose territory it departed to Antarctica, with the party under which the vessel was flagged or the party whose nationals were involved in the expedition, or with the parties whose stations were to be visited in Antarctica?

COMPREHENSIVE REGULATION – GAPS IN THE SYSTEM

The argument of some parties (and in particular the US and UK)[62] that the regulation of tourism in the Antarctic was best achieved by timely and effective implementation of the Protocol and its Annexes could, if only because of the scale of third party involvement, be construed as flawed. However, the UK also argued[63] that it was inappropriate to consider additional regulation in the absence of determining whether or not the current regime – provided by the existing recommendations and strengthened considerably by the provisions of the Protocol, was adequate. If the current regime proved inadequate, and if further regulation was needed, then what form should such regulation take? Answers to these questions required prior assessment of:

- the nature and scale of the potential problems and impacts of tourism and non-governmental activity; and
- the adequacy of the existing provisions in addressing these potential problems and impacts.

The creation of further regulation in the absence of such an assessment was likely to lead to, at best, a duplication of existing legal measures or, at worse, to provisions which failed to target the predicted problems and impacts. That analysis in 1992 concluded that there were indeed gaps in the regulatory regime and that a number of further measures needed to be introduced. Specifically,

- new unified guidelines for tour operators and tourists,
- enhanced information exchange requirements,
- an onboard observer system,

---

[61] Australia, 'Tourism and non-Governmental Activities in the Antarctic Treaty Area', p. 7.

[62] United States, 'Antarctic Tourism and the Environmental Protocol', p. 10. United Kingdom, 'The Regulation of Tourism and non-Governmental Activities in the Antarctic Treaty Area', p. 11.

[63] *Ibid.*; also United Kingdom, 'The Regulation of Tourism and non-Governmental Activities in the Antarctic Treaty Area (II)', p. 24.

- central coordination for the collection, collation, analysis and dissemination of information,
- the production of a lay guide to the Antarctic Treaty and its Environmental Protocol,
- liaison with maritime experts to elaborate details on liability and shipping.

Not all of these issues are specific to tourism and NGO activity. The question of shipping standards and matters relating to MARPOL 73/78 are for example common to all vessel activities in the Antarctic. Nor, indeed, were these issues necessarily new. Nicholson had called for enhanced information exchange as far back as 1985.[64]

Since 1992, significant progress on a number of these issues has however been made. For example, Recommendation XVIII–1 introduced new guidelines for tour operators and tourists; Resolution 3 (1995),[65] coupled with Attachment A of Recommendation XVIII–1, has updated information exchange, whilst the setting up of IDAT by ICAIR in New Zealand made some advances towards coordinating data handling, though unfortunately the IDAT initiative has since lapsed.

Further measures to enhance regulation could also be achieved through more proactive implementation of the Protocol. Negotiations on a draft Antarctic Shipping Code[66] are now the subject of Decision 2 (1999) and will be discussed at a Group of Experts Meeting, to be set up under Recommendation IV–24, in the UK in 2000.[67] The Code, once elaborated, would have implications for the Antarctic tourism industry. Other areas where regulation could be enhanced include the following:

- The inspection provisions under Article 14 of the Protocol. More widespread use of inspections, and the comprehensive inspection checklists developed by the Consultative Meeting, could target tourist vessels and facilities.
- More constructive implementation of the protected areas system; in particular ASMA designations under Annex V of the Protocol would provide an opportunity to control and manage tourist operations within, if necessary, extensive areas of the Antarctic. The ASMA mechanism could also curb the increasing trend of deep-field incursions by tourists into remote, pristine areas.

But all these possible mechanisms, and the decisions taken at Consultative Meetings since 1992, face the inherent and continuing problem that their application

---

[64] Nicholson, 'Antarctic Tourism', pp. 191–203.

[65] Resolution 3 (1995), 'Reporting of Tourism and non-Governmental Activities', in *Final Report of the Nineteenth Antarctic Treaty Consultative Meeting, Seoul, 8–19 May 1995* (Seoul: Ministry of Foreign Affairs of the Republic of Korea, 1995), p. 98.

[66] Norway, 'The International Code of Safety for Ships in Polar Waters (Polar Code)', doc. XXII ATCM/WP 17, 1998, p. 4.

[67] Decision 2 (1999), 'Guidelines for Antarctic Shipping and Related Activities', in *Final Report of the XXIII ATCM*, Annex B. See discussion by Scovazzi, Chapter 13 in this book.

does not extend formally to third states and their nationals or to third party flagged-vessels in the Antarctic.

Recognising that comprehensive regulation of tourism appears elusive in the Antarctic, some have argued that compliance with the standards of the Antarctic Treaty and its Protocol could be achieved through regulation applied outside the Treaty area, and in particular at those points where vessels and aircraft depart for Antarctica.[68]

Some Consultative Parties have incorporated the provisions of Article VII(5) of the Treaty when enacting the Protocol into their domestic law. This has been done so that national legislation relates not only to their nationals but also to expeditions – irrespective of nationality – which are organised in, or depart to Antarctica from, that state's territory. Exemption from regulation is specified if such expeditions are already the subject of appropriate authorisation from another party to the Protocol.

The issue of some form of port state control was first raised at a Consultative Meeting by the Netherlands in 1995;[69] it was inserted as a separate topic on the agenda of the following Consultative Meeting. At the 1996 Meeting, the issue was explored in more detail through papers submitted by the Netherlands and Chile.[70] Though the latter restricted itself to conventions that already incorporated provisions for inspections by port states, namely MARPOL 73/78 and SOLAS, the Netherlands went further and recommended that 'gateway ports' situated in the territories of Consultative Parties should inspect vessels for breaches, or possible breaches, of Annex IV to the Protocol. The Netherlands recognised, however, that it was 'doubtful whether action can be taken under the Antarctic Treaty against a foreign ship on its way to and from Antarctica which does not comply with the requirements of Annex IV to the Environmental Protocol'. Although the notion of control through gateway ports did not attract consensus, it was nevertheless recognised that some form of harmonisation of inspection by port states might be useful. When the parties returned to the issue a year later, however, the proposals set out in a paper by the UK on 'departure state jurisdiction'[71] were countered by Germany[72] which regarded the inspection of vessels prior to their going to Antarctica as an inappropriate mechanism, and contrary to the 1982 UN Convention on the Law of the Sea.

What then could be the next step? Consultative Parties are faced with a continuing burgeoning tourism industry which is to a substantial part operated

---

[68] See Orrego Vicuña, Chapter 3 in this book.

[69] *Final Report of the XIX ATCM*, para. 59.

[70] Netherlands, 'Inspection of Ships in Gateway Ports to Antarctica, on the Basis of MARPOL 73/78, and in Antarctic Ports under the Environmental Protocol (Annex IV) to the Antarctic Treaty', doc. XX ATCM/WP 9, 1996, p. 6. Chile, 'Enforcement of International Maritime Convention and Domestic Standards on the Inspection of Ships that Operate in the Antarctic', doc. XX ACTM/WP 17, 1996, p. 15.

[71] United Kingdom, 'Enhancing Compliance with the Protocol: Departure State Jurisdiction', doc. XXI ATCM/WP 22, 1997, p. 6.

[72] Germany, 'Inspection of Ships by Port States (Agenda Item according to the Final Report of the XXth ATCM)', doc. XXI ATCM/WP 16, 1997, p. 3.

within the purview of third party states. To cope with this situation, they have two main options: relinquish any idea of formal regulatory control (at least over that sector of the industry) and rely instead solely on industry self-regulation; or embrace more innovative means of regulation. Regulation through port state jurisdiction offers one such opportunity. This mechanism, if more widely adopted – and especially if the standards to be employed at departure ports could be harmonised between Consultative Parties – could be an effective measure. It would ensure the maintenance of high standards of environmental care in the Antarctic. A major advantage of operating such jurisdictional control is that it could be extended quite appropriately to vessels and aircraft departing to Antarctica, including those of non-Contracting Parties.

Part II

INSTITUTIONAL SUPPORT TO THE IMPLEMENTATION OF THE PROTOCOL

5

Institutional Issues for the Antarctic Treaty System with the Protocol in Force: An Overview

Jorge Berguño

Entry into force of the Protocol on Environmental Protection to the Antarctic Treaty[1] reinforces the sustained effort by the international community to develop a comprehensive response to the growing threats to the global environment. The entry into force comes at a time when the Antarctic Treaty System (ATS) finds itself at the crossroads of a number of challenges which will test its capability to implement an effective regime for the protection of the Antarctic environment and its dependent and associate ecosystems. These challenges include developing appropriate procedural arrangements and a stronger institutional framework which complies with the objectives of the Environmental Protocol. Given the special political and juridical status of the Antarctic, success or failure will inevitably reflect upon the performance of the ATS as a whole, hopefully by improving its internal operation as well as its external relations.

Internal cohesion and external projection are equally important for the appropriate functioning of the ATS as a subsystem of the global international system. The ability of the ATS to maintain peace, the protection of the unique Antarctic environment, and the free but cooperative pattern of scientific Antarctic research will remain to be severely tested in the future. If the Antarctic Treaty[2] and its supplementary norms fail to adequately perform their role in the Antarctic, it can be taken for granted that pressures on the ATS will intensify at the United Nations, and possibly in other fora, where a now dormant revisionist movement could obtain a second chance to antagonise the ATS.

The relationship of the ATS to the global international system therefore involves the issue of the further international development of the ATS, including the need to

---

[1] ILM, Vol. 30, 1991, pp. 1,461ff.
[2] UNTS, Vol. 402, pp. 71ff.

establish instrumentalities in addition to the Consultative Meetings and the Committee for Environmental Protection (CEP), such as an Antarctic Treaty Secretariat. There may also be a need for subsidiary bodies entrusted with specific tasks that demand a permanent commitment, something which at present is not well served by the ancillary meetings of the ATS (Special Consultative Meetings and Meetings of Experts).

It is incumbent upon the Consultative Parties to ensure appropriate coordination between the Treaty, its Protocol and all other Antarctic regulatory regimes. This is especially true since the Antarctic Treaty constitutes the political and legal framework for all activities undertaken in its area of application and, by virtue of the Environmental Protocol, for all impacts which the Antarctic environment and the dependent and associated ecosystems may experience as a consequence of those activities. Beyond the concerns for the systemic integrity of the ATS, the issue of effective implementation of the regime envisaged by the Environmental Protocol acquires a symbolic meaning and forcefully challenges the stewardship of the Consultative Parties in Antarctica.[3]

In order to provide a comprehensive overview of the relative adequacy or inadequacy of the existing procedural and institutional arrangements within the ATS, terms such as 'structure', 'system' and 'regime' should be defined. These terms should also be related to an appropriate theoretical framework and their practical applications to the ATS should be considered. General definitions should indeed be contrasted with their more specific Antarctic models. The usage of the above mentioned terms (structure, system and regime) in this chapter is not fundamentally different from the traditional usage in international law and international relations theory, but it also draws upon a body of literature which has as its objective the integration of the study of discrete geographical groupings of states within the systematical analysis of the international system.[4]

THEORETICAL FRAMEWORK

'Structure' has been postulated as a key concept, a peculiar reality, a thing in the secondary sense resulting from the addition to its parts of something different, namely an 'order'. The reality of that order, its organization and texture, differs from its individual components. 'System' is not a particular type of structure but any kind of structure conceived as a dynamic process, in its actual operation and functioning. 'Regime' has been defined as a set of implicit and explicit principles, norms, rules and decision-making procedures operating within a given system or subsystem.

---

[3] See Sir Arthur Watts, *International Law and the Antarctic Treaty System* (Cambridge: Grotius, 1992), pp. 73–87.

[4] See K. Dodds, *Geopolitics in Antarctica: Views from the Southern Oceanic Rim* (Chichester: Wiley and Sons, 1997); and C.C. Joyner, *Governing the Frozen Commons: The Antarctic Regime and Environmental Protection* (Columbia, SC: University of South Carolina Press, 1998).

Order, system and regimes are not synonyms. Regimes are man-made arrangements, complex social artefacts, and mechanisms governing the actions of individuals through a combination of defined 'rules of the game' and less rational compulsive social constraints which represent 'binding observances' essential for social cohesion and systemic integrity.

The rise and demise of international regimes is usually explained by the 'realists' with such parameters as self-interest and political power, while by the 'idealists' (also called 'liberalists') with principles and norms. Usage and custom, on the one hand, and theory or knowledge on the other, are not generally viewed as capable of generating a regime on their own merits. In reality, we live in an ocean of usages and customs, with patterns of behaviour based on actual or long-standing practice which cements our social environment and holds the key to its meaning and function.

Nevertheless, these constraints do not deny the power of ideas or the possible role played by epistemic communities in the origin, growth and development of a system (or sub-system) such as the ATS. The international system, as well as its subsystems, operates during periods of certainty and security with precision and automaticity. In critical times, when the international system is no longer useful to all its partners and its 'inter-national' character are questioned, or when the subsystem is in a state of crisis, the demand for a new order leads to an intellectual reappraisal of the old one. The new regime starts as an inspiration in the creative imagination, but from the moment that the vision of a new regime is conceived to the time it actually becomes a binding observance, or an institution, a significant period of time can elapse.[5]

Members of international regimes are always sovereign states; decision-making procedures at this level frequently require a two-step implementation. This is a conspicuous feature in the ATS, but less so at the domestic level or in more sophisticated regimes. The Antarctic Treaty provides that measures have the status of legally binding norms once all of the Consultative Parties have consented to be bound by them. A more definite trend towards institutional development may result in at least a limited departure from the two-step procedure, as seems to be the implication of Decision 1 (1995) adopted at the 1995 Consultative Meeting in Seoul. This Decision introduced a distinction between *measures* (legally binding once approved by all Consultative Parties), *decisions* (related to internal organisational matters and operative once adopted by a Consultative Meeting), and *resolutions* (i.e., hortatory instruments adopted by a Consultative Meeting).[6]

---

[5] J. Ortega y Gasset, 'Esquema de las Crisis', *En Torno a Galileo. Obras Completas*, Madrid, Revista de Occidente, 1947–60, Vol. V, pp. 62–63; J. Ortega y Gasset, *Man and People* (New York: Norton, 1957), pp. 258–272; M. Mann, *The Sources of Social Power. A History of Power from the Beginning to A.D. 1760* (Cambridge University Press, 1986), Vol. I.

[6] See paras. 1–3 of Decision 1 (1995), 'Measures, Decisions and Resolutions', in *Final Report of the Nineteenth Antarctic Treaty Consultative Meeting, Seoul, 8–19 May 1995* (Seoul: Ministry of Foreign Affairs of the Republic of Korea, 1995), Annex B, pp. 89–90. See also discussion by Bush, Chapter 2

Some regimes remain at the stage of spontaneous orders: natural markets, linguistic and cultural systems, habits and customs. Others are negotiated and completed through a conscious effort to agree on their substantive content. In addition to explicit consent, specific qualifications such as the ATS 'activity criterion', as provided in Article IX of the Antarctic Treaty for acquiring the consultative status, may be required. A third category of regimes, that of imposed orders, reflects the current features of the international system. In every case, even for artificial or man-made regimes either consensual or imposed hegemonically, a layer of spontaneous order sustains the more rational strata on top of the regime.

Critics of the ATS visualise it as an imposed order, hegemonically ruled by a consortia of dominant world powers. This view operates on three levels of cumulative misunderstanding: firstly, the interpretation of the Antarctic Treaty in terms of classic power politics; secondly, the understanding of the rule of consensus, intended to safeguard a non-prejudicial egalitarian *status quo*, with the veto powers awarded to permanent members of the UN Security Council; and thirdly, interpretation of international affairs in a manner detrimental to the role of ideas.

Conceptualisation of international regimes clarifies the role of *international institutions*, which are often mistakenly equated with 'regimes'. Regimes may be more or less articulately arranged, but institutional development is always a major dimension of any regime. The ATS paradigm of international cooperation is characterised by an apparently low level of institutional growth – a fact which would seem to cast doubts on the assumption that collaborative regimes are more apt to become institutionalised than coordinative regimes.

This preference for flexible accommodation and for the lack of machinery in the ATS has some enduring, historic roots. On the other hand, it is balanced by other more pervasive and coercive unconscious mechanisms. Reference to the 'degree of integration at the human level in the professional group of scientists, diplomats, lawyers and environmentalists' is pertinent to explaining the cohesive force of this group of professionals and provides a particularly eloquent example of the concept of 'diplomatic culture' defined by Hedley Bull as the 'common stock of ideas and values possessed by the official representatives of States'.[7]

THE ATS INSTITUTIONALITY: CHALLENGE AND RESPONSE

Quite often the flourishing of the Antarctic Treaty cooperation is traced to a favourable climate spawned by the International Geophysical Year (IGY). Another view suggests that the Antarctic Treaty gained impetus from an agreement nurtured

---

in this book.

[7] For a comprehensive survey of the role of international institutions, see: H.G. Schermers and N. Blokker, *International Institutional Law. Unity with Diversity* (The Hague: Kluwer Law International, 1995); Watts, *International Law and the Antarctic Treaty System,* pp. 85–86; H. Bull, *The Anarchical Society: A Study of Order in World Politics* (New York: Columbia University Press, 1977), p. 316.

by a US State Department which was eager to prevent any further escalation of Antarctic belligerence between its allies. Whenever these explanations do not amount to pure oversimplification, they represent predicaments that must be compounded by the realisation that the 1959 Antarctic Treaty aggregates a heritage of international cooperation, codifies pre-existing practices and draws on a rich experience of unfinished diplomatic negotiations. Our goal is not to rectify current interpretation of the history of the Antarctic Treaty, but rather to capture the essence of systemic practices and institutional elements which pre-dated the 1959 Treaty.

Let us, in retrospect, focus on the year 1939. A chain of events had led to the amicable solution of almost every territorial dispute in the Antarctic. The only exception was the area surrounding the Antarctic Peninsula. At the time, the US proposed to set aside this area as a condominium of American states which would be invited to a conference where they could examine the respective merits of their Antarctic titles. Precisely when an alliance between the European and Commonwealth claimants seemed imminent, the US upheld the interests of the South American claimants, which resulted in a stalemate in Antarctic politics.

An interesting question which we are not able to explore here is whether this US initiative was prompted solely by German activities in the Antarctic or whether it transpired from broader concepts such as the Monroe Doctrine, the Western Hemisphere or the short-lived American Neutrality Zone. A side-effect of the American initiative was the decision taken by Chile in 1940 and by Argentina in 1942 to proceed with the delimitation of their overlapping sectors in Antarctica, leaving the only empty, yet uninviting, space between 90° and 150° W to accommodate the US.[8]

In areas other than the 'American Quadrant', cooperation had been more successful. The US may not have been aware that Soviet officials had given assurances to the UK that they would seek advance approval for any activities undertaken in British claimed territories. The US State Department could not ignore initiatives such as the Franco–Australian–New Zealand–British Agreement on Aerial Navigation in Antarctica. Another example was the Norwegian Royal Decree 141 which placed under the sovereignty of Norway the Antarctic coast and lands stretching from the boundaries of the UK claimed sector to those of the Australian claimed territory in the Antarctic. On several instances in the year 1939, the US State Department felt compelled to reserve its position on these matters. Against this background, cancellation of Norway's offer to hold a Polar Exhibition and Conference in Bergen in 1940 can be seen as a lost opportunity for the then still unborn Antarctic community of nations.[9]

---

[8] While the unclaimed sector was accommodated by the ATS, it may add to the complexity of Antarctic maritime issues; see Vidas, Chapter 14 in this book. Chile and Argentina acted unilaterally after failing to agree on a Delimitation Treaty (1906–08).

[9] The first mention of an Antarctic conference appears in a Norwegian memorandum dated 26 October 1934 addressed to the UK; text reprinted in W.M. Bush (ed.), *Antarctica and International Law: A Collection of Inter-State and National Documents* (London: Oceana Publications, 1982–88), Vol. III,

Deadlock rather than conflict seems the most appropriate word for describing the rather confined but troublesome state of Antarctic affairs in the late 1940s. Confronted with overlapping Chilean and British claims, Argentina argued cogently that the situation created by unilateral attributions of sovereignty in Antarctica could only be solved by a conference of interested states. Once World War II was over, Chile requested in a memorandum dated 13 October 1947, the views of the US State Department on an Antarctic conference. The US procrastinated, but faced with the disrupting effects of the escalating conflict in the area of the Antarctic Peninsula, the State Department sent a special envoy to Chile and Argentina and submitted to all seven claimants proposals for a *trusteeship*, or alternatively, for an Antarctic *condominium*. [10]

Experts on Antarctic affairs consider this and other episodes to be failures in early attempts to create an Antarctic regime. These failures left a gap of some ten years before another, and this time successful, endeavour eventually led to a diplomatic conference on Antarctic problems. The diplomatic record shows otherwise a continuous effort to achieve agreement on the basic rules of the desired regime which was being deployed from 1949 to 1953. This effort was further pursued by more scattered initiatives. Early cooperative practices in demilitarisation, exchange of personnel and information, assistance to Antarctic expeditions, Southern Ocean meteorology, whaling, and postal communications, etc., acquired a multilateral perspective in the 1940s. A pattern emerged, moving away from the classic internationalist formulation and toward more informal and less institutionalised arrangements.

The US proposals of 1948 included a Commission of eight members, endowed with legislative and executive authority in the Antarctic; however, in subsequent exchanges with France, the US State Department developed a more flexible interpretation. Pursuant to a Chilean proposal, inspired from a 1908 Baltic *status quo* agreement, a different option was canvassed during several years as a *modus vivendi*. This arrangement, improved and retouched between 1949 and 1952, contained the seeds of the future Antarctic Treaty. The American negotiator, Ambassador Daniels, drew heavily on that document in his own draft treaty which was presented as the basic document for the preparatory meeting of the Antarctic Conference. Perhaps more interesting than the similarity is the outstanding discrepancy between the two texts: the Chilean *modus vivendi* entrusted to a *Committee* a coordinating function in order to avoid logistic and scientific duplication (thus anticipating Article 6 of the 1991 Environmental Protocol); in Daniel's draft treaty, management of Antarctic affairs is entrusted to a *periodic*

---

pp. 141–142.
[10] Texts of trusteeship and condominium proposals reprinted in Bush (ed.), *Antarctica and International Law*, Vol. III, pp. 461–466. See also note by the Norwegian Embassy in Washington, of 15 November 1948, in *US Foreign Relations 1948*, Vol. I, pp. 1,011–1,013.

Meeting, i.e., the periodic consultative diplomatic conference foreseen in Article IX of the Antarctic Treaty.[11]

The lack of success of the 1949–53 negotiations, which involved the US and the claimant states in a rather uneven manner, has been primarily explained as a by-product of the unsettled territorial issues. However, nations with an established interest in Antarctica were no longer exclusively claimants; they now included also nations with a basis for claim, those prohibited by Article IV of the Treaty from asserting a claim, as well as those not recognising any claim or basis for claim. M.J. Peterson writes: 'the farthest any claimant was prepared to go officially in 1948 was the Chilean proposal that territorial questions be set aside while interested States cooperated in the pursuit of scientific knowledge'.[12] Pursuant to that proposal, the litigious aspects of the Antarctic problem were effectively put aside; nevertheless, considering every laborious step towards the Antarctic Treaty, it is easy to detect that as the threshold of institutional requirements has lowered, the acceptability of the option has risen.

INSTITUTIONAL ARRANGEMENTS WITHIN THE ATS

The Antarctic Treaty epitomised a tradition of competitive emphasis and equal opportunities for Antarctic nations. It provided no mechanism, bureaucracy or administrative unit to handle the day-to-day functioning of the regime. In conformity with Article IX of the 1959 Treaty, regular Consultative Meetings of the parties are held to ensure consultation on matters of common interest, to exchange information, and to agree on measures intended to further the objectives and purposes of the Antarctic Treaty. Authorities, places and times, as well as organisational aspects, of the Consultative Meetings have continued to be decided in each instance. With the single exception of the newly created Committee for Environmental Protection, there are no permanent committees, subcommittees or subsidiary bodies monitoring the performance of the ATS.

On 1 December 1959, the day of the signature of the Antarctic Treaty, some of the representatives assembled for that purpose may have had the uneasy feeling that a process going far beyond their expectations had been triggered on that day. In his closing speech, the Chilean representative spoke of 'the birth of a new system', and back in Santiago when lecturing at the University of Chile, he stated with great enthusiasm, 'The Antarctic System is born'.[13] Four decades after the signing of the

---

[11] A 1950 draft of the *declaration* or *modus vivendi* has been published in Bush (ed.), *Antarctica and International Law*, Vol. III, pp. 470–471.

[12] M.J. Peterson, *Managing the Frozen South: The Creation and Evolution of the Antarctic Treaty System* (Berkeley, CA: University of California Press, 1988), p. 56; J.K. Moore, 'Tethered to an Iceberg: United States Policy towards the Antarctic, 1939–1949', *Polar Record*, Vol. 35, 1999, pp. 125–134.

[13] M. Mora, *Statement by the Representative of Chile*, Conference on Antarctica, doc. 25, Annex D. Roberto Guyer, adviser to the Argentine delegation at the 1959 Antarctic Conference popularised the

Antarctic Treaty, a System has developed with a significant proliferation of institutions, which is designed to ensure the observance of various conventions and supplementary instruments. A certain lack of symmetry and balance presently exists between a weaker Antarctic Treaty core and a more developed periphery of Antarctic regimes.

The 1972 Convention for the Conservation of Antarctic Seals[14] had provided for a Scientific Committee that would be activated only if commercial sealing reached significant levels. The 1980 Convention on the Conservation of Antarctic Marine Living Resources[15] is headed by a Commission assisted by a Scientific Committee, a Secretariat and other subsidiary bodies. The 1988 Convention on the Regulation of Antarctic Mineral Resource Activities,[16] now in abeyance, diversified its institutionality to include a Commission, an Advisory Committee, Regulatory Committees, the Meeting of Parties and a Secretariat. The 1991 Protocol on Environmental Protection to the Antarctic Treaty established the Committee for Environmental Protection – so far the only institutional mechanism with an overall role within the ATS.

Geographical scopes of the various Antarctic regimes do not entirely coincide; parties to those regimes are not always the same, and the generally decentralised character of the ATS has been a subject of praise, criticism and some concern. But rationalisation of the operation of those regimes is not a prior condition for the application of timely measures. The supplementary instruments have been gradually integrated into the comprehensive environmental protection regime which the Protocol introduced into the mainstream of the ATS.

Action taken to impose the supremacy of the Consultative Meeting within the ATS is noteworthy. In 1985 Recommendation XIII–2 recognised the virtue of there being a regular overview of the ATS, which included the relationships among ATS components. Reports have regularly been delivered by the Chairmen of Special Consultative Meetings, the CCAMLR Commission, SCAR, as well as by the depositaries of the Antarctic Treaty, the Environmental Protocol, the Seals Convention, by the Coordinator of the Treaty Parties at the UN, and by the Council of Managers of National Antarctic Programmes (COMNAP). The overview, however, stopped short of any advice being provided to the autonomous components of the ATS, that is until the Twenty-third Consultative Meeting held in Lima in 1999. This meeting took the unprecedented step of addressing a message to CCAMLR in its Final Report:

> It was also noted that, given the important linkages within the wider Antarctic Treaty System, CCAMLR's successful ability to combat illegal, unregulated and unreported

---

term 'Antarctic System' in academic circles; see R. Guyer, 'The Antarctic System', *Recueil des Cours de l'Académie de Droit International de la Haye*, Vol. 139, 1973/II, pp. 149–226.

[14] UNTS, Vol. 1080, pp. 175ff; and ILM, Vol. 11, 1972, pp. 251ff.

[15] ILM, Vol. 19, 1980, pp. 837ff.

[16] ILM, Vol. 27, 1988, pp. 868ff.

fishing would enhance the strength of the whole Antarctic Treaty System and the protection of the Antarctic ecosystem.[17]

Modernisation of the ATS within the framework of agenda item 'Operation of the ATS' has included the public availability of documents, circulation of Consultative Meeting reports, transmission to the UN Secretary-General of the Final Report of each Consultative Meeting, reviewing the Rules of Procedure of the Consultative Meeting to liberalise the participation of observers, inviting acceding states and NGOs to attend the Consultative Meetings, and designating SCAR, CCAMLR and COMNAP as permanent observers to the Consultative Meetings.

But other initiatives have not mustered the necessary consensus. An ATS Secretariat is accepted as a concept, pending its location. However, the fact that the Consultative Parties have ruled out most proposals[18] implies either a higher degree of institutionalisation or the unequivocal manifestation of political will, which does not give high expectations regarding their ability to agree upon a strong Secretariat.

The issue of a Secretariat arose for the first time during the interim meetings in Washington in 1958–59, in preparation for the First Consultative Meeting in Canberra. At the 1959 Antarctic Conference, two proposals involving institutional mechanisms had been rejected: a Chilean initiative to create the Antarctic Scientific Research Institute of the Treaty parties and the US idea of a Standing Committee to oversee provisional application of the Antarctic Treaty before its entry into force. However, the subsequent offer by the US State Department to coordinate in Washington preparations for the First Consultative Meeting was welcomed by all Treaty signatories.

At the time, the Australian Government informed the other parties that it wished to establish a permanent secretariat headquartered in Canberra which would be staffed by Australian diplomatic personnel. Argentina, Chile and the Soviet Union expressed their preference for a continuation of the Washington venue, privileged as it was by the availability of local expertise. The USA, as depository of the Antarctic Treaty, sought to maximise the advantages of its capital city and naturally opposed any notion of a more permanent secretariat.

The First Consultative Meeting demonstrated, however, the relative convenience of the rotating Consultative Meeting. A South African proposal for a permanent secretariat with rotating headquarters was supported by all Commonwealth countries. But improvement of the consultative and preparatory mechanisms prevented the discussion from gaining momentum. When discussion of agenda item 'Operation of the ATS' brought to life the debate on the Secretariat, the

---

[17] *Final Report of the Twenty-third Antarctic Treaty Consultative Meeting, Lima, Peru, 24 May – 4 June 1999* (Lima: Peruvian Ministry of Foreign Affairs, 1999), para. 42. At that Consultative Meeting, Resolution 3 (1999) entitled 'Support for CCAMLR' was adopted.

[18] For instance, giving the *Handbook of the Antarctic Treaty System* an official status within the ATS, issuing a 'White Book' on the achievements of the ATS, establishing a coordinating unit for public information.

concept of streamlining the existing arrangements through a coordinating 'troika' (a consultative committee composed of past, present and future host countries) won support at the time.

The entrenched resistance to a permanent secretariat was gradually waning and the 1985 Consultative Meeting, held in Brussels, gave signs of the preferences of a still silent majority. At the 1997 Consultative Meeting in Rio de Janeiro, four nations held their ground against a rising tide which demanded a strong secretariat. The rise and demise of the 1988 CRAMRA and the adoption in 1991 of the Environmental Protocol were interludes along the road toward an ATS Secretariat. Other milestones on the road to the establishment of a Secretariat include a brief discussion at the 1991 Consultative Meeting in Bonn, a basic consensus at the Consultative Meeting the following year in Venice, the 1998 Consultative Meeting held at Tromsø (with some disarray due to the politicised debate on the matter of the location), and eventually the more business-like focus at the 1999 Consultative Meeting in Lima which emphasised the need for secretarial support to the Consultative Meeting and the CEP.[19]

At the time of the more vigorous and comprehensive debate held at the Consultative Meeting in 1987, some delegations stated 'that the establishment of some more permanent infrastructure would underline the vitality of the Treaty and enhance its open character'.[20] However, the more externally directed functions of a Secretariat mentioned in the Final Report of that Meeting have been persistently omitted. These especially include:

- the provision of accurate and up-to-date information about activities in the Antarctic and the ATS,
- the preparation of information on the ATS for public dissemination,
- the attendance of meetings of, and communications with, other relevant international organisations, and
- the coordination among the components of the ATS.

THE COMMITTEE ON ENVIRONMENTAL PROTECTION

At the First Session of the Eleventh Special Consultative Meeting (Viña del Mar, Chile, 19 November to 6 December 1990), debate on the form and content of comprehensive measures for the protection of the Antarctic environment focussed on several proposals: a convention advocated by the 'Group of Four' (Australia, Belgium, France and Italy); a framework protocol, supported by drafts submitted by the UK and the US; and a comprehensive protocol submitted by New Zealand. The draft convention included a Committee for the Protection of the Environment,

---

[19] For further discussion see Francioni, Chapter 7 in this book.
[20] See *Final Report of the Fourteenth Antarctic Treaty Consultative Meeting, Rio de Janeiro, 5–16 October 1987* (Rio de Janeiro: Ministry of External Relations of Brazil, 1987).

together with a Secretariat, a Scientific and Technical Advisory Committee and an Inspectorate. New Zealand's draft protocol contained a similar layout but its proposed Committee was a consultative body.[21]

Once the deadlock was broken through a process of informal consultations, a carefully balanced text was submitted, in the form of a personal and informal contribution by the senior Norwegian diplomat Rolf Trolle Andersen. Although the Andersen text contained some unfinished provisions, including the indefinite prohibition of mining in the Antarctic, it reflected the final compromise on the nature, scope and functions of the Committee for Environmental Protection. The advisory status of the CEP was drawn from the New Zealand draft, as well as from the insistence of the advocates of a framework protocol, namely that the powers of the Consultative Meeting should remain intact, and, paradoxically, from the fact that the proposed convention also provided for a technical and scientific advisory body.

However, the Andersen text contained some elements which emerged directly from the Viña del Mar informal consultation, including the symbolism of the establishment of the CEP on the basis of the Protocol and not by the Consultative Meeting. Although its function is admittedly purely advisory, a certain amount of self-initiative is vested on the new institution called to

> provide *advice* and formulate *recommendations* in connection with the *implementation of the Protocol*, including the operation of its Annexes, for consideration at Antarctic Treaty Consultative Meetings, and to perform *such other functions* as may be referred to it by the Antarctic Treaty Consultative Meetings.[22]

The blend of conservative and activist views in this carefully balanced formulation overshadows the importance of the creation of a specific institution empowered to monitor the protection of the Antarctic environment and its dependent and associated ecosystems. The individual functions which are attributed to the CEP by Article 12(1) of the Protocol have in a rather detailed, enumerative manner concentrated more attention than is indicated by the general role in the above-cited *chapeau* on the implementation of the Protocol. This carries a political overtone as well as an environmental overtone. Other break-throughs in the institutional development of the ATS include the entrance of observers other than the 'old-timers', namely SCAR and CCAMLR, and inclusion of the newly accepted COMNAP and any other relevant scientific, environmental and technical organisation, as well as and the consultative process mandated in Article 12(2) of the Environmental Protocol.

---

[21] L.M. Elliot, *International Environmental Politics Protecting the Antarctic* (London: St. Martin's Press, 1994), pp. 187–205.

[22] Art. 12(1) of the Protocol (emphasis added).

EFFECTS OF THE PROTOCOL ON THE ATS

In conformity with a long-standing Antarctic tradition, the Final Act of the Eleventh Special Consultative Meeting recorded the desirability of ensuring, at an early date, implementation of the provisions of the Environmental Protocol. Pending its entry into force, it was agreed that it was desirable that all parties to the Antarctic Treaty individually take steps as soon as possible to apply Annexes I to IV, in accordance with their legal systems and to the extent practicable.

Resistance to proceed beyond these parameters was visible at the 1991 and 1992 Consultative Meeting. But at the 1994 Consultative Meeting in Kyoto, under the agenda item 'Committee for Environmental Protection', Australia, Chile and France seized the opportunity to present papers concerning preparatory arrangements pending the entry into force of the Environmental Protocol. That Meeting took the significant step to adjudicate all agenda items related to the implementation of Article 12 of the Protocol to a Transitional Environmental Working Group (TEWG) and requested SCAR, CCAMLR and COMNAP to cooperate fully with the TEWG. *Mutatis mutandis*, the discarded US proposal of 1959 for a Standing Committee to oversee the provisional application of the Antarctic Treaty, had been applied in the Protocol, which now related to the protection of the Antarctic environment.

The TEWG effectively prepared the ground for the CEP. Coordination with SCAR, CCAMLR and COMNAP substantially improved with the revamping of the Rules of Procedure of the Consultative Meetings, which consolidated the position of those three organisations within the ATS. Their integration into the System was further advanced by the swift approval of the CEP's own Rules of Procedure at the 1998 Consultative Meeting, which followed shortly after the entry into force of the Protocol. There are perhaps a few inconsistencies in the Rules, including the fact that CCAMLR is represented by its Commission in the Consultative Meetings while the Chairman of its Scientific Committee attends the CEP. While this division of labour appears appropriate, the CEP consultations envisaged under Article 12 of the Protocol, and Article 7 of its own Rules of Procedure, mention only the Chairman of the Scientific Committee of CCAMLR. This solution may not be efficient, in the light of Article 6 of Annex V to the Protocol ('Area Protection and Management'). This Article indicates 1) that proposed Management Plans for new protected areas shall be forwarded to the CEP, to SCAR and, as appropriate, to the CCAMLR Commission; 2) that the CEP shall take into account any comments by these organisations, including as appropriate the CCAMLR Commission; and 3) that, with regard to the provisions of Articles 4 and 5 of the Protocol, no marine area shall be designated as an Antarctic Specially Protected Area or an Antarctic Specially Managed Area without the prior approval of the CCAMLR Commission.

It is self-evident that these inconsistencies can be solved through the practice and adequate coordination between the Consultative Meeting and the CCAMLR

Commission.²³ A certain complexity also exists regarding the advice which SCAR may provide at the level of the CEP and afterwards to the Consultative Meeting. The fact is that SCAR is now effectively a component of the ATS, advising the CEP in conformity with Articles 11 and 12 of the Protocol, as well as the Consultative Parties, in conformity with both Recommendation XIII-2 and the latest Rules of Procedure of the Consultative Meeting. A suggestion to avoid duplication has been made by the current Chairman of the CEP.²⁴

The ATS faces more important challenges after the installation of the CEP and in the wake of the establishment of a Secretariat. Decisions must be made regarding the functions of such a Secretariat, duly taking into account not only the mechanics of Consultative Meetings but also the full range of issues concerning *internal* systemic coordination and the *external* relations of the ATS. At present, web sites operate on behalf of the Consultative Meetings, CCAMLR, the CEP, COMNAP and SCAR, providing useful information to the members of each.²⁵ The concept of a 'postal box' for the ATS as such seems to be only a short step away, but some legal and technical requirements must be complied with, and once established, the Secretariat should take over these tasks.

At present, the ATS is confronted with the possibility of holding Consultative Meetings every two years, along with holding an annual meeting of the CEP. The CEP Report may be received by a short Special Consultative Meeting whenever the regular Consultative Meeting is not being held. The 1991 decision to shift to annual meetings without any preparatory meeting or structured consultation was probably not wise after all, and advance approval of Consultative Meeting agendas at the preceding Meeting does not resolve all the problems in that respect. Nevertheless, these administrative issues are certainly minor compared to other matters which were already anticipated in the Final Act of the Eleventh Special Consultative Meeting. The Final Act underlined the commitment of the parties to the Protocol, stated in Article 16, to elaborate rules and procedures relating to liability for damage arising from activities taking place in the Antarctic and covered by the Protocol, including liability for damage to the Antarctic environment. However, at the same time, the Meeting agreed that an inquiry procedure should be elaborated to facilitate disputes concerning the interpretation and application of Article 3 with respect to activities undertaken or proposed to be undertaken in the Antarctic Treaty area.

These two lines of action seem to be developing separately according to the Final Act as quoted. In fact, it is extremely difficult to construct an Antarctic environmental liability regime without reference to the key Article 3.²⁶ On the other hand, decision to develop an inquiry procedure is logical since matters concerning Article 3 are substracted from the dispute settlement procedure. The future

---

[23] See also discussion by Herr, Chapter 15 in this book.
[24] See Orheim, Chapter 6 in this book.
[25] See Vidas and Njåstad, Chapter 8 in this book.
[26] See discussion by Skåre and Lefeber, Chapters 9, 10 and 11 in this book.

administration of a liability regime presents some specific challenges for the Consultative Parties: 1) the specific role of the CEP in such a future liability regime, 2) procedures to cope with unrepaired or unrepairable environmental damage, such as a fund, 3) the application of the inquiry procedure and 4) consideration of the possibility of institutionalising an Inquiry Committee as a subsidiary body of the CEP.

The most significant effect of the entry into force of the Protocol on the ATS remains to be, nevertheless, the necessity to adapt to a situation where a strong CEP requires proper attention and expeditious decision on its advice. Supported by subsidiary bodies which will entail a greater degree of institutional development, the CEP will be meeting annually and performing a wealth of intersessional work. On the other hand, the ATS itself requires that political logistics and, above all, scientific matters receive expeditious solutions and an appropriate level of attention. The Consultative Meeting may have to organise itself in the form of a plenary assembly with standing committees, perhaps becoming the pivotal organ of an Antarctic international organisation[27] which is possibly supported not only by a permanent location for its Secretariat but also by permanent subsidiary bodies located in areas accessible to, or in the vicinity of, the Antarctic continent.

CONCLUSIONS

This overview leads to the conclusion that it is necessary that the ATS be endowed with greater coordination and that additional functions be developed which will strengthen its position as the main governing framework for all Antarctic activities.

Antarctic policies are rooted in a very complex array of memories, beliefs and ideas which sometimes become embedded in a peculiar *ad hoc* institutionality, but which also carry with them an inherent mistrust of other more classical forms of institutional development. Nevertheless, grand ideas such as 'the interest of all Mankind', a 'Continent for Science', a 'Natural reserve, devoted to Peace and Science', play a role in internalising rules and linking ideologically issue-areas. Once a policy choice has been made to reinforce the organisational and normative structures, the specific model of institutional development chosen will provide the framework and incentives necessary for the participating states to make their national policies more consistent with one another. With adequate institutional development, the common stock of shared values will more successfully be promoted in the wider scenario of the international system.[28]

---

[27] See Francioni, Chapter 7 in this book. Francioni suggests transformation of the Consultative Meeting in the plenary organ of a new Antarctic Treaty Organisation.

[28] See J. Goldstein and R.O. Keohane (eds.), *Ideas & Foreign Policy: Beliefs, Institutions, and Political Change* (Ithaca and London: Cornell University Press, 1995).

6

The Committee for Environmental Protection: Its Establishment, Operation and Role within the Antarctic Treaty System

Olav Orheim

Two years have passed since the entry into force of the Protocol on Environmental Protection to the Antarctic Treaty[1] and the establishment of the Committee for Environmental Protection (CEP) under this Protocol. It is now timely to examine the operation of CEP and the development of its functions thus far, as well as to speculate on how its future relationships with other components of the Antarctic Treaty System (ATS) may evolve.

This chapter first briefly describes the history of the CEP and the reasons for its establishment[2] as reflected in the provisions of the Environmental Protocol and the Rules of Procedure for the Committee for Environmental Protection[3]. The role of the CEP within the ATS is then discussed, both with respect to policy-making and advice-giving, as well as its links to the Scientific Committee on Antarctic Research (SCAR), the Council of Managers of National Antarctic Programmes (COMNAP) and the institutions established under the Convention on the Conservation of Antarctic Marine Living Resources[4] (CCAMLR). Based on the experience gained from the initial period of the existence of the CEP, the discussion proceeds to expound on how the Committee can function in practice regarding different issues governed by the Protocol. Finally, some speculations are presented on how the CEP

---

[1] ILM, Vol. 30, 1991, pp. 1,461ff.
[2] See also Berguño, Chapter 5 in this book.
[3] See Decision 2 (1998), in *Final Report of the Twenty-second Antarctic Treaty Consultative Meeting, Tromsø, Norway, 25 May – 5 June 1998* (Oslo: Royal Norwegian Ministry of Foreign Affairs, 1998), Annex B.
[4] ILM, Vol. 19, 1980, pp. 837ff.

may develop its role within the ATS and its relationship to the other ATS components.

The following are some of the key questions considered in this chapter:

- Will the CEP make a difference to the Antarctic environmental policy? Will it work better than existing arrangements?
- What are the main internal and external challenges for the CEP?
- Do the representatives that participate in the CEP possess the necessary technical expertise?
- What roles are *not* suitable for the CEP, and how can the CEP remain non-political?
- What should be the frequency of CEP meetings?
- Are there lessons to be learned from other ATS institutions, especially from those established under CCAMLR?

THE ESTABLISHMENT OF THE CEP

The Environmental Protocol was negotiated over a very brief time period, from 1990 to 1991. This period started with conflict between the Consultative Parties that arose from the negotiations of the Convention on the Regulation of Antarctic Mineral Resource Activities[5] (CRAMRA) in the 1980s and the subsequent decision by some parties in 1989/90 not to sign or ratify CRAMRA. Consultative Parties were under considerable pressure to reduce the level of conflict and reach consensus by mid-1991 because at that time any party could in principle start a process of withdrawal from the Antarctic Treaty by requesting a review conference and thereafter *not* following the well-established consensus process.[6] In 1989 there was a strongly perceived fear that the Consultative Parties were en route to a break-up of the Treaty. Intense efforts were therefore undertaken to arrive at a consensus within the two years available.

The text of the Environmental Protocol, which emerged from 1990 to 1991, drew on numerous sources. The provisions of the Annexes to the protocol were likewise to a large extent 'lifted' from existing measures adopted at previous Consultative Meetings, such as Annex II which reflected the 1964 Agreed Measures for the Conservation of Antarctic Fauna and Flora. The concept behind the CEP itself probably originated from a combination of the positive experience with the Scientific Committee under CCAMLR, concepts developed during the CRAMRA negotiations, and proposals put forward by some of the Consultative Parties.[7] It seems reasonable to assume that many of the Protocol negotiators believed that a body like the CEP was needed because environmental protection was not only a

---

[5] ILM, Vol. 27, 1988, pp. 868ff.
[6] Art. XII(2) of the Antarctic Treaty.
[7] For the discussion on the origins of CEP, see Berguño, Chapter 5 in this book.

political and legal question but also an issue that required increasingly specialised, technical expertise. Expert consideration of environmental issues would likely require a different composition of participants from those normally taking part in the regular Consultative Meetings.

Article 11 of the Environmental Protocol forms the basis for the establishment of CEP. The article states, *inter alia*, that the CEP is to be established at the time the Protocol enters into force and that the membership in this Committee is to consist of those states that are parties to the Protocol. Article 11 reads as follows:

1. There is hereby established the Committee for Environmental Protection.
2. Each Party shall be entitled to be a member of the Committee and to appoint a representative who may be accompanied by experts and advisers.
3. Observer status in the Committee shall be open to any Contracting Party to the Antarctic Treaty which is not a Party to this Protocol.
4. The Committee shall invite the President of the Scientific Committee on Antarctic Research and the Chairman of the Scientific Committee for the Conservation of Antarctic Marine Living Resources to participate as observers at its sessions. The Committee may also, with the approval of the Antarctic Treaty Consultative Meeting, invite such other relevant scientific, environmental and technical organisations which can contribute to its work to participate as observers at its sessions.
5. The Committee shall present a report on each of its sessions to the Antarctic Treaty Consultative Meeting. The report shall cover all matters considered at the session and shall reflect the views expressed. The report shall be circulated to the Parties and to observers attending the session, and shall thereupon be made publicly available.
6. The Committee shall adopt its rules of procedure which shall be subject to approval by the Antarctic Treaty Consultative Meeting.

INITIAL CHALLENGES – RULES OF PROCEDURE FOR THE CEP

Prior to the entry into force of the Protocol, three consecutive Consultative Meetings (1995, 1996 and 1997) had each set aside one weekly session for what was termed the 'Transitional Environmental Working Group' (TEWG). The reasons for initiating the TEWG were i) to achieve the stated objectives of the Parties to implement the Protocol informally ahead of its entry into force, and ii) to begin the preparatory work that was necessary if CEP was to start its work efficiently, in particular work on rules of procedure and an agenda for the work of the future CEP.[8] At the Twenty-first Consultative Meeting in Christchurch in 1997, the TEWG examined and revised the draft Rules of Procedure for the CEP[9]. These Rules had already been drafted at the Seventeenth Consultative Meeting held in Venice in

---

[8] See *Final Report of the Eighteenth Antarctic Treaty Consultative Meeting, Kyoto, Japan, 11-22 April 1994* (Tokyo: Japanese Ministry of Foreign Affairs, 1994), para. 42.

[9] *Final Report of the Twenty-first Antarctic Treaty Consultative Meeting, Christchurch, New Zealand, 19–30 May 1997* (Wellington: New Zealand Ministry of Foreign Affairs and Trade, 1997), para. 83.

1992[10]. The work done at Christchurch, which was followed by inter-sessional correspondence, was a major contribution to the rapid resolution of the Rules of Procedure at the next Consultative Meeting in Tromsø, Norway, in May 1998. As the Tromsø Consultative Meeting was the first Meeting following the entry into force of the Protocol, it was also the arena for the first meeting of the CEP. Another reason for the rapid agreement to the Rules of Procedure was that the parties came to Tromsø with the clear intention of having substantive discussion on environmental issues. Prior to the Meeting there was a fear in some quarters that the CEP would be bogged down in a quagmire of procedural issues and not be able to commence with its functions until subsequent Consultative Meetings. That, however, did not happen. The discussion at the First Meeting of the CEP (CEP I) in Tromsø, May 1998, demonstrated the will of the parties to make substantial progress.

The Rules of Procedure for the CEP were to a large extent modelled on those for the Scientific Committee of CCAMLR. There are nevertheless some aspects of the Rules of Procedure that deserve comments here, either because they provoked substantive discussion or because they are noteworthy for other reasons.

Representatives and Experts – Rule 3

Each Party to the Protocol is entitled to be a member of the Committee and to appoint a representative who may be accompanied by experts and advisers with suitable scientific, environmental or technical competence.

A number of delegations maintained that this Rule should also describe the qualifications of the representative so as to underscore the technical and non-political nature of CEP (as is done in the corresponding rule for the Scientific Committee of CCAMLR). Others maintained that, nonetheless, there should be no restriction on any party with respect to whom it appoints as a representative, and this view prevailed.

Observers – Rule 4

Observer status in the Committee shall be open to:
a) /.../

b) the President of the Scientific Committee on Antarctic Research, the Chairman of the Scientific Committee for the Conservation of Antarctic Marine Living Resources and the Chairman of the Council of Managers of National Antarctic Programmes, or their nominated Representatives;

---

[10] *Final Report of the Seventeenth Antarctic Treaty Consultative Meeting, Venice, Italy, 11–20 November 1992* (Rome: Italian Ministry of Foreign Affairs, 1993), paras. 34–36.

COMNAP is now recognised as an observer on equal level with SCAR and CCAMLR, which reflects the observer status of COMNAP to the Consultative Meeting that was agreed in 1997. In 1991, when the Protocol was signed, Article 11(4) of the Protocol named only the President of SCAR and the Chairman of the Scientific Committee of CCAMLR as observers to the CEP.

Conduct of Business – Rules 9, 10 and 22

Rule 9

The Committee shall meet once a year, in conjunction with the Antarctic Treaty Consultative Meeting, and at the same location. With the agreement of the ATCM, and in order to fulfil its functions, the Committee may also meet between annual meetings.

The Committee may establish informal open-ended contact groups to examine specific issues and report back to the Committee.

Rule 10

The Committee may establish, with the approval of the Antarctic Treaty Consultative Meeting, subsidiary bodies, as appropriate.

Such subsidiary bodies shall operate on the basis of the Rules of Procedure of the Committee as applicable.

Rule 22

English, French, Russian and Spanish shall be the official languages of the Committee, and the subsidiary bodies referred to in Rule 10.

When read together, these three rules give the CEP flexibility in conducting its work; however, unresolved complex issues still remain which will be returned to below in this chapter. Up till now, no subsidiary bodies have been established according to Rule 10, and the practical difficulties and costs of operating in four languages mean that such bodies are unlikely to be established in the future. On the other hand, both CEP I and CEP II (the Second Meeting of CEP in Lima in 1999) established several open-ended contact groups according to Rule 9 (see discussion below). These open-ended contact groups are not required to operate in all four official languages.

Submission of Documents – Rule 13

Members of the Committee should follow the Guidelines on Circulation and Handling of CEP Documents, as set out in Annex 3 to the Report of the Committee on Environment Protection to ATCM XXII.

These Guidelines generally follow the Guidelines on circulation and handling of documents for the Consultative Meetings,[11] except that CEP documents have to be submitted earlier, namely 75 days prior to a CEP meeting. The argument for this was that the technical nature of CEP documents required more time for translation and subsequent evaluation by the parties prior to the meeting. In addition, to lay the foundation for a more rapid and efficient handling of documents, the CEP Guidelines specify that whenever feasible all documents should be electronically submitted and circulated. This is not yet a requirement for documents submitted to the Consultative Meetings.[12]

Advice and Recommendations – Rule 14

The Committee shall try to reach consensus on the recommendations and advice to be provided by it pursuant to the Protocol.

Where consensus cannot be achieved the Committee shall set out in its report all views advanced on the matter in question.

It is important to note that the CEP is not a decision-making body. The Committee gives advice to the Antarctic Treaty Consultative Meeting, and this advice should be presented in a complete and balanced manner, including minority views.

Chairperson and Vice-chairs – Rules 16 and 20

Rule 16

The Committee shall elect a Chairperson and first and second Vice-chairs from among the Consultative Parties. The Chairperson and the Vice-chairs shall be elected for a period of two years.

The Chairperson and the Vice-chairs shall not be re-elected to their post for more than one additional two-year term. The Chairperson and Vice-chairs shall not be representatives from the same Party.

The Vice-chairs to be elected at the first meeting of the Committee shall be elected for a one-year term to ensure that the terms of office of the Chairperson and Vice-chairs shall be staggered.

Rule 20

The Chairperson and Vice-chairs shall begin to carry out their functions on the conclusion of the meeting of the Committee at which they have been elected, with the exception of

---

[11] See Guideline on Pre-sessional Document Circulation and Document Handling, in *Final Report of the Twentieth Antarctic Treaty Consultative Meeting, Utrecht, 29 April – 10 May 1996* (The Hague: Netherlands Ministry of Foreign Affairs, 1997), Annex D.

[12] See discussion by Vidas and Njåstad, Chapter 8 in this book.

the Chairperson and the Vice-chairs of the first meeting of the Committee who shall take office immediately upon their election.

At CEP I, the author of this chapter was elected Chairperson for two years, while Jorge Berguño (Chile) and Gillian Wratt (New Zealand) were elected as, respectively, first and second Vice-chairs for one year. At CEP II, Jorge Berguño and Gillian Wratt were re-elected for two more years. Thus, if CEP meetings continue to be held each year, then the elected period of the current Chairperson will terminate at the end of CEP III, whereas the current Vice-chairs' term will terminate at the end of CEP IV, at which point neither Vice-chair can be re-elected. The definition of the elected period was seen in relation to the annual frequency of Consultative Meetings. Should the Parties – and the CEP – decide to adopt a lower frequency of the meetings (namely biennial instead of annual), these Rules may have to be reconsidered in order to avoid future officers being elected for only one CEP meeting, which could be detrimental to the continuity in the operation of the CEP.

Administrative Facilities – Rule 21

As a general rule the Committee, and any subsidiary bodies, shall make use of the administrative facilities of the Government which agrees to host its meetings.

The requirements for inter-sessional document handling and archives have in reality placed considerable demands on administrative facilities beyond those described above. So far these administrative facilities have been provided by the office of the CEP Chairperson. The issue of administrative functions related to inter-sessional document handling and archives will have to be resolved in the future, either by the parties offering support or through the establishment of the Antarctic Treaty Secretariat that can also carry these functions.

FUNCTIONS OF THE CEP

The Committee's functions are defined in Article 12 of the Protocol. The essence of Article 12 is that the Committee shall provide advice and formulate recommendations to the parties with respect to the implementation of the Environmental Protocol. Article 12 reads as follows:

1. The functions of the Committee shall be to provide advice and formulate recommendations to the Parties in connection with the implementation of this Protocol, including the operation of its Annexes, for consideration at Antarctic Treaty Consultative Meetings, and to perform such other functions as may be referred to it by the Antarctic Treaty Consultative Meetings. In particular, it shall provide advice on:
    a. the effectiveness of measures taken pursuant to this Protocol;
    b. the need to update, strengthens or otherwise improve such measures;

    c. the need for additional measures, including the need for additional Annexes, where appropriate;
    d. the application and implementation of the environmental impact assessment procedures set out in Article 8 and Annex I;
    e. means of minimising or mitigating environmental impacts of activities in the Antarctic Treaty area;
    f. procedures for situations requiring urgent action, including response action in environmental emergencies;
    g. the operation and further elaboration of the Antarctic Protected Area system;
    h. inspection procedures, including formats for inspection reports and checklists for the conduct of inspections;
    i. the collection, archiving, exchange and evaluation of information related to environmental protection;
    j. the state of the Antarctic environment; and
    k. the need for scientific research, including environmental monitoring, related to the implementation of this Protocol.

2. In carrying out its functions, the Committee shall, as appropriate, consult with the Scientific Committee on Antarctic Research, the Scientific Committee for the Conservation of Antarctic Marine Living Resources and other relevant scientific, environmental and technical organisations.

Obviously, it was impossible for the CEP initially to provide advice and formulate recommendations on all the items on this extensive list. Indeed, there are several items on the list that neither CEP I nor CEP II have yet considered. The main issues so far treated by those two meetings have been environmental impact assessments (EIAs),[13] area protection and management,[14] data and exchange of information,[15] and the state of the Antarctic environment[16]. Emergency response[17] and environmental monitoring[18] have also been on the agenda of the first two meetings. There seems to be an expectation that the CEP will make substantial and rapid progress on many of the issues. Thus the number of working papers on environmental issues, which generally result in substantive discussion and action, has been much larger than at the previous Consultative Meetings. At CEP I, altogether 12 working papers and 29 information papers were considered[19], while at CEP II the numbers increased to 22 working papers and 33 information papers[20].

---

[13] Protocol, Art. 12(1)(d).
[14] *Ibid.*, Art. 12(1)(g).
[15] *Ibid.*, Art. 12(1)(i).
[16] *Ibid.*, Art. 12(1)(j).
[17] *Ibid.*, Art. 12(1)(f).
[18] *Ibid.*, Art. 12(1)(k).
[19] See 'Report of the Meeting of the Committee for Environmental Protection, Tromsø, 25–29 May 1998', Annex 2, in *Final Report of the XXII ATCM*, Annex E.
[20] See 'Report of the Second Meeting of the Committee for Environmental Protection, Lima, 24–28 May 1999', Annex 1, in *Final Report of the Twenty-third Antarctic Treaty Consultative Meeting, Lima, Peru, 24 May – 4 June 1999* (Lima: Peruvian Ministry of Foreign Affairs, 1999), Annex G.

This almost doubling of working papers is an indication that the parties are putting forth considerable pre-meeting effort to deal with the subjects considered by CEP.

Based on the Draft Agenda for the Third Meeting of CEP (CEP III),[21] the Committee is likely to spend most of its time considering EIAs and protected areas, as well as the questions of data and the exchange of information, and the state of the Antarctic environment. In addition, increasing attention will probably be given to conservation of Antarctic flora and fauna[22] and emergency response and contingency planning.[23] Over time it can be expected that the CEP will be in a position to fulfil its role with respect to all items listed in Article 12 of the Protocol. Under such circumstances, there is no realistic way that an anticipated increasing substance of CEP work can be performed without the establishment of an Antarctic Treaty Secretariat. For the time being, the lack of permanent secretarial support is compensated for by the Chairperson organising the provision of secretarial support, including the CEP website; but this solution is not a tenable situation in the longer run. It is important here to note the specific requirements which are defined by the Protocol with respect to information that parties shall submit to the CEP for consideration and evaluation. These include the following:

- inspection reports (Article 14);
- annual reports from the parties (Article 17);
- draft comprehensive environmental evaluations, CEEs (Annex I, Article 3);
- information on environmental impact assessment procedures; list of initial environmental evaluations prepared (Annex I, Article 6);
- annual reports from the parties on number and nature of permits issued under Annex II (Annex II, Article 6);
- waste management plans and annual reports from the parties on implementation of these (Annex III, Article 9);
- annual reports from the parties on number and nature of permits issued under Annex V (Annex V, Article 10);
- information on actions taken in cases of emergencies (Annex I, Article 7; Annex II, Article 2; Annex III, Article 12; Annex IV, Article 7; and Annex V, Article 11).

These requirements place a significant demand on the time and resources of the CEP. If the Committee is to consider the above information in a manner suitable for giving appropriate advice to the parties, a support mechanism that can consolidate pertinent information and prepare such evaluations before the CEP meetings is essential.[24]

---

[21] See 'Report of the Second Meeting of the Committee for Environmental Protection, Lima, 24–28 May 1999', Annex 5.

[22] Protocol, Art. 12(1)(a), referring to Annex II.

[23] Protocol, Art. 12(1)(f).

[24] For further discussion on the need for an Antarctic Treaty Secretariat, see Francioni, Chapter 7 in this book.

THE ROLE OF THE CEP WITHIN THE ANTARCTIC TREATY SYSTEM AND CEP'S RELATIONSHIP TO OTHER COMPONENTS

The CEP was established to provide advice to the parties on environmental issues and its role should thus be that a of a non-political and technical body. As stated in Article 11(4) of the Protocol, SCAR and the Scientific Committee of CCAMLR participate in the CEP as observers, and the same status was later given to COMNAP through Rule 4 of the Rules of Procedure for the CEP.

In the foreseeable future, conflicts between the roles of the CEP on the one hand and the COMNAP and the Scientific Committee of CCAMLR on the other do not appear to be likely. The COMNAP has set up its own group, the Antarctic Environmental Officers Network (AEON), which provides the Council of Managers with advice on operational environmental questions. Many of the AEON members are concurrently members of the CEP. With the COMNAP and the CEP both primarily made up of representatives of governmental institutions, it seems unlikely that the representatives will have problems sorting out their dual roles of being in the CEP or in the AEON and the advising role of the COMNAP within the CEP.

With regard to the roles of the CEP and the Scientific Committee of CCAMLR, there has so far been little overlap in the subject area of the work done by the CEP. However, there are potential overlaps in issues such as marine waste and pollution in marine protected areas, none of which have so far figured prominently on the CEP agenda. In the future, a division of labour in such issues between the CEP and the Scientific Committee of CCAMLR may be necessary, but there is no reason to anticipate this leading to disagreements.[25] At CEP II, a direct dialogue was initiated between the CEP and the Scientific Committee of CCAMLR, and a general agreement was expressed regarding the advantages of fostering interaction between the two committees by participating at each others' meetings[26]. As the CEP develops its operations, it may also find it useful to draw on the experience of the Scientific Committee of CCAMLR, which has managed to establish efficient, technical sub-groups.

The potential for confusion is greater between the CEP and SCAR. The CEP's work programme has so far included protected areas and environmental impact assessments, areas where SCAR's Group of Specialists on Environmental Affairs and Conservation (GOSEAC) has until now provided advice to Consultative Meetings. Using SCAR (and COMNAP) has been a comfortable arrangement for the Consultative Parties because SCAR has performed its advisory functions without requesting financial support. As GOSEAC has coupled the necessary scientific knowledge with operational expertise from COMNAP, and as GOSEAC has had competent convenors, this arrangement has functioned well, even though some

---

[25] For the relationship between the Protocol and CCAMLR, see also Herr, Chapter 15 in this book.
[26] 'Report of the Second Meeting of the Committee for Environmental Protection, Lima, 24–28 May 1999', paras. 13–14.

delegates at Consultative Meetings at times might have perceived SCAR/GOSEAC as conveying an arrogant 'we know best' attitude.

Because the CEP at present has neither a secretariat nor a budget, it does all its inter-sessional work by e-mail. This means that the CEP is not yet equipped to carry out the technical considerations so far performed by GOSEAC. At the same time, the CEP is identified as the Consultative Parties' own advisory body for issues governed by the Environmental Protocol. This means that it will only be a matter of time before CEP takes over some roles which have until now been handled by SCAR/GOSEAC.

When the CEP is fully developed, the role of GOSEAC will be redundant for CEP purposes. However, it should be noted that present members of GOSEAC are likely to be found again as members of new CEP bodies, but will no longer give their views through SCAR. This delineation of competence and roles should improve SCAR's own role within Consultative Meetings. SCAR, with its status as observer, should continue to be the prime source for *scientific* advice to the Consultative Parties.

The CEP will have to set up one or more working groups to handle issues that require more technical consideration than is practical and possible to be handled at a CEP meeting. Whether such groups are permanent or established on a case-by-case basis from one meeting to the other will be further considered at CEP III. This question (along with many others) is presently being discussed by an open-ended e-mail correspondence group concerned with Antarctic protected areas. The e-mail group is a follow on from the workshops on protected areas that were arranged in 1998 and 1999, prior to CEP I and CEP II, respectively. For draft CEEs, the Committee has agreed to a process for starting inter-sessional considerations once a draft CEE is submitted.[27]

EXTERNAL AND INTERNAL CHALLENGES

There are at present no obvious external challenges to the CEP. Various NGOs have over the years criticised the Consultative Parties on Antarctic environmental issues, but today their attitudes are mainly non-confrontational. Thus as long as the Antarctic Treaty cooperation itself is not challenged (as it was in the United Nations in the 1980s), it does not seem likely that the CEP alone will be challenged from the outside as long as it does its work well.

The greatest foreseeable challenge to the CEP and to the whole Protocol is the maintenance of *internal* support. Without an Antarctic Treaty Secretariat, there is a real danger that issues will not be followed up and that elements of the Protocol will become paper obligations only, without follow-up action. This situation may especially develop in a few specific cases where the requirements in the Protocol are drawn up in a somewhat daunting fashion, for example the reporting obligations

---

[27] For further discussion, see below in this chapter.

outlined above. It is therefore fundamental that when parties agree on obligations prescribed by the Protocol, they also make available the means for implementing those obligations. At the Lima Consultative Meeting in 1999, there was a disturbing lack of time for discussing the report and advice of CEP II. If this reflects an attitude among the delegates that all is well now that the Protocol has entered into force, then there are some clouds on the CEP horizon. Adequate consideration of the issues brought forward by the CEP will in the long run be necessary to uphold the status of the Protocol itself.

HOW CAN THE CEP HANDLE TECHNICALLY COMPLEX ENVIRONMENTAL ITEMS?

This question raises several issues, including 1) participation in CEP meetings, 2) handling of documents, and 3) inter-sessional work, including the establishment of sub-groups.

Participation in CEP Meetings

The participants at meetings are – as described above with reference to Rule 3 of the Rules of Procedures – the representatives of each party to the Protocol accompanied by experts and advisers with suitable scientific, environmental and technical competence. At previous CEP meetings, most items to be handled by the CEP have required knowledge of the Antarctic environment but little specialised competence, for example on issues such as waste management or station construction. Judging by the composition of participants so far, it is not likely that expertise in such areas will be available at future CEP meetings either. Furthermore, taking into account the variety of types of issues to be considered by the CEP, the possibility of having all types of relevant expertise represented at the meetings seems unlikely. It can therefore be concluded that technically complex issues are in reality not likely to be handled actually at CEP meetings.

The Handling of Documents

If adequate expertise is lacking at a CEP meeting, documents on complex issues will have to be submitted early enough to involve experts in the individual countries. The Guidelines on Circulation and Handling of CEP Documents[28] state that documents requiring translation should be received by the host government at least 75 days before the CEP meeting. The documents should then be circulated in translation no later than 60 days before the meeting. In theory this may sound fine, but in reality

---

[28] See para. 1 of Guidelines, in 'Report of the Meeting of the Committee for Environmental Protection, Tromsø, 25–29 May 1998', Annex 3.

this is not what happens in many cases. Documents received after the deadline become available in translation for consideration at the meeting, and this is of course too late to allow adequate time for the examination of complex matters. Furthermore, even if documents are circulated from the host government within the 60-day deadline, there is still little time for national and international consultation. However, the time limits in the Guidelines are not likely to be extended. Experience has shown that the existing limits are difficult to adhere to, and these already have a longer time requirement than the time limits adopted for document distribution for the regular Consultative Meetings.

The CEP has been able to improve the situation somewhat with the agreement that all documents from the meeting be placed on the CEP web site as well as all submitted documents for the up-coming meeting, in a restricted area available only to CEP members and observers[29]. Thus, once a document is made available electronically by the host government, then it immediately is also available on the CEP web site, thus offering far quicker access than through the traditional means of document distribution by postal services or diplomatic channels.

Inter-sessional Work

The only realistic way for the CEP to deal with the issue of handling complex technical matters is through inter-sessional work. Rule 9 of the CEP Rules of Procedure, as quoted above, provides for this possibility; accordingly, CEP I established guidelines for contact groups working inter-sessionally.[30] From the Guidelines it follows that the convenor and terms of reference for an inter-sessional group should be decided at the CEP meeting, that the contact group should be open-ended, and that all correspondence should be circulated to all the members of the group.

Both CEP I and CEP II established inter-sessional contact groups (for example for discussions on protected areas, development of guidelines for environmental impact assessments, and consideration of a state of the Antarctic environment report), and this approach has so far proved very productive for making progress on these issues. The contact groups worked exclusively by e-mail, using only the English language. The use of one language is of course the only realistic way for such informal multilingual groups to work, even though the use of English can create a certain barrier for members of the group whose mother tongue is not English.

Although the work of the contact groups has thus far proved fruitful, these groups have not yet examined complex documents, such as a draft CEE. No draft

---

[29] See para. 3(b) of Decision 1 (1999), 'CEP Web Site', in *Final Report of the XXIII ATCM*, Annex B. See further discussion by Vidas and Njåstad, Chapter 8 in this book.

[30] See 'Report of the Meeting of the Committee for Environmental Protection, Tromsø, 25–29 May 1998', para. 9.

CEEs were transmitted to either CEP I or CEP II for consideration, but it can be safely assumed that they will arrive in the future. When such a document is sent from the originating party, that party is only obliged to provide the document in one of the four Antarctic Treaty languages,[31] and there may well be considerable delay before the document is available in all four languages.

At the same time, issues requiring in-depth discussions at a CEP meeting may be few in view of the requirement for thorough national examination in advance to submission of a draft CEE. It is recognised that a proposed activity requiring a CEE may lead to diplomatic disagreements or technical disagreements, or to both. However, it may not be necessary to have documents with long lead times in order to enable discussion on all such issues. For example, if a party proposes to establish a scientific infrastructure that would not harm the environment but would interfere with the scientific experiments of another country, this would not be an issue for the CEP, but for the Consultative Meeting itself. The parties could discuss this issue without considering the *technical* aspects of the CEE. On the other hand, if a party proposed, for example rock drilling, and the main issue of contention was the choice of an environmentally acceptable drilling fluid, then the issue should be resolved within the CEP. In such a case, the CEP needs to have the problem identified well in advance to provide sufficient time for thorough consideration, and the technical discussions would be best resolved through inter-sessional work involving appropriate expertise.

Recognition of the problems related to processing draft CEEs led to discussions at CEP II of undertaking inter-sessional work on draft CEEs and to the procedure described below. Also of importance was the thought that this could help to find a flexible way forward, should Consultative Meetings became biennial.[32]

CONSIDERATION OF DRAFT COMPREHENSIVE ENVIRONMENTAL EVALUATIONS

The activities in Antarctica that potentially might have the greatest impact on the Antarctic environment are those for which a CEE is required. Consequently, considerations of draft CEEs, and the subsequent formulation of advice to the parties, are an important task for the CEP. The issue of how to consider appropriately a draft CEE is a very complex matter. While there is a competing interest of giving a proper and independent evaluation, there is at the same time the desire to not unfairly hinder a party from proceeding with an activity for which it is responsible and which that party has judged to be environmentally acceptable. For this reason, the procedures around CEEs as laid down in Annex I of the Protocol (and indeed the whole EIA process) has a number of 'grey areas', where value

---

[31] According to Art. XIV of the Antarctic Treaty, English, French, Russian and Spanish language versions of the Treaty are all equally authentic.

[32] That issue is discussed below in this chapter.

judgements, rather than prescribed actions, will lead the way. It is therefore hoped that over time application of EIA procedures will establish common standards and practice.

For the time being, CEP II agreed on a procedure for its consideration of draft CEEs.[33] Basically, the procedure allows a party to identify an issue and ask for an inter-sessional, open-ended contact group to be established. This has further been made operational through the following provisional guidelines, which are designed to enable CEP members to evaluate whether an issue identified by another member is appropriate for the CEP to consider in an inter-sessional contact group. It is expected that any such group should normally be established to consider only the *technical* issues of a draft CEE. Operational procedures for establishing a contact group are as follows:

1. At the same time a Draft CEE is circulated to Members via diplomatic channels, the proposer should notify the CEP Chair by e-mail that a Draft CEE has been circulated, and, if available, indicate the web address on which the report can be accessed.

2. The originator of a Draft CEE should post it on its web site in the language(s) it is made available in. Links to this web site will be established on the CEP web site. If the proposer does not have a web site on which it is able to post the Draft CEE, an electronic version of the report should be forwarded to the Chair of the CEP who will immediately post it on the CEP web site.

3. The CEP Chair notifies the CEP contact points that the Draft CEE is available and of its web address. The notification should include the comment that any Party that wants CEP to consider an issue or issues concerning the draft CEE should notify the CEP Chair as soon as possible.

4. A Party that indicated that it wants the CEP to consider a draft CEE should as soon as possible indicate the issue(s) it wants examined, propose Terms of Reference (ToR) and propose their member of an open ended intersessional contact group.

5. On such notification the CEP Chair will immediately inform all contact points and indicate that an open ended intersessional contact group has been proposed. The CEP Chair will at this time suggest a convenor for the group, suggest a set of ToRs and ask for nominations of members to the group.

6. The convenor of the contact group could be the person proposed by a Party requesting an issue to be considered. It should preferably not be from the Party proposing the activity. The notification to the members should have a time limit

---

[33] See Guidelines for CEP Consideration of Draft CEEs, in 'Report of the Second Meeting of the Committee for Environmental Protection, Lima, 24–28 May 1999', Annex 4.

of 15 days for Members (Parties) to object to:
i. establishment of the group
ii. the proposed convenor
iii. the proposed terms of reference.

If the Chair does not receive a reply within 15 days it will be considered that the Member agrees to the establishment of the group, the proposed convenor and the proposed ToRs.

7. If more than one Member proposes issues to be considered by the CEP, the ToRs should be amended to reflect the additional issues at the time such issues are raised. There should be a reasonable flexibility in the ToRs to allow for consideration of related technical issues that arise in the work of the contact group.

8. The right of a Party to raise an issue on a Draft CEE at the CEP or ATCM is not affected by its action in relation to the establishment – or non-establishment – of an open ended intersessional contact group.

These proceduress are now being tested for the first time, and it is likely that they will be improved with practice. It is hoped that the procedures will provide a practical means of enabling the CEP to in the future give rapid and efficient consideration to draft CEEs. To achieve this end, however, the draft CEE must be made quickly available in English, even if this is not the original language.

THE FREQUENCY OF CEP MEETINGS

Only one formal requirement influences the time interval of CEP meetings, namely the consideration of draft CEEs.[34] This follows from Annex I of the Protocol, which in its attempt to carefully balance conflicting interests may have created a rather impractical system. The significant point here is that consideration of draft CEEs is assumed to require a CEP meeting, based on Article 3(5) of Annex I to the Protocol, which states:

> No final decision shall be taken to proceed with the proposed activity in the Antarctic Treaty area unless there has been an opportunity for consideration of the draft Comprehensive Environmental Evaluation by the Antarctic Treaty Consultative Meeting on the advice of the Committee, ...

Article 3(5) goes on to state that application of this paragraph cannot delay the decision of whether or not to proceed with the activity by more than 15 months from the date of circulation of the draft CEE.

---

[34] Other, procedural aspects are discussed above in this chapter.

Thus, when a draft CEE is submitted, the CEP must consider the draft and give its advice so that the draft CEE then can further be considered at a Consultative Meeting, all to take place within 15 months.

It can be argued that if the Consultative Parties choose not to hold a Consultative Meeting within 15 months, they have chosen not to consider the draft CEE; there is then no formal need for a CEP meeting. I suspect that most would consider such an interpretation to be in conflict with the spirit of the Protocol. However, if the Consultative Parties should decide to revert to biennial meetings (which is informally under discussion at present), it is highly likely that a procedure must be established to stay within the time limit set by Article 3(5) of Annex I. The alternative – extending the time period to something like 15 months plus one year – would as a consequence cause a strong shift in balance between the national CEE work, which is where the responsibility lies, and the international evaluation. It would also delay any work proceeding in Antarctica which is why Article 3(5) was negotiated.

Beyond the CEE requirements, the frequency of CEP meetings is laid down in Rule 9 of the CEP Rules of Procedure. Rule 9 states that the CEP shall meet once a year, in conjunction with the Consultative Meeting and at the same location. However, it goes on to state that, should it be necessary for the fulfilment of its functions, the CEP may meet more frequently, with the agreement of the Consultative Meeting.

The view of CEP members has so far been that meeting once a year is adequate. However, should the Consultative Meetings revert to a biennial cycle, several CEP members have already indicated that they believe that the CEP should in that case continue to meet annually until its working procedures are sufficiently established.[35]

WILL THE CEP MAKE A DIFFERENCE FOR ANTARCTIC ENVIRONMENTAL POLICY?

I believe the answer to this question is a clear yes. After all, the CEP has a strong basis for its work in the Protocol. The activity level within CEP meetings and the inter-sessional work is also much higher than when environmental issues were considered at the preceding Consultative Meetings.

What has kept the Antarctic so relatively pristine up to now is the low human presence and the recognition by scientists, tourists and support personnel that the Antarctic is unique and worth preserving. Today, the rapidly increasing human presence in Antarctica presents a new challenge, especially with regard to tourism

---

[35] 'Report of the Second Meeting of the Committee for Environmental Protection, Lima, 24–28 May 1999', paras. 25–28.

and adventure expeditions. Another challenge is the global impact through long-distance transported pollutants. With these issues in mind, it is more important than ever that the CEP take its recognised role in leading the Consultative Parties and the environmental community to the right courses of action.

7

Establishment of an Antarctic Treaty Secretariat: Pending Legal Issues

Francesco Francioni

The 1959 Antarctic Treaty[1] made no provision for permanent institutions, for fixed headquarters for treaty organs or for an administrative infrastructure with secretariat functions. As originally conceived, the Antarctic Treaty was intended primarily to be a security arrangement designed to guarantee stability and cooperation in an area marked by competing territorial claims and rivalries among superpowers. In this context, there was no provision for the establishment of international institutions. Thus, rather than establish a form of international administration, the Treaty envisaged a forum of inter-governmental consultation and cooperation that would guarantee continuous and peaceful access to the Antarctic. This original intent is reflected in the rather limited number of the original parties to the 1959 Treaty: seven claimants – Argentina, Australia, Chile, France, New Zealand, Norway and United Kingdom; the two 'quasi-claimants' – the USA and the USSR, both asserting an inchoate title; and three non-claimants – Belgium, Japan and South Africa. The original intent is also reflected in the type of solution applied to the difficult issue of territorial claims. Rather than taking a position with regard to the basis of such claims in international law, Article IV of the Antarctic Treaty preserved a *status quo* and foreclosed the assertion of new claims while the Treaty was in force.

Consistently with this political matrix, the evolution of the Antarctic Treaty System over a period of almost 40 years has confirmed the minimalist approach toward institutional development, thus differing from the general trend that has characterised the great majority of treaty regimes in contemporary international law.[2] The resulting institutional 'deficit' reflects the continuing reluctance with

---

[1] UNTS, Vol. 402, pp. 71ff.
[2] Most multilateral environmental treaties concluded in the past 25 years include a provision for either the establishment of a secretariat or for the provision of secretarial functions by an international organisation to which the treaty is attached. See, for example, the 1973 Convention on International

which the Consultative Parties, or at least a substantial number of them, have looked at the prospect of transforming the Antarctic Treaty into some sort of international organisation. What distinguishes the Antarctic Treaty System from an 'international organisation' is that the Antarctic Treaty neither constitutes an international subject that is endowed with international 'personality' nor does it have an independent legal entity capable of representing the corporate will of the Antarctic Treaty parties even in domestic law.[3]

At present there seems to be no general interest in a repositioning of the Antarctic Treaty Parties on this matter. On the contrary there is ample evidence that the establishment of international institutions within the Antarctic Treaty is still looked upon with suspicion as a step toward internationalisation of the Antarctic and the eventual weakening of the individual countries' positions on the question of exclusive territorial claims. Furthermore, the process of institutionalisation of an international treaty would normally entail the development of a collective will of the entity through majority decision-making. However, this would not be consistent with the rule of consensus, which has been one of the pillars of Antarctic Treaty cooperation since the Treaty's emergence. Finally, financial concerns with regard to the inevitable costs of maintaining an international bureaucracy have played a role in keeping alive the affection for the present 'anorganic' structure of the ATS.

INSTITUTIONAL ELEMENTS IN THE ANTARCTIC TREATY SYSTEM

Despite this background, it would be naive to believe that the ATS has been entirely unaffected by the seeds of institutional development during its evolution over the past four decades. The regular convening of the Antarctic Treaty Consultative Meetings, 23 of which have been held thus far, has led to the consolidation of a plenary body which meets and works on the basis of a precise agenda, and according to its rules of procedure. The Meetings are held in locations chosen in accordance with a preordained principle of rotation among Consultative Parties as host states, following their alphabetical order in the English language.[4] Some basic secretarial functions are provided in a decentralised form by the host country of the

---

Trade in Endangered Species of Wild Fauna and Flora, Art. XII (ILM, Vol. 12, 1973, pp. 1,085ff; also available at <www.wcmc.org.uk/CITES>); 1985 Vienna Convention for Protection of the Ozone Layer, Art. VII (ILM, Vol. 26, 1987, pp. 1,516ff) and 1987 Montreal Protocol on Substances that Deplete the Ozone Layer, Art. XII (ILM, Vol. 26, 1987, pp. 1,541ff; also available at <www.unep.chlozone>); 1989 Basel Convention on the Transboundary Movement of Hazardous Waste, Art. XVI (ILM, Vol. 28, 1989, pp. 657ff; also available at <www.unep.ch/basel>); 1992 UN Framework Convention on Climate Change, Art. VIII (ILM, Vol. 31, 1991, pp. 849ff; also available at <www.unfccc.de>).

[3] Some practical ramifications of this deficit are discussed below, in this chapter.

[4] Recommendation I–XIV, text reprinted in J.A. Heap (ed.), *Handbook of the Antarctic Treaty System*, 8th edition (Washington, DC: US Department of State, 1994), p. 22.

Consultative Meeting. In particular, the host country has the responsibility for 1) preparing the Meeting, 2) circulating documents prior to the Meeting, 3) providing administrative and secretarial services during the Meeting, 4) preparing and timely publication of the Final Report from the Meeting, 5) providing certified copies of the Final Report to the Treaty parties as well as to the Secretary-General of the United Nations,[5] and 6) providing authentic texts of the measures adopted at the Meeting.[6]

This decentralised mechanism has improved through the years owing to the establishment, pursuant to Recommendation XIII–1,[7] of the national 'contact points', and to the introduction, pursuant to Recommendation XIII–2,[8] of the practice of presenting reports by other components of the ATS at each Consultative Meeting. This has greatly facilitated communication among the Parties and exchange of information among the various components of the System.

In addition to the basic secretarial functions provided by the host country, the depository government of the Antarctic Treaty (the Government of the United States) has provided secretarial functions in other instances, for example for the preparation of the latest (eighth) edition of the *Handbook of the Antarctic Treaty System* (published by the United States Department of State in 1994) and the undertaking of a study on the public availability of documents and information about the ATS.[9] In some exceptional circumstances, administrative functions have been provided on an *ad hoc* basis by individual governments delegated to perform specific institutional activities. This was the case of the 'convenor' of Treaty parties for the 'Question of Antarctica' at the UN General Assembly from the mid-1980s to the early 1990s.[10]

In addition to these surrogate forms of institutional infrastructure, a true secretariat was at that time contemplated in two instruments that were developed as part of the ATS: the 1980 Convention on the Conservation of Antarctic Marine Living Resources (CCAMLR)[11] and the 1988 Convention on the Regulation of

---

[5] Recommendation XII–6, text reprinted in Heap (ed.), *Handbook of the Antarctic Treaty System*, pp. 49–50.

[6] Measures adopted in accordance with Art. IX(1) of the Antarctic Treaty. These were, until the XIX Consultative Meeting in 1995, treated as a single category of instruments, known as 'Recommendations'; at the XIX Consultative Meeting, a Decision was adopted to distinguish between three different types of instruments: 'Measures', 'Decisions' and 'Resolutions'. See also discussion by Bush, Chapter 2, and Berguño, Chapter 5, in this book.

[7] Recommendation XIII–1, text reprinted in Heap (ed.), *Handbook of the Antarctic Treaty System*, pp. 51-52.

[8] Recommendation XIII–2, text reprinted in Heap (ed.), *Handbook of the Antarctic Treaty System*, p. 52.

[9] Recommendation XII–6, para. 5.

[10] Australia began the coordinating work as the 'convenor'. The role was later performed on a rotating basis by host states of the Antarctic Treaty Consultative Meeting.

[11] ILM, Vol. 19, 1980, pp. 837ff. CCAMLR, Art. XVII provides that an Executive Secretary shall be appointed by the Commission, the plenary organ established under Art. VII of CCAMLR, and endowed with legal personality pursuant to Art. VIII.

Antarctic Mineral Resource Activities (CRAMRA).[12] Indeed, a secretariat was actually established at the headquarters of the CCAMLR Commission in Hobart, Tasmania.[13] With regard to CRAMRA, the establishment of a secretariat, as contemplated by its Article 33, has remained a dead letter as a consequence of the Consultative Parties' decision to adopt a ban on mineral activities, contained in the Protocol on Environmental Protection to the Antarctic Treaty.[14]

SHORTCOMINGS OF THE PRESENT SYSTEM

Although the present system of a decentralised performance of secretarial functions by individual Consultative Parties has served the Antarctic cooperation reasonably well until now, the disadvantages of the system cannot be ignored. One disadvantage is the lack of centralised archives of the Antarctic Treaty, a lacuna which until recently might have appeared tolerable in view of the rather small number of parties to the Treaty and the limited scope of Antarctic activities. However, with today's growing number of states parties and the expansion of the ATS, especially with regard to environmental protection, the need for a centralised system of Antarctic Treaty records has become more urgent. For instance, a state that intends to proceed with the approval of a large number of measures adopted at earlier Consultative Meetings is not able to obtain certified copies of these instruments from a single source. Through diplomatic channels, the state must address every country that has hosted the Consultative Meeting at which the relevant measures were adopted. The *Handbook of the Antarctic Treaty System* does not contain certified texts; it indeed contains occasional errors as well as certain gaps.[15] At any rate, the *Handbook* is neither complete nor up-to-date any longer, with the latest revised edition published in 1994.

Another disadvantage of the present system is the necessity of setting-up administrative and secretarial support from scratch at every subsequent Consultative Meeting. Although this process gives every host country the opportunity of going through a learning experience, the high cost has caused some Consultative Parties to forego the opportunity to host the Meeting basically because of the financial burden involved.[16]

---

[12] ILM, Vol. 27, 1988, pp. 868ff. Similarly to CCAMLR, the establishment of a secretariat would have been the responsibility of the Commission provided for under Art. 18 of CRAMRA.

[13] CCAMLR Art. XVII.

[14] ILM, Vol. 30, 1991, pp. 1,461ff. Protocol, Arts. 7 and 25.

[15] The latest, eighth edition of the *Handbook*, for instance, omits one entire paragraph from the Antarctic Treaty, namely Art. III(2); see Heap (ed.), *Handbook of the Antarctic Treaty System*, pp. 11–12.

[16] In 1991–92 India declined to host the Consultative Meeting and the responsibility was then passed on to Italy. Other countries may well also 'pass on' the responsibility in the coming years, as is illustrated by the most recent case: The XXIV Consultative Meeting should have been held in 2000 – yet, as of 29 December 1999, neither a date nor a host country for the Meeting was adopted.

The recent entry into force of the Environmental Protocol has given rise to new arguments in favour of the establishment of a centralised secretariat. With the Protocol in force, the complexity and scope of Antarctic regulation and management has increased significantly. Moreover, the Protocol has led to the establishment of a new permanent institution, the Committee for Environmental Protection (CEP).[17] The CEP has introduced new requirements for communication and the circulation of documents, especially with regard to environmental impact assessment, and in the future these requirements may be extended to liability for environmental harm.[18] This new situation requires a substantial strengthening of continuity between Consultative Meetings and the timely circulation of documents, requirements which would be better fulfilled by a permanent secretariat than the existing decentralised system.

THE FORMAL RECOGNITION OF THE NEED FOR A SECRETARIAT

Increasing awareness about these difficulties and the continuing expansion of the scope of regulatory powers within the ATS has led to the gradual recognition by Consultative Parties of the need for a permanent secretariat. This need was first fully evaluated at the XIII Consultative Meeting held in Brussels in 1985. The Final Report of that Meeting stated the following:

> The way in which the work of the Consultative Meetings had developed in recent years led many delegations to the conclusion that there was a growing need for some kind of permanent infrastructure to be established, and that there may well soon be no alternative to doing so. Such infrastructure, which in the view of some delegations might take the form of a small secretariat, was envisaged as having functions which should include the documentary preparation and follow-up of Consultative Meetings as well as certain functions relating to the preparation and dissemination of up-to-date records and information about the Antarctic Treaty and the Consultative Meeting process. Some delegations noted that a secretariat might in time come to have more extensive functions. Emphasis was, however, laid by some on the need for any permanent infrastructure to be proportional to the real needs, which in the immediate future called only for something modest in scope and cost.[19]

In spite of the favourable orientation expressed in this Report by 'many delegations', consensus on the establishment of a secretariat did not develop quickly. At the subsequent Consultative Meeting held in Rio de Janeiro in 1987, the

---

[17] On the CEP see Orheim, Chapter 6 in this book.
[18] A secretariat will of course become indispensable in the event of the establishment, in a future liability annex or annexes, of a compensation fund. But even without such a fund, the environmental liability regime will increase the need for communication and an administrative infrastructure. On the Antarctic liability regime, see chapters by Skåre and Lefeber in Part III of this book.
[19] *Final Report of the Thirteenth Antarctic Treaty Consultative Meeting, Brussels, 7–18 October 1985* (Brussels: Ministry of Foreign Affairs, External Trade and Co-operation in Development, 1986), para. 25.

question of a permanent infrastructure was again discussed, this time on the basis of several papers presented, *inter alia*, by Argentina, Australia, China and the United States. Although the majority of the Consultative Parties were again in favour of a permanent secretariat, several delegations maintained the view that the System did not need any new permanent infrastructure:

> The role of host Governments in the preparation and follow-up of Consultative Meetings (including preparation and publication of the Final Report), the role of the Depositary Government of the Antarctic Treaty, the national contact points established pursuant to Recommendation XIII-1, and the provision for reports by the other components of the Antarctic Treaty System at each Consultative Meeting pursuant to Recommendation XIII-2, were all cited as examples of how the Consultative mechanism has evolved and could continue to evolve in the future in an effective fashion to meet the growing needs. These delegations emphasized that the establishment of a permanent infrastructure to carry out some or all of the functions mentioned was not necessary and *expressed doubts about what effective contribution such infrastructure could make*.[20]

It was not until the XVII Consultative Meeting, held in Venice in 1992, that we can find the first unambiguous expression of a general consensus among Consultative Parties as to the need for the establishment of an Antarctic Treaty Secretariat. The Final Report of that Meeting stated that 'consensus existed that a secretariat should be established to assist the Antarctic Treaty Consultative Meeting and the Committee for Environmental Protection in performing their functions'.[21]

The Venice Meeting also marked the beginning of serious discussions on the form of establishment, composition, functions and sharing of costs of the new structure. An *ad hoc* contact group, which benefited from the active contributions of many Consultative Parties, was established and a document containing the essential elements of an Antarctic Treaty Secretariat was attached to the Final Report.[22]

After the 1992 Venice Meeting, the subject of the secretariat came up regularly at every Consultative Meeting: Kyoto (1994), Seoul (1995), Utrecht (1996), Christchurch (1997) and Tromsø (1998); the same subject was also deliberated during inter-sessional meetings of the Group of Legal Experts discussing the question of liability.[23] At the latest, XXIII Consultative Meeting, held in Lima, 24 May – 4 June 1999, the Consultative Parties again addressed the question of the secretariat, confirming their consensus as to the need for a permanent and cost-effective secretariat. New arguments were articulated by several Consultative Parties to highlight the importance of a timely decision on this matter. One such argument was the increasing discrimination that the present rotating system unwittingly produces between countries that can afford the financial burden of hosting a

---

[20] *Final Report of the Fourteenth Antarctic Treaty Consultative Meeting, Rio de Janeiro, 5–16 October 1987* (Rio de Janeiro: Ministry of External Relations, 1988), para. 28 (emphasis added).
[21] *Final Report of the Seventeenth Antarctic Treaty Consultative Meeting, Venice, Italy, 11–20 November 1992* (Rome: Italian Ministry of Foreign Affairs, 1992), para. 43.
[22] *Ibid*.; Annex E.
[23] For an overview of meetings of that Group, see Skåre, Chapter 9 in this book.

Consultative Meeting and countries for whom such a burden presents a real hardship or even insurmountable difficulties. In the long run, this differentiation was felt to be a potentially divisive element in the Antarctic Treaty System. The Final Report of that Meeting stated the following on the matter:

> With regard to the ever-increasing cost that organising an ATCM represents for the host country, concern was expressed that a *de facto* division might develop in the Antarctic Treaty System between those countries that could assume the burden and others that could not. It was noted that the establishment of a secretariat would be one way of diffusing that burden.[24]

Another argument put forward at the 1999 Consultative Meeting in support of a permanent secretariat was the possibility that future Meetings would be held biennially rather than annually, again due to the increasing cost of hosting the Consultative Meeting. In the event of biennial Meetings, the System will be in even more need of a permanent secretariat to ensure continuity during the prolonged inter-sessional period. This is especially true with regard to the pending work on liability and the administrative support required by the newly established CEP, which has to develop its work programme, manage the information and data provided by the Parties, and, most important, provide advice for the Consultative Meeting on draft comprehensive environmental evaluations within the terms and time limits provided by Article 5 of Annex I of the Environmental Protocol. This problem is addressed in the Final Report of the 1999 Consultative Meeting:

> Some delegations were prepared to consider the possibility of holding ATCMs every other year provided that there would be a guarantee for the continuity of the work of the CEP, on the issue of liability and on issues relating to the establishment of a permanent Secretariat. In this connection, it was expressed that meeting every other year would entail a practical problem in respect of the application of paragraphs 3 and 5 of Annex I to the Madrid Protocol, with regard to CEEs and other ATCMs' responsibilities.[25]

After seven years of discussions on the matter of an Antarctic Treaty Secretariat, the lack of substantial results is due less to the technical and legal problems involved in its establishment than to the political issue of the secretariat's future location. This issue has proven difficult to deal with and has effectively delayed the process of creating a secretariat. Argentina has offered Buenos Aires as the secretariat seat and the proposal has been supported by most Consultative Parties. Argentina's offer has the advantage of contributing to a fair geographic distribution of Antarctic institutions, as none are currently located in South America. On the other hand, the United Kingdom has so far expressed its reservations with regard to placing the secretariat in any country that maintains overlapping claims in the Antarctic.[26]

---

[24] See *Final Report of the Twenty-third Antarctic Treaty Consultative Meeting, Lima, Peru, 24 May – 4 June 1999* (Lima: Peruvian Ministry of Foreign Affairs, 1999), para. 33.
[25] *Ibid.*, para. 35. See also discussion by Orheim, Chapter 6 in this book.
[26] See especially *Final Report of the Twenty-second Antarctic Treaty Consultative Meeting, Tromsø,*

It is not the purpose of this chapter to enter into the merits of this sensitive political issue. Its solution may be facilitated by bilateral consultations between Argentina and the United Kingdom, as was noted in the Final Report of the 1999 Consultative Meeting.[27] Our purpose in the remaining part of this chapter is rather to focus on the legal issues surrounding the establishment of the secretariat. Clarification of these issues is not likely to accelerate the process of political accommodation among potential candidates for the seat of the secretariat. However, it is hoped that after available options have been evaluated and the issue of the secretariat seat is finally resolved, clarification of these issues will remove the technical obstacles to the establishment of the secretariat and make possible the timely adoption of its constitutive act.

LEGAL QUESTIONS INVOLVED IN THE ESTABLISHMENT OF A SECRETARIAT

There are basically three major legal questions involved in the creation of an Antarctic Treaty Secretariat: 1) the form of the legal instrument by which it is to be established, 2) the status of the secretariat, i.e., its personality and legal capacity, and 3) its privileges and immunities. To these, further substantive questions may be added that relate to the range of functions, the criteria for sharing the costs among the Parties, and perhaps the advisability of a dispute settlement mechanism.

These questions are examined individually below, although with a caveat that there is a high degree of interdependence among them.

What Legal Instrument for the Constitution of a Secretariat?

In principle, one can envisage three different means, by which the establishment of a secretariat can be achieved in the present legal setting of the Antarctic Treaty. The first is the adoption of an *ad hoc* deliberation by the Antarctic Treaty Consultative Meeting. The second is the conclusion of a headquarters agreement with the prospective host state. The third is the conclusion of an additional protocol to the Antarctic Treaty, supplemented by a host country agreement to deal especially with the issue of privileges and immunities.

The first option might appear rather attractive because of its simplicity and the strong connotation it would attach to the establishment of the secretariat as part of the structure and process of the Consultative Meetings. However, this avenue is not problem free. Firstly, it is not clear in what 'type' of act such deliberation should be embodied. As already mentioned above, at the XIX Consultative Meeting, held in

---

Norway, 25 May – 5 June 1998 (Oslo: Royal Norwegian Ministry of Foreign Affairs, 1998), Appendix 1, p. 27.

[27] See *Final Report of the XXIII ATCM*, para. 27.

Seoul in 1995, the Consultative Parties decided that measures adopted pursuant to Article IX(1) of the Antarctic Treaty were to be divided into three categories: measures, decisions and resolutions.[28] Under this new typology, it is clear at the outset that one should exclude the possibility of resorting to a resolution. Its non-binding character would not provide an adequate legal basis for the establishment of a new institution whose functioning must be secured by effective obligations of all contracting parties. The question, therefore, is whether such deliberation should take the form of a measure or a decision. Surprisingly enough, remarkably little attention has been given to this subject during the six years of discussions within the *ad hoc* contact group. In the view of the present author, the more suitable instrument would be a *measure* because the area of application of a decision is intended to remain confined to 'internal organisational matters'. This narrow definition can hardly encompass the setting up of a new infrastructure that would be capable of influencing the operation of the ATS in 'external' relations. In any event, a decision could only proclaim the 'birth' of a secretariat whose actual existence and functioning would only depend on the appropriate follow-up by the Consultative Parties, in terms of financial support and implementing measures. This consideration again leads to the conclusion that the adoption of a measure would be a more suitable instrument than a decision in the constitution of a secretariat.

Another problem arising from the use of an Article IX(1) instrument for the creation of the secretariat is the requirement that all the Antarctic Treaty Consultative Parties would have to give their consent and subsequent approval before the instrument becomes legally binding. This is a notoriously slow process. However, it is doubtful that this process would be slower than that of an ordinary international agreement because the new institution would at any rate only become established with the consent and the full participation of all the Consultative Parties.

The second option available for setting up the secretariat would be an international agreement with the host country. This agreement would at the same time have the character of a constitutive instrument of the secretariat and of a headquarters agreement that would provide for the seat, legal capacity and privileges of the new organ. Such an international agreement would underlie the early work laid down by the contact group of the 1992 Venice Consultative Meeting. This solution is also expressly mentioned in Annex D attached to the Final Report of the 1994 Kyoto Consultative Meeting.[29] Although the advantage of this solution would be its simplicity, there are two disadvantages. Firstly, the uncertainties surrounding the legal status of the secretariat outside the host country could be critical when the secretariat needs to leave its permanent seat to assist the rotating Consultative Meeting. Secondly, there would be the lack of a unified international subject,

---

[28] For a more detailed examination of the legal nature and content of these three types of instruments see Bush, Chapter 2, and Berguño, Chapter 5, in this book.

[29] *Final Report of the Eighteenth Antarctic Treaty Consultative Meeting, Kyoto, Japan, 11–22 April 1994* (Tokyo: Japanese Ministry of Foreign Affairs, 1994), Annex D.

supposedly an Antarctic Treaty Organisation, which would have the legal capacity and competence to enter into the headquarters agreement with the host state. The latter deficiency could be overcome by the conclusion of an identical agreement between the host country and all the Consultative Parties acting jointly as participants in the Antarctic Treaty Consultative Meeting.

Lastly, a third option would be the adoption of a protocol in addition to the Antarctic Treaty, designed especially for the establishment of the secretariat, and the conclusion of a separate host state agreement that provided for the accommodation, services, privileges and immunities to be enjoyed by the secretariat and its staff within the territory of the host country. This is probably the most traditional approach that could be followed. It would have the advantage of facilitating recognition of the secretariat's legal capacity, and perhaps of privileges and immunities, in the territory of all the Consultative Parties.

This approach, however, raises concerns with regard to two main issues. The first is the risk that this option may lead to the creation of a secretariat with a separate international status, independent of a 'parent' international entity for whose services the secretariat would be created. This risk is amplified by the specific features of the Antarctic Treaty, under which the Consultative Meeting is not personified as an international subject, nor does it possess legal capacity as such independently of the joint will of the Consultative Parties. The secretariat would thus be an inter-governmental institution that is supposed to act as an organ of an international body that, in legal terms, does not exist. This conclusion, quite correctly, is not accepted by some Consultative Parties.[30] The second issue is operational. If the secretariat is endowed with international personality, and thus for all intents and purposes becomes a small international institution in its own right, what assurance is there that it will remain responsive to and effectively controlled by the Consultative Meeting? The most radical, yet legally sound way to resolve these issues would be the transformation of the present Consultative Meeting into the plenary organ of a new Antarctic Treaty Organisation. Such a step, although technically possible, would require a modification or amendment of the Antarctic Treaty by unanimous vote pursuant to Article XII(1). However, such a decision is quite unlikely, at least in the near future. Indeed, the Antarctic Treaty Consultative Parties confirmed on the occasion of the 30th anniversary of the entry into force of the Treaty that they remain 'convinced of the continued effectiveness'[31] of the present structure of the ATS. There is no indication that a new philosophy is now emerging in the Consultative Meetings.

Short of such a radical step, a more realistic approach would appear to be the creation of a secretariat with functions limited to providing administrative and organisational support to the Consultative Meeting and its subsidiary bodies, such as

---

[30] See UK paper on file with the author.

[31] *Final Report of the Sixteenth Antarctic Treaty Consultative Meeting, Bonn, Germany, 7–18 October 1991* (Bonn: German Federal Ministry of Foreign Affairs, 1991).

the CEP. In this perspective, the constitutive instrument should contain provisions that would ensure that the Consultative Parties are able to exercise effective control over the secretariat's functions and its acts. In this way, the secretariat would be conceived as an 'organ' of the collective will of the Consultative Parties, expressed through the Consultative Meeting, rather than an inter-governmental entity capable of its own will.

Legal Status of the Secretariat

The question of the secretariat's legal status has been discussed for quite some time within the Consultative Meeting's contact group. The discussion has revolved essentially around the dilemma of whether the secretariat ought to have legal personality only in the domestic law of the host country or whether it should enjoy full international personality. The Consultative Parties continue to remain divided on this.

In the opinion of the present author, the choice between these two options should be based on the pragmatic evaluation of the object, purpose and functions of the future secretariat rather than on pre-conceived ideological models.

So far, a fairly broad consensus has emerged that the secretariat should be rather small and cost effective. From the work of the Consultative Meetings held from 1992 to 1996, the following set of well-identified functions has been defined:

- assistance in the preparation of Consultative Meetings and of any other meetings under the Antarctic Treaty;
- coordination of information under the Antarctic Treaty and the Environmental Protocol;
- maintaining the records of the Consultative Meeting and of other meetings under the Antarctic Treaty;
- preparing reports of Consultative Meetings and other official meetings;
- circulating information received from Treaty parties of activities in the Antarctic by non-parties.

Although the Consultative Meeting may determine that other technical or administrative functions could be attributed to the secretariat, it is clear from the above list that there is no need to create a secretariat with a high profile, with the attribute of international personality that even the Consultative Meeting does not possess. What is in reality needed is an executive body which has the limited contractual capacity to ensure the hiring of staff, maintain and transfer funds, acquire and dispose of property, and if necessary the capacity to commence legal proceedings in the host state, possibly with the approval of the Consultative Meeting. At this stage, proposals that have been advanced to the effect of giving the secretariat the capacity to conclude international agreements[32] are not convincing. Given the instrumental character of the secretariat, such agreements could only reflect the will of the Consultative Meeting of which the secretariat was only an organ. Consequently, if the need arises for an international agreement, such as the headquarters agreement, the agreement ought to be concluded by the Consultative Parties acting jointly or by the secretariat acting in an agency relationship with the Consultative Meeting and pursuant to an *ad hoc* delegation of powers by the Consultative Parties.

The main argument put forward in support of an international personality of a future secretariat is that the secretariat should be able to act in the territory of all Consultative Parties, especially in relation to the need to provide support to the rotating venues of the Consultative Meetings. Although this need is certainly genuine, it is hardly conceivable that it ought to justify the granting of an international legal personality. For this reason, organs are set up by international agreements that, for the fulfillment of their functions, convene in a different host country every year without enjoying or claiming an international personality. An example is the World Heritage Committee, an organ of the 1972 UNESCO Convention Concerning the Protection of the World Cultural and Natural Heritage.[33] The World Heritage Committee has a broad range of powers and an annual budget of several million US dollars. To the best knowledge of this writer, in the more than 25 years of practice of the World Heritage Committee, the need for attributing an international legal personality to this organ has on no occasion ever arisen or even been proposed.[34]

Within the family of the Antarctic instruments, we can find another example of an executive organ which, although it is required to perform some functions outside the host state, is neither for this sole reason endowed with an international personality nor with an independent legal status. This is the case of the CCAMLR

---

[32] 'The Secretariat shall have full legal personality and such legal capacity to carry out its functions. In particular, the secretariat shall have the capacity ... to conclude, subject to prior approval of the Antarctic Treaty Consultative Meeting, international agreements'; Art. 3(iv) of the draft presented at the informal contact group on the secretariat convened in Tromsø, during the XXII Consultative Meeting in 1998. Document on file with the author.

[33] 1972 Convention Concerning the Protection of the World Cultural and National Heritage; ILM, Vol. 11, pp. 1,358ff, also available at <www/unesco.org/whc/nwhc/>.

[34] Art. 14 of the World Heritage Convention provides that 'the World Heritage Committee shall be assisted by a secretariat appointed by the Director General of the United Nations Educational, Scientific and Cultural Organisation'. From 1992, the Committee has been served by a special unit of the UNESCO Secretariat that has been named the 'World Heritage Centre'. The question of the legal capacity of the Secretariat/Centre to enter into international agreements – especially agreements with member states and inter-governmental and non-governmental organisations – has been a recurrent theme of discussion at the World Heritage Committee's meetings. At the 1996 meeting in Merida, Mexico, the World Heritage Committee decided that the capacity to enter into international agreements should be reserved to the Committee itself and the competence to enter into such agreements would be conferred to the Chairperson of the Committee. See UNESCO, *Report of the World Heritage Committee*, Merida, 1996. Paper on file with the author.

Secretariat. Under Article XVII of the CCAMLR, the Secretariat is appointed by the Commission, which is the sole body under that Convention to be endowed with legal personality and to 'enjoy in the territory of each of the Contracting Parties such legal capacity as may be necessary to perform its functions'. This wording makes it clear that the legal capacity referred to is of a domestic law nature and does not necessarily entail an international personality.

To conclude on this point, it is the view of the present author that the Consultative Meeting does not need to set up a secretariat with an international legal personality. Such a step is not required by the nature of the activities and functions to be performed by the secretariat. Further, it is believed that the establishment of such an international legal personality would not be appropriate in the specific circumstances of the Antarctic Treaty where the plenary body itself – the Consultative Meeting – is not an international subject. The best solution would be the adoption of a functional approach, similar to that found in Articles VIII and XVII of the CCAMLR Convention. In analogy with the CCAMLR Commission, the Antarctic Treaty Secretariat would simply be attributed 'legal personality', without further specification, and would enjoy legal capacity, first of all, in the host country, and secondarily in the territory of the other Treaty parties, but only to the extent that such capacity would be indispensable for the performance of the functions contemplated in the constitutive instrument (measure, protocol or headquarters agreement) or provided, if need arises, by a specific agreement or a Consultative Meeting decision.

Privileges and Immunities

The efficient and unimpeded performance of the secretariat's functions would require that a certain number of privileges and immunities be granted to the secretariat as an institution as well as to its staff. There is no disagreement that such immunities ought to be accorded in the territory of the host state where the secretariat is permanently located and where its functions normally are to be performed. The question arises as to whether immunities should also be extended to the territory of other parties to the Antarctic Treaty. This point has attracted extensive debate in the Consultative Meetings, especially after the introduction of several papers (by the United Kingdom and Australia) at the Consultative Meetings in 1995 and 1996. In a UK paper presented at the Utrecht Meeting in 1996, a strong case was made for the granting of a broad range of privileges and immunities *by all* Consultative Parties, in view of the need to guarantee the undisturbed performance of the secretariat functions in the territory of all states that would be potential hosts of the Consultative Meeting or of other meetings for which the assistance by the secretariat is required.[35] It is still unclear to what extent this position is a

---

[35] Paper on file with the author.

consequence of the theory that favours the granting of international personality to the secretariat. Such a consequential link would appear to be inherent to the view whereby the obligation to grant immunities to international organisations rests on a norm of customary international law.[36] This view, however, is contested by those who maintain that the sole basis for the obligation to accord immunities to international institutions is the treaty that provides these immunities.[37] If this view is correct, nothing would prevent the Consultative Parties from regulating the question of immunities independently of the international personality of the secretariat and in a way that would restrict such immunities to the territory of the host country.

Undoubtedly, this theoretical dispute is relevant to the correct orientation of the ongoing debate over the status and immunities of the future Antarctic Treaty Secretariat. However, for the purpose of this chapter, it is not believed to be necessary to take a firm position on the issue of the legal foundation of immunities of international organisations because to give the future secretariat the status of an international organisation is not only unnecessary but also inappropriate.[38]

On this point, the need to avoid solutions based on the mechanical imitation of existing models of international organisations should be recalled. In the case of Antarctic cooperation, the choice should rather be inspired by the specific nature of the ATS and by the pragmatic evaluation of the object and functions of the secretariat. It is hardly necessary to repeat that, short of an amendment of the Antarctic Treaty, the future secretariat cannot be conceived as an organ of an *international organisation* – which the Consultative Meeting is *not*. Moreover, the secretariat must not be established as an independent, international entity capable of operating autonomously since its sole purpose is to serving the Consultative Meeting and the other subsidiary organs of the ATS such as the CEP. The secretariat should therefore be constituted as an organ of the collective will of the Consultative Parties as expressed in the organisational structure of the Consultative Meeting. From this perspective, what is necessary and sufficient is the delineation of a set of *functional immunities* designed to protect the undisturbed exercise of the secretariat's functions rather than the independence of an imaginary international organisation. Such immunities should primarily include traditional tax exemptions, immunity from local jurisdiction and execution, inviolability of the seat, archives and official communications. They would primarily be relevant in the relations between the secretariat and the host country. They should therefore be provided for either in an

---

[36] C. Dominicé, 'L'immunité de juridiction et d'exécution des organisations internationales', *Recueil des Cours de l'Académie de Droit International de la Haye*, Vol. 187, 1984/IV, pp. 145–238; J.-F. Lalive, 'L'immunité de juridiction des Etats et des Organisations internationales', *Recueil des Cours de l'Académie de Droit International de la Haye*, Vol. 84, 1953/III, pp. 205–396; B. Conforti, *Diritto Internazionale*, 5th edition (Naples: Editoriale Scientifica, 1997); G. Weissberg, *The International Status of the United Nations* (New York and London: Oceana Publications, 1961).

[37] Sir Ian Sinclair, in *Yearbook of the International Law Commission*, 37th Session, Vol. 1, 1985, p. 289.

[38] See the discussion above, in this chapter.

ad hoc headquarters agreement between the host country and the Consultative Parties or in the constitutive instrument of the secretariat (be it a protocol or a Consultative Meeting measure). If the need arises during the actual life of the secretariat for the extension of privileges and immunities to the territory of other parties to the Antarctic Treaty, this result could be achieved by including an *ad hoc* provision in a separate protocol of agreement or in a measure to be adopted by the Consultative Meeting with subsequent approval and ratification in the respective domestic legal systems of parties to the Treaty.

This functional approach is in accordance with several precedents in international practice. In addition to the already mentioned CCAMLR Articles VIII and XVII, under which the privileges of the secretariat are 'parasitic' on the Commission's own privileges, we may also refer to the GATT whose secretariat was established in the absence of an international organisation that had privileges and immunities. In this case, the functional approach was underscored by the fact that immunities were conferred by a decree of the Swiss Government, extending unilaterally to the GATT Secretariat the guarantees provided to the United Nations in the headquarters agreement of 19 April 1946.[39]

Other Issues

Besides the three key legal issues discussed above, the establishment of an Antarctic Treaty Secretariat entails other questions of predominantly political and financial character, for example the questions of the scope of functions of the secretariat and cost sharing.

The first question has been addressed since the Venice Consultative Meeting in 1992, and there is now a broad consensus on the range of functions, as outlined by the contact group at the 1994 Consultative Meeting in Kyoto (see discussion above).

Regarding the second question, uncertainties still exist as to whether the future secretariat should be financed by equal shares of all the Consultative Parties or by a system of assessed contributions that should reflect the different economic strength of the participants or the different scale and importance of their activities in the Antarctic. Resistance to the latter solution comes from countries who wish to preserve the special character of the Antarctic Treaty as an instrument based on the formal equality of all Consultative Parties and on consensus. However, even if the approach based on equal shares were to prevail, it would be difficult to maintain an unadulterated consensus system with regard to the adoption of a budget necessary to meet the expenses of the secretariat. A system based on a large majority, such as a three-quarters majority, would avoid the risk of paralysing the secretariat in the event of a veto by one or a small minority of the Consultative Parties. However, even the introduction of such a majority would constitute a departure from the

---

[39] J. Jackson, *World Trade and the Law of GATT* (Indianapolis, Ind.: Bobbs-Merrill, 1969), pp. 119ff, 145ff.

Antarctic Treaty, Article IX(1) and (4), and could raise the issue of the necessity of an amendment pursuant to Article XII.

CONCLUSIONS

This chapter has argued that developments that occurred in the ATS over the past 40 years give strong evidence in favour of the timely establishment of an Antarctic Treaty Secretariat. The entry into force of the Environmental Protocol in 1998 and the formal establishment of the CEP the same year represented new important developments in the System, which now calls for corresponding institutional upgrading.

Having recognised this, it has been maintained here that the setting up of a new secretariat infrastructure should not occur in such a manner that would transform the Antarctic Treaty into an international organisation. The secretariat should be conceived and set up as a support structure for the Consultative Meeting. It should have a functional capacity that is limited to the performance of the tasks entrusted to it by the Consultative Meeting. Its privileges and immunities should reflect the intergovernmental nature of the body – the Consultative Meeting – of which it would be an emanation. In this perspective, it would not be necessary, and perhaps it would be legally inappropriate, to endow the secretariat with international personality.

As to the choice of the constitutive instrument, preference should go to either the adoption of a 'measure' under Article IX of the Antarctic Treaty or of an *ad hoc* protocol additional to the Treaty. The choice between one or the other ultimately depends on how high a profile the Consultative Parties wish to give to the future secretariat.

Beyond these issues, there remains the political problem of how to determine the location of the future secretariat. This decision is an important one, not so much because of the prestige and national pride that are at stake, but rather because hosting the first institution of the overall Antarctic Treaty carries with it a strong symbolic value and legacy: the host country's uncompromising commitment to further the spirit of mutual trust and international cooperation that has so far preserved the Antarctic as a 'natural reserve, devoted to peace and science' – as the Environmental Protocol states as its chief objective.

8

The ATS on the Web: Introducing Modern Information Technology in Antarctic Affairs

Davor Vidas and *Birgit Njåstad*

When the Twenty-second Antarctic Treaty Consultative Meeting was hosted by Norway in the Spring of 1998, only few months after entry into force of the Environmental Protocol,[1] an excerpt from the Final Report from that meeting stated the following:

> Norway introduced Working Paper (XXII ATCM/WP 25) proposing that an ATCM home page be established on the World Wide Web, with the purpose of assisting future ATCM Host Governments in pre-sessional circulation of documents, as well as enabling easier access to information about the Antarctic and the Antarctic Treaty system to the general public. The Meeting expressed support for the Norwegian initiative and adopted Resolution 5 (1998).[2]

This wording reflected well the two-fold purpose behind the Norwegian initiative. In relation to ATS internal affairs, the purpose was to facilitate the distribution of documents for Consultative Meetings, in the absence of an Antarctic Treaty Secretariat. In relation to ATS external presentation, Norway wanted to contribute to enhanced transparency of the ATS operation, by providing easy access to information about the ATS, including publicly available documentation, at one location. The advantages of modern information technology offered the unique

---

[1] Protocol on Environmental Protection to the Antarctic Treaty, which entered into force on 14 January 1998; text reproduced in ILM, Vol. 30, 1991, pp. 1,461ff.

[2] Para. 22 of the *Final Report of the Twenty-second Antarctic Treaty Consultative Meeting, Tromsø, Norway, 25 May – 5 June 1998* (Oslo: Royal Norwegian Ministry of Foreign Affairs, 1998). The matter was discussed under agenda item 5(c), 'Operation of the Antarctic Treaty System: Consequences of the Entry into Force of the Protocol on Environmental Protection and Related Issues'.

possibility of achieving both purposes through the same media – the Internet. Two host countries for the regular Consultative Meetings have thus far taken advantage of Internet services to achieve the same two purposes named above – Norway when it hosted the XXII Meeting in 1998 and Peru when it hosted the XXIII Meeting in 1999.[3]

Further to the initiative for the establishment of a website for the Consultative Meetings, the advantages of using Internet were at the same time considered by the Committee for Environmental Protection (CEP) at its first meeting, held in conjunction with the Tromsø Consultative Meeting.[4] The CEP had agreed that

> there is a need to simplify the means for information exchange and that the use of electronic mechanisms would be valuable, including the establishment of an Internet home page. The format of the home page, however, still remained to be finalised, as were the modalities for protecting documents in an electronic exchange system. The Committee therefore suggested that these and other related issues be considered by ATCM XXII.[5]

This chapter discusses the reasons behind the introduction of the websites for the Consultative Meeting and the CEP, the principal procedural questions that had to be solved before setting up the websites, and the resulting differences in the status of the two websites in the overall ATS. Some practical information on the design and usage of the websites also is included.

THE CONSULTATIVE MEETING WEBSITE

Ideally, the website of an international cooperative arrangement would be maintained by its own secretariat. This is not least the case for multilateral cooperative arrangements involving a large number of states, such as the 1959 Antarctic Treaty[6] which currently numbers 44 states parties. Notwithstanding an acute need for a permanent secretariat, no Antarctic Treaty Secretariat is as yet available.[7] While not perceived as a significant deficiency in the early years of the ATS, the lack of a secretariat has increasingly become an impediment to achieving smooth cooperative process among Antarctic Treaty parties and in the parties' relationship to the wider international community. This impediment has been

---

[3] The website established by Norway was located at <www.tromso.npolar.no/atcm> and maintained by the Norwegian Polar Institute. The website established by Peru for the XXIII Consultative Meeting is, as of 8 June 2000, still operative, though not at the original location (<www.antartica-rcta.com.pe>); it is currently located at <www.rree.gob.pe/conaan/meeting1>.

[4] The CEP held its first meeting in Tromsø, 25–29 May 1998; see 'Report of the Meeting of the Committee for Environmental Protection, Tromsø, 25–29 May 1998', in *Final Report of the XXII ATCM*, Annex E (hereinafter Report of the CEP I). On the CEP see Orheim, Chapter 6 in this book.

[5] Para. 38 of the *Final Report from the XXII ATCM*. See also paras. 24 and 56 of the Report of the CEP I.

[6] UNTS, Vol. 402, pp. 71ff.

[7] For a discussion on an Antarctic Treaty Secretariat, see Francioni, Chapter 6 in this book. See also Berguño, Chapter 5 in this book.

manifested in various ways, from the practical need for a permanent body to manage the 'logistics' of preparing and carrying through Consultative Meetings, as well as of the follow-up tasks, to the need of having one central location from which information concerning cooperation within the ATS can be distributed to third parties and to the general public who are interested in ATS affairs.[8] In this situation, the initiative for a Consultative Meeting website can actually be seen as an improvisation aimed at a temporary — pending the establishment of the Secretariat — filling in of some of the gaps in the operation of the ATS.

Facilitating the Preparation of a Consultative Meeting

The lack of permanent secretarial support has resulted in many practical, 'internal' deficiencies in the operation of the Antarctic Treaty cooperation. In accordance with Article IX of the Antarctic Treaty, the representatives of the Consultative Parties are to meet 'at suitable intervals and places, for the purpose of exchanging information, consulting together on matters of common interest pertaining to Antarctica, and formulating and considering, and recommending to their Governments, measures in furtherance of the principles and objectives of the Treaty'. Those meetings, known as the Antarctic Treaty Consultative Meetings,[9] do not have any permanent venue. The Antarctic Treaty provided only the place and time of the First Consultative Meeting, which in accordance with Article IX(1) of the Treaty was held 'at the City of Canberra within two months after the date of entry into force of the Treaty'. Subsequent practice introduced the hosting of Consultative Meetings by *rotation* among the Consultative Parties, in principle by following the alphabetical order of the names of the countries in the English language. Until 1991, regular Consultative Meetings were mainly held biennially. As follow-up action on the adoption of the Protocol, it was agreed at the XVI Consultative Meeting held in Bonn in 1991 that 'Consultative meetings should in future be held annually and in the first half of the year'.[10] The efficiency of this agreement has come under questioning at the Lima Consultative Meeting in 1999. Based on experiences from several annual Consultative Meetings held in the course of the 1990s, some delegations 'were prepared to consider the possibility of holding ATCMs every other year',[11] whereas

---

[8] Certain components of the ATS, such as the CCAMLR, do have a secretariat and maintain their own website; for CCAMLR website see <www.ccamlr.org>.

[9] The meetings are named so in accordance with Rule 1 of the Rules of Procedures; the latest revision of the Rules was made in 1997, and is published attached to Decision 1 (1997) in the *Final Report of the Twenty-first Antarctic Treaty Consultative Meeting, Christchurch, New Zealand, 19–30 May 1997* (Wellington: New Zealand Ministry of Foreign Affairs and Trade, 1997), Annex B.

[10] Para. 134 of the *Final Report from the Sixteenth Antarctic Treaty Consultative Meeting, Bonn, 7–18 October 1991* (Bonn: German Federal Ministry of Foreign Affairs, 1992).

[11] See para. 35 of the *Final Report of the Twenty-third Antarctic Treaty Consultative Meeting, Lima, Peru, 24 May – 4 June 1999* (Lima: Peruvian Ministry of Foreign Affairs, 1999).

other delegations 'stressed that holding meetings every other year would allow for a more structured preparation of the issues and more meaningful outcomes'[12].

The government that hosts a Consultative Meeting must have adequate resources at its disposal to facilitate the organisation of the Meeting; it must, among other tasks, undertake the handling of all the documents for the Meeting, organise pre-sessional translation, and distribute the translated documents to the other parties. Recommendation I–XVI, which was adopted at the First Consultative Meeting, sets forth the following requirements for document distribution:

> [Documentation]which any participating government may desire to place before the next consultative meeting, shall be forwarded through diplomatic channels so as to reach all governments entitled to participate in that consultative meeting, at least one month prior to the meeting, except in circumstances of urgency.

When this recommendation was adopted, the Antarctic Treaty numbered altogether 12 states parties; today, there are 44 states parties to the Treaty, 27 of which with the consultative status. As a rule, documents prepared for the Consultative Meetings must be translated into four official languages.[13] The practicalities of timely document distribution therefore require the host country for the Meeting to employ significant resources if it is to meet the agreed deadlines for document distribution. The documentation submission and distribution deadlines are specified in the 1996 Guideline on Pre-sessional Document Circulation and Document Handling, with the main rule as follows:[14]

> [A]ll Working Papers prepared by Consultative Parties and Observers referred to in Rule 2, and Information Papers which a Representative of a Consultative Party requests be translated, should be received by the Host Government no later than 45 days before the meeting. The Host Government should circulate these papers in translation no later than 30 days before the meeting through diplomatic channels.[15]

While the number of documents distributed at the early Consultative Meetings was relatively limited, the amount of documentation at recent Meetings has become quite voluminous. The 1999 Meeting in Lima, for instance, required that 180 official papers (comprising both working and information papers) be distributed to 44 states parties; most of the documentation had to be available in all four official languages[16]. Disregarding the documentation distributed to observers and other entities attending the Meetings, as well as documents other than working and

---

[12] *Ibid.*, para. 37.

[13] According to Rule 22 of the Rules of Procedure, English, French, Russian and Spanish are the official languages of the Consultative Meeting.

[14] Annex D to the *Final Report of the Twentieth Antarctic Treaty Consultative Meeting, Utrecht, 29 April – 10 May 1996* (The Hague: Netherlands Ministry of Foreign Affairs, 1997).

[15] Para. 1 of the Guideline on Pre-sessional Document Circulation and Document Handling.

[16] According to para. 6 of the Guideline, 'No Working Paper or Information Paper submitted to the ATCM will be used as a basis for a discussion at the ATCM unless it has been translated into four official languages'.

information papers, these 180 official papers alone, if made available in four official languages and distributed to 44 states parties, amount to 31,680 pieces of documentation which must be circulated. Many of the documents are extensive in length, and the 1996 Guideline included a suggestion that 'Information Papers for which translation has been requested ... should ordinarily be limited to 30 pages';[17] there is, however, no suggested page limit for the working papers.

As John Heap noted already some time ago, developments in the ATS have 'presented hosts to Consultative Meetings with formidable challenges'.[18] Indeed, it is because of these challenges that there have been recent examples of countries who have questioned their capacity to host a Consultative Meeting, or even forego their turn for hosting a Meeting. A recent illustration of this occurred at the XXIII Consultative Meeting in Lima in 1999; no country was able to accept the hosting of the next Consultative Meeting, which would have normally been held in 2000, and it is currently (as of 2 April 2000) still uncertain where the next regular Consultative Meeting will be held. The Final Report of the Lima Consultative Meeting states, *inter alia*, the following concerning this matter:

> With regard to ever-increasing cost that organising an ATCM represents for the host country, concern was expressed that a *de facto* division might develop in the Antarctic Treaty System between those countries that could assume the burden and others that could not.[19]

To ameliorate part of these problems, Norway offered at the XXI Consultative Meeting, held in 1997 in Christchurch, New Zealand, to explore the possibility of document distribution by electronic means. As the host country for the XXII Consultative Meeting in 1998, Norway established a World Wide Web (WWW) homepage for that specific meeting. Norway's primary aim in establishing the website for the Tromsø Consultative Meeting was to assess the potential for using the WWW as a tool for simplifying document distribution. Because of the potential sensitivity of the contents of working papers and the fact that electronic access to Consultative Meeting documents had not yet been adopted as a formal mechanism for document distribution, it was decided that, in this initial pilot project, only information papers would be posted on the website. Furthermore, only those documents were posted on the website for which the submitting parties specified that these could be made accessible through the website. The documents were posted on a username and password protected page, which ensured restricted access. Information about the website and procedures for procuring a username and password was distributed to the parties through traditional diplomatic channels. The secretariat for the Tromsø Consultative Meeting ensured that parties submitted their

---

[17] Guideline, para. 1.

[18] See J. Heap (ed.), *Handbook of the Antarctic Treaty System*, 8th edition (Washington, DC: US State Department, 1994), p. 21.

[19] Para. 33 of the *Final Report of the XXIII ATCM*.

documents electronically in addition to the traditional hard copy submission; the secretariat also queried parties as to whether or not they would permit the posting of their papers on the website. The Meeting secretariat thus ensured that the website for the XXII Consultative Meeting was utilised even though no formal decision had been made beforehand with respect to its establishment within the ATS. Although only one third of the information papers were posted on the website before the beginning of the Meeting, the pilot project did indicate to the parties that there was a great potential in distributing documents through a protected website. The project demonstrated furthermore that such a mechanism would not necessarily entail major efforts or costs for the hosting government; rather, the use of the website could greatly contribute in optimising the efficient use of the resources available for preparing a Consultative Meeting.

Based on this first experience of the potentials of electronic exchange of and access to documents for the Consultative Meetings through the Internet, Norway proposed at the XXII Consultative Meeting in Tromsø that a decision be adopted by the Consultative Parties which would, with respect to document distribution among the Parties, require the following:

1. An Antarctic Treaty Consultative Meeting Internet Home Page (*ATCM homepage*) shall be established at the World Wide Web (WWW).

2. The *ATCM homepage* shall *inter alia* contain:

    a) in an area freely accessible to the general public:
    - general information on Antarctica and the Antarctic Treaty System;
    - within the resources available and in accordance with paragraph 3 of this Decision, an archive of official documentation from the Antarctic Treaty Consultative Meetings, containing working and information papers submitted to the Meetings, as well as the Final Reports of the Meetings with text of annexes.

    b) in a password protected area accessible only to the Antarctic Treaty Parties and the Antarctic Treaty System Observers, experts invited by the ATCM and any other entities that the ATCM decides should have such access:
    - official documents submitted electronically to the Host Government in advance of an Antarctic Treaty Consultative Meeting;
    - any practical information related to the Antarctic Treaty Consultative Meeting, which the Host Government may wish to communicate this way.[20]

In the absence of a permanent Antarctic Treaty Secretariat, the arrangement proposed by Norway cannot be a permanent one. However, with reliance on the 1996 Guideline for Pre-sessional Document Circulation and Document Handling, the arrangement does offer possibility for ensuring *continuity* in the maintenance of an ATCM homepage. According to paragraph 7 of the Guideline, the host government should, in principle within three months of the end of the Consultative Meeting it has hosted, circulate the Final Report from that Meeting and a

---

[20] See Norway, 'ATCM Homepage', doc. XXII ATCM/WP 25, 1998, pp. 5–6.

comprehensive list of official papers from the Meeting. It could thus be understood that the reasonable time during which the host government continues to perform certain secretarial tasks is a period of three months after the conclusion of the Meeting it has hosted. The following arrangement was accordingly proposed to ensure the continuity of a Consultative Meeting homepage:

> Norway, as the Host Government of the XXII Antarctic Treaty Consultative Meeting shall, based on the Home Page prepared in advance of the XXII Meeting, establish the *ATCM homepage*, and shall maintain it until three months after the closure of ATCM XXII.
>
> Thereafter, within the resources available and pending any more permanent solution agreed upon by the Antarctic Treaty Consultative Parties, the *ATCM homepage* should be maintained by any subsequent Host Government of the Antarctic Treaty Consultative Meeting, from three months after the closure of the previous Meeting until three months after the closure of the Meeting it hosts.[21]

Enhancing Transparency in the Operation of the Antarctic Treaty System

The matter of public availability of documents from the Consultative Meetings, rather than being a practical matter such as document distribution, deals rather with the perception of the cooperation within the ATS in the wider international community.[22]

During the 1960s and early 1970s, documents from the Consultative Meetings were generally not available to the public. Since that time, the ATS has passed through a gradual process of evolution towards transparency. In the early 1980s, the public availability of documents became a regular agenda item at Consultative Meetings with the aim of making information about Antarctica and the ATS more widely available to the interested public. The Consultative Parties' decisions and the consistent practice of the Parties in the course of the 1990s enabled the public availability of all official documents of all the Consultative Meetings held so far.[23]

---

[21] *Ibid.*

[22] See the discussion in D. Vidas, 'The Antarctic Treaty System in the International Community: An Overview', in O.S. Stokke and D. Vidas (eds.), *Governing the Antarctic: The Effectiveness and Legitimacy of the Antarctic Treaty System* (Cambridge University Press, 1996), pp. 35–60, and especially on demands for increased transparency in the operation of the ATS, at pp. 55–58.

[23] For the development of the ATS towards public availability of documents from the Consultative Meetings, see relevant paragraphs in the Final Reports of the Seventh (1972), Ninth (1977), Tenth (1979) and Eleventh (1981) Consultative Meetings. Further see Recommendation XII-6 (1983), paras. 29–33 of the *Final Report of the Thirteenth Antarctic Treaty Consultative Meeting, Brussels, 7–18 October 1985* (Brussels: Ministry of Foreign Affairs, External Trade and Cooperation in Development, 1985), paras. 61–67 of the *Final Report of the Fourteenth Antarctic Treaty Consultative Meeting, Rio de Janeiro, 5–16 October 1987* (Rio de Janeiro: Brazilian Federal Ministry of Foreign Relations, 1987) and Recommendation XIV-1 adopted at that Meeting. Further developments towards public availability of documents have reflected the practice of the Consultative Parties, as expressed

Although the official documents submitted by the Consultative Meetings are now, in principle, available to the public from the closure of the Meeting, practical obstacles to the transparency of the ATS still exist. The papers, though available in theory, are still hardly accessible in practice. While the Consultative Parties have, through their decisions and consistent practice, allowed for public availability of documents from the Consultative Meetings, they have still failed to ensure easy access to these documents.

Considering the current status of the public availability of documents and in the interests of the Parties' succeeding in ensuring the transparency of the ATS, Norway proposed at the Tromsø Consultative Meeting that all official documents from Consultative Meetings (including both working and information papers) should not only be *offered available* but also *made accessible*, as far as possible. In the light of modern-day electronic communication, one way of remedying this situation is to establish an archive on the Internet, which would be accessible through the Consultative Meeting website.

It was accordingly proposed by Norway that all the official documents submitted to the host country in the preparation of a Consultative Meeting be placed at a password protected website. At the *closure* of the Consultative Meeting, all password restrictions on the protected document page for that Meeting would be eliminated, thus making all official documents available to the public. When submitting official documents for the Meeting in question, Parties should indicate whether certain documents should not be made public. The proposal made by Norway was as follows:

> As from the closure of an Antarctic Treaty Consultative Meeting, and if no Delegation has indicated its intention to the contrary when submitting a document, the Host Government shall enable free public access to all the official documents which have been placed on the *ATCM homepage*, by removing the password protection from these.[24]

Save for the replacement of the word 'shall' with 'should', this proposal was incorporated in the text of Resolution 5 (1999). The consequence of the change from the imperative to the conditional was that no strict requirement was being established for the host government of a Consultative Meeting regarding enabling public access to the documents through the Consultative Meeting website upon the closure of the Meeting.

---

in: paras. 35–39 of the *Final Report of the Fifteenth Antarctic Treaty Consultative Meeting, Paris, 9– 20 October 1989* (Paris: French Ministry of Foreign Affairs,1989), paras. 44–47 of the *Final Report of the Sixteenth Antarctic Treaty Consultative Meeting, Bonn, 7–18 October 1991* (Bonn: German Federal Ministry of Foreign Affairs, 1991), para. 52 of the *Final Report of the Seventeenth Antarctic Treaty Consultative Meeting, Venice, Italy, 11–22 November 1992* (Rome: Italian Ministry of Foreign Affairs,1993), para. 71 of the *Final Report of the Eighteenth Antarctic Treaty Consultative Meeting, Kyoto, Japan, 11–22 April 1994* (Tokyo: Japanese Ministry of Foreign Affairs, 1994); and para. 69 of the *Final Report of the Nineteenth Antarctic Treaty Consultative Meeting, Seoul, 8–19 May 1995* (Seoul: Ministry of Foreign Affairs of the Republic of Korea, 1995).

[24] Norway, 'ATCM Homepage', pp. 5.

Norway also expressed the view that there are no real obstacles for a large number of official documents submitted to the Consultative Meetings to be made publicly available already *in advance* of a Meeting. It was argued that this, in fact, could only increase the general public's involvement in and awareness of Antarctic management issues in the contemporary international community, which in itself would be a desirable development. It was suggested that, in practice, this could be accomplished by establishing a page in addition to a protected page for net-accessible documents which are not password protected. The following was proposed by Norway:

> Should any Delegation, when submitting a document to the Host Government indicate that the document can be placed in the *ATCM homepage* area freely accessible to the general public in advance of the Meeting, and if no other Antarctic Treaty Consultative Party objects within 30 days from the placement of that document on the *ATCM homepage* password protected area, the Host Government shall remove the password protection from that document.[25]

However, as no consensus was reached among the Consultative Parties on this matter at the Tromsø Meeting, this proposal was not reflected in the text of the adopted Resolution 5 (1999);[26] and the proposed paragraph was altogether omitted from the final text of the resolution.

The Current Legal Status of the Consultative Meeting Website and Practical Implications

The proposal submitted by Norway at the 1998 Consultative Meeting was to establish the Consultative Meeting website by a *decision*,[27] operative at the adoption (and, as an internal organisational matter, legally binding for the Consultative Parties). Though the Norwegian initiative for the establishment of a Consultative Meeting website received wide support at the XXII Consultative Meeting in Tromsø, not all delegations favoured the establishment of such a website by a decision. Eventually, Resolution 5 (1998) was adopted. That resolution stated that the host governments of Consultative Meetings 'be *encouraged to consider* the establishment of ... ATCM Home Page',[28] rather than be required to establish such a website. Regarding the public availability of documents after the closure of a Consultative Meeting, the same resolution stated that the host government '*should*

---

[25] *Ibid.*

[26] Resolution 5 (1998), 'ATCM Home Page', published in the *Final Report of the XXII ATCM*, Annex C

[27] See Norway, 'ATCM Homepage', pp. 5–6.

[28] See para. 1 of Resolution 5 (1998); emphasis added.

enable free public access' to these documents,[29] using the 'should enable' wording rather than 'shall enable' as was originally proposed by Norway[30].

Thus, in the absence of permanent secretarial support to the Consultative Meetings, the Consultative Parties still remained somewhat conservative about introducing internal organisational changes, such as the establishment of a Consultative Meeting website, as a matter of legal procedure in the preparation of a Consultative Meeting. On the other hand, the practical need for the more efficient use of the resources available to the host country of a Consultative Meeting, as well as quicker access to documents and enhanced speed in document distribution provided through the Internet, anticipates that the use of a Consultative Meeting website will most likely be accepted in practice by future host countries as a facilitating rather than hampering tool in their effort to use secretarial resources efficiently and effectively. This practice would also be consistent with the need in the ATS cooperation for a collective memory in the form of archives, which at present does not exist.[31] However, with the current regulation of a Consultative Meeting website based on a resolution, there is no certainty for maintaining continuity of the website or of its possible development toward the achievement of an instrument of 'collective memory'.

Design and Practical Usage of the First Website for the XXII Consultative Meeting Hosted by Norway

The Norwegian Polar Institute (NP) hosted and operated the website for the Tromsø Consultative Meeting. The development of the site was a cooperative effort between the NP and the Norwegian Ministry of Foreign Affairs. A consultant was hired to assist in the design of the site.

Norway had multiple aims in establishing the site. In addition to facilitating document distribution, a website was considered a suitable means by which practical information about the meeting could be distributed (for example information about the conference city, the meeting venues, the meeting programme and receptions, and other practical matters). Norway furthermore held that a website could be an excellent avenue for distributing information about the Antarctic and the ATS to the general public. Information that was considered to have the potential of contributing to and increasing the public's awareness and understanding of the Antarctic Treaty System was carefully selected, including a description of the functions of the Treaty, an overview of parties to the Treaty, and an overview of previous Consultative Meetings and decisions adopted at these. These overarching goals are reflected in the structure of the website, which is shown in Figure 1.

---

[29] *Ibid.*, para. 3; emphasis added.
[30] See Norway, 'ATCM Homepage', pp. 5–6.

The ATS on the Web 151

Figure 1: ATCM XXII Website Structure

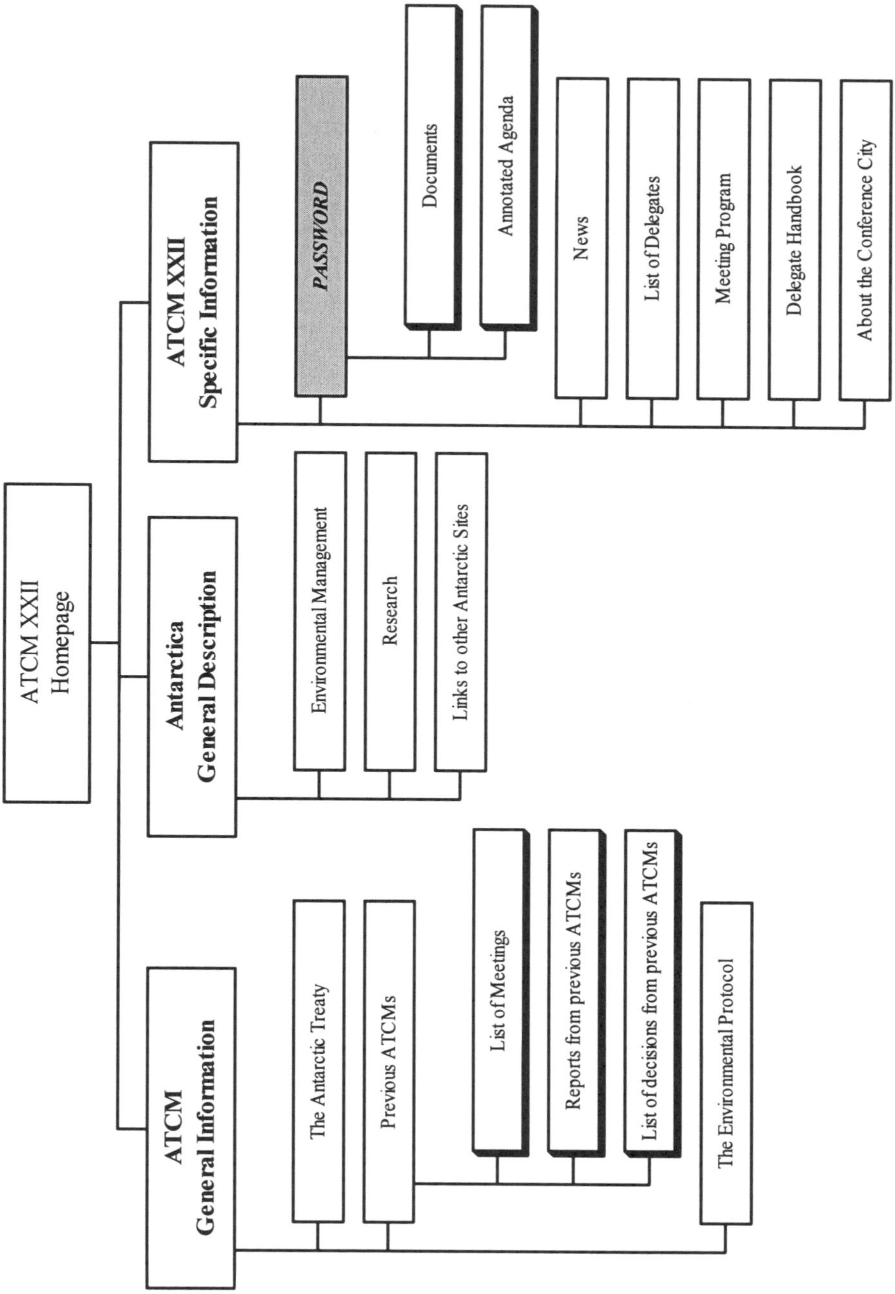

```
                              ATCM XXII
                              Homepage
                                  |
        ┌─────────────────────────┼─────────────────────────┐
        |                         |                         |
   ATCM                      Antarctica                 ATCM XXII
   General Information       General Description        Specific Information
        |                         |                         |
   ┌────┼────┬────────┐      ┌────┼────┐            ┌───────┼────────┐
   |    |    |        |      |    |    |            |       |        |
 The   Previous  List of  The    Env.  Research  Links   PASSWORD  Documents  Annotated
Antarctic ATCMs  Meetings Env.   Mgmt.           to other                     Agenda
Treaty           |        Protocol               Antarctic
                 |                                Sites
                 ├──────────┐
                 |          |
            Reports from   List of decisions
            previous ATCMs from previous ATCMs

                                                  ┌────┬────┬────┬────┬────┐
                                                  News  List  Meeting  Delegate  About the
                                                        of    Program  Handbook  Conference
                                                        Delegates                City
```

Norway considered the website for the XXII Consultative Meeting a success even though only a limited number of the submitted information papers were posted on the protected page of the website. One week in advance of the Meeting, only 15 of the 45 information papers that had been received were posted on the website. This limited posting was likely a result of the novelty of the concept and the fact that document distribution by Internet had not been formalised by the Consultative Meeting. Already at the following Consultative Meeting, held in Lima in 1999, the trend had changed significantly; in advance of that meeting almost all working and information papers were posted in all four Treaty languages as soon as translations were available.

The statistics[32] for the use of the Tromsø Meeting website show that there seems to be a large potential for the use of the WWW as a mechanism for distribution of information in the ATS. From the time of its conception in February 1998 until the start of the XXII Consultative Meeting in May the same year, approximately 1500 visitor sessions were registered for the website, averaging 14 visitor sessions per day. Not surprisingly, there was a peak in visitor sessions immediately before the start of the Meeting. Further examination of the statistics indicate that the website was visited in the main by delegates[33] who were searching for both practical and document information.

At the conclusion of the XXII Consultative Meeting, Norway updated the website by posting a number of additional working and information papers, having first cleared this with the parties that submitted the papers. The password protection was removed and the documents were thereby made available to the public in accordance with Resolution 5 (1998). In early September 1998, three months after the closure of the XXII Consultative Meeting, Norway forwarded the website to the Peruvian government for it to be used as the foundation for the website of the XXIII Consultative Meeting hosted by Peru in 1999 in Lima.

THE CEP WEBSITE

A year after proposing the establishment of a Consultative Meeting website in 1998, Norway introduced another proposal for placing the 'ATS on the Web'. The 1999 document discussed the operational aspects of the CEP website, which was established on a provisional basis by Norway as the country of the current Chair of the CEP.[34] Upon consideration of this initiative, the CEP 'recommended that Decision 1 (1999) be approved by XXIII ATCM'.[35] According to that decision,

---

[32] Statistics for the use of the ATCM website has been prepared by the NP utilising the statistical program WebTrends.

[33] Users from a large number of countries representing all continents accessed the website, although North America clearly stands out as the top geographic user region.

[34] See Norway, 'CEP Home Page', doc. XXIII ATCM/WP 26, 1999; also para. 5 of the 'Report of the Second Meeting of the Committee for Environmental Protection, Lima, 24–28 May 1999', in the

the home country of the Chair of the Committee for Environmental Protection (CEP) shall, within the resources available to it, and only for as long as it provides the CEP Chair, operate a CEP web site on the World Wide Web on an interim basis.[36]

This Decision introduced the CEP website as an 'internal organisational matter' of the ATS, rather than a practice which the parties may be 'encouraged to consider'.

Facilitating the Work of the CEP

The CEP meetings, which are normally held in conjunction with the Antarctic Treaty Consultative Meetings,[37] clearly face many of the same challenges as the Consultative Meetings with respect to document handling and distribution. The experience gained at the first and second meetings of the Committee indicated that the volume of papers to be considered by the Committee may become quite large. The first meeting of the CEP (CEP I) considered 12 working papers and 29 information papers, while the second meeting of the CEP (CEP II) considered 22 and 33, respectively. The Environmental Protocol's extensive provisions on exchange of information and submission of documents[38] make it reasonable to assume that in the future there will be as many, and quite likely more, papers submitted to the Committee. The host country for the CEP meeting is responsible for handling all documentation for the meeting. At the first meeting of the Committee, the Guidelines for Circulation and Handling of CEP Documents were adopted[39] stating, *inter alia*, that:

> All Working Papers prepared by Parties and Observers referred to in Rule 4-a and -b of the CEP Rules of Procedure and Information Papers which a Representative of a Party requests be translated, should be received by the Host Government no later than 75 days before the meeting. The Host Government should circulate these papers in translation no later than 60 days before the meeting.[40]

Final Report of the XXIII ATCM, Annex F (hereinafter Report of the CEP II).

[35] 'Report of the CEP II', para. 12.

[36] Para. 1 of Decision 1 (1999), 'CEP Web Site', in the *Final Report of the XXIII ATCM*, Annex B.

[37] Rule 9 of the CEP Rules of Procedure, published as Decision 2 (1998) in the *Final Report of the XXII ATCM*. Note that the Committee, with the agreement of the Consultative Meeting, and in order to fulfil its functions, may also meet between annual meetings. Furthermore, the Committee may establish subsidiary bodies, which could conduct separate meetings, operating on basis of the CEP Rules of Procedures (Rule 10). See further Orheim, Chapter 6 in this book.

[38] See, *inter alia*, Article 17, specifying provisions for annual reporting; Article 3 of Annex I, which specified that any draft Comprehensive Environmental Evaluation is to be submitted to the CEP; Article 9 of Annex III, which states that parties shall send copies of their waste management plans, and reports on their implementation and review, to the CEP.

[39] The Guidelines for Circulation and Handling of CEP Documents are referred to in Rule 13 of the CEP Rules of Procedure, and set out in Annex 3 to the Report of the CEP I.

[40] Para.1 of the Guidelines for Circulation and Handling of CEP Documents.

As for Consultative Meeting documents, there is a suggested length limit of 30 pages for information papers,[41] but no limit has been specified for working papers. Unlike the specifications for the submission of the Consultative Meetings documents, the Guidelines for circulation and handling of CEP documents clearly states that 'the submission and circulation of all documents should be done by electronic means whenever feasible'[42]. The CEP has consequently been given a foundation that should enable a more efficient distribution of documents than the traditional diplomatic exchange mechanism.

Other aspects of the operation of the CEP make it imperative that efficient mechanisms for communication and information exchange are established. For example, the Environmental Protocol specifies that information on activities undertaken in cases of emergencies shall be circulated to the Committee *immediately*.[43] In such cases, the traditional methods for dissemination of information through diplomatic channels may not be adequate. Furthermore, in addition to the formal meetings of the Committee, or of subsidiary bodies should such be established, the Committee may establish informal open-ended contact groups to examine specific issues intersessionally;[44] these contact groups report back to the Committee[45]. Such informal open-ended contact groups operate mainly through electronic communication (e-mail)[46] and will likely need access to the archive of past communications. At both CEP I and CEP II meetings, informal contact groups were established.[47] At CEP II, the meeting furthermore adopted procedures that allow the CEP Chair to establish, in-between sessions, contact groups for the consideration and review of any draft CEE submitted.[48] The fact that the CEP is operative in-between session indicates that the CEP will have to find suitable communication channels for conducting its work.

Already after only two meetings of the CEP, it has become clear that the Committee will require more sophisticated communication channels than the Consultative Meetings ever needed during the first 40 years of the operation of the ATS. The CEP will require both an efficient data management and an information

---

[41] *Ibid*.

[42] *Ibid*., para. 7.

[43] See, *inter alia*, Article 2 of Annex II to the Environmental Protocol. Similar provisions exist in Annexes III, IV and V.

[44] See discussion by Orheim, Chapter 6 in this book.

[45] CEP Rules of Procedure, Rule 9.

[46] See para. 9 of the Report of the CEP I.

[47] At CEP I, informal contact groups were established to prepare a draft guide on EIA procedures and to further consider issues related to the development of a State of the Antarctic Environment Report (see paras. 28 and 60 of the Report of the CEP I). At CEP II, contact groups were established to prepare an initial report on matters arising from the Workshop on Diseases on Antarctic Wildlife and to further develop issues related to Antarctic protected area management (see paras. 59 and 80 of the Report of the CEP II).

[48] The Guidelines for CEP Considerations of Draft CEEs are published as Annex 4 to the Report of the CEP II.

exchange system. The CEP must have efficient tools available for the distribution of documents in advance of a meeting, for easy access to documents from previous meetings, for the exchange of information required by the Environmental Protocol, and for easy access to documents related to intersessional work. In light of these needs, at the CEP I meeting in Tromsø, Norway as the home country of the CEP Chair offered to perform the following:

> ... to develop and establish a temporary CEP Home Page, to be ready for use in advance of the next meeting of the Committee. Furthermore, Norway would provide the CEP, at its next meeting, with a working paper discussing various aspects of the operations of a CEP Home Page that would need to be clarified before a permanent CEP Home Page is established.[49]

In December 1998, the Norwegian Polar Institute, the home organisation of the CEP Chair, established the CEP website.[50] The CEP website was to a large degree directed toward the CEP Members only, the main aim being to establish a CEP database and an efficient data management system for the operation of the CEP. The parts of the CEP website related to the on-going operation of the CEP are username and password protected, restricting access to the site to members only.

After having established the website and having operated it over a period, Norway submitted a working paper to CEP II[51] that considered the unresolved issues related to the operation of the CEP website. Issues considered included the questions of the costs for operating and maintaining the website, the language, and who should have access to the restricted areas of the website. Although some issues, such as the language issue,[52] remained unresolved,[53] the Consultative Parties adopted Decision 1 (1999), by which it was decided that the home country of the CEP Chair shall, within the resources available to it, operate a CEP website on an interim basis. It was specified that the website should, *inter alia*, contain the following:

a) in an area freely accessible to the general public:
- general information about the Committee on Environmental Protection and environmental issues in Antarctica;
- an archive of official documentation from the previous meetings of the Committee, containing Working and Information Papers submitted to its meetings, as well as the final reports of its meetings;
- links to related web sites (e.g. SCAR, COMNAP, CCAMLR)

---

[49] See para. 24 of the Report of the CEP I.

[50] The NP operated the website on the address <www.npolar.no/cep>.

[51] See Norway, 'CEP Home Page'.

[52] See para. 6 of the Report of the CEP II: 'The present situation of the CEP web site not carrying its material in all four official languages of the Antarctic Treaty System should be considered an interim arrangement'.

[53] It is acknowledged that most of these issues will remain unresolved until a permanent secretarial structure is in place for the ATS. CEP II noted that, ideally, the website should be operated by the Antarctic Treaty Secretariat, should it be established (see para. 5 of the Report of the CEP II).

b) in a password protected area accessible only to the Members of the CEP, Observers to the CEP, and other experts as appropriate that the Committee decides should have such access [sic]:

- official documents submitted electronically to the Host Country and the CEP Chair in advance of a CEP Meeting;
- any other documents that have been provided to the CEP Chair for consideration at the Meeting.[54]

The Meeting supported the need for an official contact point to be designated in each country to regulate access and the submission of information and documents to the CEP website.[55] An official list of CEP Contact Points was established to further facilitate electronic communication among CEP Members.[56]

Design and Practical Usage of the CEP Website

At the time of the establishment of the CEP website, it was decided to use an uncomplicated structure and design in order to minimise the resources needed to maintain and operate the website. This should be beneficial for other CEP members that will in turn operate and maintain the website upon the election of a new CEP Chair.

The structure of the website (Figure 2) reflects the specifications of Decision 1 (1999) with respect to the content of the site. The website has one gateway to general information about the Environmental Protocol and the CEP. This gateway is specifically aimed at the general public with the intention of creating a greater awareness about the work of the Committee and environmental protection in Antarctica in general. The website also has a gateway to an archive of documents and reports from previous meetings. Both members and the interested public have easy access to the documents, which also thus enhances the transparency of the operation of the CEP. The CEP website has a third gateway to a member-restricted site containing, *inter alia*, documents submitted to the next CEP meeting and any document relevant for on-going intersessional work. The CEP website also contains links to other relevant Antarctic websites, including the website for the Consultative Meeting, provided this is established by the host country and in operation.

The statistics[57] for the use of the CEP website show that use of the Internet has excellent potential for the operation of the Committee. During the first year of its existence the website had approximately 3000 visitor sessions, averaging 7 visitor sessions per day. Immediately before the start of the CEP meeting in Lima, May 1999, there was a boost in the number of visitor sessions, but following the meeting

---

[54] Para. 3 of Decision 1 (1999).
[55] Para. 7 of the Report of the CEP II.
[56] See Annex 2 to the Report of the CEP II.
[57] Statistics have been prepared by the NP utilising the statistical website program WebTrends.

The ATS on the Web 157

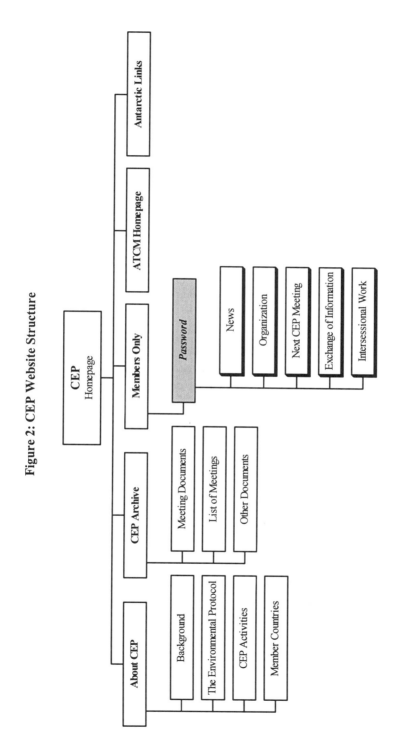

Figure 2: CEP Website Structure

to the end of March 2000, the number of visitor sessions increased even further.[58] Closer examination of the statistics shows that all the underlying pages of the website seem to have also drawn visitors and that the CEP reports are the files most downloaded. Visitors from the United States presently comprise the most active user group of the website, but the statistics also show a large number of visitors from other countries.

Parallel to the establishment of the CEP website, the Norwegian Polar Institute also established an e-mail address to which any website visitor could send queries. Judging by the number and types of queries received, it can be deduced that the CEP website has a potentially large user audience in a segment of the public with interests in the operation of the Antarctic Treaty System, the Committee for Environmental Protection and the Antarctic issues in general. As such, the CEP website seems to have potential both as a communication mechanism for the Committee itself and as an information resource for a public interested in Antarctic affairs in general.

FINAL REMARKS

Some final remarks about prospects for future developments are added here, based on the experiences gained from the use of websites for the Consultative Meeting and the CEP. Modern information technology and Internet communication have already afforded beneficial assistance in the operation of the ATS, not least while it still lacks a permanent secretarial infrastructure. Once established, the Antarctic Treaty Secretariat would most likely assume the maintenance of the websites for the Consultative Meeting and the CEP and perhaps also integrate these into an overall ATS website. However, it may still be some time before a Secretariat is established; at present it is indeed impossible to predict with certainty whether, where and when an Antarctic Treaty Secretariat will be established.

Meanwhile, it would seem to be of joint interest to the individual host governments and all the Consultative Parties that the resources needed in the preparation of a Consultative Meeting be rationalised in order to maximise the efficiency of meeting arrangements.

The experience thus far derived from the establishment and operation of the CEP website indicate that the adoption of the CEP website through Decision 1 (1999) has had positive results, and the gains for the host government and the ATS as a whole outweigh the burden of maintaining the website. However, the continuation of the Consultative Meeting website and its exact contents is at the discretion of each subsequent host government.[59] Worth noting is an operational

---

[58] The statistics available in the time of finalisation of this chapter are up-dated only as of end-March 2000.

[59] Compare the differences between the contents of the websites for the Tromsø (1998) and Lima (1999) Consultative Meetings.

aspect of this disparity between the websites for the CEP and the Consultative Meeting. At the CEP II meeting, 'members noted the need for a close link between the ATCM web site and the CEP web site, and that there should be no inconsistency between the two'.[60] Decision 1 (1999) requires that the CEP website 'be operated in close co-operation with the Host Country of the ATCM operating the ATCM Home Page'.[61] However, whereas the operation of the CEP website is required as an 'internal organisational matter', operation of the Consultative Meeting website is based on no more than 'encouragement' to the host country to 'consider' establishing it.

Pending the establishment of the Secretariat, it seems in the collective interest of responsible and efficient ATS operation to enhance the status of the Consultative Meeting website from a volunteer action to an internal organisational matter, based on a Decision. This would guarantee continuity in the preparation of the Consultative Meetings and would help avoid the need for each subsequent host government to go through the process of preparing for the meeting 'from scratch'.

---

[60] Para. 6 of the Report of the CEP II.
[61] Para. 2 of Decision 1 (1999).

Part III

NORMATIVE SUPPORT TO THE IMPLEMENTATION OF THE PROTOCOL: AN ANTARCTIC LIABILITY REGIME

9

Liability Annex or Annexes to the Environmental Protocol: A Review of the Process within the Antarctic Treaty System

Mari Skåre

Under Article 16 of the Protocol on Environmental Protection to the Antarctic Treaty,[1] the parties are obliged to elaborate rules and procedures relating to liability for damage arising from activities taking place in the Antarctic Treaty[2] area, with the view to adopting one or more annexes to the Protocol on such rules and procedures.[3] The XVII Antarctic Treaty Consultative Meeting in Venice in 1992 convened a Group of Legal Experts to undertake such elaboration. The Group presented its report to the XXII Consultative Meeting in 1998 and thereafter was dissolved. The first round of negotiations on the issue of liability for environmental damage was held at the XXIII Consultative Meeting in 1999.

The purpose of this chapter is to give an overview of the process to date of developing rules and procedures on liability under the Protocol and of the major outstanding issues involved.

In developing rules on liability for environmental damage in international law in general, the focus has primarily been on civil liability for high-risk activities, such as transport of oil and other hazardous substances and nuclear related activities. Progress has, however, been slow because of the complex inter-relationship of technical, legal and policy considerations. Antarctica is extremely remote from populated areas. Consequently, the continent and the surrounding ocean provide an ideal template for scientific studies, particularly those related to global environmental changes. The political regime governing the Antarctic is centred on

---

[1] ILM, Vol. 30, 1991, pp. 1,461ff.
[2] UNTS, Vol. 402, pp. 71ff.
[3] On the nature of the obligation contained in Art. 16, see discussion by Lefeber, Chapter 10 in this book.

international scientific cooperation. The management regime remains delicately balanced given that the sovereignty questions in the Antarctic are not resolved. In this respect, the crucial provision of the Antarctic Treaty is Article IV, whereby the parties basically agree to disagree regarding sovereignty claims. Thus, this huge continent is co-managed by states that claim territorial rights, which in some cases overlap, in the Antarctic and by non-claimant states that do not recognise these claims, including the USA and Russia who assert that they have the basis for a claim to sovereignty in the Antarctic. In this complex situation, questions concerning jurisdiction cause specific concerns for the Antarctic Treaty parties.

The issue of liability for environmental damage in the Antarctic is one of these concerns. Moreover, the absence of any decision-making bodies other than the Consultative Meeting and of a permanent secretariat which can be available to attend to administrative matters magnifies the challenge of developing rules on liability. Because of this situation, the technical, legal and policy issues that need to be solved regarding liability in the Antarctic context are even more complicated than under other international regimes.

TYPES OF HUMAN ACTIVITIES CARRIED OUT IN THE ANTARCTIC AND THEIR IMPACT ON THE ENVIRONMENT

The principal environmental concerns in the Antarctic relate to changes occurring at the global level (for example depletion of the ozone layer and climate change) rather than changes originating from human activities within the region itself. Such global changes and their potential impact on the Antarctic are not addressed by the Protocol, which is a regional instrument.

To date no comprehensive risk-analysis regarding the potential environmental impact caused by human activities taking place in the Antarctic Treaty area has been undertaken. The 1998 Consultative Meeting requested the Council of Managers of National Antarctic Programmes (COMNAP) to undertake an assessment of the risks of environmental emergencies arising from activities in Antarctica. COMNAP's report was delivered at the 1999 Consultative Meeting, and concluded that there are only minor risks for such environmental emergencies.[4]

Major types of human activities in Antarctica relate to 1) fishing, 2) science and environmental monitoring, 3) support and logistic activities, and 4) tourism. Mineral activities are prohibited under the Protocol (Article 7). Likewise, commercial whaling is held under a moratorium under the 1946 International Convention for the Regulation of Whaling,[5] whilst commercial sealing is regulated and *de facto* prohibited by the 1972 Convention for the Conservation of Antarctic Seals[6]. The

---

[4] COMNAP, 'An Assessment of Environmental Emergencies Arising from Activities in Antarctica', doc. XXIII ATCM/WP 16, 1999.
[5] UNTS, Vol. 161, pp. 74ff.
[6] UNTS, Vol. 1080, pp. 175ff; and ILM, Vol. 11, 1972, pp. 251ff.

dominant human activity in the late 1800s and though the first half of the twentieth century was whaling, an activity that caused the near extinction of some species and subsequent major perturbations in the marine environment. There are no reports of commercial sealing and whaling taking place in the Antarctic Treaty area today.

Fishing

Fishing in sub-Antarctic and Antarctic waters is governed by the Convention on the Conservation of Antarctic Marine Living Resources[7] (CCAMLR) whose area of application is greater than the Antarctic Treaty itself – extending northward approximately to the Antarctic Convergence (CCAMLR, Article I). CCAMLR was negotiated in the late 1970s to bring unregulated fishing under control. The Commission under CCAMLR has adopted numerous conservation measures, in order to deal with all commercial, new and exploratory fisheries. In the 1996/97 season, fisheries targeted krill, squid and Patagonian toothfish, with the latter accounting for 97 per cent of the 10,562 tonnes of finfish taken. However, despite these regulations, illegal and unregulated fishing has re-emerged as a serious problem. The problem is two-fold, illegal fishing carried out by vessels of CCAMLR parties and unregulated fishing by non-CCAMLR, i.e., third–party vessels.[8]

Not only is the stock of Patagonian toothfish under threat but populations of seabirds (Albatrosses and Petrels) are also in jeopardy. These birds are caught and killed incidentally during long-line operations. In 1998 an estimated 50,000 to 89,000 birds were killed as a result of fishing.[9]

Fishing activities under CCAMLR are outside the scope of the Protocol, and thus so is liability under the Protocol resulting from these activities. It is, however, apparent that fishing is currently creating the greatest impact to the Antarctic environment.

Science and Environmental Monitoring

The Antarctic flora and fauna, characterised by a low number of species and a high number of populations, attracts scientific interest. The continent of Antarctica and the surrounding Southern Ocean play a critical role in the global environmental system and are particularly appropriate for monitoring and studying climate change.

---

[7] ILM, Vol. 19, 1980, pp. 837ff.
[8] See a brief overview in D. Vidas, 'Emerging Law of the Sea Issues in the Antarctic Maritime Area: A Heritage for the New Century?', *Ocean Development and International Law*, Vol. 31, 2000, pp. 201–205.
[9] CCAMLR, *Report of the Seventeenth Meeting of the Commission, Hobart, Australia, 26 October – 6 November 1998* (Hobart: CCAMLR, 1999), para. 6(21).

As local contamination in the Antarctic is very low, the area is an excellent laboratory for monitoring and studying long-range pollutants.

Today science and its logistic support activities comprise the predominant human activity in the continent of Antarctica. In 1997 there were 42 year-round research stations in Antarctica and, in addition, some 40 summer stations. The summer population (January) was approximately 4,115 persons.[10] Since 1997, the number of year-round stations has been declining.

The science activities are taking place in the continent of Antarctica, which is slightly less than 1.5 times the Europe (over 14 millions square kilometres). The highest concentration of research activities and permanent stations is on the Antarctic Peninsula, which is not covered with ice all year round. This land area is roughly 200,000 km^2.

Support and Logistic Activities

Of the activities that fall within the scope of the Protocol, it is not science itself that causes the greatest impact on the Antarctic environment today, but rather the logistic support to scientific activities. Logistic support activities are the primary cause for generating waste; they result in fuel spills, increase the levels of traffic by aircraft, vessels, vehicles and personnel, and create the need for infrastructure on ice and land. As there are no ports and harbours in Antarctica, the ships have to anchor offshore. In 1997, there were 42 landing facilities, ten of which have runways – mostly on the ice.[11]

COMNAP's report on environmental emergencies submitted to the latest, Twenty-third Consultative Meeting in 1999 states that the most common type of logistic support incident over that past 10 years involves liquid spills, mainly of petroleum products. The majority of the spills were relatively small (under 1,000 litres) and occurred on land. COMNAP concludes that most spills will be small and unlikely to threaten any wildlife. COMNAP further concludes that although fuel spills in the marine environment have a low probability, they pose a greater magnitude of threat than terrestrial or ice-sheet spills.

In 1989, an Argentine ship (*Bahia Paraiso*) carrying both supplies and tourists ran aground in Arthur Harbour and caused a spill of some 600,000 litres of diesel fuel in the Southern Ocean. An immediate recovery of 65,000 litres was effected, and two years after the spill, approximately 148,500 litres were pumped out from the tanks of the ship. The spill had immediate effects on the local sea and wildlife, killing penguins and krill.[12] However, seven years after the accident, it appeared that the majority of the species had recovered.[13]

---

[10] Central Intelligence Agency, *Yearbook 1997*, available at <www.us.gov/cia>.
[11] *Ibid.*
[12] Environmental Defence Fund web site, 1989 Letter, available at <www.edf.org>.
[13] COMNAP, 'An Assessment of Environmental Emergencies Arising from Activities in Antarctica';

Open burning of garbage and dumping of waste has taken place in the past, but is now prohibited by the Protocol. COMNAP reports that local air quality is not impacted significantly by normal station operations. There have, however, been some transport-related accidents where it has been impossible to remove a vehicle or aircraft from the site of the accident.

The vast majority of research programmes or projects are conducted in cooperation between two or more states, *inter alia*, to ensure efficient use of logistic support.[14] Resources are pooled to allow joint transportation of goods and personnel to Antarctica as well as the joint use of pipelines and storage facilities. These cooperative efforts also extend to response actions in cases where environmental damage threatens to occur, as in the case of the *Bahia Paraiso* spill.

Tourism[15]

Tourism in the Antarctic is mostly concentrated around the South Shetland Islands and the western side of the Antarctic Peninsula during the austral summer. The overwhelming majority of the tourism is ship-born (around 90 per cent) and the actual time spent ashore by tourists is very limited. Ship-born tourism to Antarctica increased in the 1990s. During the austral summer season of 1997–98, approximately 9,400 persons travelled to Antarctica on touring vessels. One of the landing facilities in Antarctica is run by a commercial tourist organisation. Over-flights carrying tourists is a recent re-addition in the tourist business; one airline flew 3,146 passengers over Antarctica in the 1997–98 season, only a small decline from the previous season. Land-based tourism is extremely modest. There is, however, one lodging facility located in Chile's base at King George Island that may accommodate tourists.

Tourism has the potential for environmental impact particularly from a shipping accident. However, the few shipping accidents that have occurred in the Antarctic, and that released pollutants into the sea, do not seem to have resulted in any lasting environmental damage. Accidents involving passenger ships may, however, cause severe personal loss and injury.

---

also Untied States, 'Implications of the Current Draft Liability Annex to Activities among Treaty Members', doc. XXI ATCM/IP 104, 1997.

[14] COMNAP 'Information Paper on Scientific and Operational Co-operation', doc. XXII ATCM/IP 7, 1998.

[15] All data used in the preparation of this section has been collected from the website of the International Association of Antarctic Tour Operators, at <www.iaato.org> and from IAATO 'Information Paper to ATCM XXII', doc. ATCM XXII/IP 88, 1998. For further details see Richardson, Chapter 4 in this book.

REGULATING NON-EXISTING ACTIVITIES: LIABILITY FOR ENVIRONMENTAL DAMAGE UNDER CRAMRA

Under the Convention on the Regulation of Antarctic Mineral Resource Activities[16] (CRAMRA), the Consultative Parties had constructed a management regime for any mineral resource activity in the Antarctic. However, when France and Australia decided in 1989 not to sign CRAMRA, and New Zealand signed but decided not to ratify the Convention, its entry into force was effectively blocked.

A basic consideration for the negotiators was the protection of the Antarctic environment and its dependent and associated ecosystems.[17] The result of negotiations was a carefully balanced decision-making system, establishing a new set of institutions as well as laying down environmental rules and standards.[18] The merits of CRAMRA have, however, been disputed. While some view it as a negotiating success, setting up a regulatory regime to ensure environmental protection,[19] others have characterised it as a disaster to the environment if ever put in force.[20]

In assessing the merits of the rules on liability under CRAMRA and the value of transferring such rules to a liability regime under the Protocol, it is important to note that the scopes of application of these two instruments differ substantially. The Protocol addresses activities in the Antarctic in general, which – apart from some tourism – today basically relate to science and associated logistic activities conducted for the benefit of mankind. CRAMRA, on the other hand, limited its attention to mineral resource activities. The potential environmental impacts from these types of activities vary considerably.

During the negotiations of CRAMRA, the item of liability was discussed in a legal contact group chaired by Professor Rüdiger Wolfrum. Liability proved to be a difficult issue, and it remained unsolved until the very final session of the Special Consultative Meeting.

Article 8 of CRAMRA provides obligations on the operator to take response action if any activity threatens to result in environmental damage. The standard of

---

[16] ILM, Vol. 27, 1988, pp. 868ff.

[17] On the basic considerations see: R.T. Andersen, 'Negotiating a New Regime: How CRAMRA Came into Existence', in A. Jørgensen-Dahl and W. Østreng (eds.), *The Antarctic Treaty System in World Politics* (London: Macmillan, 1991), pp. 94–109; C.C. Joyner 'The Effectiveness of CRAMRA', in O.S. Stokke and D. Vidas (eds.), *Governing the Antarctic: The Effectiveness and Legitimacy of the Antarctic Treaty System* (Cambridge University Press, 1996), pp. 152–173; and *Final Report from the Eleventh Antarctic Treaty Meeting, Buenos Aires, Argentina, 23 June – 7 July, 1981* (Buenos Aires: Argentinean Ministry of Foreign Affairs, 1981), Recommendation XI–1, para. 5(c).

[18] For more information on CRAMRA see Joyner, 'The Effectiveness of CRAMRA'.

[19] R.T. Scully, 'The Antarctic Treaty as a System', in R.A. Herr, H.R. Hall and M.G. Haward (eds.), *Antarctica's Future: Continuity or Change?* (Hobart: Tasmanian Government Printer, 1990), pp. 95–102.

[20] J.N. Barnes, 'Protection of the Environment in Antarctica: Are Present Regimes Enough?', in

liability is strict, but the operator may invoke defences (e.g., natural disasters, and armed conflicts). States may only be held liable for damage that would not have occurred or continued if the sponsoring state had carried out its obligations. Such liability is limited to that portion of liability not met by the operator. [21]

The parties decided that further elaboration of liability under CRAMRA should take place through the development of a separate protocol. This protocol should have, *inter alia*, covered issues such as limits, claims tribunal and a fund. No application for a mineral exploration or development permit would be permitted until such a protocol had entered into force. At the time of the negotiations there was, however, no commercial interest in mineral activities. Rolf Trolle Andersen, the chairman of numerous international meetings under the ATS, describes the driving forces behind the decision to initiate the CRAMRA negotiations:

> It was believed that precisely the *lack* of knowledge regarding possible occurrences of minerals and the *absence* of a commercial interest in activity might make it possible to reach an agreement in this very complicated field.[22]

LIABILITY UNDER THE PROTOCOL

The Protocol supplements the Antarctic Treaty and neither modifies nor amends the Treaty that applies to the area south of 60° south latitude, including all ice shelves (Article VI). Nothing in the Treaty prejudices the rights of any state under international law with regard to the high seas within that area.

In the Final Act of the Eleventh Special Consultative Meeting, the parties stated that nothing in the Protocol shall derogate from the rights and obligations under CCAMLR, the Seals Convention and the International Whaling Convention – this formulation being understood that activities undertaken within the Antarctic Treaty area regulated by those conventions are not covered by the Protocol.

Article 16 refers only to activities taking place in the Antarctic Treaty area *and covered by the Protocol*. This means that fishing activity regulated by CCAMLR, which perhaps is the activity causing the greatest impact on the environment, falls outside the scope of rules on liability under the Protocol. Associated activities, such as pollution arising from fishing vessels, may, however, be included, since such activities are not regulated by CCAMLR.

At the start of the negotiation of the Protocol, it was held that rules on liability would constitute an important element in the environmental regime for the Antarctic. However, the actual shape and content of the regime on liability was not

---

Jørgensen-Dahl and Østreng (eds.), *The Antarctic Treaty System in World Politics*, pp. 186–228.

[21] For more details on the solutions in CRAMRA, see R. Wolfrum 'The Unfinished Task: CRAMRA and the Question of Liability', in Jørgensen-Dahl and Østreng (eds.), *The Antarctic Treaty System in World Politics*, pp. 120–132.

[22] R.T. Andersen, 'Negotiating a New Regime: How CRAMRA Came into Existence', p. 96.

clear. There were differing views as to how strict liability should be, and the parties realised that developing rules on liability would be a challenging and time-consuming process.

The failure of CRAMRA provided the Consultative Parties with the need to demonstrate to the 'outside world' their ability to manage the Antarctic in a responsible manner. The Parties were under pressure to complete the negotiations of the Protocol quickly, even though it meant cutting some corners. The result was a Protocol providing general environmental standards but weak institutional structures. The Protocol relies for a large part on national implementation.[23] The matter of liability was, in Article 16 of the Protocol, again postponed for future elaboration.

In international cooperation, consensus is often sought and found using so-called 'constructive ambiguity' in the drafting of documents. Disagreements that prove difficult to bridge may be set aside for later considerations. Ambiguities provide room for different interpretations. Different views on the obligations embedded in the Protocol have already surfaced. There are for example different opinions as to whether and to what extent the Protocol requires the parties to undertake remedial measures if environmental harm occurs.[24]

At the final session of the Eleventh Special Consultative Meeting in Madrid, October 1991, the Consultative Parties underlined their commitment to elaborate rules and procedures relating to liability; they also expressed their wish that the elaboration could begin at an early stage. Since the entry into force of the Protocol, implementation and enforcement has become a priority. Unfinished tasks may now begin to receive increased attention. One such task is the obligation, embedded in Article 16, to elaborate rules and procedures related to liability.

THE GROUP OF LEGAL EXPERTS ON LIABILITY: GOALS AND OUTCOMES

The XVI Consultative Meeting, held in 1991, received a report from the Eleventh Special Consultative Meeting on the negotiation of the Protocol. The 1991 Consultative Meeting considered that the next Consultative Meeting should decide on how and when the matter of liability under the Protocol should be dealt with.[25]

That Consultative Meeting, held in Venice in 1992, agreed to convene a Group of Legal Experts to undertake the elaboration of rules and procedures relating to liability as called for under Article 16 of the Protocol. However, no time limit for

---

[23] See chapters in Part V of this book.
[24] 'Liability – Report of the Group of Legal Experts', doc. XXII ATCM/WP 1, 1998, para. 25. See also United States 'Negotiation on an Annex or Annexes on Liability', doc. XXII ATCM/IP 126, 1998.
[25] *Final Report of the Sixteenth Antarctic Treaty Special Consultative Meeting, Bonn, Germany, 7–18 October 1991* (Bonn: German Federal Ministry of Foreign Affairs, 1991).

achieving a final result was set. Although the Meeting received substantial input from several delegations (especially from the Netherlands, Chile and Germany), it did not offer any policy guidance on the work to be undertaken by the Group of Legal Experts, except a reference to Article 16 of the Protocol. Thus, the purpose and the intended effect of the future rules on liability were not made clear. To ensure adequate preparation for the work, proposals were to be prepared and exchanged through diplomatic channels.[26] The Group's Chairman, Professor Wolfrum, convened the first meeting of the Group in November 1993. To prepare for this meeting, he circulated a questionnaire to which most Consultative Parties responded.[27]

The Group of Legal Experts met nine times during the period 1993–98,[28] in conjunction with Consultative Meetings and inter-sessionally. In addition, there were informal meetings between parties. The deliberations took place on the basis of draft 'offerings' prepared by the Chairman. Starting with the 1995 meeting of the Group in Brussels, drafting suggestions from legal experts taking part were included more systematically in the subsequent fifth, sixth, seventh and eighth 'Offerings'.[29] At the 1996 meeting of the Group in Utrecht, the USA tabled an alternative draft on liability which followed a less comprehensive approach than that of the Chairman's 'Offerings'.[30] The US text focussed only on liability that would result from a failure by an operator or party to carry out appropriate emergency response action.

The Group as such was never in a position to put forward or recommend a draft negotiating text or texts simply because it was never possible to attain agreement on key issues. Throughout the period during which the Group met, the Chairman provided oral reports on the progress to the Consultative Meetings. Except for the report prepared for the XXII Consultative Meeting in 1998[31] and the recordings of the Consultative Meetings from 1994 and thereon, there are no public reports or records from the Group's deliberations. However, the Chairman's summaries have been circulated to the members of the Group.[32]

---

[26] *Final Report of the Seventeenth Antarctic Treaty Consultative Meeting, Venice, Italy, 11–20 November 1992* (Rome: Italian Ministry of Foreign Affairs, 1993).

[27] The questionnaire is contained in doc. XVIII ATCM/WP 2, 1994, Annex I; text reproduced in W.M. Bush (ed.), *Antarctica and International Law: A Collection of Inter-State and National Documents* (Dobbs Ferry, NY: Oceana Publications, 1992–), Binder II, Booklet AT92C, pp. 83–85.

[28] The meetings of the Legal Expert Group were held: 18–20 November 1993 in Heidelberg (Germany), 11–15 April 1994 in Kyoto (Japan), 7–9 November 1994 in The Hague (the Netherlands), 8–12 May 1995 in Seoul (Republic of Korea), 27–30 November 1995 in Brussels (Belgium), 29 April – 3 May in Utrecht (the Netherlands), 7–11 October 1997 in Cambridge (United Kingdom), 19–23 May 1997 in Christchurch (New Zealand) and 27–29 May 1998 in Tromsø (Norway).

[29] The last, the eighth, of the 'offerings' is annexed to the 'Liability – Report of the Group of Legal Experts'.

[30] United States, 'Annex VI to the Protocol on Environmental Protection to the Antarctic Treaty – Liability for Emergency Response Action', doc. XX ATCM/IP 43, 1998; text also annexed to 'Liability – Report of the Group of Legal Experts'.

[31] 'Liability – Report of the Group of Legal Experts'.

[32] At the 1998 Consultative Meeting only a limited set of issues were discussed by the Group. The

As mentioned, the Group prepared a written report to the XXII Consultative Meeting in 1998, on results achieved, major outstanding problems, alternative approaches and key issues to be considered by the Consultative Parties. That report covered issues such as the scope of application, the question of uniform rules for all activities, definition of damage, standard of liability, imposing liability on operators, response action and remedial, irreparable damage, exemptions from liability, insurance, limits to liability, environmental protection fund, dispute settlement, and procedural matters regarding future approaches.

The Goal and How to Get There

The Group's report of 1998 states that 'there is agreement that the aim of an annex or annexes is a liability regime which would cover all categories of activities undertaken in the Antarctic Treaty area and which are covered by the Protocol'.[33] The report reveals, however, considerable disagreement on how exactly this goal should be achieved.

The approach taken by the Group of Legal Experts has been the so-called 'comprehensive approach', whereby all the elements of a liability regime should be included in one instrument, i.e., a single annex to the Protocol rather than multiple annexes. However, a common understanding of the substantive implications of the pursuit of the 'comprehensive approach' was never developed. It was argued that this approach would provide for a strict liability regime and an expedient result, while it was feared that a piecemeal approach could result in a weak liability regime. Some parties were of the view that due to the complexity of the issue of liability, a 'step-by-step' approach, dividing the issues into two or more annexes, would be more appropriate. As mentioned, the USA has to that effect tabled a proposal to an annex on liability for emergency response action.[34] The US has criticised the 'comprehensive approach' as a means of redefining the Protocol and then attaching liability to newly created obligations. Others have held that the US proposal does not meet the parties' obligations under Article 16 of the Protocol. Consequently, it has been held that it was not within the mandate of the Group to discuss the US proposal.

The parties' differing views regarding the content of the obligations provided for by the Protocol are reflected in the Final Report of the XXII Consultative Meeting held in Tromsø, Norway. These views address the matters as to whether or

---

issues addressed at the Group's meeting in Tromsø were 1) the definition of operator and how liability may be imposed on an operator, 2) joint and several liability, 3) insurance, and 4) dispute settlement; see 'Summary from the Meeting of the Group of Legal Experts on Liability', 27–29 May 1998 (document on file with the author).

[33] 'Liability – Report of the Group of Legal Experts', para. 7.

[34] United States, 'Liability for Emergency Response Action'.

not an annex on liability should contain obligations for the operator to take 1) precautionary measures, 2) response action, or 3) remedial measures:

> Some Delegations considered that at least some of those obligations are not covered by the Protocol and that elaboration of such obligations should not be included in an annex or annexes on liability. The greatest concern reflected inclusion of obligations to take remedial measures, to which some Delegations had significant objections. Other Delegations disagreed and concluded that all three issues should be included in an annex on liability. Some Delegations, while speaking in favour of elaboration rules on these issues, preferred to deal with them in a separate annex.[35]

Those favouring the 'comprehensive approach' from the outset of negotiations (Germany, the Netherlands, New Zealand, Australia, and France) have also stressed the need to move forward expeditiously. However, as time has passed and the Group of Legal Experts has made little progress, the voices calling for a more pragmatic, feasible approach have become louder and stronger.

Some parties (articulated by Italy) suggested at the 1998 Consultative Meeting that in future deliberations initial focus should be given to environmental damage resulting from activities *in violation* of the Protocol.[36]

Salient Questions Raised by the Group in 1993–98 and Reported as Key Issues in the Group's Report to the XXII Consultative Meeting in 1998

Scope of Application. Before the XVII Consultative Meeting decided in 1992 to convene a Group of Legal Experts, a number of parties pointed out basic considerations that would have to be made in developing rules on liability for environmental damage in the Antarctic. One such statement came from Germany:

> Of course existing rules on international law on liability and experiences with these instruments will have to be taken into our consideration. However, prior international conventions on liability, especially on international liability, are dealing with certain special activities and risks. Thus, it seems to us necessary to examine very well whether – with view to the specific situation and needs of Antarctica – these conventions offer adequate solutions or whether and to what extent special rules will have to be developed.[37]

The questionnaire sent to the parties in preparation for the first meeting of the Group of Legal Experts was accompanied by extensive background information including copies of conventions and other documents that might be of relevance for

---

[35] *Final Report of the Twenty-second Antarctic Treaty Consultative Meeting, Tromsø, Norway, 25 May – 5 June 1998* (Oslo: Royal Norwegian Ministry of Foreign Affairs, 1998), para. 75.
[36] See also discussion by Lefeber, Chapter 11 in this book.
[37] 'Statement on Basic Elements for an Annex on Liability', submitted by Germany, 16 November 1992, para. 3; reprinted in Bush (ed.), *Antarctica and International Law* (1992–), Binder II, Booklet AT1992A, p. 114.

the Group's work. No further analysis or study, as called for in the German statement has, however, ever been made.[38] Existing international rules on liability were not discussed extensively in the Group of Legal Experts and their relevance was not commented on in its 1998 report to the XXII Consultative Meeting.

The Chairman's January 1994 summary of the responses to the questionnaire pointed out that there is disagreement as to whether or not the liability regime should cover loss or damage to persons and property.[39] The Group did not preclude the possibility of including such damage, but decided at that stage to concentrate its deliberations on damage to the environment.

Another issue not discussed extensively by the Group was whether or not the environmental harmful impact of *lawful activities* should attract liability. Some parties believed that no liability should attend environmental impacts incurred from permitted activities. Others, however, have held that not *all* permitted activities should be exempted from liability; in deciding whether or not the activity should incur liability, such exclusion should depend on the type of activity and by whom the activity was being undertaken.

Uniform Rules for All Types of Activities. A key question raised by the Group of Legal Experts was whether all activities in the Antarctic Treaty area should be addressed by an annex on liability in a uniform way, or whether scientific and associated logistic activities should be given preferential treatment. In addition, it has been asked what considerations should be given to any effects of a liability regime on cooperation among parties or their national programmes.

The Protocol designated the Antarctic as a 'natural reserve devoted to peace and science'. The necessity to take into consideration the fact that scientific research is a desired and valuable task has been expressed by many parties, for instance by the Netherlands in 1992:

> It seems appropriate to recognise that certain activities are central to the operation of the Antarctic Treaty System (ATS) and are of global importance (i.e. scientific research), while others are of fundamental different nature (for example tourism) ... This is to be reflected in a future liability regime.[40]

Considerable concerns have been raised as to how a liability regime would affect science activities. At the meeting of the Group of Legal Experts in Cambridge in 1996, the Scientific Committee on Antarctic Research (SCAR) made the following statement:

> SCAR's overall concern is that the latest draft ... deals with science on the same basis as activities which aim to utilise the Antarctic for commercial gain ... However, it is a critical issue for SCAR that the practical operation of any future liability annex should not

---

[38] See discussion by Lefeber, Chapter 10 in this book.
[39] See text reprinted in Bush (ed.), *Antarctica and International Law* (1992–), Binder II, Booklet AT94A, pp. 1–4.
[40] *Ibid.*, Binder II, Booklet AT92B, pp. 1–3.

unnecessarily prejudice the pursuit of scientific knowledge. ... SCAR considers it important that once a research permit has been granted, based on an accepted EIA procedure, this research activity can be pursued from a secure legal position.[41]

Scientists have in particular expressed concern regarding the possible effect of a liability regime on cooperative scientific programs or activities and response actions to environmental emergencies. Because of the remoteness from larger emergency facilities, the personnel in the Antarctic will typically provide for cooperative response actions such as were undertaken during the *Bahia Paraiso* spill and in other emergency situations.[42]

The approach followed thus far has been to elaborate uniform rules and procedures on liability in relation to all activities covered by the Protocol. The prevailing view expressed has been that the basic issue is whether damage has been caused *to* the environment, not whether damage has been caused *by* a particular type of activity. Provided that concerns raised by the scientific community are met, uniform rules may, however, result in less strict norms for all activities, including commercial activities carrying higher risks for environmental damage than the present science activities.

The Definition of Damage – Exemptions. The question on how to define damage has been subject to protracted discussions. There are different views on both the formulation of the general threshold and on exemptions (excuses). The issue raised for discussion at the XXII Consultative Meeting in 1998 was whether environmental impacts resulting from activities found acceptable by national authorities following environmental impact assessment procedures should be excluded from a liability regime. Again, parties have been expressing different views.

The general opinion appears to be that environmental impacts resulting from terrorists acts, armed conflicts and natural catastrophes should be exempted from liability. The question of whether lawful activities, such as permitted discharges from vessels, should attract liability has also been raised in the discussion on exemptions. The Group's report on this question reads:

> A different approach would endeavour to avoid some of the complex issues by imposing liability only for activities which violate the Protocol. It was maintained that by adopting this approach there would then be no need to exclude assessed activities from the notion of damage.[43]

---

[41] Statement from XXIV SCAR Delegates Meeting, Cambridge, 12–16 August 1996.

[42] Scientists were specifically invited to express their views under the 1997 meeting of the Legal Experts Group in Christchurch, New Zealand. Two information papers were written in that connection: United States, 'Implications of the Current Draft Liability Annex to Activities among Treaty Members', doc. XXI ATCM/IP 104, 1997; and Russia, 'Implications of the Draft Liability Annex to Scientific Research Activities in Antarctica', doc. XXI ATCM/IP 115, 1997. See also a recent working paper by New Zealand, 'Joint and Several Liability and International Collaborative Science', doc. XXIII ATCM/WP 10, 1999.

[43] 'Liability – Report of the Group of Legal Experts', para. 19.

It has been argued, however, that the concept of 'violation' of the Protocol is too vague, especially since its provisions will have to be implemented in domestic legislation.

The Standards of Liability. All participants in the Group of Legal Experts favoured strict liability (i.e., liability not requiring the proof of fault) as the appropriate standard. Some parties opted for imposing joint and several liability on the responsible operators. It is, however, recognised that the activities in the Antarctic are often of a cooperative nature, in contrast to many of the activities regulated by other international agreements. Concern has been raised as to whether joint and several liability would create a barrier to cooperative scientific and associate logistic activities.

Who Should be Held Liable for Environmental Damage and How Should Liability be Imposed? Most operators in the Antarctic carrying out scientific activities are 'state operators', i.e., they are governmental agencies or funded by governments. The prevailing view of the Group appeared to be that the operator of an activity is the one who should be held responsible for damage caused by that activity rather than the state under whose jurisdiction the operator is operating. In consequence, states should only be responsible if the damage would neither have occurred nor continued had a state party carried out its obligations under the Protocol – and this only to the extent that the operator did not meet liability. The intention of the Group was to present a solution that would not go beyond state responsibility under general international law.

The issue on how liability should effectively be imposed on the operator raised controversial questions, compounded by the complexities of what would be the proper basis for jurisdiction. An issue raised was whether operators of a state party should be subject to the jurisdiction of 'departure states',[44] and if so, whether 'departure state' should be subject to residual liability in case they fail to enforce the Protocol through their port state controls.[45]

The following example may illustrate the challenge: A Canadian owned cruise ship, flying the flag of a state not party to the Protocol, its passengers being of different nationalities, leaves the port of Christchurch (New Zealand) to sail for Antarctica. If an accident should occur within the Antarctic Treaty area, New Zealand would be the only possible link to any enforcement of liability under the Protocol. However, for political reasons to non-coastal states and for financial and

---

[44] The term 'departure state' was introduced in the Group by the United Kingdom, and it originates in the United Kingdom, 'Enhancing Compliance with the Protocol: Departure State Jurisdiction', doc. XXI ATCM/WP 22, 1997. The United Kingdom holds the view that a vessel voluntarily seeking admission to the port of a foreign state implicitly accepts the jurisdiction of that state even in matters arising outside the exclusive economic zone of that state.

[45] Further on this subject see Orrego, Chapter 3, and Richardson, Chapter 4, in this book.

practical reasons to port states close to the Antarctic, such a solution might prove to be difficult if not unacceptable.

Obligations Comprised by Liability. As already indicated, there was considerable disagreement on this point. The parties hold different views regarding the obligations of the Protocol, and in particular the meaning of liability, as provided for in Article 16.

However, the parties are in agreement that liability – at least – must imply an obligation to cover the costs related to emergency response action so that an operator who fails to carry out appropriate response action would be obliged to cover the costs incurred by others.

Regarding cases where the operator has failed to undertake required response action, the Group – without reaching a common understanding – discussed whether and on what conditions other entities might undertake response action on behalf of the operator.

It was also discussed whether it would be appropriate to include liability for irreparable damage within a liability regime. Differing views remained on this issue. One view expressed was that an operator that has caused irreparable damage should not be in a more advantageous position than an operator that has caused reparable damage. However, some parties were against including irreparable damage in an annex on liability because of the considerable difficulties in identifying proper means of measuring compensation in such cases. They argued that liability should be limited to calculable, economic losses, and suggested that the issue of irreparable damage should be addressed at a later stage. The introduction of punitive elements in a liability regime was also cautioned against.

Limits to Liability – Insurance. Some parties considered the necessity of setting limits to liability in order to enable operators to acquire affordable insurance. Others held that limits to liability would not be appropriate in all cases, for instance when the damage has been caused deliberately.

The parties sought advice from insurance institutions on the issue of liability, and the International Group of Protection and Indemnity Clubs[46] (P&I Clubs) responded in May 1998, stating that it is generally accepted that no regime of compulsory insurance can be introduced without a monetary limit.

The question of insurance and how costs can be internalised in the budgets for various activities taking place in the Antarctic has not been thoroughly assessed. State operators in the Antarctic would probably for the large part be subject to self-insurance. Concerns were expressed that if strict liability were to be introduced, the budgets of national Antarctic science programmes would have to take severe cuts,

---

[46] The International Group represents 14 members who are shipowners, mutual insurers who cover among them the third party liabilities of more that 90 per cent of the world's ocean-going tonnage.

which could in some cases threaten to terminate their activities.[47] The Group of Experts did not offer any assessment of the financial and practical consequences of a legal regime structured in the lines of the 'Eight Offering'.

Environmental Protection Fund. An issue raised for discussion in the Group was whether rules on liability should provide for an environmental protection fund. There are no industries in Antarctica which are in the position to provide any financial resources for such a fund, nor are the Consultative Parties themselves likely to provide financial resources for that purpose. The question of the fund is therefore very much linked to the question of liability for irreparable damage because any environmental fund would most likely receive its assets from compensation paid for such damage. The establishment of such a fund raised several complex questions on, for instance, legal personality and administration. The practical merit of establishing such a fund was questioned because the fund was regarded not likely to receive any contributions if liability for irreparable damage was not agreed upon and no other sources for repairing the damage were found.

Procedural and Other Issues. Because of the unresolved sovereignty in the Antarctic, all issues dealing with jurisdiction remain sensitive. In the context of liability, who should be regarded a claimant and who should decide on claims for compensations? What procedures should be laid down for calculating the damages – in particular irreparable damage? What should be the role of the Consultative Meeting, of the Arbitral Tribunal provided for under the Protocol, and of national courts? Is there a need to establish new institutions? Although the Group has examined these issues, major disagreements remain.

Outcome of the Group's Work

As stated in the Final Report of the XXII Consultative Meeting in 1998, Consultative Parties concluded that the Group of Legal Experts had fulfilled its tasks and agreed that, under the Consultative Meeting, Working Group I would in the future be the appropriate forum for continuing the work on liability.[48] In reality, it was felt that the Group of Legal Experts had come to a dead end, and a procedural move was made to try to get the process on a more practical and speedy track. The Consultative Parties recognised the need to bring more factual data into future deliberations and therefore asked COMNAP and other expert bodies to provide input on the practical aspects of liability. The deliberations continued in a policy rather than in a legal forum – Working Group I of the Consultative Meeting – starting from the XXIII Consultative Meeting, held in 1999 in Lima, Peru.

---

[47] Russia, 'Implications of the Draft Liability Annex to Scientific Research Activities in Antarctica'.
[48] *Final Report of the XXII ATCM*, para. 65.

THE FIRST ROUND OF NEGOTIATIONS IN WORKING GROUP I

There were no inter-sessional meetings preparing the first round of negotiations in Lima in 1999. However, several parties prepared working papers for discussion at the XXIII Consultative Meeting. In addition to the report of the Group of Legal Experts (including 'the Eight Offering' and the proposal from the US), the delegates had received draft annexes on liability proposed by the UK, Chile and the Netherlands. Argentina, Brazil, Chile, Ecuador, Peru and Uruguay submitted jointly a paper on basic definitions and considerations, Australia submitted a paper on principles for liability, New Zealand a paper on joint and several liability and Germany a paper on how to proceed.[49]

The mandate given by the XXII Consultative Meeting in 1998 was that Working Group I, at that Meeting, should seek to elaborate texts, based on submissions by parties, for further considerations at the XXIV Consultative Meeting. The Meeting in Lima focussed on the key policy issues, as earlier raised by the Group of Legal Experts. During the negotiations, informal contact groups were convened to work on specific topics. The Chairman for the liability discussion, Don MacKay of New Zealand, summed up the first round of negotiation:

> [t]he future elaboration of a liability regime remains bedevilled by a fundamental difference of approach between delegations. This is over the basic question of whether we should be seeking to elaborate a so-called 'comprehensive' (or single) annex, or a so-called 'limited' annex (which could be the first in a series of annexes focusing on particular aspects of the liability problem). Until this fundamental issue is resolved there will inevitably be limits as to the further progress that will be possible.[50]

This does not mean that there was no movement at all. The need to seek other avenues is clearly recognised and other approaches actively sought. Recorded in the Final Report from the Lima Meeting are some 'areas of convergence' which indicate a will to continue to negotiate with a view to reach consensus. The next round of negotiations on liability will take place at a Special Consultative Meeting, to be held in The Hague, 11–15 September 2000.

CONCLUDING REMARKS

Since the negotiations on CRAMRA, rules on liability have generally been regarded as an important element of any environmental protection regime in the Antarctic. During the negotiations of CRAMRA, the Swedish representative stated, 'It is of crucial importance that we are able to demonstrate to the outside world that all activities on Antarctica clearly are connected to the strongest possible rules of

---

[49] See an analysis of the main issues involved, as discussed by Lefeber, Chapter 11 in this book.

[50] 'Personal Report from the Chairman of the Liability Discussion in Working Group I', doc. XXIII ATCM/WP 41, 1999, para. 6.

liability'.[51] It appears that, when CRAMRA failed to enter into force, the sentiment expressed was carried over to the negotiations of the Protocol.

Antarctica remains the cleanest continent on Earth, a remote place in which – with the exception of fishing – no important commercial interests are vested. Perhaps it was the vision of imposing the most stringent environmental rules that spurred the majority of the parties to take on the task of elaborating rules and procedures on liability with the ambition of setting high standards of liability in a single comprehensive annex. This would indeed have demonstrated the ability of the Consultative Parties to manage Antarctica, not only in coherence with emerging international environmental law, but even in the forefront of international developments. The goals were laudable at the outset, but the outcomes so far have proved modest, at best.

The Group of Legal Experts commenced its *legal* deliberation with no specific *policy* guidance as to what would be the purpose of the rules of liability, no analysis of the *activities* taking place in the Antarctic Treaty area including the *environmental risks* involved, and no analysis on rules on liability under *other international regimes* already in force. The lack of these inputs has contributed to making the elaboration of an Antarctic liability regime an abstract, theoretical exercise for several years. Liability for environmental damage is a complex legal issue that can indeed provide enough intellectual 'food' to keep any Group of Legal Experts meeting for years; already at the first meeting of the Group, in Heidelberg, in November 1993, the observation was made that the Group was operating in a theoretical vacuum. Throughout the process, a *practical* perspective has constantly been called for. Input from COMNAP and from SCAR has been provided and parties have been encouraged to include scientific experts in their delegations. Basically, in inter-governmental cooperation the lawyers are there to provide legal assessments and legal language for expressing the common intentions of the parties. As the purpose of developing rules on liability has not been clear from the outset, it should now be no wonder why the Antarctic Treaty parties have not been able to take full advantage of available legal and scientific expertise.

The decision of the XXII Consultative Meeting in 1998 to delegate liability to Working Group I of the Consultative Meeting has put the matter on another track. Policy and practical matters have now come more to the forefront – a development that has been much needed indeed. Future deliberations would benefit from focussing on the *purpose* of the rules on liability.

---

[51] The statement is annexed to a letter to the Norwegian Ministry of Foreign Affairs from the Swedish Embassy dated 6 July 1987; available at archive of the Royal Norwegian Ministry of Foreign Affairs, Oslo.

10

The Legal Need for an Antarctic Environmental Liability Regime

René Lefeber

The already legally complex issue of the establishment of liability for damage to the environment is, in the Antarctic Treaty area, further complicated by the uncertain jurisdictional situation in the Antarctic and the fragility of its environment and dependent and associated ecosystems.[1] This may be illustrated by the *Bahia Paraiso* oil spill which to date has been the most serious case of damage to the Antarctic environment.[2] The *Bahia Paraiso* was a polar transporter and tourist vessel which was owned by the Argentine government and flew the Argentine flag. The vessel ran aground on 28 January 1989, 1.5 km from Bonaparte Point, near Palmer Station. Large slicks remained for two weeks after the spill, while small slicks persisted in coastal areas until March 1989. All components of the ecosystem were contaminated to varying degrees by the spill, including sediments, limpets, macroalgae, clams, bottom-feeding fish, and seabirds. A reported, possible indirect effect of the spill is a reproductive failure in a population of South Polar Skuas which has been ascribed to parental neglect. The effects of the spill were, however, limited in time and space.

---

[1] On liability for damage caused by activities in the Antarctic Treaty area, see also, e.g., Germany, 'Environmental Protection through the Law of International Liability', doc. XXIII ATCM/IP 37, 1999; S. Blay and J. Green, 'The Development of a Liability Annex to the Madrid Protocol', *Environmental Policy and Law*, Vol. 25, 1995, pp. 24–37; H.C. Burmester, 'Liability for Damage from Antarctic Mineral Resource Activities', *Virginia Journal of International Law*, Vol. 29, 1989, pp. 621–660; F. Francioni, 'International Responsibility for Damage to the Antarctic Environment', in F. Francioni (ed.), *International Environmental Law for Antarctica* (Milan: Giuffrè, 1992), pp. 233–257; M. Poole, 'Liability for Environmental Damage in Antarctica', *Journal of Energy and Natural Resources Law*, Vol. 10, 1992, pp. 246–266; Sir Arthur Watts, 'Liability for Activities in Antarctica – Who Pays the Bill to Whom?', in R. Wolfrum (ed.), *Conflicting Interests, Cooperation, Environmental Protection, Economic Development* (Berlin: Duncker & Humblot, 1985), pp. 147–161.
[2] For an account of the oil spill and its impact, see the 'Initial Environmental Evaluation, Associated with the Salvage Operation of the Remaining Oil of the Bahia Paraiso', prepared by the Instituto Antartico Argentino, Argentina & Rijkswaterstaat – Tidal Waters Division, the Netherlands, 1992.

This may be attributed to the high-energy environment, the relatively small volume of the materials released, and the volatile nature of the materials released. The most effective removal processes were evaporation, dilution, wind, and currents. Only a minor portion of the spill was removed through sedimentation, biological uptake, microbial oxidation, and photo-oxidation. The effects of the spill were still present in several areas one year after the spill, but these have now disappeared. In order to minimise the risk of further contamination, the fuel and oil remaining in the hull of the ship were salvaged from the wreckage. Incidents like the *Bahia Paraiso* oil spill produce numerous legal questions, such as the following: Should liability for damage be channelled to a state as an international person under international law or to the state as the owner of the vessel under municipal law, or to both? Who is entitled to claim reparation for the damage to the environment on behalf of Antarctica? Where can such reparation be claimed? What forms of damage can be subject of a claim, and should general international law or municipal law be applicable, or should special rules and procedures be developed?

THE OBLIGATION TO DEVELOP AN ANTARCTIC ENVIRONMENTAL LIABILITY REGIME

During the negotiations on the 1991 Protocol on Environmental Protection to the Antarctic Treaty[3] (Protocol), several negotiating states insisted on the adoption of rules and procedures relating to liability for damage to the Antarctic environment and dependent and associated ecosystems. Faced with time constraints and diverging views on the issue, the negotiating states agreed that the adoption of such rules and procedures was desirable, but deferred it to the implementation phase of the Protocol. This has been laid down in Article 16 of the Protocol, which reads:

> Consistent with the objectives of this Protocol for the comprehensive protection of the Antarctic environment and dependent and associated ecosystems, the Parties undertake to elaborate rules and procedures relating to liability for damage arising from activities taking place in the Antarctic Treaty area and covered by this Protocol. Those rules and procedures shall be included in one or more Annexes to be adopted in accordance with Article 9 (2).

This provision imposes an *obligation* on the parties to the Protocol to develop *special* rules and procedures relating to liability for damage to the Antarctic environment and dependent and associated ecosystems. From a legal point of view, one can therefore no longer question the need to develop a special liability regime. A risk analysis of activities in the Antarctic Treaty area may nevertheless provide useful guidance for the development of the modalities of a special Antarctic environmental liability regime.

---

[3] ILM, Vol. 30, 1991, pp. 1,461ff.

A *special* liability regime supersedes or concurs with the rules and procedures relating to liability that would apply in the absence of such a regime. Several reasons can be given for developing a special liability regime. One might be the wish to harmonise municipal rules and procedures relating to liability in order to prevent the application of diverging rules and procedures to identical cases or to ensure the availability of legal remedies to plaintiffs on a non-discriminatory basis. Another reason may be that the substantive or procedural rules of such *ordinary* liability regimes do not provide prompt, adequate and effective compensation or any other form of reparation for damage caused.

Article 16 of the Protocol provides that the parties to the Protocol 'undertake to elaborate' special rules and procedures relating to liability. Although this provision is principally a procedural obligation, it nevertheless remains *binding* for the parties to the Protocol. However, Article 16 is not an obligation of result, but an obligation of conduct.[4] It only requires that the parties to the Protocol cooperate, albeit in good faith and with the aim of developing a special Antarctic environmental liability regime. Article 16 of the Protocol does not, however, require the parties to the Protocol to carry such cooperation to the desired end-result.[5] In fact, Article 16 of the Protocol is complied with if the parties to the Protocol make serious work of its implementation.[6] If the parties to the Protocol do not succeed in developing a special Antarctic environmental liability regime, they continue to be bound by the obligation to cooperate, unless they release themselves by deleting this obligation from the Protocol.

Because the development of a special Antarctic environmental liability regime may follow divergent scenarios, the parties to the Protocol will have to make several strategic choices before they can work on specific texts. Fundamental is whether they wish to develop a state liability regime or a civil liability regime, or a combination.[7] Another issue that the parties to the Protocol must take a stand on concerns the relation of an Antarctic environmental liability regime to other special liability regimes that may be applicable in the Antarctic Treaty area. If such other special liability regimes are applicable to an incident in the Antarctic Treaty area, a decision must be made as to whether it is desirable to replace these special liability regimes with an Antarctic environmental liability regime or whether the parties to the Protocol should be stimulated to join these regimes.[8]

---

[4] See also R. Lefeber, *Transboundary Environmental Interference and the Origin of State Liability* (The Hague: Kluwer Law International, 1996), pp. 74–81.
[5] *Ibid.*, pp. 39–41.
[6] See Skåre, Chapter 9 in this book.
[7] See the following discussion on international liability law, containing a concise analysis of contemporary international liability law and an assessment of the legal need to develop a special Antarctic environmental liability regime in view of ordinary and special liability regimes now in existence or in the process of being developed.
[8] See the discussion on the applicability of other special liability regimes which addresses this issue, below in this chapter.

Only if the above-mentioned strategic choices have been made is there a basis to proceed with the elaboration of the modalities of an Antarctic environmental liability regime. In this context, it becomes necessary to look at the common features of special liability regimes, including functional scope, definition of damage, channelling of liability, standard of liability, limitation of liability, establishment of financial security to cover the liability, residual liability, and settlement of claims. These issues will be addressed in a subsequent chapter,[9] together with an assessment of the work done so far on an Antarctic environmental liability regime.

INTERNATIONAL LIABILITY LAW

State Liability

The general rules and procedures with respect to the liability of states for their internationally wrongful acts (state responsibility or liability *ex delicto*) are incorporated in customary international law. Accordingly, these rules and procedures are applicable to the acts of states in the Antarctic Treaty area. Furthermore, the parties to the Protocol seem to agree that an Antarctic environmental liability regime should not be prejudicial to the application of the law of liability *ex delicto* to damage arising from activities in the Antarctic Treaty area.[10] However, in view of the self-imposed obligation to develop a special Antarctic environmental liability regime, the parties apparently are of the opinion that the general rules and procedures relating to liability *ex delicto* do not provide for prompt, adequate and effective compensation or other form of reparation for damage to the Antarctic environment and dependent and associated ecosystems. If this presumption is correct, there is indeed a legal need for the development of a special Antarctic environmental liability regime; yet a review of this presumption is necessary.

In order to hold a state liable *ex delicto*, the following conditions must be fufilled:1) the attribution of an act to the state, 2) the wrongfulness of that act, and 3) the presence of an injured state.[11] This means that a state can only be held liable *ex delicto* for damage to the Antarctic environment and dependent and associated ecosystems if the damage-causing conduct can be attributed to that state, the failure of that state to prevent the occurrence of damage is internationally wrongful, and another state, or one of its nationals, has been injured. With respect to damage to the

---

[9] See Lefeber, Chapter 11 in this book.

[10] See proposals of Chile, 'Draft Annex on Environmental Liability to the Madrid Protocol', doc. XXIII ATCM/WP 34, 1999, Art. 13(3); and the Netherlands, 'Liability', doc. XXIII ATCM/WP 18, 1999, Art. 9(2); see also 'Chairman's Eighth Offering – Annex on Environmental Liability', reproduced in 'Liability – Report of the Group of Legal Experts', Appendix 1 (EO), doc. XXII ATCM/WP 1, 1998, Art. 7(3) .

[11] See also Arts. 3, 23 and 40 of the 1996 ILC Draft Articles on State Responsibility, reprinted in ILM, Vol. 37, 1998, pp. 442ff.

Antarctic environment and dependent and associated ecosystems, proof of these three conditions may become an insurmountable burden for the plaintiff.

Firstly, the act-of-the-state condition *prima facie* means that a state can only become liable *ex delicto* for damage caused by the governmental activities it carries out in the Antarctic Treaty area. However, although a state cannot be held liable merely for the reason that a person falling within its jurisdiction or control has caused damage,[12] international law requires states to supervise non-governmental activities with the aim of ensuring that such activities do not violate the rights of other states under international law. Activities in the Antarctic Treaty area require prior authorisation under the Protocol. The authorising state cannot, therefore, claim that it was unaware that a certain activity was being carried on. However, the effective supervision of non-governmental activities in the Antarctic Treaty area is much more difficult than 'at home', and this provides authorising states with an opportunity to evade liability. It has even been suggested to incorporate in the Antarctic environmental liability regime an explicit provision that a state is not liable for the damage that a person has caused merely because the person was within the jurisdiction or control of that state.[13] Such a provision states the obvious but also carries the risk of abuse as a state may try to invoke the provision in order to evade liability, even if it has not adopted appropriate measures to prevent and abate damage to the Antarctic environment and dependent and associated ecosystems.[14]

Secondly, liability *ex delicto* only arises if the damage-causing conduct is internationally wrongful, in other words contrary to international law. This requires the identification of the conventional and extra-conventional obligations that are applicable to an incident. It will then be necessary to proceed with an analysis of these obligations in order to establish whether one or more obligations have been violated. As for damage to the Antarctic environment and dependent and associated ecosystems, the discussion will in many cases focus on the obligation of states to adopt appropriate measures to prevent and abate such damage. This obligation can be derived both from the Protocol and extraconventional law.[15] It applies to the conduct of governmental activities as well as to the supervision of non-governmental activities in the Antarctic Treaty area. However, obligations of this

---

[12] See Lefeber, *Transboundary Environmental Interference and the Origin of State Liability*, pp. 56–60; see also the 'Explanatory Note to Article 9 (State Liability and Responsibility)' to the Netherlands proposal of 28 May 1999, drafted by the Netherlands at the XXIII Consultative Meeting, reproduced in 'Personal Report of the Chairman of the Liability Discussion in Working Group 1', submitted by New Zealand, doc. XXIII ATCM/WP 41, 1999.

[13] See e.g. Art. 7(1) EO; Art. 13(1) in Chile, 'Draft Annex on Environmental Liability'.

[14] This proposal is derived from Art. 8(3) of the 1988 Convention on the Regulation of Antarctic Mineral Resource Activities, ILM, Vol. 27, 1988, pp. 868ff.

[15] See Arts. 3(4) and 13(1) of the Protocol on Environmental Protection to the Antarctic Treaty; and Francioni, 'International Responsibility for Damage to the Antarctic Environment', pp. 238–242; L. Pineschi, 'The Antarctic Treaty System and General Rules of International Environmental Law', in F. Francioni and T. Scovazzi (eds.), *International Law for Antarctica* (Milan: Giuffrè, 1987), pp. 187–246.

kind are generally not considered absolute obligations, but due diligence obligations.[16] The mere occurrence of damage is therefore not sufficient to establish the violation of international law. In each and every case, the substance of the action taken by the state in question must be reviewed to find a breach of the applicable norm. The action actually taken by a state must be compared with the action a 'good government' would have taken to prevent and abate the damage in similar circumstances. Only if the action taken falls short of this objectified standard, is the damage-causing conduct internationally wrongful.

Thirdly, liability *ex delicto* only arises *vis-à-vis* states of which a legally protected interest has been infringed. With respect to damage to the Antarctic environment and dependent and associated ecosystems, it is of course not difficult to identify the injured state if a third state or nationals of that state have taken response measures or remedial measures in the aftermath of an incident. However, the situation is different if damage to the Antarctic environment and dependent and associated ecosystems cannot be repaired. As for the disputed areas in the Antarctic Treaty, the presence of an injured state depends on the validity of a claim to territorial sovereignty. If no valid claim can be presented related to a part of the Antarctic Treaty area that has not been claimed or cannot be claimed, no state exists that has an individual legal interest in presenting a claim on behalf of Antarctica. In this case, a claim for damage to the Antarctic environment and dependent and associated ecosystems can only be brought by a state in the collective interest. In particular, a Consultative Party could point to the intention of the Consultative Parties to administer Antarctica in the collective interest.[17] However, the right of a state to bring a claim in the collective interest has not been recognised by the International Court of Justice, and only in exceptional cases has it been recognised by the International Law Commission.[18]

If either one of these conditions has not been fulfilled, a state cannot be held liable *ex delicto* and damage to the Antarctic environment and dependent and associated ecosystems will not be repaired. In view of these legal gaps, the law of state liability *ex delicto* does not provide for a liability regime that would provide for

---

[16] See Francioni, 'International Responsibility for Damage to the Antarctic Environment', pp. 245–246; and Lefeber, *Transboundary Environmental Interference and the Origin of State Liability*, pp. 61–74, especially at p. 62.

[17] See Preambles to the Antarctic Treaty and to the Protocol; see also Blay and Green, 'The Development of a Liability Annex to the Madrid Protocol', pp. 32–33; Francioni, 'International Responsibility for Damage to the Antarctic Environment', p. 254.

[18] See Art. 40(2) of the 1996 ILC Draft Articles on State Responsibility; see also Blay and Green, 'The Development of a Liability Annex to the Madrid Protocol', p. 32; J. Charney, 'Third State Remedies for Environmental Damage to the World's Common Spaces, in F. Francioni and T. Scovazzi (eds.), *International Responsibility for Environmental Harm* (The Hague: Kluwer Law International, 1991), pp. 149–177; Francioni, 'International Responsibility for Damage to the Antarctic Environment', pp. 252–257; and Lefeber, *Transboundary Environmental Interference and the Origin of State Liability*, pp. 113–128.

prompt, adequate and effective compensation or any other form of reparation for damage to the Antarctic environment and dependent and associated ecosystems.

In addition to its liability for internationally wrongful acts, a state may also be held liable for the injurious consequences arising out of acts not prohibited by international law (liability *sine delicto*). However, as yet this form of liability has not been incorporated into positive extraconventional international law.[19] Only one multilateral, special state liability regime provides for this form of liability, the 1972 Convention on Liability for Damage Caused by Objects Launched into Outer Space. The question of whether or not this Convention is applicable to incidents in the Antarctic Treaty area is addressed below in this chapter, under the discussion on the applicability of other liability regimes.

The main reason that a special state liability regime for outer space activities was concluded was the high degree of governmental involvement in these activities at the time of the drafting and conclusion of the above-mentioned Convention.[20] In view of the increasing number of non-governmental activities in the Antarctic Treaty area, this reason cannot alone be advanced to support the development of a special state liability regime for activities in that area. In addition, the development of a comparable special state liability regime is not a realistic option in view of the prevailing views of the international community on the imposition of state liability if a state has not committed an internationally wrongful act. In the absence of an internationally wrongful act, the majority of states are not prepared to accept state-to-state liability for damage caused by activities within their jurisdiction or control, irrespective of the governmental or non-governmental nature of these activities.[21] In the negotiations on an Antarctic environmental liability regime, the development of such a state liability regime has also not been seriously considered.

Civil Liability

The rules and procedures with respect to civil liability are incorporated in municipal law. They are applicable to activities in the Antarctic Treaty area subject to a state's right to exercise jurisdiction over activities – including ships, aircraft, or natural or legal persons engaging in such activities. This may be illustrated by the case that was brought before a court in the United States in the aftermath of the Mount Erebus disaster. On 28 November 1979, a DC-10 aircraft, owned by Air New Zealand and engaged on a sight-seeing trip to the Antarctic, crashed into Mount Erebus, killing

---

[19] See Germany, 'Environmental Protection Through the Law of International Liability', p. 11; Francioni, 'International Responsibility for Damage to the Antarctic Environment', pp. 245–246; and Lefeber, *Transboundary Environmental Interference and the Origin of State Liability*, pp. 145–187.

[20] See Lefeber, *Transboundary Environmental Interference and the Origin of State Liability*, pp. 161–162.

[21] See also Francioni, 'International Responsibility for Damage to the Antarctic Environment', p. 247; and Lefeber, *Transboundary Environmental Interference and the Origin of State Liability*, pp. 178–181.

all the 237 passengers and 20 crew members aboard. Families of the passengers not only claimed compensation from Air New Zealand, but also from the United States Government for negligence of the United States air traffic controllers at McMurdo Station. In court, the United States Government pleaded immunity from jurisdiction.[22] Under the Federal Tort Claims Act, such a plea can only be upheld if Antarctica would fall under the foreign-country exception to the waiver of sovereign immunity in this Act. The Court concluded that, for the purposes of this case, Antarctica was not a foreign country and allowed the case to proceed.[23]

The Mount Erebus disaster is illustrative of some of the procedural legal difficulties that plaintiffs or defendants may face if damage arising from activities in the Antarctic Treaty area is governed by an ordinary civil liability regime. Apart from states invoking immunity from jurisdiction and forum shopping by plaintiffs, other complex procedural legal issues may impede the adjudication of claims for damage, such as the right of access to courts, the choice of an applicable law, and the recognition and enforcement of judgements in third states.

Regarding the right of access to courts, the legal systems of many states permit domestic and foreign plaintiffs to bring tort actions if the tort has been committed within the jurisdiction of the state to whose legal system access is sought. It is, however, not clear where the tort has been committed if neither the place where the harm-causing activity is situated nor the place where the harm has been suffered falls within the jurisdiction of a single state. In such cases, actions may usually be brought before the courts of the place where the harm-causing activity is situated or the place where the harm was suffered, or the courts of both places. In the case of damage to the Antarctic environment and dependent and associated ecosystems, the question of access to courts is further complicated by the uncertain jurisdictional situation in the Antarctic Treaty area. The municipal courts of a claimant state will be inclined to accept jurisdiction if the harm-causing activity is situated in the part of Antarctica that is claimed by that claimant state or if the damage has been suffered in that part. This rule is, however, not very helpful if access is sought to the courts of non-claimant states. Nevertheless, the courts of those states may decide to exercise jurisdiction because the defendant, or the plaintiff, has his habitual residence in that state. The plaintiff may thus have a wide choice of forum, but there is also the risk that the court to which access is sought will decline to exercise jurisdiction. This risk is considerable if an activity is being carried on by a state because the state may claim immunity from jurisdiction before municipal courts. The recognition of such claims by municipal courts is governed by the applicable law that may provide for absolute or limited immunity from jurisdiction.

---

[22] See Beattie v. United States, 765 F.2d 91 (DC Cir. 1984).

[23] In a case that was brought before a New Zealand court, the claim of the New Zealand Government for reimbursement of part of the cost of the investigation into the cause and circumstances of the incident was rejected for technical legal reasons. See Mahon v. Air New Zealand Ltd., Privy Council, 1 AC 808, 3 All ER 201 (1984) (Opinion by Lord Diplock).

The choice of the applicable law is governed by the conflict-of-laws rules of the forum. None of the Consultative Parties has a municipal law with a statutory conflict-of-laws rule on environmental torts. The substance of environmental torts is therefore subject to the general conflict-of-laws rules with respect to torts of the forum. However, the practice of states is not uniform in this respect. Examples of conflict-of-laws rules that are used to select the applicable law include: 1) the law of the forum (*lex fori*), 2) the law of the place where the tort has been committed (*lex loci delicti*), 3) the law with which the facts have the most significant relationship, and 4) the law chosen by the litigants. The possible application of different laws by different fora is likely to stimulate 'forum shopping' by the plaintiffs. Some crucial elements in the choice of forum of plaintiffs will be the differences in the standard of liability, the concept of environmental damage, and the limitations to liability of different laws. Forum shopping increases the options for the plaintiff to obtain compensation or any other form of reparation, but it will also increase the uncertainty for the defendant. This can be avoided by the development of uniform liability rules and procedures. The development of uniform liability rules and procedures is also to the advantage for the plaintiff because the objective of such rules is normally to secure prompt, adequate and effective compensation or any other form of reparation for damage caused. For these purposes, special liability regimes have been developed with respect to damage from space-borne and airborne sources, the use of radio-active substances, the transport of hazardous substances, and the exploration and exploitation of mineral resources.[24]

Once a favourable judgement has been obtained, it will be meaningless for the plaintiff to enforce this judgement in the state where it has been rendered, unless the defendant has assets in that state. If the defendant does not have sufficient assets in that state or refuses to accept the judgement rendered, the judgement will have to be recognised and enforced in the state where the defendant does have assets. The recognition of a judgement is not likely in the absence of a treaty to that effect between the states in question, and even if such a treaty is in force, it is likely to provide for exceptions to the rule.

Although the use of civil law remedies may result in an equitable settlement of claims in one case, it may not in others. The legal difficulties to bring and effectuate claims in a multi-jurisdictional context, or the wish to secure prompt, adequate and effective compensation or other form of reparation for damage caused, necessitate the development of uniform procedural and substantive rules for the settlement of claims arising from damage to the Antarctic environment and dependent and associated ecosystems. This is particularly the case in view of the uncertain jurisdictional situation in the Antarctic Treaty area. The Consultative Parties also seem to hold this view because the emerging Antarctic environmental liability regime is a special civil liability regime that provides for such uniform procedural and substantive rules.

---

[24] See further discussion below in this chapter.

THE APPLICABILITY OF OTHER SPECIAL LIABILITY REGIMES

Introduction

The analysis above has shown that ordinary state and civil liability regimes do not provide for prompt, adequate and effective compensation or other form of reparation for damage caused to the Antarctic environment and dependent and associated ecosystems. The analysis also confirms the legal need for a special Antarctic environmental liability regime and the propriety of the incorporation of Article 16 in the Protocol. However, several existing and proposed special liability regimes already cover activities that may be carried on in or outside the Antarctic Treaty area and may cause damage to the Antarctic environment and dependent and associated ecosystems. The point of departure of these special liability regimes, which are with one exception special civil liability regimes, is the source of pollution. Whether or not these 'source-oriented' liability regimes can fill the legal need must therefore be examined.

As special liability regimes that are confined to a specific geographic area outside the Antarctic Treaty area do not fill the legal need, these will not be addressed here. Furthermore, it would be premature to discuss special liability regimes that have not yet been developed and are also not in *statu nascendi*, such as the regimes that are to be developed under Article 235 of the 1982 UN Convention on the Law of the Sea, Article X of the 1971 London Convention on the Prevention of Marine Pollution by Dumping of Wastes and Other Matter, or Article 15 of the 1996 Protocol to that Convention.

Source-Oriented Liability Regimes

Air-Borne and Space-Borne Sources. Liability for damage caused by aircraft is regulated by the 1952 Convention on Damage Caused by Foreign Aircraft to Third Parties on the Surface.[25] This Convention provides for a civil liability regime. Strict liability is imposed on the operator of the aircraft (Articles 2(1) and 11). The Convention covers damage suffered by natural or legal persons on the ground and, hence, in the Antarctic Treaty area, provided that such damage is the direct consequence of an incident involving civil aircraft (Articles 1(1) and 26). Irreparable damage to the Antarctic environment and dependent and associated ecosystems is thus not covered by the Convention, nor are the costs of response measures and remedial measures because these costs are not likely to be considered a direct consequence of an incident. Moreover, only damage caused in the 'territory' of a party to the Convention is covered (Article 23(1)). Accordingly, even if damage to the environment is covered by the Convention, the award of reparation depends on

---

[25] See 1952 Convention on Damage Caused by Foreign Aircraft to Third Parties on the Surface, published in UNTS, Vol. 310, pp. 181ff.

the recognition by the competent court of territorial sovereignty over the damaged environment in Antarctica.

Liability for damage caused by spacecraft is regulated by the 1972 Convention on Liability for Damage Caused by Objects Launched into Outer Space.[26] This Convention provides for a state liability regime. Strict liability rests with states that launch or procure the launching of spacecraft and states from whose territory or facility spacecraft is launched (Article II). The Convention covers damage caused on the surface of the earth and, hence, in the Antarctic Treaty area. The definition of damage only refers to damage to persons and property of states or of natural or juridical persons (Article I(a)). However, according to the accepted interpretation, this definition does not only cover physical harm to persons and property but also cover the reasonable costs of response measures.[27] It is, however, less clear whether the definition can also be interpreted to cover the reasonable costs of remedial measures and irreparable damage. Even if it would, the award of the reimbursement of the reasonable costs of remedial measures and the compensation for irreparable damage will depend on the presentation of a valid claim to territorial sovereignty over the damaged environment, because the Convention does not cover damage to property that does not belong to states or to natural or legal persons.

The Use of Radioactive Substances. Some nuclear activities cannot be lawfully carried on in the Antarctic Treaty area, namely the conducting of nuclear explosions and the disposal of nuclear waste (Article V(1)). It is, furthermore, not likely that other nuclear activities will be carried on in the Antarctic Treaty area or cause damage in that Treaty area. An exception may be nuclear ships that navigate within the Antarctic area. Liability for damage caused by nuclear ships is covered by the 1962 Convention on the Liability of Operators of Nuclear Ships, but this civil liability regime has not entered into force nor is it expected to enter into force.[28] The use of nuclear energy in the Antarctic Treaty area for the generation of electricity is not prohibited, but at present no nuclear reactor is employed in the Antarctic Treaty area for such purpose.[29] Nuclear installations outside the Antarctic Treaty area, however, may cause long-range nuclear damage in that area. This damage and also the damage caused by the transport of nuclear substances is covered by two special civil liability regimes that are linked, the 1960 Paris Convention on Third Party Liability in the Field of Nuclear Energy and the 1963 Vienna Convention on Civil

---

[26] See 1972 Convention on Liability for Damage Caused by Objects Launched into Outer Space, text reprinted in ILM, Vol. 10, 1971, pp. 965ff, and ILM, Vol. 11, 1972, pp. 250ff.

[27] See, e.g., C.Q. Christol, 'International Liability for Damage Caused by Space Objects', *American Journal of International Law*, Vol. 74, 1980, pp. 346–371, at pp. 355–365; and B.A. Hurwitz, *State Liability for Outer Space Activities* (Dordrecht: Martinus Nijhoff Publishers, 1992), pp. 12–18.

[28] See *American Journal of International Law*, Vol. 57, 1963, pp. 268–278.

[29] In the 1960s and 1970s, a nuclear reactor was employed at McMurdo station by the United States. See J. May, *The Greenpeace Book of Antarctica – A New View of the Seventh Continent* (London: Dorling Kindersley, 1988).

Liability for Nuclear Damage.[30] In principle, these Conventions also apply to nuclear damage that is caused beyond the limits of national jurisdiction and therefore do not require the presentation of a valid claim to territorial sovereignty over the damaged environment (Articles 2 and IA respectively). The Conventions provide for strict liability of the operators of nuclear installations (Articles 3(a) and II(1) respectively). The definition of nuclear damage of the 1960 Paris Convention, however, is limited to damage to persons and property (Article 3(a)), but this Convention is in the process of being revised. The 1963 Vienna Convention was revised in 1997, *inter alia*, to include in the definition of nuclear damage the costs of response measures and the costs of remedial measures, but the definition does not extend to irreparable damage to the environment (Article I(k)).

The Transport of Hazardous Substances. Liability for damage caused by the transport of oil is regulated by the 1969 IMCO International Convention on Civil Liability for Oil Pollution Damage, as amended. Liability for damage caused by the transport of other hazardous and noxious substances, with the exception of the transport of nuclear substances, is regulated by the 1996 IMO Convention on Liability and Compensation for Damage in Connection with the Carriage of Hazardous and Noxious Substances by Sea.[31] Pursuant to these civil liability regimes, strict liability is imposed on the ship owner (Articles III(1) and 7(1) respectively). The definition of damage does not cover damage to the environment outside the territory or the exclusive economic zones of states with the exception of the costs of 'preventive measures', i.e. 'any reasonable measures taken by any person after an incident has occurred or to prevent or minimise damage' (Articles II and I(7), and 3 and 1(7) respectively). As a result, the award for the reimbursement of the reasonable costs of remedial measures will depend on the presentation of a valid claim to territorial sovereignty over the damaged environment. The award of compensation of irreparable damage is not possible at all because the definition of damage of these Conventions does not extend to such damage.

The transport of hazardous wastes and other wastes is regulated by the 1989 UNEP Convention on the Control of Transboundary Movements of Hazardous Wastes and Their Disposal.[32] The Convention prohibits the export of hazardous

---

[30] See 1960 Paris Convention on Third Party Liability in the Field of Nuclear Energy, as amended by protocols of 1962 and 1982 (UNTS, Vol. 956, pp. 251ff); 1963 Vienna Convention on Civil Liability for Nuclear Damage (UNTS, Vol. 1063, pp. 265ff), as amended by protocol of 1997 (ILM, Vol. 36, 1997, pp. 1,462ff); and 1988 Joint Protocol Relating to the Application of the Vienna Convention and the Paris Convention, doc. IAEA/GOV/2326 (1988), Annex I.

[31] See 1969 IMCO Convention on Civil Liability for Oil Pollution Damage (UNTS, Vol. 973, pp. 3ff), as amended by protocols of 1976 (ILM, Vol. 16, 1977, pp. 617ff), of 1984 (ILM, Vol. 23, 1984, pp. 195ff), and of 1992 (doc. IMO/LEG/CONF.9/15, of 1992); and 1996 IMO Convention on Liability and Compensation for Damage in Connection with the Carriage of Hazardous and Noxious Substances by Sea (ILM, Vol. 35, 1996, pp. 1,406ff).

[32] See 1989 UNEP Convention on the Control of Transboundary Movements of Hazardous Wastes and Their Disposal, text reprinted in *Environmental Policy and Law*, Vol. 19, 1989, p. 68.

wastes and other wastes to the Antarctic Treaty area for disposal (Article 4(6)). Further to Article 12 of the Convention, the parties have entered into negotiations with a view to adopting a protocol on liability and compensation for damage resulting from the transboundary movements of hazardous wastes and other wastes and their disposal. With respect to transport, the proposed civil liability regime imposes liability on the person who notifies the authorities of a transboundary movement of hazardous wastes and other wastes or, if the state of export is the notifier or if no notification has taken place, on the exporter (draft Article 4(1)).[33] The definition of damage is comparable to the definition of the Conventions mentioned in the previous paragraph. It does not cover damage to the environment beyond the limits of national jurisdiction, with the exception of the costs of 'preventive measures' (draft Article 3(2)(c)), nor does it cover irreparable damage to the environment. Also in this case the award for the reimbursement of the reasonable costs of remedial measures will depend on the presentation of a valid claim to territorial sovereignty over the damaged environment. In this context, the declarations of Chile and the United Kingdom upon their ratification of the 1989 UNEP Convention on the Control of Transboundary Movements of Hazardous Wastes and Their Disposal are telling.[34] Chile has declared that the scope of application of the Convention covers 'both the continental territory of the Republic and its area of jurisdiction situated south of latitude 60°S'. The United Kingdom has explicitly stated that its ratification extends to its Antarctic territories.

The Exploration and Exploitation of Mineral Resources. The exploration and exploitation of mineral resources is prohibited by the Protocol (Article 7). Since illegal exploration and exploitation of mineral resources is not likely, there does not seem to be a need for the development of special liability rules and procedures for mineral resource activities. It is, however, disputed whether the mining ban applies to all areas south of 60° south latitude.[35] According to some Consultative Parties, it does. Other Consultative Parties argue, however, that mineral resource activities on the deep seabed in the Antarctic Treaty area can be lawfully carried on because the 1959 Antarctic Treaty is not prejudicial to the rights of states on the high seas (Article VI).[36] If the mining ban does not apply to the deep seabed in the Antarctic Treaty area, mineral resource activities on the deep seabed are governed by the 1982 UN Convention on the Law of the Sea. Liability for damage caused by mineral resource activities on the deep seabed south of 60° south latitude will then be

---

[33] See, for the text of the draft protocol referred to in this paragraph, doc. UNEP/CHW.1/G.1/10/L.2, 1999.
[34] See *Tractatenblad* [Dutch Treaty Series] 1993, No. 72, at 5, and *Tractatenblad* 1996, No. 81, at 2, respectively.
[35] See D. Vidas, 'The Southern Ocean Seabed: Arena for Conflicting Regimes?', in D. Vidas and W. Østreng (eds.), *Order for the Oceans at the Turn of the Century* (The Hague: Kluwer Law International, 1999), pp. 291–314, especially at pp. 303–307.
[36] See UNTS, Vol. 402, pp. 72 ff.

covered by the relevant provisions of the Mining Code that is being developed under the auspices of the International Seabed Authority.[37] It is notable that the draft Mining Code provides for fault liability of the contractor (Section 12.1). The contractor can be held liable for 'the actual amount of any damage, including damage to the marine environment', including 'the costs of reasonable measures to prevent or limit damage to the marine environment'. Although the text is not explicit, this provision could be interpreted so as to include irreparable damage to the environment.

Conflicting Special Liability Regimes

The patchwork of 'source-oriented' liability regimes leaves gaps and does not provide for prompt, adequate and effective compensation or other form of reparation for damage to the Antarctic environment and dependent and associated ecosystems. Firstly, these liability regimes only cover incidental damage arising out of certain, defined hazardous activities. They do generally not cover non-incidental damage arising out of hazardous activities, or non-incidental damage or incidental damage arising out of non-hazardous activities. Secondly, because these liability regimes are generally tied to damage to persons and property, they do not adequately cover liability for damage to the environment. Thirdly, even if a source-oriented liability regime applies, plaintiffs will usually not be able to obtain compensation or full compensation if the damage-causing activity cannot be identified, if no causal connection can be established between the damage and the damage-causing activity, or if a limitation of liability can be validly invoked by the person to whom liability is channelled. Fourthly, a certain instance of damage may not be covered by a source-oriented liability regime as the result of limited participation of Consultative Parties and other states in these regimes. Moreover, the application of source-oriented liability regimes that are linked to damage suffered within the limits of national jurisdiction in a given case of damage in the Antarctic Treaty area may unleash sensitive jurisdictional disputes.

Given the legal gaps left by source-oriented liability regimes, these regimes do not rule out the legal need for an Antarctic environmental liability regime. However, the place of such a liability regime within the system of international special liability regimes remains to be examined. In particular, the Consultative Parties will have to decide whether to adopt a complementary approach, which would take into account existing source-oriented liability regimes, or a comprehensive approach, which would not take into account such existing regimes. A complementary approach would involve: 1) the accession of Consultative Parties and other states to relevant source-oriented liability regimes, and 2) the development of an Antarctic environmental liability regime for damage not covered by such liability regimes.

---

[37] See 1997 Draft Standard Terms of Exploration Contract, doc. ISBA/LTC/WP.2.

However, a complementary approach would seem to be of limited value. The primary objective of an Antarctic environmental liability regime is to impose liability for a certain manifestation of damage irrespective of the nature of the damage-causing activity. The fragility of the Antarctic environment and dependent and associated ecosystems has been the reason to incorporate in the Protocol an obligation to develop a special liability regime. In contrast, source-oriented liability regimes focus on the hazardous nature of a certain activity. They have been developed with the aim of affording special protection for damage suffered by natural or legal persons and do generally not adequately cover damage to the environment, in particular irreparable damage.[38]

Furthermore, a complementary approach would only be useful if damage to the Antarctic environment and dependent and associated ecosystems is already covered by source-oriented liability regimes, and overlap between an Antarctic environmental liability regime and such source-oriented liability regimes would prove to be considerable. However, an Antarctic environmental liability regime will only be applicable to activities in the Antarctic Treaty area that are covered by the Protocol. It will not apply if damage to the Antarctic environment and dependent and associated ecosystems is caused by activities not covered by the Protocol, irrespective of whether these activities are carried on inside or outside the Antarctic Treaty area. For example, the operation of spacecraft above the Antarctic Treaty area is not covered by the Protocol and damage caused on the surface of Antarctica should therefore not be brought within the scope of the Antarctic environmental liability regime. Likewise, the exploitation of living resources in the Antarctic Treaty area is not covered by the Protocol and damage caused by it will not be covered by the Antarctic environmental liability regime.[39] The exploitation of mineral resources is not admissible under the Protocol and will therefore probably not be covered by the Antarctic environmental liability regime. However, this does not exclude that damage to Antarctic environment and dependent and associated ecosystems is covered by a source-oriented liability regime. In contrast, the operation of aircraft and ships as well as the use of nuclear energy in the Antarctic Treaty area are covered by the Protocol, and damage caused by such activities could be covered by an Antarctic environmental liability regime. This regime will only cover damage to the Antarctic environment and dependent and associated ecosystems and not damage to other environments or to persons or property. However, this does not exclude that such damage to other environments is covered by a source-oriented liability regime. Hence, an activity covered by the Protocol may be subject to different special liability regimes.

In view of the different foci of source-oriented liability regimes, and to avoid sensitive jurisdictional issues in a given case of damage, the development of a comprehensive Antarctic environmental liability regime for damage to the Antarctic

---

[38] See Lefeber, Chapter 11 in this book.
[39] *Ibid.*

environment and dependent and associated ecosystems has an added value. With respect to damage to the Antarctic environment and dependent and associated ecosystems, the parties to an Antarctic environmental liability regime would then have to exclude the application *inter se* of other special liability regimes – reasonable costs of response measures and remedial measures included – if such regimes had less-favourable provisions for the compensation or any other form of reparation of such damage. Some proposals for an Antarctic environmental liability regime contain a conflict-of-treaties rule that precisely has that effect.[40]

CONCLUSION

From a legal point of view, the need to develop a special Antarctic environmental liability regime cannot be questioned. Article 16 of the Protocol imposes an unambiguous obligation on the parties to develop such a regime. The foregoing discussion in this chapter has shown that there were also good legal reasons for the incorporation of Article 16 in the Protocol. As was also shown, neither ordinary state and civil liability nor special 'source-oriented' liability regimes appear to provide for prompt, adequate and effective compensation or any other form of reparation for damage to the Antarctic environment and dependent and associated ecosystems.

Nevertheless, the need for a special Antarctic environmental liability regime continues to be questioned. It is said that no serious incident with non-intentional and lasting damage to the environment has ever occurred in the Antarctic Treaty area. Moreover, it is said that it is not likely that such a serious incident will occur. The risk of damage to the environment is also said to be low, because the activities being carried on in the Antarctic Treaty area are not hazardous, for example scientific research, or are not likely to cause lasting damage, for example oil spills.[41] Yet, if damage occurs, legal remedies may not be available for reparation of the damage to the environment. But then, it is said that the Consultative Parties will take the necessary action to repair the damage to the environment. The *Bahia Paraiso* incident is cited as a precedent in which other Consultative Parties assisted Argentina in taking the necessary response measures to repair the damage to the environment.[42]

If the added reparative value of an Antarctic environmental liability regime would be thus limited in practice, the development of an Antarctic environmental liability regime may be justified by its corrective function or its preventive function.

---

[40] See Art. 3(2) in Chile, 'Draft Annex on Environmental Liability'; Art. 3(2) in the Netherlands, 'Liability'; but see Art. 2(3) in EO, that gives priority to other international agreements in force within the Antarctic Treaty System.

[41] See, e.g., COMNAP, 'An Assessment of Environmental Emergencies Arising from Activities in Antarctica', doc. XXIII ATCM/WP 16, 1999; see also Skåre, Chapter 9 in this book.

[42] See also Lefeber, Chapter 11 in this book.

The correction of wrongful conduct through the imposition of civil liability may be a useful side-effect of a liability regime, but the imposition of criminal or administrative liability would be more appropriate to effectuating such correction. Likewise, the prevention of damage to the environment through the risk of facing civil liability may be a useful side-effect of a liability regime, but the regulation of activities would be more appropriate to achieving the same result. In particular, an Antarctic environmental liability regime should not result in the *de facto* prohibition of an activity by the imposition of prohibitive insurance coverage. Hence, the performance of the reparative function should remain the primary objective of an Antarctic environmental liability regime.[43]

The question of whether the opportunity costs of the negotiations actually justify the development of such regime has been answered by the Consultative Parties, which have negotiated the Protocol. It would require an amendment of the Protocol to reverse this answer, as is currently contained in Article 16 of the Protocol.

---

[43] See Blay and Green, 'The Development of a Liability Annex to the Madrid Protocol', p. 26 ('combination will require caution, because they can lead to complex results').

11

The Prospects for an Antarctic Environmental Liability Regime

René Lefeber

In view of the discussions at Consultative Meetings and in the 'Group of Legal Experts on Liability' convened under Article 16 of the 1991 Protocol on Environmental Protection to the Antarctic Treaty[1] (Protocol),[2] the Consultative Parties seem to have agreed on the development of a special civil liability regime for damage to the Antarctic environment and dependent and associated ecosystems. This is also reflected in the draft annex on environmental liability, proposed by the chairman of the Group of Legal Experts on Liability (Eighth Offering or EO),[3] and in the proposals of Australia, of Chile, of the Netherlands, of South American Consultative Parties jointly, and of the United Kingdom – that all were on the agenda of the 1999 Consultative Meeting in Lima.[4] This chapter will analyse, in the light of the aforementioned proposals, the need and desirability of specific procedural and substantive rules of a special civil liability regime for damage to the Antarctic environment and dependent and associated ecosystems.

---

[1] ILM, Vol. 30, 1991, pp. 1,461ff.
[2] See an overview provided by Skåre, Chapter 9 in this book.
[3] See 'Chairman's Eighth Offering – Annex on Environmental Liability', reproduced in 'Liability – Report of the Group of Legal Experts', Appendix 1 (EO), doc. XXII ATCM/WP 1, 1998.
[4] See Australia, 'Principles for an Antarctic Liability Regime', doc. XXIII ATCM/WP 15, 1999; the Netherlands, 'Liability', doc. XXIII ATCM/WP 18, 1999; United Kingdom, 'Liability', doc. XXIII ATCM/WP 21, 1999; Chile, 'Draft Annex on Environmental Liability to the Madrid Protocol', doc. XXIII ATCM/WP 34, 1999; Argentina, Brazil, Chile, Ecuador, Peru and Uruguay, 'Basic Definitions and Considerations for the Annex on the Liability Regime', doc. XXIII ATCM/WP 35, 1999.

THE DEVELOPMENT OF AN ANTARCTIC ENVIRONMENTAL LIABILITY REGIME

One controversial issue is whether an Antarctic environmental liability regime should be laid down in a single annex covering damage to the Antarctic environment and dependent and associated ecosystems arising from all activities covered by the Protocol[5] or step by step in different annexes. Although Article 16 of the Protocol does not exclude the development of two or more annexes, the idea of a step-by-step approach originates in a United States proposal, originally submitted at the 1996 Consultative Meeting in Utrecht, to focus on the implementation of Article 15 of the Protocol (on emergency response action).[6] One could have sympathy for the priority attached by the United States to the further regulation of emergency response action in view of the events following the *Bahia Paraiso* oil spill.[7] The *Bahia Paraiso* was a vessel which was owned by the Argentine government and flew the Argentine flag. However, in the absence of adequate action taken by Argentina in the aftermath of the oil spill, it was the United States, Chile, Korea and Spain that incurred a substantial part of the costs for the immediate response measures and the Netherlands for the further response measures, namely the costs connected with the salvaging of the fuel and oil remaining in the hull of the wreckage.[8]

Although a step-by-step approach is not excluded by the text of Article 16 of the Protocol, it cannot encroach on the obligation to ultimately develop a comprehensive special liability regime for damage to the Antarctic environment and dependent and associated ecosystems. The Protocol applies to environmental impacts from *all* human activities conducted in the Antarctic Treaty[9] area with the exception of activities regulated by the 1946 International Convention for the Regulation of Whaling[10], the 1972 Convention for the Conservation of Antarctic Seals[11], and the 1980 Convention on the Conservation of Antarctic Marine Living Resources[12].[13] The damage to the Antarctic environment and dependent and associated ecosystems that

---

[5] See, e.g., Art. 2 in EO; para. 3 in Australia, 'Principles for an Antarctic Liability Regime'; Art. 2 in Chile, 'Draft Annex on Environmental Liability to the Madrid Protocol'; and Art. 2 in the Netherlands, 'Liability'.

[6] See 'Report of the Group of Legal Experts on the Work Undertaken to Elaborate an Annex or Annexes on Liability for Environmental Damage in Antarctica', reproduced in doc. XXII ATCM/WP 1 (1998), 'Liability – Report of the Group of Legal Experts', paras. 41–45, and 'Appendix 2 – Proposal of the United States Delegation (April 1996)'.

[7] See also Lefeber, Chapter 10 in this book.

[8] See the 'Initial Environmental Evaluation, Associated with the Salvage Operation of the Remaining Oil of the Bahia Paraiso', prepared by the Instituto Antarctico Argentino, Argentina & Rijkswaterstaat – Tidal Waters Division, the Netherlands, 1992.

[9] UNTS, Vol. 402, pp. 71ff.

[10] UNTS, Vol. 161, pp. 74ff.

[11] UNTS, Vol. 1080, pp. 175ff; and ILM, Vol. 11, 1972, pp. 251ff.

[12] ILM, Vol. 19, 1980, pp. 837ff.

[13] See Art. 4(2) of the Protocol; see also Final Act of the Eleventh Antarctic Treaty Special Consultative Meeting, reproduced in ILM, Vol. 30, 1991, pp. 1,455ff.

results from the harvesting of species covered by these Conventions, in other words the possible biodiversity loss or the application of inhumane catch methods, is therefore excluded from the scope of an annex on environmental liability. However, other damage to the Antarctic environment and dependent and associated ecosystems that is caused in connection with the harvesting of such species does not seem to be excluded, for instance damage caused by oil pollution from fishing vessels or in the process of removing the wreckage of a fishing vessel.[14]

A step-by-step approach still requires a general framework, for example for the adjudication of claims, to be developed in order to make the first step work. It is difficult to envisage how the establishment of such framework is to be apportioned over two or more liability annexes. However, a step-by-step approach could leave open the drawing up of a list of activities to which, and the circumstances under which, an annex on environmental liability is applicable. Moreover, a step-by-step approach could facilitate the development of different rules and procedures on liability for different activities or certain manifestations of damage to the Antarctic environment and dependent and associated ecosystems, including the possible preferential treatment of scientific activities and associated logistic activities.[15] Since some of the moot issues relate to the scope of such a list, a step-by-step approach could considerably speed up the negotiation process on other parts of an annex on environmental liability.

THE FUNCTIONAL SCOPE OF AN ANTARCTIC ENVIRONMENTAL LIABILITY REGIME

A special feature of Article 16 of the Protocol is its focus on a certain manifestation of damage irrespective of the nature of the damage-causing activity. This feature evidences that it has not been the hazardous nature of human activities in the Antarctic but the fragility of the Antarctic environment and dependent and associated ecosystems that has prompted the incorporation in the Protocol of an obligation to develop a special liability regime. The Antarctic annex on environmental liability will therefore be different from the majority of special liability regimes that focus on the hazardous nature of a certain activity.[16] These special liability regimes have been developed with the aim of affording special protection to natural and legal persons rather than to the environment. Special liability regimes for damage caused by hazardous activities have proved to be indispensable because hazardous activities may cause considerable, if not enormous, damage for which an individual natural or legal person cannot be reasonably held responsible.

---

[14] See also Art. 2(2) in EO; Art. 2(2) in Chile, 'Draft Annex on Environmental Liability to the Madrid Protocol'; Art. 2(2) in the Netherlands, 'Liability'.
[15] See Art. 3(3) of the Protocol.
[16] For examples of such regimes see Lefeber, Chapter 10 in this book.

Since the origin of the Antarctic annex on environmental liability is the wish to preserve the fragile Antarctic environment, proposed texts do not cover damage other than damage to the Antarctic environment and dependent and associated ecosystems. The Eighth Offering, however, provides that the development of an annex on liability for loss or impairment of an established use, or loss or damage to property of a third party arising directly out of damage to the Antarctic environment and dependent and associated ecosystems is not necessarily precluded.[17] One may add personal injury to these manifestations of damage. Although the regulation of liability for these manifestations of damage seems desirable, the scope of an annex on environmental liability is limited by the objectives of the Protocol, explicit reference to which is made in Article 16 of the Protocol. The Protocol's objective is the 'comprehensive protection of the Antarctic environment and dependent and associated ecosystems' (Article 2 of the Protocol). To bring damage other than damage to the Antarctic environment and dependent and associated ecosystems within the scope of the Protocol, and hence within an annex developed under Article 16 of the Protocol, would require a stretched interpretation of the cited objective of the Protocol. Unfortunate as this may seem, it is therefore submitted that the Protocol does not provide a legal basis for an annex covering manifestations of damage other than damage to the Antarctic environment and dependent and associated ecosystems itself. Although Article 16 of the Protocol does not provide a legal basis for these manifestations of damage, liability for these could be laid down in a measure under Article IX(1) of the 1959 Antarctic Treaty. Since the origin of special civil liability regimes is the wish to afford protection to the victims of damage caused by hazardous activities, it could be argued that such a measure, if developed, should not afford special protection to victims of damage caused by non-hazardous human activities in the Antarctic Treaty area. In view of the controversial legal status of the Antarctic region and related jurisdictional issues, there is, however, a need to develop special procedural safeguards in order to ensure that victims can avail themselves of domestic legal remedies.[18] This may require the development of special procedural rules on the establishment of jurisdiction of courts, the identification of an applicable law, and the recognition and enforcement of judgements. However, in cases of damage caused by non-hazardous activities, it would not seem necessary to afford special protection to victims by substantive safeguards, such as the introduction of a strict standard of liability or rules on the establishment of financial security to cover the liability. In these cases, there is no reason not to apply the general rule of contemporary international and municipal law that provides for damage to lie where it falls, unless the occurrence of damage is imputable, in the sense of wrongful conduct, to the source of the damage.[19]

---

[17] Art. 1 in EO, second sentence.
[18] See also *ibid*.
[19] See further discussion below in this chapter.

THE DEFINITION OF DAMAGE

The Protocol only provides a legal basis for an annex that covers damage to the Antarctic environment and dependent and associated ecosystems. The phrase 'damage to the Antarctic environment and dependent and associated ecosystems' can only mean: 1) the costs of immediate response measures to minimise and contain damage; 2) the costs of further response measures to clean up damage; 3) the costs of remedial measures to restore damaged components of the Antarctic environment and dependent and associated ecosystems or to return it to a comparable state; and 4) compensation for irreparable damage if remedial measures are not possible, feasible, or desirable. In the discussions on the definition of damage to the Antarctic environment and dependent and associated ecosystems, two proposals have been advanced that seriously limit the scope of the definition.

One proposal excludes from the definition of 'damage' impacts that have been considered and judged acceptable on the basis of an environmental impact assessment conducted in accordance with the Protocol, be it an initial environmental evaluation (IEE) or a comprehensive environmental evaluation (CEE).[20] This idea, for which no precedent can be found in international law, is fundamentally misconceived. The *admissibility* of an activity should not be confused with *liability* for damage *resulting* from such an activity. According to the municipal law of many countries, the authorisation of an activity by the competent authorities does not prejudice the existence of third-party liability. The adoption of such a definition of damage would result in a significant and undesirable limitation of liability that does not find support in international liability law. The idea apparently originates in the limitation of the scope of a liability annex, or annexes, to damage to the Antarctic environment and dependent and associated ecosystems. However, taking this into account, a case for exemption from liability can be made for foreseeable, non-accidental damage to the Antarctic environment and dependent and associated ecosystems.[21] In such a case, a financial contribution to the environmental protection fund, if established, could – and should – be considered before an activity can be admitted. However, it is difficult to see how this idea could work in cases of accidental damage or damage that could not be foreseen at the time of the environmental impact assessment.

---

[20] See, e.g., Art. 3(1) in EO; Art. 4(a) in Chile, 'Draft Annex on Environmental Liability to the Madrid Protocol' (on the assumption that the indents are meant to be alternative conditions and not cumulative conditions); first indent of Argentina, Brazil, Chile, Ecuador, Peru and Uruguay, 'Basic Definitions and Considerations for the Annex on the Liability Regime'; Art. A in United Kingdom, 'Liability'; see also S. Blay and J. Green, 'The Development of a Liability Annex to the Madrid Protocol', *Environmental Policy and Law*, Vol. 25, 1995, pp. 24–37, at p. 28. See, however, para. 1 in Australia, 'Principles for an Antarctic Liability Regime'.

[21] This is also reflected in the Contact Group Text of 28 May 1999, drafted by the United Kingdom at the XXIII Consultative Meeting, 'Personal Report of the Chairman of the Liability Discussion in Working Group 1', doc. XXIII ATCM/WP 41, 1999; see also Art. 1(a) in the Netherlands, 'Liability' (that contains cumulative conditions).

A second proposal suggests that liability should only arise if the damage is 'of a more than minor and more than transitory nature' or, alternatively, is 'significant and lasting' or 'significant given its nature or scale and its duration'.[22] A similar element can be found in many international environmental obligations and must be proved to establish a state's liability *ex delicto*, or state responsibility, for acts not in compliance with such obligations.[23] In international environmental law, this qualification has its origin in the balance between the sovereign right of states to develop activities within their jurisdiction, on the one hand, and the duty to prevent transboundary environmental damage, on the other. Apart from the difficulties encountered in establishing what is 'minor' and what is 'transitory' or what is 'significant' and particularly what is 'lasting', this does not justify the incorporation of a threshold of damage in a special civil liability regime.[24] However, although such thresholds cannot be found in many special civil liability regimes, they have found their way, albeit in different wordings, in a few recently concluded special civil liability regimes, including the 1988 Convention on the Regulation of Antarctic Mineral Resource Activities (CRAMRA).[25] This Convention only covers damage to the Antarctic environment or dependent and associated ecosystems 'beyond that which is negligible or which has been assessed and judged to be acceptable pursuant to this Convention' (Article 1(15)).

THE ADOPTION OF PRECAUTIONARY MEASURES, RESPONSE MEASURES, AND REMEDIAL MEASURES

The incorporation in an annex on environmental liability of obligations for the operator to take precautionary measures, response measures, and remedial measures has proved to be another controversial issue. Firstly, it is not clear whether such obligations are already covered by the Protocol, in particular Article 15 which provides that 'each Party *agrees* to: (a) provide for prompt and effective response action to [environmental] emergencies ... and (b) establish contingency plans for response to incidents' (emphasis added). Response measures, including the

---

[22] See Art. 3(1) in EO ('of a more than minor and more than transitory nature' or 'significant and lasting'); para. 1 in Australia, 'Principles for an Antarctic Liability Regime' ('substantial and lasting'); Art. 1(a) in the Netherlands, 'Liability' ('significant'); first indent in Argentina, Brazil, Chile, Ecuador, Peru and Uruguay, 'Basic Definitions and Considerations for the Annex on the Liability Regime' ('significant and lasting'); Art. A in United Kingdom, 'Liability' ('significant and lasting'); and Contact Group Text of 28 May 1999 ('significant and lasting' or 'significant given its nature or scale and its duration').
[23] See R. Lefeber, *Transboundary Environmental Interference and the Origin of State Liability* (The Hague: Kluwer Law International, 1996), pp. 24–25 and 86–89.
[24] See also Blay and Green, 'The Development of a Liability Annex to the Madrid Protocol', p. 28 ('potential for dispute').
[25] ILM, Vol. 27, 1988, pp. 868ff. See also 1993 Council of Europe Convention on Civil Liability for Damage Resulting from Activities Dangerous to the Environment that excludes 'pollution at tolerable levels under local relevant circumstances' (Art. 8(d)); reprinted in ILM, Vol. 32, 1993, pp. 1,228ff.

preparation of contingency plans, are clearly covered by Article 15 of the Protocol. It is, however, debated whether obligations in this respect can be imposed on operators on the basis of Article 13(1) of the Protocol, which provides that '[e]ach Party shall take appropriate measures within its competence ... to enforce compliance with this Protocol'. Some Consultative Parties argue that Article 15 of the Protocol has been meant to impose obligations on states only. Although their contention finds some support in the placement of Article 15 in the Protocol and in its text, the Protocol does not prohibit the imposition on operators of obligations with respect to response measures. The object and purpose of the Protocol rather lend support to the contention that obligations in this respect can – and should – be imposed on operators by national legislation (Articles 3(3)(vi), 3(4) and 13(1) of the Protocol).[26] In contrast, precautionary measures and remedial measures would not seem to be covered by Article 15 of the Protocol, which only deals with 'emergency response action'. Remedial measures would also not seem to come within any other provision of the Protocol. However, precautionary measures relate to prevention, and obligations in this respect can – and should – be imposed on operators by national legislation (Articles 3(4) and 13(1) of the Protocol). Although obligations to adopt precautionary measures and response measures would thus seem to be covered by the Protocol, the incorporation in an annex on environmental liability of obligations to adopt precautionary measures, response measures, and remedial measures would strengthen the system of environmental protection in the absence of unambiguous provisions in this respect in the Protocol. If states would be required, either by the Protocol or an annex on environmental liability, to take appropriate measures within their competence to ensure that the operator adopts precautionary measures, response measures, and remedial measures, they will have to impose criminal or administrative liability in case of non-compliance.[27] However, the imposition of criminal or administrative liability in case of non-compliance with obligations related to precautionary measures, response measures, and remedial measures would be a strange phenomenon in a special civil liability regime.

The primary objective of a special civil liability regime is the reimbursement of costs or the payment of compensation. Special civil liability regimes normally fail to provide for other forms of relief, notably a right to bring a claim against the operator with the aim of forcing the operator to take precautionary measures, response measures, and remedial measures. However, it is worth considering whether the admissibility of civil actions under the annex should not be extended to include the adoption of such measures.[28] In view of their special interest in the protection of the

---

[26] The municipal legislation of South Africa and the United States reflect this view; see Dodds, Chapter 21 in this book, and Joyner, Chapter 22 in this book. The municipal legislation of Norway does not; see Njåstad, Chapter 20 in this book.

[27] This seems to be the intention of paras. 4, 5 and 7 in Australia, 'Principles for an Antarctic Liability Regime' and Arts. 6, 9 and 10 in Chile, 'Draft Annex on Environmental Liability to the Madrid Protocol'.

[28] This is the intention of Art. 4(1–3) in the Netherlands, 'Liability'. See also Blay and Green, 'The

Antarctic environment and dependent and associated ecosystems, the endowment to an Antarctic environmental protection fund – or to non-governmental organisations – of the right to bring such a claim should particularly be considered. The award of such a claim as a means of reparation would be a relative novelty, but one which seems to become increasingly accepted in international instruments.[29] However, if awarded as a means of reparation, the obligation of the operator to adopt precautionary measures, response measures, and remedial measures is distinct from obligations imposed on operators on the basis of Article 13(1) of the Protocol. As a means of reparation, the obligation to adopt such measures can only be enforced in civil proceedings, and not in criminal or administrative proceedings.

A related issue concerns the question whether and under what circumstances the operator can be held liable if a person other than the operator takes response measures or remedial measures in the aftermath of an incident. It is particularly debated whether the need to adopt such measures should be determined, or at least, whether the adoption of such measures should be authorised, by a state; and in addition whether that state should be the state to whose jurisdiction such other person is subject or the state to whose jurisdiction the operator is subject.[30] In order to prevent the adoption of premature or excessive measures by third parties, some state involvement seems inevitable. However, in this respect a distinction should be made between immediate response measures, further response measures, and remedial measures.[31] In order to enable the timely adoption of immediate response measures, the third party cannot be expected to wait for state authorisation; rather, a mere notification of its intention to adopt such measures should suffice. In contrast, the adoption of further response measures or remedial measures is already subject to state authorisation and does not need further regulation. Since an operator is only liable for damage that transgresses a certain threshold (see the definition of damage above in this chapter), the adoption of further response measures and remedial measures by the operator or a third party is only permitted if the Protocol's obligatory procedures on environmental impact assessment are complied with. These procedures apply to all activities covered by the Protocol, unless the activity has less than a minor or transitory impact on the Antarctic environment and dependent and associated ecosystems (Article 8 and Annex I of the Protocol).

Development of a Liability Annex to the Madrid Protocol', p. 34.

[29] See also Lefeber, *Transboundary Environmental Interference and the Origin of State Liability*, p. 293.

[30] See Art. 9 in Chile, 'Draft Annex on Environmental Liability to the Madrid Protocol' (state of the operator determines the need for 'further response action', including remedial measures); seventh and eighth indents in Argentina, Brazil, Chile, Ecuador, Peru and Uruguay, 'Basic Definitions and Considerations for the Annex on the Liability Regime'. See, however, Art. 4(5–6) in the Netherlands, 'Liability' (no state authorisation required).

[31] This is also reflected in the text of informal Contact Group of 28 May 1999, drafted by Germany at the XXIII Consultative Meeting, reproduced in 'Personal Report of the Chairman of the Liability Discussion in Working Group 1', submitted by New Zealand, doc. XXIII ATCM/WP 41, 1999.

THE CHANNELLING OF LIABILITY

The annex on environmental liability is likely to become a special civil liability regime that imposes primary liability on operators,[32] in other words on a legal or natural person who organises an activity in the Antarctic Treaty area and not on states. The annex does not envisage the imposition of liability on states with the exception of: 1) cases in which states are operators, and 2) the possible imposition of residual state liability. Evidently, an annex on environmental liability cannot prejudice the application of customary international law in relevant cases; hence it cannot exclude a state's liability for an internationally wrongful act (liability *ex delicto*).[33] The channelling of primary liability to the operator is in accordance with the polluter-pays principle and also has a preventive function in addition to its corrective and reparative functions.

A surprising and paralysing discussion has, however, emerged on the establishment of jurisdiction of states parties over operators as a prerequisite for the imposition of liability on an operator.[34] I doubt whether it is really necessary or desirable to limit the imposition of liability for damage to operators that have a jurisdictional link with a state party. This would only be necessary if the emerging annex would impose criminal or administrative liability, but not if it would provide for civil law remedies. If one legal instrument could implement the obligation of the parties to the 1959 Antarctic Treaty to ensure that 'no one engages in any activity in Antarctica contrary to the principles or purposes' of the Treaty (Article X) or the Protocol (see Article 13(2)), that instrument would be an annex to the Protocol. Liability for damage to the Antarctic environment and dependent and associated ecosystems should be imposed on operators purely because the damage has occurred in the Antarctic Treaty area. The location of the organiser of an activity or the establishment of other jurisdictional grounds should therefore not become criteria for imposing liability. For the effectuation of civil liability, in particular the enforcement of a judgement, it suffices that the operator has assets in a state party to the annex on environmental liability. The establishment of jurisdiction would then only be relevant in the context of the establishment and verification of the operator's financial security to cover his liability or the possible imposition of residual state liability.[35]

---

[32] See Art. 3 *quater* in EO; Arts. 11 and 12 in Chile, 'Draft Annex on Environmental Liability to the Madrid Protocol'; Art. 5 in the Netherlands, 'Liability'; and Art. C in United Kingdom, 'Liability'.

[33] See Lefeber, Chapter 10 in this book.

[34] See Art. 3 *quater* in EO; Arts. 4(h) and 6 in Chile, 'Draft Annex on Environmental Liability to the Madrid Protocol'; Art. 1(g) in the Netherlands, 'Liability'; and Arts. C and D in United Kingdom, 'Liability'.

[35] See on this further discussion below in this chapter.

THE QUESTION OF JOINT AND SEVERAL LIABILITY

The principle of joint and several liability is a common element of special civil liability regimes and there would *prima facie* not seem to be any reason for not including it in an Antarctic environmental liability regime.[36] However, at the Meeting of the Group of Legal Experts on Liability in Tromsø, objections were raised against the imposition of joint and several liability because it may discourage the undertaking of joint activities.[37] Joint activities, in particular in the field of scientific research, are common in the Antarctic, and such activities should also be stimulated because they produce environmental economies of scale. However, it is submitted that objections against joint and several liability can easily be overcome by the conclusion of an indemnification agreement between the partners of a joint activity, or, if no such agreement has been concluded, by adequate legal remedies for redress.[38] The imposition of joint and several liability will only stimulate the evaluation of the financial security of the partners and, hence, secure the availability of funds for the reimbursement of costs or the compensation of damage should this be necessary.

THE STANDARD OF LIABILITY

It seems beyond discussion that the standard of liability for damage to the Antarctic environment and dependent and associated ecosystems should be strict.[39] The concept of 'strict liability' is a common element of special civil liability regimes that seek to ensure relief to innocent victims for damage arising out of hazardous activities, even if no fault of a legal or natural person can be proved. This, however, is a deviation from general international and municipal law, which generally provide for damage to lie where it falls, unless the occurrence of damage is imputable, in the sense of wrongful conduct, to the source of the damage: *ubi jus, ibi remedium* (only a violation of a right creates a right to redress). Since the Antarctic environmental liability regime is also meant to serve a special purpose, namely the protection of the Antarctic environment and dependent and associated ecosystems, a case can be made for a similar encroachment upon the maxim *ubi jus, ibi remedium*. In order to

---

[36] See also Art. 3 *ter* in EO; Art. 8 in Chile, 'Draft Annex on Environmental Liability to the Madrid Protocol'; and Art. 6 in the Netherlands, 'Liability'.

[37] See 'Summary from the Meeting of the Group of Legal Experts on Liability, 27–29 May 1998 in Tromsø, Norway'.

[38] See also New Zealand, 'Joint and Several Liability and International Collaborative Science' doc. XXIII ATCM/WP 10, 1999; and Art. 5(2) in EO; Art. 8(2) in Chile, 'Draft Annex on Environmental Liability to the Madrid Protocol'; and Art. 6(3) in the Netherlands, 'Liability'.

[39] See Art. 3 *bis* in EO; Art. 7 in Chile, 'Draft Annex on Environmental Liability to the Madrid Protocol'; Art. 5 in the Netherlands, 'Liability'; third indent in Argentina, Brazil, Chile, Ecuador, Peru and Uruguay, 'Basic Definitions and Considerations for the Annex on the Liability Regime'; Art. C(2) in United Kingdom, 'Liability'.

promote the timely implementation of immediate response measures by third parties, the reimbursement of the reasonable costs of the adoption of such measures requires a strict standard of liability. Yet, it is not so obvious why such a strict standard of liability should also apply to the reimbursement of the costs of further response measures and remedial measures or to the payment of compensation for irreparable damage. In these cases, there is sufficient time to evaluate, if need be in court: 1) whether the conduct of the operator justifies the imposition of liability, and 2) if damage can be cleaned up or damaged components of the environment can be restored or returned to a comparable state to obtain a court order against the operator with the aim of forcing him to take the necessary further response measures and remedial measures. The principal reason for special civil liability regimes to provide for strict liability is the compelling need to compensate innocent victims, even if the person that has caused the damage has not been at fault. Since the origin of the Antarctic annex on environmental liability is the wish to preserve the fragile Antarctic environment, the need to deviate from general international law and municipal law may not be so compelling.

LIABILITY FOR IRREPARABLE DAMAGE

At the 1998 Consultative Meeting in Tromsø, most delegations spoke in favour of an article providing for the payment of compensation for irreparable damage to the Antarctic environment and dependent and associated ecosystems.[40] The adoption of an article on irreparable damage would break new ground but likewise be a threat to the successful completion of an annex on environmental liability, as it requires a solution of the problem of the monetary valuation of irreparable damage to the environment.[41] It should therefore not come as a surprise that several proposals attribute to the Consultative Meeting the task of developing guidelines in this respect and of determining the amount to be paid by the operator.[42]

Common to these proposals is the provision that the operator should pay compensation to an Antarctic environmental protection fund if: 1) damage has not been repaired because the operator or a third party has not taken response measures or remedial measures within a certain time frame, or 2) damage is irreparable because such measures are not possible, feasible, or desirable.[43] It does not seem to

---

[40] See *Final Report of the Twenty-second Antarctic Treaty Consultative Meeting, Tromsø, Norway, 25 May–5 June 1998* (Oslo: Royal Norwegian Ministry of Foreign Affairs, 1998), para. 79.

[41] See also Lefeber, *Transboundary Environmental Interference and the Origin of State Liability*, pp. 136–138 and 291–294.

[42] See Art. 19(2) in Chile, 'Draft Annex on Environmental Liability to the Madrid Protocol'; Art. 7(2) in the Netherlands, 'Liability', that delegates the determination of the amount to be paid to specially appointed experts; and Art. G in United Kingdom, 'Liability'.

[43] See Art. 5 *bis* in EO; Art. 12 in Chile, 'Draft Annex on Environmental Liability to the Madrid Protocol'; Art. 7(1) in the Netherlands, 'Liability'; and Art. E(1)(b) in United Kingdom, 'Liability' (only 'unreparable' damage).

be effective to impose these obligations on the operator because doing so would eventually require state involvement to enforce these obligations in criminal or administrative proceedings. If damage is reparable and neither the operator nor a third party has taken response measures or remedial measures, the Antarctic environmental protection fund – or non-governmental organisations – should be entitled to bring a civil action against the operator with the aim of forcing him to adopt such measures. Only if damage is really irreparable, should it be considered to endow the Antarctic environmental protection fund, which is after all to become the beneficiary of compensation for irreparable damage, with the right to bring financial claims for such damage.[44]

THE LIMITATION OF LIABILITY

Special civil liability regimes normally limit the liability of the person liable by financial limits, time limits, and exemptions. The introduction of such limitations is necessary to keep activities insurable and, hence, economically viable. If no financial products are available to cover the risks, the costs for *bona fide* natural or legal persons engaging in activities subject to a strict liability regime are likely to become prohibitive. It is therefore generally considered necessary to limit strict liability and to consider the introduction of residual liability schemes to provide additional means to repair the damage caused.[45]

It must be recognised that the compulsory coverage of environmental risks associated with activities in the Antarctic Treaty area will be a new challenge for the insurance industry. This is particularly true for the coverage of irreparable damage. It therefore seems inevitable to provide, at least initially, for a limitation of the liability of operators in an annex on environmental liability. However, such limitations, in particular financial limits, could be increased gradually if the insurance market developed the necessary financial products to cover the risks associated with activities in the Antarctic Treaty area. To facilitate such increases, an annex should provide a flexible decision-making procedure for adjusting the limits.

The limitation of liability in amount is part of all proposed texts for an annex on environmental liability.[46] The proposals are rudimentary and do not specify financial limits. The establishment of ceilings is likely to become a tedious issue because current proposals extend to all activities covered by the Protocol. Since the magnitude of damage is related to the nature of the damage-causing activity, it will

---

[44] See Arts. 18(3)(b), 18(5), 19(2) and 19(3) in Chile, 'Draft Annex on Environmental Liability to the Madrid Protocol'. See, however, Art. F(2) in United Kingdom, 'Liability', that attributes such right to two or more states parties. See also discussion below in this chapter.

[45] See further discussion below in this chapter.

[46] See, e.g., Art. 9 in EO; Art. 15 in Chile, 'Draft Annex on Environmental Liability to the Madrid Protocol'; and Art. 11 in the Netherlands, 'Liability'.

be necessary to establish different ceilings for different activities. Another approach to this matter may be not to limit the liability, but to permit states to fix per activity the amount for which the operator has to establish financial security.[47] However, in such a case, liability would not be limited in amount and the operator would remain liable for the surplus if the financial security is not sufficient to cover all damage. This approach may discourage the development of activities in the Antarctic Treaty area.

The limitation of liability in time has somewhat been neglected in the discussions but has found its place in a few, recently proposed texts which, however, do not specify the time limits.[48] In this case too, it will be necessary to establish different time limits for different activities because the point in time at which damage manifests itself is also related to the nature of the damage-causing activity. Some activities may cause latent damage that manifests itself only many years after the incident. For this reason it does not seem appropriate to establish a uniform time limit. Existent special civil liability regimes have shown that liability does not extend further than 30 years after the event from which the damage resulted.[49]

In accordance with other special civil liability regimes, it has been proposed that operators be exempted from liability if damage is caused as a result of an extreme natural disaster, armed conflict, or an act of terrorism.[50] However, it has also been proposed that operators be exempted from liability for activities in accordance with the Protocol and its annexes, such as the disposal of waste or the discharge of oil or garbage from ships in case of emergency in accordance with Annexes III and IV to the Protocol.[51] To the extent that this latter exemption originates in the occurrence of

---

[47] See, e.g., 1993 Council of Europe Convention on Civil Liability for Damage Resulting from Activities Dangerous to the Environment (Art. 12), ILM, Vol. 32, 1993, pp. 1,228ff.

[48] See, e.g., Art. 16 in Chile, 'Draft Annex on Environmental Liability to the Madrid Protocol'; and Art. 12 in the Netherlands, 'Liability'.

[49] See, e.g., 1996 IMO International Convention on Liability and Compensation for Damage in Connection with the Carriage of Hazardous and Noxious Substances by Sea (Art. 37(3): 10 years) (ILM, Vol. 35, 1996, pp. 1,406ff); 1993 Council of Europe Convention on Civil Liability for Damage Resulting from Activities Dangerous to the Environment (Art. 17(2): 30 years); 1989 UN/ECE Convention on Civil Liability for Damage Caused During Carriage of Dangerous Goods by Road, Rail and Inland Navigation Vessels (Art. 18(2): 10 years) (doc. UN/ECE/TRANS/79 (1989)); 1969 IMCO International Convention on Civil Liability for Oil Pollution Damage (Art. VIII: 6 years) (UNTS, Vol. 973, pp. 3ff), as amended by protocols of 1976 (ILM, Vol. 16, 1977, pp. 617ff), of 1984 (ILM, Vol. 23, 1984, pp. 195ff), and of 1992 (doc. IMO/LEG/CONF.9/15 (1992)); 1963 IAEA Convention on Civil Liability for Nuclear Damage (Art. VI(1)(a): 30 years (loss of life and personal injury) or 10 years (other damage) (UNTS, Vol. 1063, pp. 265ff), as amended by protocol of 1997 (ILM, Vol. 36, 1997, pp. 1,462ff); 1960 OECD Convention on Third Party Liability in the Field of Nuclear Energy (Art. 8(a): 10 years), as amended by protocols of 1962 and 1982 (UNTS, Vol. 956, pp. 251ff).

[50] See Art. 6(2) in EO; para. 2 in Australia, 'Principles for an Antarctic Liability Regime'; Art. 14(a–b) in Chile, 'Draft Annex on Environmental Liability to the Madrid Protocol'; fourth indent in Argentina, Brazil, Chile, Ecuador, Peru and Uruguay, 'Basic Definitions and Considerations for the Annex on the Liability Regime'; Art. 8(a–b) in the Netherlands, 'Liability'; and Art. H(2) in United Kingdom, 'Liability'.

[51] See Art. 6(1) in EO; Art. 14(c) in Chile, 'Draft Annex on Environmental Liability to the Madrid

an emergency that is covered by the aforementioned exemption, there is no need for a specific provision. It may be doubted, however, whether all these cases of emergency are the result of an extreme natural disaster, armed conflict, or an act of terrorism and whether operators should be exempted from liability in all cases. Since the admissibility of an activity is not the same as liability for damage resulting from such an activity (see also the definition of damage above), there does not seem to be a compelling need to provide for these additional exemptions from liability.

It has furthermore been suggested to exclude from the annex on environmental liability damage caused by ships that are solely used for governmental non-commercial service.[52] Similar provisions can be found in other special civil liability regimes, but its insertion in an Antarctic environmental liability regime would significantly affect its scope as most ships used in the Antarctic Treaty area are on governmental non-commercial services, or are said to be so.[53]

THE ESTABLISHMENT OF FINANCIAL SECURITY

The establishment of financial security to cover the risk of liability is an indispensable instrument to effectuate special civil liability regimes. In view of the risk that the operator's financial security is not sufficient to cover that operator's liability in a particular case, it is important to look at the obligations of states in this respect. The Eighth Offering merely provides that states 'shall ensure ... that non-State operators ... be required to have and maintain' financial security.[54] It is submitted that the phrase 'be required to' limits the use of the provision. This wording merely obliges states to promulgate regulations that require operators to establish financial security in accordance with the annex on environmental liability. However, it does not provide that a state must supervise that such financial security be actually obtained by the operator or that it incurs residual state liability for the operator's failure to obtain such financial security. In view of the difficulties in ascertaining which state has jurisdiction over an operator in the Antarctic, it cannot be expected that residual state liability be imposed for the mere failure of an operator to establish and maintain sufficient financial security (absolute obligation).[55] However, states could be required to take 'appropriate measures' to

---

Protocol'; Art. 8(c) in the Netherlands, 'Liability'; fourth indent in Argentina, Brazil, Chile, Ecuador, Peru, and Uruguay, 'Basic Definitions and Considerations for the Annex on the Liability Regime'; and Art. H(1) in United Kingdom, 'Liability'. See, however, Blay and Green, 'The Development of a Liability Annex to the Madrid Protocol', p. 29 ('[b]y its very nature it is inappropriate to subject any remedial liability such as immediate response action to any defences').

[52] See Art. 6(3) in EO.

[53] See also Orrego Vicuña, Chapter 3 in this book.

[54] Art. 8 in EO; see also Art. 17 in Chile, 'Draft Annex on Environmental Liability to the Madrid Protocol'.

[55] See, however, para. 6 in Australia, 'Principles for an Antarctic Liability Regime' and Art. 10(1) in the Netherlands, 'Liability' that could be read in this way; and Blay and Green, 'The Development of

ensure that operators establish and maintain sufficient financial security to cover the risk of liability (due diligence obligation). If this obligation would be qualified in this way, there will be no need to establish an exclusive jurisdictional link between an operator and a state regarding the establishment and maintenance of financial security. All states parties that at some point have a jurisdictional link with an operator could be required to enforce this obligation, including the state of departure to or arrival from the Antarctic Treaty area. If compliance with the obligation to establish and maintain financial security only requires the exercise of due diligence, it would also become acceptable to require departure states and arrival states to supervise the establishment and maintenance of financial security by operators from third states, including operators from non states parties.

THE INTRODUCTION OF RESIDUAL LIABILITY SCHEMES

The limitation of the liability of the operator means that part of the damage will not be repaired. This disadvantage can be overcome by the introduction of residual state liability or collective compensation arrangements. From existing and proposed special civil liability regimes, it appears that, with the exception of nuclear liability regimes, states are extremely reluctant to accept liability on a residual basis.[56] Yet, several proposals envisage the imposition of residual state liability if liability has not been satisfied by the operator or otherwise.[57] However, these proposals only envisage residual state liability if a state has not complied with its obligations under the Protocol or its annexes (liability *ex delicto*). Accordingly, a state can only be held liable if it has not diligently enacted or implemented appropriate regulations to prevent, minimise, and contain damage to the Antarctic environment and associated or dependent ecosystems, including rules and procedures to ensure prompt, adequate, and effective relief for any such damage caused. This provision would seem to state the obvious as did CRAMRA (Article 8(3)(a)) from which it is derived. Hence, these proposals do not seem to go beyond existing international law and carry the risk that a state will not accept liability for an internationally wrongful act, unless civil law remedies have been exhausted.[58] This would only be different if the person seeking reimbursement would be entitled to bring a 'residual' claim against a state in a municipal court.[59] In such a case, the lawfulness of *acta jure*

---

a Liability Annex to the Madrid Protocol', p. 30 (in relation to shortfalls in insurance cover: 'the most effective mechanism is that of 'residual liability' of the state of origin').

[56] See Lefeber, *Transboundary Environmental Interference and the Origin of State Liability*, pp. 299–311.

[57] See Art. 7(2) in EO; Art. 13(2) in Chile, 'Draft Annex on Environmental Liability to the Madrid Protocol'; Art. 9 in the Netherlands, 'Liability'; ninth indent in Argentina, Brazil, Chile, Ecuador, Peru and Uruguay, 'Basic Definitions and Considerations for the Annex on the Liability Regime'; and, less clearly, Art. I in United Kingdom, 'Liability'.

[58] See also Lefeber, Chapter 10 in this book.

[59] See Art. 9 in the Netherlands, 'Liability', and especially the Explanatory Note to Article 9 (State

imperii of states would come within the jurisdiction of municipal courts. The acceptance of such transnational settlement of claims would be a novelty in international liability law.

A more realistic alternative to residual state liability is the introduction of a collective compensation arrangement. Examples of such arrangements can be found in several special civil liability regimes.[60] The objective of such arrangements is normally to make available supplementary funds if not all damage can be satisfied by the operator. Thus, such arrangements generally enable the reimbursement of reasonable costs if: 1) the operator cannot be held liable because he can avail himself of an exoneration or cannot be identified, 2) the operator is financially incapable to meet his liability up to the financial limit, 3) the total amount of damage exceeds the financial limit, or 4) relief is provided on an interim basis. Several texts propose the establishment of an Antarctic environmental protection fund, but these range from the attribution to such a fund of mere financial tasks to the attribution of operational tasks, notably the adoption of response measures and remedial measures together with a right of recourse against the operator.[61] As argued in previous sections, consideration should also be given to attributing to an Antarctic environmental protection fund the right to bring actions against an operator before municipal courts or before other dispute-settlement mechanisms provided for in the annex with the aim of claiming compensation for irreparable damage or with the aim of forcing the operator to take precautionary measures, response measures, and remedial measures. After all, the protection of the Antarctic environment and dependent and associated ecosystems is of common interest to the parties to the Protocol that claim to act 'in the interest of mankind as a whole' (see Preamble). It would seem that the interest of mankind is best served by endowing a relatively independent Antarctic environmental protection fund with the necessary powers to act as a trustee on behalf of the international community.

The costs of operating collective compensation arrangements are normally charged to the sectors of industry that benefit from the activities that are subject to a special civil liability regime. The proposed Antarctic environmental protection fund is to be financed by the payment of compensation for irreparable damage[62] and by

---

Liability and Responsibility) of 28 May 1999, drafted by the Netherlands at the XXIII Consultative Meeting, reproduced in 'Personal Report of the Chairman of the Liability Discussion in Working Group I', submitted by New Zealand, doc. XXIII ATCM/WP 41, 1999.

[60] See 1996 IMO International Convention on Liability and Compensation for Damage in Connection with the Carriage of Hazardous and Noxious Substances by Sea (Arts. 13–36); 1971 IMCO International Convention on the Establishment of an International Fund for Compensation for Oil Pollution Damage (ILM, Vol. 22, 1972, pp. 284ff), as amended by protocols of 1976 (ILM, Vol. 16, 1977, pp. 621ff), of 1984 (ILM, Vol. 23, 1984, pp. 177ff), and of 1992 (doc. IMO/LEG/CONF.9/16 (1992)).

[61] See Art. 10 in EO; Arts. 18 and 19(3) in Chile, 'Draft Annex on Environmental Liability to the Madrid Protocol'; Art. 13 in the Netherlands, 'Liability'; and Art. G in United Kingdom, 'Liability'.

[62] See Arts. 10(2)(a) and 5 *bis* (1) in EO (only financial tasks); Art. 19(2) and 12 in Chile, 'Draft Annex on Environmental Liability to the Madrid Protocol' (including operational tasks); Art. 13(6) and 7(1) in the Netherlands, 'Liability' (only financial tasks); and Art. G(1) in United Kingdom,

voluntary contributions.[63] It goes without saying that such an arrangement will not guarantee a secure and regular income for the fund. The imposition of special liability levies on activities conducted in the Antarctic Treaty, for example, a per capita levy on tourists, deserves attention in the interest of ensuring the solvability of an Antarctic environmental protection fund.[64]

THE SETTLEMENT OF CLAIMS

The settlement of claims for damages is a complex issue. Plaintiffs may bring claims for: 1) the reimbursement of the costs of response measures, 2) compensation of irreparable damage, and 3) possibly other forms of relief, such as the enforcement of the operator's obligation to take precautionary measures, response measures, and remedial measures. It must be decided whether to endow an international tribunal or municipal courts with jurisdiction to entertain all or some of these claims. The solution chosen may be different depending on whether the plaintiff or the defendant is a state. The discussion has been volatile and no clear consensus has yet emerged on the issue; the discussion concentrates on several alternatives. One proposal provides for a traditional approach according to which cases involving states are to be brought before an international tribunal. Other cases are to be brought before the domestic court where the defendant or the plaintiff has his habitual residence or place of business.[65] Another proposal is innovative as it allows the plaintiff to choose between the adjudication of claims before an international tribunal or a domestic court.[66] If a case is brought before a municipal court, it is presumed that states have waived their right to claim immunity from jurisdiction. A third proposal suggests that all claims, irrespective of whether or not the plaintiff or the defendant is a state, should be brought before the municipal court of the state in which the operator has his habitual residence or place of business.[67]

Except for the last, these proposals are not clear regarding vesting exclusive jurisdiction in an international tribunal or a domestic court in respect of a single incident. This has some disadvantages if several plaintiffs bring claims in respect of a single event. One reason for endowing a single court with exclusive jurisdiction is the limitation of liability in amount because this may require the equitable distribution of limited funds over the plaintiffs. Other reasons are forum shopping by

---

'Liability'.

[63] See Art. 10(3) in EO; Art. 18(6) in Chile, 'Draft Annex on Environmental Liability to the Madrid Protocol'; and Art. 13(7) in the Netherlands, 'Liability'.

[64] See commentary to Art. 13 in the Netherlands, 'Liability', arguing that 'possible further sources of contribution should be considered'.

[65] See Art. 11 in EO (alternative A); Art. 29(2) in Chile, 'Draft Annex on Environmental Liability to the Madrid Protocol'; Arts. 14(1) and 15(2) in the Netherlands, 'Liability' (unless a state involved consents to the referral of the case to domestic courts).

[66] See Art. 11 in EO (alternative B).

[67] See Art. F(1) in United Kingdom, 'Liability'.

plaintiffs who have the right to choose the forum, the problems of *lis pendens* and related actions, the risk of diverging case law, and considerations of transparency. The problems involved can best be overcome by endowing an international tribunal or the municipal court of the habitual residence or place of business of the defendant, be it a state or not, with exclusive jurisdiction to hear the claim. However, if the defendant does not have his habitual residence or place of business in a state party, for example, a defendant who has his habitual residence or place of business in a third state, another jurisdictional link should suffice. This link could, for example, be the habitual residence or place of business of the plaintiff.

Competent, domestic courts or tribunals are required to apply the annex on environmental liability as the uniform law of the states parties with respect to the adjudication of claims for damage to the Antarctic environment and dependent and associated ecosystems. Domestic courts, however, can avail themselves of their *lex fori* to deal with legal issues not regulated in the annex, for instance, the establishment of a causal link between the activity and the damage. In contrast, an international tribunal will have to build its own, but uniform, rules in such a case on the basis of the general principles of law of the states parties to the annex.

At the end of the day, the effectiveness of any settlement-of-claims procedure depends on the recognition and enforcement of judgements in the state where the defendant has assets. Safe for generally accepted exceptions, such as considerations of public policy, the annex must therefore unequivocally provide for the recognition and enforcement of judgements in other states parties to the annex.[68]

CONCLUSION

It is difficult to draw any meaningful conclusions from the on-going negotiations on the Antarctic environmental liability regime. Although several Consultative Parties endeavour to bring the negotiations to a fruitful conclusion, others have remained silent and do not even voice their concerns on the emerging Antarctic environmental liability regime. It is the personal assessment of the author of this chapter that consensus among the Consultative Parties on an Antarctic environmental liability regime can only be achieved if the negotiations move away from a strict liability standard towards a fault liability standard. The strict liability standard should only be maintained in respect of the reimbursement of the reasonable costs of immediate response measures. This is not as bad as it may sound. The risk that activities carried on in the Antarctic Treaty area will cause lasting damage to the Antarctic environment and dependent and associated ecosystems is said to be relatively low.[69] Furthermore, the objective of the Antarctic environmental liability regime is the protection of the fragile environment, not the protection of natural or legal persons

---

[68] See only Art. 14(7) in the Netherlands, 'Liability'.
[69] See COMNAP, 'An Assessment of Environmental Emergencies Arising from Activities in Antarctica', doc. XXIII ATCM/WP 16, 1999; see also Skåre, Chapter 9 in this book.

that have suffered damage. In the absence of precedents, it will be difficult to justify the imposition of strict liability by mere reference to the fragility of the Antarctic environment and dependent and associated ecosystems. Since appropriate precautionary measures will also have to be adopted to avoid the imposition of fault liability, the automatic imposition of strict liability is also not likely to strengthen the preventive function or corrective function of an Antarctic environmental liability regime.[70]

In the opinion of the present author, it would also be more useful to focus on the civil enforcement of the adoption of precautionary measures, response measures, and remedial measures. Serious thought should, furthermore, be given to the idea of endowing an independent fund with the right to bring claims with the aim of enforcing the adoption of necessary precautionary measures, response measures, and remedial measures. The establishment of a civil enforcement mechanism would also contribute to the reparative function of an Antarctic environmental liability regime and enhance its practical use.

---

[70] See, however, Blay and Green, 'The Development of a Liability Annex to the Madrid Protocol', p. 28 ('likely to encourage parties undertaking activities in Antarctica to take greater care to avoid harm in the first place').

Part IV

RELATIONSHIP WITH OTHER INTERNATIONAL INSTRUMENTS AND ARRANGEMENTS

12

Relationship between the Environmental Protocol and UNEP Instruments

Donald R. Rothwell

One of the characteristics of the Antarctic Treaty System (ATS) is that it has generally sought to remain beyond the reach of a number of important international institutions. The ATS of course operates within the framework of international civil society created by the United Nations system. With the exception of Switzerland, all parties to the Antarctic Treaty[1] are members of the UN. The ATS has also been subject to debate within the UN. During the second half of the 1980s and some years of the 1990s, the ATS was required to 'defend' itself against criticism from within the UN General Assembly.[2] However, notwithstanding that incidence, the ATS has never actively sought to engage the UN system. This is to be contrasted with the more active engagement between the ATS and other international institutions.[3] This engagement has partly been facilitated through the mechanisms of the Antarctic Treaty itself, especially through Articles II and III, which have provided an avenue for active exchange between the ATS and a number of 'international organisations having a scientific or technical interest in Antarctica'[4]. The most prominent of these organisations is the Scientific Committee on Antarctic Research (SCAR). In

---

[1] UNTS, Vol. 402, pp. 71ff.
[2] Peter Beck has undertaken an extensive and ongoing study of the debate within the United Nations concerning Antarctica, see especially P. Beck, 'Antarctica, Viña del Mar and the 1990 UN Debate', *Polar Record*, Vol. 27, 1991, pp. 211–216; P. Beck, 'The 1991 UN Session: The Environmental Protocol Fails to Satisfy the Antarctic Treaty System's Critics', *Polar Record*, Vol. 28, 1992, pp. 307–314; and P. Beck, 'The United Nations and Antarctica, 1992: Still Searching for the Elusive Convergence of View', *Polar Record*, Vol. 29, 1993, pp. 313–320.
[3] For a general overview of the ATS in the international community, see D. Vidas, 'The Antarctic Treaty System in the International Community: An Overview', in O.S. Stokke and D. Vidas (eds.), *Governing the Antarctic: The Effectiveness and Legitimacy of the Antarctic Treaty System* (Cambridge University Press, 1996), pp. 35–60.
[4] Art. III(2) of the Antarctic Treaty.

addition, institutions such as the Intergovernmental Oceanographic Commission, the International Civil Aviation Organisation (ICAO), the International Hydrographic Organisation (IHO), the International Maritime Organisation (IMO) and the World Meteorological Organisation (WMO) have all been engaged in the ATS process, principally through invitations to attend Consultative Meetings.[5]

During the past 40 years, the ATS has evolved from the relatively simple 1959 Antarctic Treaty into a 'system' which now, in addition to the Treaty itself, includes three international agreements[6] and over 250 other international instruments (measures, decisions, resolutions and recommendations) adopted at altogether twenty-three Antarctic Treaty Consultative Meetings held thus far.[7] As a result of this evolution of the ATS, new institutions have been created along the way. In the case of the CCAMLR, a Commission was provided for collecting catch data and monitoring Southern Ocean fishing activities.[8] In the case of the Environmental Protocol, the Committee for Environmental Protection (CEP) was established to provide advice and formulate recommendations to the parties in relation to the implementation of the Protocol.[9]

The evolution of the ATS and expansion of its focus beyond just peaceful scientific cooperation in the Antarctic have inevitably forced the Antarctic Treaty Consultative Parties to consider the merits of developing linkages with a broader range of international organisations that may have expertise in areas that the Treaty parties are now in need of. To this end, the ATS has during the 1990s begun to develop more linkages with the United Nations Environment Programme (UNEP). Since its establishment in 1972, UNEP has increasingly become a focus for environmental programs within the UN system. UNEP became engaged with the ATS in the early 1990s during the negotiation of the Environmental Protocol, and the first consultative meeting it attended was the Eighteenth Consultative Meeting in 1994. At this Meeting, UNEP indicated its willingness to cooperate and to offer experience and assistance to Antarctic Treaty parties.[10] In the following year, an international symposium of Antarctic experts met to consider the future of the ATS

---

[5] For background, see P. Gautier, 'Institutional Developments in the Antarctic Treaty System', in F. Francioni and T. Scovazzi (eds.), *International Law for Antarctica*, 2nd edition (The Hague: Kluwer Law International, 1996), pp. 31–47.

[6] These are the 1972 Convention for the Conservation of Antarctic Seals (Seals Convention) (ILM, Vol. 11, 1972, pp. 251ff); the 1980 Convention on the Conservation of Antarctic Marine Living Resources (CCAMLR) (ILM, Vol. 19, 1980, pp. 837ff); and the 1991 Protocol on Environmental Protection to the Antarctic Treaty (Environmental Protocol) (ILM, Vol. 30, 1991, pp. 1,461ff). Note that the Environmental Protocol defines the 'Antarctic Treaty System' in Art. 1(e) as follows: 'the Antarctic Treaty, the measures in effect under that Treaty, its associated separate international instruments in force and the measures in effect under those instruments'.

[7] The first Antarctic Treaty Consultative Meeting was held in 1961.

[8] CCAMLR, Art. VII.

[9] Protocol, Art. 12. See discussion by Orheim, Chapter 6 in this book.

[10] *Final Report of the Eighteenth Antarctic Treaty Consultative Meeting, Kyoto, Japan, 11–22 April 1994* (Tokyo: Japanese Ministry of Foreign Affairs, 1994), Annex C (v).

and recommended that '[C]onsideration should be given to closer cooperation with relevant United Nations specialised agencies and programs, in particular UNEP'.[11] Since that time, UNEP has repeated its offer to the ATS that avenues for cooperation be further explored.[12]

It is clear that a range of opportunities exists for enhanced interaction between UNEP and the ATS at an institutional level. However, with the entry into force of the Environmental Protocol, the more imminent question may be the level of interaction that will take place at the normative level, i.e., between the Protocol and UNEP instruments. Essentially this raises the question as to how the Protocol will interact with international environmental law, in particular that part of it over which UNEP has oversight. This interaction may prove beneficial to the ATS but could also result in conflict because of possible overlap between the Protocol and UNEP instruments. The Consultative Parties have been aware of the various implications of this interaction ever since the adoption of the Protocol. At the Eighteenth Consultative Meeting in 1994, a working paper was submitted by Chile which raised for consideration the interaction of the Protocol with international environmental agreements.[13] Among the agreements identified for assessment were the following:

- the 1985 Vienna Convention for the Protection of the Ozone Layer (Ozone Layer Convention);[14]
- the 1989 Basel Convention on the Control of Transboundary Movements of Hazardous Wastes and Their Disposal (Basel Convention);[15]
- the 1992 United Nations Framework Convention on Climate Change (FCCC);[16]
- the 1992 Convention on Biological Diversity (CBD).[17]

All of these Conventions are UNEP Conventions. Reference was also made to the 1972 Convention on the Prevention of Marine Pollution by Dumping of Wastes and Other Matter and the 1973/78 International Convention for the Prevention of Pollution from Ships,[18] and attention was drawn to the 1982 UN Convention on the Law of the Sea.[19] However, all that was agreed upon at the Consultative Meeting was that 'it was important to ensure proper coordination between global

---

[11] R. Guyer and H. Wyndham, 'Chairmen's Summary of the Symposium', in A.W. Jackson (ed.), *On the Antarctic Horizon: Proceedings of the International Symposium on the Future of the Antarctic Treaty System* (Hobart: Australian Antarctic Foundation, 1996), p. 113.
[12] *Final Report of the Nineteenth Antarctic Treaty Consultative Meeting, Seoul, 8–19 May 1995* (Seoul: Ministry of Foreign Affairs of the Republic of Korea), Annex G (v).
[13] *Final Report of the XVIII ATCM*, para. 51.
[14] ILM, Vol. 26, 1987, pp. 1,529ff.
[15] ILM, Vol. 28, 1989, pp. 657ff.
[16] ILM, Vol. 31, 1992, pp. 849ff.
[17] *Ibid.*, pp. 818ff.
[18] *Final Report of the XVIII ATCM*, para. 52.
[19] On the aspects of the relationship with the Law of the Sea Convention, see discussion by Vidas,

environmental agreements and the operation of the Antarctic Treaty system, in particular, of the Protocol on Environmental Protection to the Antarctic Treaty'.[20] The relations of the ATS with other environmental treaties continued to be discussed at both the Nineteenth and Twentieth Consultative Meetings in 1995 and 1996. At the Nineteenth Consultative Meeting specific note was made of the 'importance of examining the potential overlap between the Protocol and other international treaties'.[21]

In this chapter, these issues will be explored by first assessing the role of UNEP in Antarctic affairs, then reviewing the UNEP instruments and their role in Antarctica, and finally considering the existing relationship between the Environmental Protocol and UNEP instruments.

UNEP AND ANTARCTICA[22]

The United Nations Environment Programme

The United Nations Environment Programme was established in 1972, following the United Nations Conference on the Human Environment held that year in Stockholm. UNEP reports to the UN General Assembly through the Economic and Social Council of the United Nations. The role of UNEP has been stated as to 'catalyse, coordinate and stimulate environmental action within the UN system'.[23] Originally operating only with a small secretariat and an Executive Director, UNEP has gradually expanded considerably and currently maintains several regional offices and a headquarters located in Nairobi.[24]

UNEP's activities are numerous. Significantly for international environmental law, they include sponsoring the negotiation of global environmental treaties through UNEP's Environmental Law and Institutions Programme Activity Centre. At the global level, prominent examples of treaties concluded under UNEP's sponsoring include the 1985 Ozone Layer Convention with its 1987 Montreal Protocol[25] and the 1989 Basel Convention. At the regional level, the UNEP Regional Seas Programme has resulted in the negotiation of 13 regional seas conventions, while UNEP was also responsible for negotiation of the African regional version of the 1973 Convention on International Trade in Endangered Species (CITES)[26]

Chapter 14 in this book.

[20] *Final Report of the XVIII ATCM*, para. 55.

[21] *Final Report of the XIX ATCM*, para. 52.

[22] For more detailed discussion of UNEP and ATS interaction, see D.R. Rothwell, 'UNEP and the Antarctic Treaty System', *Environmental Policy and Law*, Vol. 29, 1999, pp. 17–24.

[23] UNEP, *Executive Director's Report* (Nairobi: United Nations Environment Programme, 1985).

[24] For more details see the UNEP website <www.unep.org>.

[25] ILM, Vol. 26, 1987, pp. 1,550ff.

[26] UNTS, Vol. 993, pp. 243ff.

entitled the 1994 Lusaka Agreement on Cooperative Enforcement Operations Directed at Illegal Trade in Wild Flora and Fauna.[27]

UNEP has also been engaged in developing a range of guidelines or principles to guide international environmental law. Prominent examples include the 1987 Cairo Guidelines and Principles for the Environmentally Sound Management of Hazardous Wastes,[28] the 1987 London Guidelines for the Exchange of Information on Chemicals in International Trade,[29] the 1987 Goals and Principles of Environmental Impact Assessment,[30] and the 1985 Montreal Guidelines for the Protection of the Marine Environment Against Pollution from Land-Based Sources.[31] Other UNEP activities include administering or supervising international and regional agreements, such as the UNEP Regional Seas Programme, CITES, and the 1971 Ramsar Convention on Wetlands of International Importance Especially as Waterfowl Habitat[32] administered on behalf of UNEP by IUCN. Furthermore, as a complementary function, UNEP coordinates a number of monitoring and information systems such as the Global Environment Monitoring System, the International Register of Potentially Toxic Chemicals and the International Programme on Chemical Safety.[33] As the lead UN environmental agency, UNEP also advises governments on the development of national environmental legislation, administrative programs and institutional arrangements.[34]

In the 1990s, UNEP provided, at the request of the UN Secretary-General, assistance in the preparation of Antarctic 'State of the Environment' reports in 1994[35] and 1996.[36] In addition, UNEP has expressed interest in matters such as ecosystems, environmental monitoring and management, oceans and the development of environmental law, all of which have direct relevance for Antarctic environmental protection. At the institutional level, notwithstanding the competence and the authority of UNEP in relation to environmental matters, UNEP has not been a regular attendee at Consultative Meetings. With the adoption of the Environmental

---

[27] UNEP doc. No. 94/7929.

[28] UNEP Governing Council Decision 14/30, 1987.

[29] UNEP Governing Council Decision 15/30, 1987.

[30] UNEP Governing Council Decision 14/25, 1987.

[31] UNEP Governing Council Decision 13/1811, 1985.

[32] UNTS, Vol. 996, pp. 245ff.

[33] List drawn from UNEP, *Environmental Law in UNEP* (Nairobi: United Nations Environment Programme, 1991).

[34] See D. Kaniaru, M. Iqbal, E. Mrema and S. Chowdhury, 'UNEP's Programme of Assistance on National Legislation and Institutions', in S. Lin and L. Kurukulasuriya (eds.), *UNEP's New Way Forward: Environmental Law and Sustainable Development* (Nairobi: United Nations Environment Programme, 1995), pp. 153–170; and D. Kaniaru and L. Kurukulasuriya, 'Capacity Building in Environmental Law', in *ibid.*, pp. 171–184.

[35] *Final Report of the XVIII ATCM*, Annex C (v).

[36] See *Question of Antarctica: State of the Environment in Antarctica. Report of the Secretary-General*, UN doc. A/51/390, 1996. See also the 1994 UN Secretary-General Report on the 'Question of Antarctica', in UN doc. A/49/370, 1994.

Protocol and the Protocol's dedication of Antarctica as a 'natural reserve, devoted to peace and science',[37] the Consultative Parties' interest in developing linkages between the ATS and organisations with an environmental mandate has increased. Against this background, UNEP has both attended a number of recent Consultative Meetings and submitted its reports at a number of these.[38]

The UNEP Conventions

Since its establishment, UNEP has been responsible for the negotiation of a number of significant global international environmental conventions. While these initiatives have seldom sought to deal specifically with Antarctic matters, many have implications for the Antarctic environment because of their global geographic outreach. Below, some of the principal UNEP instruments will be assessed with comments on their application in the Antarctic.

CITES. The Convention on International Trade in Endangered Species is a well-established international environmental regime that operates to place limitations on the international trade of species which are or which may become threatened with extinction.[39] Relying on import and export controls, the Convention places obligations upon the parties to control the trade in species that have been listed under the Convention. CITES does not seek to protect habitats, to prevent pollution, or to protect over-exploitation of species.[40] Notwithstanding its limited scope, CITES is considered to have been reasonably successful in regulating trade in some endangered species. For present purposes, however, it is important to note that CITES did not seek to override any existing or future conventions dealing with certain trade in endangered species or specimens.[41]

CITES has potential application to trade in Antarctic species which may be considered endangered. Currently, some Antarctic species of birds, penguins, seals and whales are listed for protection. However, CITES does not provide for in situ protection; the regime can only provide an additional mechanism for the protection of any Antarctic species being traded. Given near unanimous acceptance of CITES amongst Antarctic Treaty parties, it is possible for the Treaty parties to individually or collectively seek to have a variety of endangered Antarctic species placed on either of the Appendix I or II CITES lists. This applies not only for species which

---

[37] Protocol, Art. 2.
[38] See, for instance, report submitted by UNEP in *Final Report of the Twenty-second Antarctic Treaty Consultative Meeting, Tromsø, Norway, 25 May – 5 June 1998* (Oslo: Royal Norwegian Ministry of Foreign Affairs, 1998), Annex G, pp. 301–302.
[39] CITES, Art. II.
[40] P.W. Birnie and A.E. Boyle, *International Law and the Environment* (Oxford: Clarendon Press, 1992), p. 475.
[41] CITES, Art. XIV.

are found on the continent of Antarctica but also especially for the marine living resources found in the Southern Ocean, many of which are facing a range of threats to their existence. The most prominent threatened species in Antarctic waters are whales, but a number of fish species have also become threatened because of recent overfishing.[42] Listing these species under CITES would further enhance the international protection of endangered Antarctic species both in situ (restricting or prohibiting trade in the species and thereby removing the incentive for commercial exploitation) and ex situ, and at the same time provide further global support for existing ATS instruments.

The Ozone Layer Convention and the Montreal Protocol. The 1985 Vienna Convention on the Protection of the Ozone Layer and the subsequent 1987 Montreal Protocol were negotiated as a result of the growing international concern for the effect of CFCs and other airborne pollutants on the Earth's ozone layer. Since its negotiation and implementation, the Ozone Layer Convention has been subject to revision through the Montreal Protocol and at subsequent review meetings of the Conference of Parties. The principal obligation stipulated in the Ozone Layer Convention is that the parties take measures 'to protect human health and the environment against adverse effects resulting, or likely to result from human activities which modify or are likely to modify the ozone layer.'[43] The Convention creates a framework under which the states parties can develop specific obligations by way of subsequent instruments. This was the approach adopted with the Montreal Protocol in 1987 under which the parties committed themselves to specific obligations to limit or phase out the use of ozone-depleting chemicals.[44] The goals set under the Montreal Protocol for eventual phase-out have been continually reviewed and adjusted by the Conference of Parties, and new commitments have been added with respect to ozone-depleting substances.[45] The Montreal Protocol was also significant in that it recognised the need to distinguish between developed and developing countries in regard to imposing differentiated targets for substance phase-out.[46] Notwithstanding some problems in implementation, the Ozone Layer Convention and the Montreal Protocol have generally been considered successful in terms of limiting the use of CFCs and other ozone-depleting substances. However, it

---

[42] Extensive recent overfishing of the Patagonian toothfish may suggest that this Antarctic fish stock is a suitable candidate for CITES listing; see G.L. Lugten, 'The Rise and Fall of the Patagonian Toothfish – Food for Thought', *Environmental Policy and Law*, Vol. 27, 1997, pp. 401–407.

[43] Ozone Layer Convention, Art. 2(1).

[44] These included CFCs, Halons, Carbon Tetrachloride and Trichloroethane, Hydrochlorofluorocarbons, Hyrdobromofluorocarbons and Methyl Bromide. See Montreal Protocol, Art. 2A-2H.

[45] See, e.g., the 1990 Adjustments and Amendments to the 1987 Montreal Protocol on Substances that Deplete the Ozone Layer (ILM, Vol. 30, 1991, pp. 537ff); and the 1992 Adjustments and Amendments to the 1987 Montreal Protocol on Substances that Deplete the Ozone Layer (ILM, Vol. 32, 1993, pp. 874ff).

[46] See Montreal Protocol, Art. 5, and comment in Birnie and Boyle, *International Law and the Environment*, pp. 407–408.

is now understood that it will be some years before the regime has any noticeable impact in terms of reducing the depletion of the ozone layer – a process which commenced some time ago and is continuing because of the impact of chemicals that have already been released into the atmosphere.

With the existence of the ever-expanding 'hole in the ozone layer' which appears over the Antarctic during the (Southern Hemisphere) summer, the Ozone Layer Convention regime has a direct impact on the Antarctic. This impact on the Antarctic environment is difficult to quantify. However, it is known that the continued depletion of the ozone layer has global consequences for both the climate and the environment. Because of this global threat, the Antarctic Treaty Consultative Parties have a genuine interest in the issues that arise from meeting obligations under the Ozone Layer Convention. While the depletion of the ozone layer presents new opportunities for Antarctic science, this development needs to be balanced against the loss of opportunity to conduct science in a completely 'pure' Antarctic environment, which was one of the fundamental objectives of the Antarctic Treaty.[47]

Basel Convention. The 1989 Basel Convention on the Control of Transboundary Movements of Hazardous Wastes and their Disposal was negotiated in order to place controls on the unregulated movement of hazardous wastes. The Basel Convention seeks to place limitations on the export of such wastes between states of export and import; and the Convention also introduces safeguards for transit states.[48] Article 4(6) of the Convention, which deals directly with the Antarctic, states the following provision:

> The Parties agree not to allow the export of hazardous wastes or other wastes for their disposal within the area south of 60° South latitude, whether or not such wastes are subject to transboundary movement.

The effect of this provision is to ensure that transboundary wastes cannot be exported to any area in the Antarctic, irrespective of whether a state of import exists. The seven Antarctic territorial claimants are therefore precluded under this Convention from seeking to export waste from their mainland territories to their claimed Antarctic territories. The Convention would also control the export of hazardous wastes from Antarctica,[49] but only in situations where there was a distinctive state of export and state of import. If a claimant state did seek to export

---

[47] See the Antarctic Treaty, Arts. II and IX(1).

[48] For discussion of the Convention see K. Kummer, 'The Basel Convention: Ten Years On', *Review of European Community and International Environmental Law*, Vol. 7, 1998, p. 227. For a review of how the Convention applies in Antarctica, see V. Bou, 'Waste Disposal and Waste Management in Antarctica and the Southern Ocean', in Francioni and Scovazzi (eds.), *International Law for Antarctica*, pp. 366–373.

[49] The Basel Convention, Arts. 1 and 2 are broad enough to cover both hazardous domestic waste and industrial wastes, which would include wastes generated on Antarctic scientific bases and associated activities.

waste from an Antarctic scientific facility to its mainland territory, notwithstanding there not being transboundary movement, the Convention would still apply if the export of waste took place through a state of transit.[50]

Convention on Biological Diversity. The CBD developed out of the 1992 United Nations Conference on Environment and Development (UNCED). The Convention entered into force in 1993 and meetings of the Conference of Parties have been held since 1994. The Convention is the first attempt to adopt common global standards for the conservation of biological diversity and sustainable use of components of biological diversity.[51] To this end, state parties to the Convention accept obligations for the conservation and sustainable use of biological diversity, identification and monitoring, in situ and ex situ conservation, and the sustainable use of components of biological diversity.[52] Other relevant aspects of the Convention deal with general principles for impact assessment and minimising adverse impacts and access to genetic resources.[53] The CBD has a clear application to Antarctica and the Southern Ocean, and it complements a number of instruments adopted by the ATS, especially the Environmental Protocol.

Framework Convention on Climate Change. The 1992 Framework Convention on Climate Change was also adopted as part of the UNCED process. The Convention entered into force in March 1994 and since then three meetings of the Conference of Parties have been held. The Convention was negotiated in response to global scientific concerns over climate change as a result of a range of industrial and agricultural activities by various states around the world. As noted by Birnie and Boyle, the FCCC 'seeks to establish a framework for the elaboration of further measures to address the causes of climate change and is an important example of the principles of common but differentiated responsibility and of precautionary action endorsed in the Rio Declaration on Environment and Development'.[54] Article 2 notes that the ultimate objective of the Convention is to achieve 'stabilisation of greenhouse gas concentrations in the atmosphere at a level that would prevent dangerous anthropogenic interference with the climate system'. The Convention's main commitments are outlined in Articles 3 and 4. To date, the meetings of the Conference of Parties have generated considerable debate over the obligations upon state parties to meet greenhouse gas emission targets and the utility of the 'common but differentiated responsibilities' approach. These issues dominated the 1997 Kyoto meeting of the Conference of Parties, and the result was a Protocol that provides for

---

[50] See *ibid.*, Art. 6(4).
[51] CBD, Art. 1.
[52] *Ibid.*, Arts. 6 to 10.
[53] *Ibid.*, Arts. 14 and 15.
[54] P.W. Birnie and A.E. Boyle (eds.), *Basic Documents on International Law and the Environment* (Oxford: Clarendon Press, 1995), p. 252.

wide disparity in commitments to reduce greenhouse gas emissions amongst developed states.[55] While the FCCC presently only has a limited impact upon activities in the Antarctic, the long-term impact of its commitments for the protection of the Antarctic are substantial, given the vulnerability of the polar ice cap to global warming and the consequences of such an event for the polar ecosystem. To date, however, there is divided scientific opinion as to the real impact that global warming is having on Antarctica.[56]

THE ENVIRONMENTAL PROTOCOL AND UNEP INSTRUMENTS

International Law and Treaty Interaction[57]

Since the beginning of the United Nations era, treaties have become the dominant source of international law. In addition, there has been a substantial growth in regional international law, the Antarctic Treaty System being one of the most prominent examples. There is an increasing recognition of 'treaty congestion'; as more treaties interact, newly developed treaties will need to take account of existing international instruments operating in related fields. The difficulties posed by the interaction between global and regional international legal regimes are increasing, and as regional legal regimes become more complex, the potential for conflict and overlap also increases. Notwithstanding attempts to ensure that no overlap or conflict occurs, it is inevitable that a regional regime that is sufficiently dynamic will eventually interact with the global regime. However, the legal rules concerning treaty interaction and conflict are not comprehensive or sophisticated enough to deal with the issues that currently exist between global and regional regimes, and this also applies in the context of the interaction between the ATS and UNEP instruments. In response to these issues, some commentators have suggested that the appropriate time has come for reassessing the global state of international environmental treaties and rationally determining where the law should be developed.[58]

---

[55] See the 1997 Kyoto Protocol to the United Nations Framework Convention on Climate Change (ILM, Vol. 37, 1998, pp. 22ff); for an assessment, see F. Yamin, 'The Kyoto Protocol: Origins, Assessment and Future Challenges', *Review of European Community and International Environmental Law*, Vol. 7, 1988, pp. 113–127.

[56] See, e.g., C. Harris and B. Stonehouse (eds.), *Antarctica and Global Climatic Change* (London: Belhaven Press, 1991).

[57] This section reflects the views of the author developed elsewhere; see B. Boer, R. Ramsay and D.R. Rothwell, *International Environmental Law in the Asia Pacific* (London: Kluwer Law International, 1997), Ch. 15; D.R. Rothwell 'The Relationship Between Global and Regional Legal Regimes', in B. Davis (ed.), *Overlapping Maritime Regimes: An Initial Reconnaissance*, Antarctic CRC Monograph No. 2 (Hobart: Antarctic CRC and Institute of Antarctic and Southern Ocean Studies, 1995), pp. 27–59.

[58] See the comment by E.B. Weiss, 'Global Environmental Change and International Law: The Introductory Framework', in E.B. Weiss (ed.), *Environmental Change and International Law* (Tokyo:

The potential for overlap and conflict between global and regional environmental law has long been recognised.[59] This was a matter which occupied the International Law Commission in preparing the original drafts of the 1969 Vienna Convention on the Law of Treaties (Vienna Convention).[60] The early rule of treaty law which applied when two treaties conflicted was that the latter of the two conflicting treaties was void.[61] However, while this rule may have been appropriate at a time when states accepted relatively few treaty-based obligations, its suitability became questionable once multilateral treaty-making took over from customary international law as the principal source of international obligations accepted by states. Today, the Vienna Convention provides the legal rules dealing with treaty interaction. It largely contains a codified law of treaties. While its application can often depend on when the relevant treaty was concluded and which states are also parties to the Vienna Convention, many of its provisions are considered to reflect customary international law.

In regard to treaties that are in conflict, the Vienna Convention only has one provision that directly deals with the matter. Article 59 provides that a treaty will be terminated if the parties conclude a later treaty that relates to the same subject matter and

- it appears from the later treaty or is otherwise established that the parties intended that the matter should be governed by that treaty, or
- the provisions of the later treaty are so far incompatible with those of the earlier one that the two treaties are not capable of being applied at the same time.

While this provision clearly indicates that the later treaty shall prevail in the case of conflict, its potential application is rather narrow. The provision will only apply when the parties make it clear, either by the terms of the treaty itself or by their subsequent actions, that their relationship be governed by the later treaty. In this instance, the parties are clearly intending that their international legal relationship be governed by the later treaty. In the second instance, there is a requirement that the provisions of two treaties be so incompatible that they are incapable of being applied simultaneously.

United Nations University Press, 1992), p. 12.

[59] See the comments by C.W. Jenks, 'The Conflict of Law-Making Treaties', *British Yearbook of International Law*, Vol. 30, 1953, pp. 403–404. For a more contemporary assessment, see G. Binder, 'The Dialectic of Duplicity: Treaty Conflict and Political Contradiction', *Buffalo Law Review*, Vol. 34, 1985, p. 340, where treaty conflict is defined to occur 'when a state concludes a treaty which creates international obligations the performance of which would be inconsistent with the performance of an international obligation to a third state under a previously concluded treaty'.

[60] UNTS, Vol. 1155, pp. 331ff. This has not been the only effort to solve the problem of treaty conflict; in 1980, the UN Secretary-General referred to the problem of 'accidental conflicts' in treaty-making and the matter was placed on the agenda of the Sixth Committee of the General Assembly; S. Rosenne, *Breach of Treaty* (Cambridge: Grotius, 1985), pp. 94–95.

[61] See the discussion in Binder, 'The Dialectic of Duplicity', pp. 363–384.

The Vienna Convention also deals with the case in which an existing treaty comes into conflict with an emerging peremptory norm of international law, i.e., *jus cogens*. Article 64 of the Vienna Convention provides that, in such case, 'if a new peremptory norm of general international law emerges, any existing treaty which is in conflict with that norm becomes void and terminates'. This provision seeks to terminate treaties, or provisions of treaties, that are in conflict with fundamental principles of international law. While this provision does not deal with conflict between treaties as such, it could apply in instances where a new fundamental norm of international law has been crystallised in a global convention.[62]

The Vienna Convention also sought to deal with successive treaties that deal with the same subject matter. Article 30 creates several rules:

– in cases where potential treaty conflict has been recognised and a treaty indicates that it is subject to or not to be considered to be incompatible with another treaty, the provisions of the other treaty prevail;
– when all the parties to the earlier treaty are also parties to the later treaty but the earlier treaty remains operative, it will only apply if its provisions are compatible; and
– when there is a difference in the parties to the relevant treaties, the earlier treaty only applies when it is compatible with the later treaty that both states are parties to. In all other cases the treaty to which both states are parties governs the relationship.

These provisions will have rather limited operation because of the requirement that treaties be compatible, which is unlikely in the case of large multilateral treaties. The provisions may also create some legal uncertainty because they have the potential to create considerable variation in the legal rights and obligations that states owe to each other depending on whether the states were parties to the earlier or later treaty.

Interaction between the Environmental Protocol and UNEP Instruments

In assessing the interaction between the Environmental Protocol and UNEP instruments, it is useful to note that the Protocol provides that it does not derogate from rights and obligations under other ATS instruments, and that it should be interpreted in such a fashion so as to avoid interference with the objectives and principles of those instruments.[63] The Protocol was therefore clearly designed to not interfere with the obligations created under other ATS instruments, notwithstanding

---

[62] See also Vienna Convention, Art. 53.
[63] Protocol, Arts. 4 and 5.

the potential for interaction between them.[64] With respect to non-ATS instruments, apart from an express reference in Annex IV to the 1973/78 International Convention for the Prevention of Pollution from Ships,[65] no other non-ATS international law instruments are directly referred to in the Protocol. It would be remiss, however, to not comment that the Antarctic Treaty in Article VI provides that its area of application is not to infringe on high seas freedoms, and this provision also has relevance for the application of the Protocol south of 60° south.[66] Nevertheless, this 'limitation' has in practice not greatly constrained the ATS from regulation of traditional high seas activities, with the possible exception of whaling in the Southern Ocean which has traditionally been regulated under the 1946 International Convention for the Regulation of Whaling.[67]

What then is the relationship between the Antarctic Environmental Protocol and UNEP environmental instruments? Are they to be read as interacting in a complementary fashion, or do they contain conflicting obligations? Given the clear potential application in and importance of some UNEP conventions for Antarctica, it seems remiss that in the Protocol no reference is made to relevant, existing UNEP conventions.[68] This is especially the case with respect to Annex III of the Protocol which deals with matters directly relevant to the Basel Convention.

One of the more notable features of the Protocol is its designation of Antarctica as a 'natural reserve, devoted to peace and science'.[69] This has created some debate as to the consequences of the designation of an area as a 'natural reserve' and the precedent, if any, for such a designation. While a linkage may be seen to exist between such a designation of Antarctica and the listing of 'World Heritage' sites under the 1972 Convention Concerning the Protection of the World Cultural and Natural Heritage,[70] a more contemporary connection can be seen with the provisions of the CBD. In its Preamble, the Convention singles out the 'fundamental requirement' of the in situ conservation of ecosystems and natural habitats and goes further to provide that each contracting party shall 'establish a system of protected areas or areas where special measures need to be taken to conserve biological diversity' (Article 2). The designation of Antarctica as a 'natural reserve' would most certainly be in the spirit of the CBD obligations, though as some commentators

---

[64] See discussion in D.R. Rothwell, 'The Madrid Protocol and Its Relationship with the Antarctic Treaty System', *Antarctic and Southern Ocean Law and Policy Occasional Papers*, Vol. 5, 1992, pp. 8–18.

[65] ILM, Vol. 12, 1973, pp. 1,319ff (Convention); and ILM, Vol. 17, 1978, pp. 546ff (Protocol).

[66] See D.R. Rothwell, *The Polar Regions and the Development of International Law* (Cambridge University Press, 1996), pp. 81–83.

[67] UNTS, Vol. 161, pp. 74ff.

[68] It needs to be recalled that the Protocol was negotiated prior to the conclusion of the CBD and FCCC; however, it should be noted that the Protocol does contain mechanisms for adjustment and amendment by which account could be taken of these later instruments. See Protocol, Art. 25 and Annex III, Art. 13.

[69] Protocol, Art. 2.

[70] UNTS, Vol. 1037, pp. 151ff.

have noted the designation may carry with it more political than legal weight.[71] Likewise, the prohibition on mining in Article 7 of the Protocol complements the approach of in-situ conservation in the CBD and the theme of sustainable development and in-situ conservation.[72] Another feature of the Protocol which complements the CBD is the broad ecosystem approach taken to ensure protection of the 'Antarctic environment and dependent and associated ecosystems'.[73] This approach fits broadly within the notion of 'biological diversity' which recognises the interconnection and diversity between and amongst organisms.

The Environmental Protocol lists a series of 'Environmental Principles'.[74] These principles are designed to provide the parties with a broad framework within which the Protocol's mechanisms are to operate. Given that the focus is naturally upon protection of the Antarctic environment, it is not surprising that a great many of these principles under the Protocol directly complement and in some instances duplicate similar principles under which the UNEP conventions operate under. Examples of some of these linkages include

- activities avoiding adverse effects on climate or weather patterns (FCCC),[75]
- activities avoiding significant changes in the atmosphere (Ozone Layer Convention and Montreal Protocol),[76]
- activities not having detrimental changes on species or populations of species of fauna and flora (CBD),[77]
- activities not jeopardising endangered or threatened species (CITES/CBD).[78]

In addition, references made in Article 3 of the Protocol, to the need to conduct environmental impact assessment and monitoring, are also broadly consistent with the spirit of the Convention on Biological Diversity.

The potential also exists for interaction between the Protocol's Annexes and UNEP instruments. Annex II, dealing with the conservation of Antarctic fauna and flora, and Annex V, which established a revised Antarctic protected area system, are clearly within the spirit of both CITES and especially the CBD. Indeed it could be argued that the 1964 Agreed Measures for the Conservation of Antarctic Fauna and Flora (the predecessor of Annex II) was one of the first attempts to comprehensively provide for in-situ conservation as now provided for in the CBD. However, while these respective provisions are broadly consistent, there are some inevitable clashes.

---

[71] See especially Sir Arthur Watts, *International Law and the Antarctic Treaty System* (Cambridge: Grotius, 1992), p. 277.
[72] See CBD, Art. 8.
[73] See Protocol, Arts. 2 and 3.
[74] See *ibid.*, Art. 3.
[75] See in particular Protocol, Art. 3(2)(b)(i) and FCCC, Arts. 1 and 3(1).
[76] See in particular Protocol, Art. 3(2)(b)(iii) and Ozone Layer Convention, Art. 2.
[77] See in particular Protocol, Art. 3(2)(b)(iv) and CBD, Art. 8.
[78] See in particular Protocol, Art. 3(2)(b)(v) and CITES, Art. II, and CBD, Art. 8.

For example, the CBD provides for measures to be taken to prevent the introduction of alien species,[79] and while the Protocol does address this issue in Annex II, it is perhaps not as all encompassing as it could be, preferring to identify only certain sources and carriers of disease as requiring control.[80] Annex V to the Environmental Protocol, dealing with area protection and management, is also not as far reaching as the CBD's provisions for 'environmentally sound and sustainable development in areas adjacent to protected areas'.[81]

It seems clear from the above review that one of the closest relationships between the Protocol and a UNEP instrument exists in the case of the CBD. For example, one of the key environmental principles of the Protocol is to ensure that activities in the Antarctic Treaty area are planned so as to avoid 'detrimental changes in the distribution, abundance or productivity of species or populations of species of fauna and flora'.[82] Likewise, Annexes I, II and V to the Protocol all contain features which ultimately seek to conserve and protect elements of Antarctic biodiversity. However, notwithstanding the apparent complementarity of the two instruments, it must be recognised that the CBD deals with a range of other matters, especially with regard to genetic resources, which the Protocol does not directly address. Given the ongoing scientific research in the Antarctic, and the questionmarks that exist with respect to sovereign rights, some important issues may arise in the future as to who may claim exclusive access to Antarctic genetic resources. In this scenario, the interaction of rights and duties under the CBD and the Protocol may prove decisive.

Interaction of UNEP Instruments with Other Legal Regimes

The UNEP instruments have developed in parallel with regional environmental law, the two most prominent examples being the European Union and the ATS. It is therefore not surprising that some of the UNEP instruments make direct reference to how they may interact with other instruments, both global and regional.

Convention on Biological Diversity. The most prominent UNEP instrument that refers to other legal regimes is CBD, which in Article 22 provides the following:

9. The provisions of the Convention shall not affect the rights and obligations of any Contracting Party deriving from any existing international agreement, except where the exercise of those rights and obligations would cause a serious damage or threat to biological diversity.

---

[79] CBD, Art. 8(h).
[80] See Protocol, Annex II, Art. 4.
[81] CBD, Art. 8(e).
[82] Protocol, Art. 3(2)(b)(iv).

10. Contracting Parties shall implement this Convention with respect to the marine environment consistently with the rights and obligations of States under the law of the sea.

One interpretation of Article 22 is that the Environmental Protocol is 'immune' from the impact of the CBD, providing that nothing takes place under the Protocol to cause damage or threat to biological diversity. While this may provide some 'exemption' for the Antarctic Treaty parties, it should however alert them that if a major environmental disaster should occur in Antarctica (such as the introduction of disease which results in serious and widespread impact on various Antarctic species), a clear conflict in legal obligations could arise between the CBD and the Protocol.

The CBD implicitly accepts the legitimacy of other environmental conventions that deal with the protection and preservation of biodiversity and are compatible with its general principles. Nevertheless, some uncertainties remain as to the relationship of the Convention with already existing international environmental regimes.[83] For example, there is the potential that, through subsequent treaty interpretation, pivotal terms such as 'biological diversity', 'biological resources' and 'ecosystem' in Article 2 of the CBD may be defined more broadly than equivalent terms and provisions in the Environmental Protocol. Such a result may not, however, necessarily be seen as a conflict between the global and regional regimes but rather as a strengthening of the international law framework for the protection of biological diversity.

A more likely area of conflicting interaction between the CBD and the Environmental Protocol is the impact of CBD provisions that deal with access to genetic resources and the access to and transfer of technology. These matters are not dealt with in the ATS regime. A further area of uncertainty is the impact of the CBD upon sub-regional and bilateral resource management agreements, especially dealing with fisheries. The operation of these agreements may in the terms of Article 22 of CBD cause 'serious damage or threat to biological diversity' so as to conflict with the Convention's obligations. While this is not a matter that would directly create a conflict with the Environmental Protocol, it especially could become an issue under the CCAMLR regime with respect to fishing in the Southern Ocean.

CITES. A feature of CITES are the detailed allowances for other international legal regimes operative both prior to and subsequent to entry into force.[84] While the focus is largely on trade agreements, the scope is still broad enough to include other environmental agreements, including the Environmental Protocol. Article XIV of CITES, in particular, makes allowance for provisions dealing with 'trade, taking,

---

[83] P. Sands, *Principles of International Environmental Law I* (Manchester University Press, 1995) p. 387.
[84] See CITES, Art. XIV.

possession, or transport of specimens' including 'any measure pertaining to customs, public health, veterinary or plant quarantine fields'. This would permit measures taken under the Environmental Protocol on these matters to be exempt from any contrary controls imposed under CITES.

ASSESSING INTERACTION OF ATS AND UNEP INSTRUMENTS

Conflicting Interaction

The growth of both the global and regional legal regimes and the increasing interaction that takes place among these regimes has resulted in an increased potential for overlap and conflict; however, neither international law nor state practice has yet developed adequate rules for dealing with this development. The question can be posed as to whether this is truly a major issue that really matters. Should states be concerned about the potential for overlap and conflict between global and regional legal regimes?

One response could be that because the protection and preservation of the Antarctic environment requires as comprehensive a legal regime as possible, and if through the interaction of global and regional legal regimes such a regime can be created, then this is a positive development. While some conflicts may inevitably occur between these legal regimes, the state parties will eventually make choices as to which regimes they accept or reject and accordingly modify their behaviour to be consistent with their international obligations. However, it must also be acknowledged that with the current, rapid development of the law, the high potential for interaction and conflict between differing legal regimes is so great that the problem cannot be ignored. The view by Alvarez that global and regional international law are correlated can only apply when the regimes are complementary.[85] Such a characterisation may be appropriate in the case of a framework global regime, such as the CBD and the Environmental Protocol being broadly consistent with the CBD framework; however, this will not always be the case.

The consequence of overlap and conflict between regional and global conventions is the lack of legal certainty which could result in the breakdown of the legal regime or in conflict between the state parties. This matter requires greater attention both at global and regional levels. In many cases, there may be some scope for the operation of the rules found in the Vienna Convention. Article 59 of the Vienna Convention could apply in instances where states that originally were parties to a regional agreement could become parties to a global agreement. In these instances the regional agreement would be considered terminated. There is also some scope for the operation of Article 30 of the Vienna Convention; however,

---

[85] See the comments by Judge Alvarez in the *Asylum Case* (*Columbia* v. *Peru*), ICJ Reports 1950, p. 294.

operation of Article 30 seems to be rather limited in cases in which global and regional treaties are competing and in which the state parties to the competing treaties are different. While international law, particularly treaty law, provides some assistance in resolving the relationship between legal regimes, it is not conclusive. In particular, the Vienna Convention clearly does not directly address all of the issues which arise from treaty interaction. To this end it would certainly be more helpful if newly adopted conventions embodied express treaty interaction provisions such as Article 22 of the CBD.

It may also be possible to argue that one effect of the ATS has been to create an area of special regional customary international law which applies in Antarctica and the Southern Ocean.[86] International law recognises the capacity of a group of states to create local custom,[87] which results in an exception to the general nature of customary international law.[88] On this basis, it could be argued that the special regional customary rules which apply in the Antarctic operate to exclude contrary customary rules which operate more generally. A key principle of such a regional customary regime would clearly be environmental protection which has been in the course of development since the adoption of the Antarctic Treaty through to the contemporary implementation of the Protocol.[89] However, for this proposition to be of any consequence, it requires evidence of the existence of customary international law rules which have developed from the UNEP conventions. While there may be grounds for asserting that the UNEP conventions have established some general customary international law principles,[90] it still remains difficult to argue that the rules established are of such a precise nature that they conflict with any similar rules which may have been established for the Antarctic.

Positive Interaction

It can be seen from the above review that UNEP has secretariat responsibility for a number of important global environmental conventions that also have an application in Antarctica and the Southern Ocean. While it is not often directly recognised that these instruments have relevance for the Antarctic, no doubt due to the dominance of the ATS within the region, a review of the level of acceptance of the above

---

[86] For discussion of this issue, see J.I. Charney, 'The Antarctic System and Customary International Law', in Francioni and Scovazzi (eds.), *International Law for Antarctica*, pp. 62–100.

[87] See the discussion in the *Asylum Case* (*Columbia* v. *Peru*), ICJ Reports 1950, p. 266 and the *Right of Passage Case* (*Portugal* v. *India*), ICJ Reports 1960, p. 6.

[88] M. Shaw, *International Law*, 4th edition (Cambridge University Press, 1997), p. 73.

[89] Charney, 'The Antarctic System and Customary International Law', p. 96.

[90] The large number of parties to the conventions could be one basis for asserting that the conventions reflect customary international law. For example, parties to the conventions numbered 150 for CITES (on 27 April 2000); 173 for the Ozone Layer Convention and Montreal Protocol (on 29 March 2000); 134 for the Basel Convention (on 2 March 2000); 177 for CBD (on 14 February 2000); and 181 for the FCCC (on 10 December 1999).

instruments by Consultative Parties shows near unanimous acceptance of these instruments.[91] A similar pattern emerges for non-Consultative Parties.[92] As such, the Consultative Parties have a further layer of international obligation imposed upon them by these international instruments. While it is true that only the Basel Convention makes specific reference to the Antarctic, it is clear that all of the UNEP instruments further support the essential obligations to protect and preserve the Antarctic, as provided for by the Environmental Protocol. Moreover, in some instances the UNEP instruments advance environmental protection to a level not addressed in the Protocol, especially in the case of global issues such as climate change and ozone depletion. The UNEP Conventions are thus clearly relevant to the Antarctic.

To date, in spite of the above, the linkages between the Protocol and these UNEP instruments have not been fully explored. It would seem to be more productive if the relationship could be developed. It is clear that the Consultative Parties are prepared to accept the terms of other global instruments and either implicitly[93] or expressly[94] acknowledge their application in the Antarctic. It may therefore be fruitful to explore whether UNEP conventions may be more actively applied and adopted within the ATS in order to not only supplement the ATS but also to extend its potential operation by providing a firmer legal regime. Would it be possible, for example, for the ATS and UNEP to explore the merits of amending some of the UNEP conventions so that they would not only apply more directly to the Antarctic but also contain provisions for placing obligations on the international community to provide more complete protection for the Antarctic? Such an initiative would assist in extending obligations of Antarctic environmental protection beyond the parties to ATS instruments, to the wider international community. Particularly relevant in this regard would be the provisions of CITES and their capacity to enhance the international protection of Antarctic endangered species. Consultative Parties could attempt to make more active use of CITES and work with UNEP to ensure the enhanced conservation of Antarctic species through that regime. In addition, the Basel Convention could be made applicable to the export of all hazardous wastes from the Antarctic continent. The effect of this prohibition would be not only to place a further international legal obligation upon the Consultative Parties in regard to hazardous wastes but also to extend the application of this prohibition to the state parties to the Basel Convention. This would in effect provide an enhanced means for Antarctic environmental protection through the operation of

---

[91] Of the current 27 Consultative Parties, 26 are simultaneously a party to CITES, the Ozone Layer Convention and Montreal Protocol, Basel Convention, CBD and FCCC. The remaining Consultative Party, the United States, is a party to all of the above with the exception of the Basel Convention and CBD.
[92] Of the non-Consultative Parties, only the Democratic People's Republic of Korea and Turkey are not parties to all of the five UNEP conventions which have been considered.
[93] See CCAMLR, Art. VI.
[94] See Protocol, Annex IV.

the more readily accepted global regime – the Basel Convention – and also side-step the sovereignty issue which exists in Antarctica. More extensive consideration could also be given to whether the terms of the CBD would assist in further developing the principles of the Environmental Protocol, particularly those dealing with protected areas and environmental impact assessment. This is an area which the Consultative Parties have to date neglected but which, following the recent entry into force of the Protocol, is ripe for further consideration given the emphasis which the ATS has traditionally given to the question of protected area management.[95]

Given the widespread acceptance of the UNEP conventions[96] and the general reflection of the some of the provisions of these conventions in customary principles of environmental protection which have now developed, it can in addition be argued that an overlay of international environmental law also applies in the Antarctic to support the initiatives which have been specifically adopted by the Antarctic Treaty parties. This line of reasoning has two consequences. Firstly, in some instances the gaps in the ATS environmental regime can be filled so that Antarctic Treaty parties not only would be mindful of their ATS obligations but also of their wider international environmental law obligations. Secondly, for states not participating in the ATS, the wider principles of international environmental law reflected in treaty commitments of the UNEP conventions and in customary international law ensure the extension of environmental obligations towards the Antarctic to the global international community. Given the limitations on the operation of the Antarctic regime beyond the Antarctic Treaty parties (and, *mutatis mutandis*, the Environmental Protocol regime to the parties to the Protocol), this represents an additional, significant level of environmental protection for the Antarctic.

CONCLUSION

Global and regional legal regimes are a relatively recent phenomena in international law. Prior to the United Nations era, few conventions had either the number of state parties or the breadth of global coverage that are seen in the multilateral conventions which are now being negotiated. While regional legal regimes have existed for some time, during the late twentieth century both the complexity and the number of these regimes have increased. Considering these developments, it is surprising that greater attention has not been given to the interaction between global and regional legal regimes. Under the UN Charter, regional legal regimes are not seen as competing global regimes but as regimes working in cooperation. This point is reinforced by Judge Alvarez in the *Asylum Case* when he noted, in referring to the relationship

---

[95] See the discussion in B.M. Clark and K. Perry, 'The Protection of Special Areas in Antarctica', in Francioni and Scovazzi (eds.), *International Law for Antarctica*, pp. 293–318.

[96] See discussion above, in this chapter.

between American international law and international law, that 'Such systems of law are not subordinate to universal international law, but correlated to it'.[97]

This chapter has demonstrated that a high degree of interaction exists between the provisions of the Environmental Protocol and the UNEP conventions. To a large degree these regimes seem to have been developed without mutual recognition, and this calls into question whether there is any correlation between the regimes, and indeed whether there is scope for treaty conflict. While the Consultative Parties have been mindful of the existence of other legal regimes that deal with matters such as whaling and marine pollution, this has not stopped them in addressing similar or identical problems. However, it is important to recall that the Antarctic has unique legal problems that are not comparable with those of any other region. As a result, states tend to rely upon regional international law as a means of regulating activity rather than on national laws because of the limitations placed upon both territorial claimants and other parties by the Antarctic Treaty. In addition, the Antarctic legal regime is not universally accepted by all states.

This situation may help in explaining why there has been so little actual engagement between the ATS and the UNEP regimes. It must also be recalled that, with few exceptions, the Antarctic Treaty parties have actively sought to keep Antarctic affairs off international agendas of non-ATS fora, which somewhat explains why so few other global international instruments make precise reference to Antarctic issues. The UNEP Conventions do, however, enhance Antarctic environmental protection with their wider global membership, providing also an additional overlay of international law. The regional ATS and global UNEP legal systems have not developed parallel with each other. Although there are no clear conflicts, there are overlaps, and the potential certainly does exist for engagement among the legal regimes that may not be in harmony with what each is attempting to achieve. It would therefore seem advisable that in the future greater harmony be developed among these legal regimes. This can only further Antarctic environmental protection.

---

[97] *Asylum Case* (*Columbia* v. *Peru*), ICJ Reports 1950, p. 294.

13

Towards Guidelines for Antarctic Shipping: A Basis for Cooperation between the Antarctic Treaty Consultative Parties and the IMO

Tullio Scovazzi

THE PROPOSAL FOR A POLAR CODE

In December 1997, a document entitled 'Development of a Code on Polar Navigation – The International Code of Safety for Ships in Polar Waters (Polar Code)' was submitted by Canada to the Sub-Committee on Ship Design and Equipment of the International Maritime Organisation (IMO).[1] The basic aim of the draft Polar Code, which constituted Annex 1 to the document, was to establish an international framework governing polar shipping. The draft Polar Code was the only bi-polar instrument ever proposed. In particular, it aimed at the harmonisation of different legal regimes by meeting

> the need for special measures to protect the safety of life, and to preserve the quality of the environment in the world's circumpolar seas and oceans. These waters pose a number of unique challenges to shipping, due to their remoteness, extreme climatic conditions and unusual navigation hazards. Hitherto, safety has been managed in a piecemeal manner, with a large number of incompatible national systems of navigational control coupled with equally varied construction, equipment and crewing standards. The Polar Code harmonises all of these to ensure that all future operations in polar waters meet satisfactory and internationally-recognised minimum standards.[2]

---

[1] IMO doc. DE 41/10, 1997. The Polar Code was drafted by an Outside Working Group of technical experts set forth by the IMO and including representatives from a number of administrations, classification societies, operators and experts on polar issues. See an overview and analysis by L.W. Brigham, 'The Emerging International Polar Navigation Code: Bi-polar Relevance?', in D. Vidas (ed.), *Protecting the Polar Marine Environment: Law and Policy for Pollution Prevention* (Cambridge University Press, 2000), pp. 244–262.

[2] IMO doc. DE 41/10, para. 1(2).

On the one hand, the Polar Code was an application of the 'precautionary approach', which aimed at preventing ship casualties in the particularly difficult polar waters and averting the consequences that these could have for the safety of human life and the preservation of the environment.[3] On the other hand, the Polar Code should have avoided that differences in domestic legislation by states having an interest in polar navigation become an obstacle to the further development of shipping routes and other activities.[4]

Regarding its structure, the Polar Code was composed of three opening sections ('Preamble', 'Guide to the Code' and 'General') and three parts devoted respectively to 'Construction Requirements' (Part A, setting out general requirements for ice-capable ships in seven Polar Classes[5]), 'Equipment' (Part B, relating, *inter alia*, to lifesaving, fire-fighting, navigation and communication systems) and 'Operational' (Part C, focusing on the presence of personnel trained and certified for ice navigation). A number of annexes and appendices complemented the Polar Code.[6]

The Polar Code followed an integrated approach, 'which covers the design and outfitting of ships for the conditions which they will encounter, their crewing by adequate numbers of suitably trained personnel and their operation in a planned and prudent manner with adequate liability provisions'.[7]

Far from being a stand-alone document, the Polar Code should have *supplemented* other IMO conventions which are presently in force among many states, such as the 1974 International Convention for the Safety of Life at Sea (SOLAS), the 1973/78 International Convention for the Prevention of Pollution from Ships (MARPOL), the 1978 International Convention on Standards of Training, Certification and Watchkeeping for Seafarers (STCW).[8] As regards the relationship with the requirements contained in those conventions, the Polar Code expressed additional requirements to mitigate the increased risks imposed on

---

[3] 'The precautionary approach has guided development of the Code. Existing systems have, to date, largely prevented catastrophic ship casualties. There have, however, been a large number of less serious incidents which indicate a continuing high level of risk'; *ibid.*, para. 1(3). It can be noted that the expression 'precautionary principle' is avoided, and the wording 'precautionary approach' preferred.

[4] 'Meanwhile the differences between existing systems have also imposed cost and regulatory barriers to exploitation of polar shipping routes and offshore resource developments within national Exclusive Economic Zones (EEZs)'; *ibid.*

[5] These are the ships whose strength, propulsion machinery and other features allow them to transit ice-covered waters with reasonable safety.

[6] Polar Ship Safety Certificate (Annex I), Permit to Operate in Polar Waters (Annex II), Breathing Apparatus (Annex III), Life-Saving Appliances and Survival Equipment (Annex IV); Draft Amendment to SOLAS Convention (Appendix I), Draft Amendment to the STCW Code (Appendix II), Draft IMO Resolution on Equivalencies for Existing Ships (Appendix III), Draft IMO Resolution on Equivalencies for Existing Ice Navigators (Appendix IV), Comparison of SOLAS and New Polar Code Requirements (Appendix V).

[7] Para. 2(2) of the 'Preamble' to the Polar Code.

[8] Within the IMO framework there are many treaties whose area of application in general covers all seas and oceans, including implicitly the Antarctic seas. SOLAS, MARPOL and STCW are among

shipping in polar regions.⁹ For example, along with meeting the requirements arising from the applicable treaties, it was envisaged that all ships operating in polar waters should have been certified as 'Polar Class' ships and have a 'Permit to Operate in Polar Waters'.

It is impossible to analyse here the various technical requirements set forth in the Polar Code.[10] They related, *inter alia*, to double hulls, hull stress indicators, intact stability requirements, emergency response training and effective damage control. Suffice it to say that five key provisions, as contained in paragraphs from 2(2) to 2(6) of the Guide to the Code, characterised the instrument as a whole, namely

> 2(2) The combination of hull structural design, material quality, subdivision, and segregation measures prescribed in the Code and supporting requirements, should be adequate to reduce the risks of pollution incidents or ship losses to acceptably low levels of probability, during prudent operations in Polar Waters.

> 2(3) No pollutants are to be carried directly against the shell in areas at significant risk of ice impact. Operational pollution of the polar environment should be minimised by equipment selection and operational practice.

> 2(4) Key safety-related, survival, and pollution control equipment should be rated for the temperatures and other conditions which may be encountered in the service intended.

> 2(5) Navigation and communications equipment should be suitable to provide adequate performance in high latitudes, areas with limited infrastructure, and unique information transfer requirements.

> 2(6) Regular audit, inspection and re-certification procedures should be applied to both ships and crew members covered by the Code.

THE SCOPE OF APPLICATION OF THE DRAFT POLAR CODE

The Problem

One of the main problems posed by the Polar Code was its proposed bipolar scope of application which included polar waters belonging to both the Northern Hemisphere (approximately the greater part of those waters north of 60° N, beyond the coastline of circumpolar countries, but with a more northerly boundary in the

---

relevant examples.
[9] See para. 2(3) of the Preamble to the Polar Code.
[10] Some outstanding issues of technical nature are mentioned in paras. 5(3) to 5(5) of IMO doc. DE 41/10. They include issues of the grandfathering approach for existing ships, the damage stability requirements, and the carriage of pollutants in the double bottom tanks; see further Brigham, 'The Emerging International Polar Navigation Code', pp. 250–254.

North Atlantic) and the Southern Hemisphere (south of latitude 60° S).[11] The Polar Code generally applied to ships operating in polar waters[12] while engaged in international voyages (para. 1(1)(1)).[13]

As was clearly stated in a discussion paper submitted by Canada on the application and impact of the Polar Code,[14] the initiative for the Code had a 'northern' origin:

> Much of the technical work has come from the northern experience, as it is in the north that a number of circumpolar nations have developed distinctly different ship classification systems, navigational systems, and operating constraints. These have been tailored for each nation's particular Arctic experience, its national interest in Arctic resource development and expansion of related shipping capability, and its need to support northern settlements, and indigenous peoples' offshore hunting and fishing.[15]

The discussion paper did not ignore the differences that existed between Arctic and Antarctic regions and waters:

> The different shipping patterns in the two Polar areas are based more on geography and seasonality than on ice severity. The Arctic Ocean is first of all an ocean, capping the world at the northern apex with deep, ice-encrusted waters. These floating ice-fields and ice-infested waters are ringed by the circumpolar nations, with their seasonally-frozen archipelagoes and coastlines. In this environment, shipping routes are typically long circular routes, through extensive and varied ice regimes, around the outer edges of the Arctic Ocean, close to the coastlines of, or within the internal waters of, the circumpolar nations. The season is roughly from June to October.
>
> In contrast, the Antarctic is a continent that covers the southern apex and is almost completely covered in turn by massive, snow-encrusted glaciers. This bitterly cold, windy continent is surrounded by the Southern Ocean. Waters close to the continent are very cold, out to a distance at which there is a sudden patterned exchange with warmer waters, called the Antarctic Convergence. The huge expanses of ocean, the convergence of waters, and the bitter temperature at the southern apex contribute to the tempestuous ocean swells, cyclonic winds and wind-driven snow squalls along the hostile coastline. Here, ship traffic patterns, operative in the December to March period, generally have a

---

[11] The content of the relevant provisions may vary depending on the hemisphere. 'The requirements of the Code, and the means in which they will be applied, reflect the different circumstances of the Arctic and Antarctic seas and oceans, where different legislative regimes now apply. In the Arctic, the emphasis is on the responsible development and exploitation of the sea routes and offshore resources; while in the Antarctic, development affecting the pristine environment is more stringently constrained'; para. 1(4) of IMO doc. DE 41/10.

[12] Exceptions are envisaged. The Polar Code does not apply to ships that are not subject to any existing IMO Convention (para. 1(1)(3)). Neither does it apply to warships, troopships and the other categories of ships listed in para. 1(1)(4), namely, ships operating independently, powered solely by oars, sails or other non-mechanical means; wooden ships of primitive build; stationary ships permanently anchored or moored in a single location.

[13] See the definition of 'Polar Waters' given in para. 3(19) of the 'Guide to the Code'.

[14] IMO doc. DE 41/INF. 8, 1997.

[15] *Ibid.*, para. 3(2).

north-south orientation as they approach the outer edge of the continent and its overflowing ice shelves.[16]

The above differences significantly influence the volume and type of shipping:

> Traffic volumes and shipping routes differ markedly. The Arctic waters are host to many transits, many convoys in the summer; Russia's Northern Sea Route has been, at its peak, one of the busier routes, reporting millions of tonnes of freight. Traffic volumes in Canada and along the Alaskan coast of the United States include major re-supply routes for aboriginal communities and government sites, with smaller numbers of tug/barge fleets, tankers, cruise ships, and commercial ore carriers, often supported by icebreakers. Canada and the United States, like Russia, have spent some decades seeking to develop the hydrocarbons and minerals of their Arctic territories and coastal waters. The specialized and generally highly ice-capable ships used in these developments have added dozen of transits in the peak years of the 1970's and early 1980's. In comparison, the Antarctic continent, with no indigenous peoples or residents save transient scientific populations and support personnel, sees only those cargo ships required to deliver fuel and supplies, an increasing number of cruise ships, marine science vessels, support icebreakers, and, well offshore in open or near-open waters, an increasing number of fishing vessels from as far away as Japan and Russia.[17]

After the differences were addressed, the discussion paper analysed the similarities which existed between the two polar regions and their seas. These similarities included the high navigational risks due to the presence of ice fields, wind-driven ice forces, cold, isolation and lack of full support services. Another common element was that the impacts of incidents in polar waters are often graver than elsewhere, both for human life and the environment. For example, lives are at risk from exposure to air temperatures in addition to water temperatures, oil spilled on or under moving ice is not amenable to containment and cleanup technologies that are normally employed in temperate waters.[18] While flora and fauna are basically different in the two polar regions, 'what is common, however, is that both areas provide precarious conditions for survival of their peculiar ecology, and each provides to the world an image of uniqueness of species, of a human heritage of value beyond any commercial estimation'.[19]

One of the basic assumptions of the discussion paper by Canada was that linkages and similarities were more important than differences. The Polar Code was intended to provide a better defined shipping regime 'in a framework which can handle common issues in a consistent and rational manner, while having the flexibility to treat any challenges unique to Arctic or Antarctic waters appropriately. The International Maritime Organisation is the logical sponsor for such an instrument, given its mandate covering international shipping issues'.[20]

---

[16] *Ibid.*, paras. 4(2) and 4(3).
[17] *Ibid.*, para. 4(4).
[18] *Ibid.*, paras. 5(2) and 5(3).
[19] *Ibid.*, para. 5(4).
[20] *Ibid.*, para. 6(3).

During the session of the IMO Maritime Safety Committee held in May 1999, the Canadian delegation reiterated its support for the draft Polar Code. Referring to the perceived conflicts with other international instruments, it stated that 'while the Antarctic waters were protected by the Antarctic Treaty regime and its Protocols, there was a need to provide harmonised technical standards for ships working or transiting all polar waters'.[21]

The purpose of the draft Polar Code was very far-reaching. If adopted, such an instrument could substantially promote sustainable development in polar waters and improve cooperation among all who are involved in polar shipping. Suffice it to mention that the draft Polar Code also aimed at making marine insurance easily available for polar operation at costs that are predictable and affordable[22], which is not the case today. Among other so-called 'Polar Code efficiencies', as mentioned in the Canadian discussion paper,[23] it deserves to be noted that the Code would significantly contribute to meeting the current shipping constraints.

Nevertheless, the content and the aim of the draft Polar Code might have seemed too optimistic for the time being. In particular, it is difficult to predict to what extent the innovative bipolar approach represented a feasible solution in the light of the current political and legal conditions of the two polar areas. Irrespective of the merits of the bipolar approach, the fact remains that the polar regions are governed by two completely different political and legal frameworks.[24] While there are difficult problems to be faced in both cases, they are not comparable in nature and have so far been discussed in different fora.

North Polar Waters

The question of sovereignty over land (continental territories or islands) in the Arctic region is today generally settled.[25] It follows that the state that exercises sovereignty over a certain territory is also entitled to exercise the corresponding rights in the adjacent jurisdictional zones, be these internal waters, territorial seas, exclusive economic zones or continental shelves. The high seas begins beyond the limit of national jurisdiction, that is, beyond the 200-mile limit in respect to the water column.

---

[21] IMO, 'Outcome of Discussion at the 71st Session of the Maritime Safety Committee', doc. XXIII ATCM/IP 111, 1999.

[22] Under para. 1(6)(1) of the Polar Code, all ships operating in polar waters should maintain insurance or other financial security. However, the paragraphs relating to liability insurance were deleted by the IMO Sub-Committee on Ship Design and Equipment.

[23] Paras. 8(1) to 8(7) of IMO doc. DE41/INF. 8.

[24] For a general discussion, see D. Vidas (ed.), *Protecting the Polar Marine Environment*, especially pp. 3–15 and 78–103.

[25] For the specific status of Svalbard, see the 1920 Treaty concerning Spitsbergen; text published in *League of Nations Treaty Series*, Vol. 2, pp. 7ff.

Still disputed in Arctic waters, however, are the rights of innocent passage, transit passage and freedom of navigation that third states can exercise along the circumpolar routes or through straits between the continent and the islands (or between the islands themselves). The questions of disputed historic claims, the legality of straight baselines systems and the contested international use of maritime straits are still open in Arctic waters. Two well-known instances are briefly commented below.

In recent years, Canada made innovative steps in the field of Arctic navigation with the adoption in 1970 of the Arctic Waters Pollution Prevention Act[26] and the establishment of a straight baselines system under the Territorial Sea Geographic Coordinates (Area 7) Order of 10 September 1985.[27] These measures were made under the pressure of planned developments of navigation through the Northwest Passage (also involving nuclear-powered super-tankers) that could cause serious hazards to the environment in the region.[28] The United States protested against the Canadian enclosure of the Arctic Archipelago and the consequences of the enclosure on the freedom of navigation and passage through international straits.[29]

Measures on polar navigation have also been adopted by another circumpolar state. On 14 September 1990, the Minister of Merchant Marine of the former Soviet Union enacted the Regulations for Navigation of the Seaways of the Northern Sea Route.[30] As stated in Article 2,

> [T]he Regulations shall, on the basis of non-discrimination for vessels of all States, regulate navigation through the Northern Sea Route for the purpose of ensuring safe navigation and preventing, reducing, and keeping under control marine environment pollution from vessels, since the specifically severe climatic conditions that exist in the Arctic Regions and the presence of ice during the larger part of the year bring about

---

[26] 'Where no law exists, or where law is clearly insufficient, there is no international common law applying to the Arctic seas, we are saying somebody has to preserve this area for mankind until the international law develops. And we are prepared to help it develop by taking steps on our own and eventually, if there is a conference of nations concerned with the Arctic, we will of course be a very active member in such a conference and try to establish an international regime. But, in the meantime, we had to act now' (press statement made in 1970 by the Prime Minister of Canada, Mr. Trudeau, in IML, 1970, Vol. 9, pp. 600–604, where the Act is also reproduced).

[27] Canada, *Gazette*, Part II, of 2 October 1985.

[28] For further discussion, see D.R. Rothwell and C.C. Joyner, 'Domestic Perspectives and Regulations in Protecting the Polar Marine Environment: Australia, Canada and the United States', in Vidas (ed.), *Protecting the Polar Marine Environment*, pp. 150–161.

[29] See United States Department of State, Bureau of Oceans and International Environmental and Scientific Affairs, *Limits in the Seas*, No. 112, *United States Responses to Excessive National Maritime Claims*, Washington, DC, 1992, p. 70.

[30] Russian text published in *Izveshcheniya Moreplavatelyam* (Notices to Mariners), No. 29, 18 June 1991; English translation published in *Guide to Navigating through the Northern Sea Route* (St. Petersburg: Head Department of Navigation and Oceanography, Russian Ministry of Defence, 1996), pp. 81–84. English translation also available, with minor deviations, in *International Challenges*, Vol. 12, 1992, pp. 121–126. See discussion by D. Brubaker, 'Regulation of Navigation and Vessel-Source Pollution in the Northern Sea Route: Article 234 and State Practice', in Vidas (ed.), *Protecting the Polar Marine Environment*, pp. 221–243, especially at pp. 226–231.

obstacles, or increased danger, to navigation while pollution of sea, or the northern coast of the USSR might cause great harm to the ecological balance, or upset it irreparably, as well as inflict damage on the interests and well-being of the peoples of the Extreme North.

Several precise requirements are established for navigation through the Northern Sea Route. The owner or master of the vessel must submit a notification and request for leading to the Administration of the Northern Sea Route. The Administration may carry out an inspection of the vessel while it navigates the route if unfavourable ice, navigational, hydrographic, weather and other conditions occur that might endanger the vessel, or where there is a threat of polluting the marine environment or the coast. Compulsory icebreaker-assisted pilotage is provided to ensure safe navigation through the straits of Vil'kitski, Sokal'skii, Sannikov and Dimitri Laptev because of the adverse navigational and ice conditions. In the other regions, the Administration shall prescribe one of the following types of leading as determined by the circumstances, namely leading along recommended routes, aircraft-assisted leading, conventional pilotage, icebreaker leading, or icebreaker-assisted pilotage. The master of the vessel must maintain contact with the radio centre of the appropriate marine operations headquarters. Vessels must have on board a certificate of due financial security for civil liability of the owner for any inflicted damage which pollutes the marine environment and the coast. The Administration provides vessels with navigational information and leading and rescuing services. It is also envisaged that 'when navigating the Northern Sea Route, payments for the services rendered to vessels by the Marine Operations Headquarters and the Administration shall be collected in accordance with the rates duly adopted' (Article 8(4)).

Again, the conformity with international law of the measures adopted by the former Soviet Union (and today confirmed by Russia) has been questioned by the United States, especially regarding its incidence on passage through maritime straits.[31]

An Arctic international forum could be the best instrument for discussing the regime of shipping in Arctic waters and addressing the complex navigational issues which mainly involve, but are not limited to, two circumpolar Arctic states (Canada and Russia) on the one hand, and at the same time another circumpolar Arctic state

---

[31] 'While the United States is sympathetic with efforts which have been made by the Soviet Union in developing the Northern Seaway Route and appreciates the importance of this waterway to Soviet interests, nevertheless, it cannot admit that these factors have the effect of changing the status of the waters of the route under international law. With respect to the straits of the Kara Sea described as overlapped by Soviet territorial waters it must be pointed out that there is a right of innocent passage of all ships through straits used for international navigation between two parts of the high seas and that this right cannot be suspended (...). In the case of straits comprising high seas as well as territorial waters there is of course unlimited right of navigation in the high seas areas' (note of 22 June 1965, in US Department of State, Bureau of Oceans and International Environmental and Scientific Affairs, *Limits in the Seas*, No. 112). See J.A. Roach and R.W. Smith, *United States Responses to Excessive Maritime Claims*, 2nd edition (The Hague: Kluwer Law International, 1996), pp. 328–339. See also discussion by Brubaker, 'Regulation of Navigation and Vessel-Source Pollution in the Northern Sea Route: Article 234 and State Practice', pp. 239–243.

and the major maritime power (the United States) on the other. To a certain extent, an appropriate forum to discuss the issue could also be the Arctic Council[32] and its bodies, including the Working Group on Protection of the Arctic Marine Environment.

It may also be added that, as regards this most important circumpolar navigation route, institutions in Russia, Norway and Japan have, from 1993 to 1999, carried out the comprehensive 'International Northern Sea Route Programme' (INSROP) aimed at evaluating all the various factors influencing prospects for commercial shipping along the Northern Sea Route.[33]

As can be seen, negotiations on international Arctic shipping might develop in different fora, including of course the IMO. The prospects of success could be more promising if negotiations were carried out in a prudent manner, avoiding as far as possible contentious legal issues and focussing on the technical aspects of the case. But what could the prospects be if the questions unique to the Antarctic were added to the Arctic navigational problems, which are in themselves difficult to face?

South Polar Waters

Unlike the sovereignty question in the Arctic, the question of sovereignty over both land and sea areas in the Antarctic region is extremely controversial. The conflicting positions of states, which in the 'Antarctic jargon' are often called 'claimants', 'non-claimants' and 'basis-for-claim' states,[34] determined the well-known 'agreement to disagree' embodied in Article IV of the 1959 Antarctic Treaty,[35] which 'freezes' the question on sovereignty in the Antarctic Treaty area. The agreement to disagree was developed and strengthened by the subsequent ATS instruments, including the 1972 Convention for the Conservation of the Antarctic Seals,[36] the 1980 Convention on

---

[32] The Arctic Council is a 'high-level forum' established to promote cooperation and political action and to most effectively address the wide range of issues common to its members (the Arctic States, namely Canada, Denmark, Finland, Iceland, Norway, Russia, Sweden and the United States). The Council was established by a Declaration signed by the eight arctic states in Ottawa on 19 September 1996 (text reprinted in ILM, Vol. 35, 1996, pp. 1,387ff).

[33] INSROP was sponsored by Central Marine Research & Design Institute, Russia, the Fridtjof Nansen Institute, Norway, Ship & Ocean Foundation, Japan, with the Secretariat based at the Fridtjof Nansen Institute. It was divided into four sub-programmes, namely 1) natural conditions and ice navigation, 2) environmental factors, 3) trade and commercial shipping aspects, and 4) political, legal and strategic factors. For integration of findings resulting from the work carried out by INSROP, see W. Østreng (ed.), *The Natural and Societal Challenges of the Northern Sea Route – A Reference Work* (Dordrecht: Kluwer Academic Publishers, 1999).

[34] See further Orrego Vicuña, Chapter 3 in this book.

[35] UNTS, Vol. 402, pp. 71ff.

[36] UNTS, Vol. 1080, pp. 175ff; and ILM, Vol. 11, 1972, pp. 251ff.

the Conservation of Antarctic Marine Living Resources[37] and the 1991 Protocol on Environmental Protection to the Antarctic Treaty[38] (Environmental Protocol).

While the language initiating in the Antarctic cooperation may seem rather mysterious (especially where 'Pandora's boxes' or 'bifocal approaches' are evoked), the essence of the agreement to disagree on coastal jurisdictional zones is based on the different perspectives under which the core question of sovereignty is understood. On the one hand, the seven claimant states do not renounce their rights to coastal zones (territorial sea, exclusive economic zone, continental shelf and, perhaps, presential sea) which ensue from their claimed sovereignty over portions of Antarctic territory. On the other hand, states that do not recognise any claim of territorial sovereignty consider that there are no coastal jurisdictional zones off Antarctica, simply for the reason that there are no coastal states whatsoever in this continent.[39]

No doubt, the determination not to enter into the difficult field of claims, counterclaims, rejections of claims and rejections of counterclaims has been an able legal device and a wise political step upon which a progressive and fruitful cooperation has been developed among the Antarctic Treaty parties. Nevertheless, the ghost of frozen sovereignty is still present in Antarctic affairs.[40] In every aspect

---

[37] ILM, Vol. 19, 1980, pp. 837ff.

[38] ILM, Vol. 30, 1991, pp. 1,461ff.

[39] On certain problems in application of the UN Convention on the Law of the Sea in the Antarctic context, see Vidas, Chapter 14 in this book. On various issues involved in the application of the law of the sea to Antarctic waters, see: F. Orrego Vicuña and M.T. Infante, 'Le droit de la mer dans l'Antarctique', *Revue Générale de Droit International Public*, Vol. 54, 1980, pp. 340–350; A. Van der Essen, 'The Application of the Law of the Sea to the Antarctic Continent', in F. Orrego Vicuña (ed.), *Antarctic Resources Policy. Scientific, Legal and Political Issues* (Cambridge University Press, 1983), pp. 231–242; B.H. Oxman, 'Antarctica and the New Law of the Sea', *Cornell International Law Journal*, Vol. 19, 1986, pp. 211–247; V.S. Safronchuk, 'The Relationship between the ATS and the Law of the Sea Convention of 1982', in A. Jørgensen-Dahl and W. Østreng (eds.), *The Antarctic Treaty System in World Politics* (London: Macmillan, 1991), pp. 328–334; P. Gautier, 'The Maritime Area of the Antarctic and the New Law of the Sea', in J. Verhoeven, P. Sands and M. Bruce (eds.), *The Antarctic Environment and International Law* (London: Graham and Trotman, 1992), pp. 121–137; C.C. Joyner, *Antarctica and the Law of the Sea* (Dordrecht: Martinus Nijhoff, 1992); T. Scovazzi, 'The Application of the Antarctic Treaty System to the Protection of the Antarctic Marine Environment', in F. Francioni (ed.), *International Environmental Law for Antarctica* (Milan: Giuffré, 1992), pp. 113–148; Sir Arthur Watts, *International Law and the Antarctic Treaty System* (Cambridge: Grotius, 1992), pp. 141–163; T. Scovazzi, 'The Antarctic Treaty System and the New Law of the Sea: Selected Questions', in F. Francioni and T. Scovazzi (eds.), *International Law for Antarctica*, 2nd edition (The Hague: Kluwer Law International, 1996), pp. 377–394; D. Vidas, 'The Antarctic Treaty System and the Law of the Sea: A New Dimension Introduced by the Protocol', in O.S. Stokke and D. Vidas (eds.), *Governing the Antarctic: The Effectiveness and Legitimacy of the Antarctic Treaty System* (Cambridge University Press, 1996), pp. 61–90; and D. Vidas, 'Emerging Law of the Sea Issues in the Antarctic Maritime Area: A Heritage for the New Century?', *Ocean Development and International Law*, Vol. 31, 2000, pp. 197–222.

[40] For instance, Art. 234 of the UN Convention on the Law of the Sea allows coastal states, under certain conditions, to adopt and enforce non-discriminatory laws and regulations for the prevention, reduction and control of marine pollution from vessels in ice-covered areas within the limits of the coastal states' exclusive economic zone. At first sight, the adoption of such measures does not seem to

of Antarctic cooperation, including navigation safety, the interested states have to deal with the unresolved sovereignty problem. They are called to display political goodwill in order to overcome all the legal intricacies through the elaboration of imaginative solutions. The development of the concept of port state jurisdiction could be a promising idea in this direction.[41]

The considerations above explain why some of the definitions that were proposed in the Polar Code are likely to become very controversial if they are related to Antarctic waters. This is the case with the definitions of 'coastal state'[42] or 'port state',[43] which are hardly compatible with the principles underlying the ATS. For example, the agreement to disagree on the sovereignty question is difficult to reconcile with the assumption that 'specific limitations and conditions associated with the operation of ships of all classes and types may be applied to voyages falling in whole or in part under the jurisdiction of a Port or Coastal State' (para. 1(4)(3) of the Polar Code). Even more questionable is the assumption that in polar waters 'coastal states, port states or Antarctic Authorities exercise rights of control under international law with regard to the high seas'.[44] This seems to conflict with the principle of the non-prejudice of high seas rights that is explicitly provided by Article VI of the Antarctic Treaty. Inspection is also a very sensitive issue under the ATS, considering that under Article VII(3) of the Antarctic Treaty the inspection of ships is limited at points of discharging or embarking cargoes or personnel in Antarctica and cannot be made at sea.

Regarding institutional matters, the references to an 'Antarctic Authority' (or 'Antarctic Authorities' in plural), which appeared in the Polar Code,[45] raised a number of doubts. Presently, it is hardly possible to identify any such 'Authority', not least if it is meant as an international institution.

---

be in contradiction with any of the instruments of the ATS. The claimant states that have already established an exclusive economic zone in Antarctic waters, namely Argentina, Australia, Chile and the United Kingdom (the UK as regards South Georgia and the South Sandwich Islands) could thus avail themselves of this right. Protests from non-claimant states are, however, likely to follow.

[41] See especially Orrego Vicuña, Chapter 3 in this book. Also Richardson, Chapter 4, and Bush, Chapter 2, in this book.

[42] Meaning 'the state whose Exclusive Economic Zone includes the area of operation of the ship'; para. 3(1) of the 'Guide to the Code'.

[43] Meaning a 'state whose Exclusive Economic Zone includes any destination port of a ship where such port lies within Polar Waters'; para. 3(21) of the 'Guide to the Code'.

[44] Para. 14(2)(4) of the Polar Code. How exactly this provision should be interpreted does not seem quite clear.

[45] See, in this sense, para. 14(2)(2) of the Polar Code, according to which specific items to be considered by the flag state in issuing a 'Permit to Operate in Polar Waters' include 'the arrangements made by the ship for complying with operational control according to the method adopted by the Port State, Coastal State or Antarctic Authority as appropriate'.

The Developments in 1999

As was expected, the bipolar sphere of application of the draft Polar Code was subject to strong criticism by several states. In the session of the IMO Maritime Safety Committee (MSC) held in May 1999, the United States, supported by many other delegations, expressed concern at the proposed Code, while recognising the need to protect polar waters and safeguard the lives of the travellers in those waters. The United States pointed out that 1) the draft Polar Code 'failed to distinguish between the conditions and nature of shipping in the Arctic and those in Antarctica, which is unique in its geography and governance', and 2) that 'there were conflicts between the draft Code and the Antarctic Treaty'.[46] In light of the criticism, the MSC agreed, as a basis for further work, that hortatory guidelines be developed for areas north of 60° N,[47] and that 'Antarctic waters are to be excluded from the application of the guidelines, unless Antarctic Treaty members decide otherwise'.[48]

However, the ideas behind the draft Polar Code were not completely lost. At the Twenty-third Antarctic Treaty Consultative Meeting, held in Lima, 24 May – 4 June 1999, the United Kingdom submitted a paper in which four possible options were set for the polar shipping code:[49]

i. The status quo, i.e. unified IMO guidelines which do not distinguish between Arctic and Antarctic conditions in relation to shipping;

ii. IMO guidelines which differentiate in two sections between the Arctic and the Antarctic;

iii. Two distinct IMO guidelines dealing separately with the Arctic and the Antarctic;

iv. IMO guidelines dealing only with the Arctic.

The United Kingdom shared the views expressed at MSC that the draft Polar Code neither adequately recognised the nature of Antarctic shipping operations nor took sufficient account of the unique environmental conditions in the Antarctic and the international system of governance for the Antarctic Treaty area. Nevertheless, the United Kingdom expressed its preference for option (ii) above, believing that the area of application of the guidelines should include Antarctic waters for a number of reasons:

> Article 10 of Annex IV to the Protocol places an obligation on Parties to take account of the objectives of the Annex in the design, construction, manning, and equipment of ships engaged in or supporting Antarctic operations.

---

[46] See IMO, 'Outcome of Discussion at the 71st Session of the Maritime Safety Committee'.

[47] Moreover, these hortatory guidelines should be developed only for SOLAS ships operating in ice-covered waters, with the exclusion of ice-free waters north of 60° N.

[48] See IMO, 'Outcome of Discussion at the 71st Session of the Maritime Safety Committee'.

[49] United Kingdom, 'Polar Shipping Code', doc. XXIII ATCM/WP 40, 1999.

For reasons of safety and environmental protection, it is wholly appropriate that Parties establish adequate minimum standards for vessels operating in Antarctic waters. This issue is of particular relevance at a time when the trend in Antarctic maritime traffic, particularly in respect of tourist ships, is towards the use of larger vessels. /.../

Existing international maritime instruments (e.g., MARPOL, SOLAS, UNCLOS) and the Environmental Protocol to the Antarctic Treaty do not address in a comprehensive fashion all elements of polar ship design, construction, operation, manning and equipment, intended to be covered by the Polar Code/guidelines.

COMNAP (Council of Managers of National Antarctic Programs) has identified that shipping operations, particularly those relating to tourist operations, represent the most significant risk to the Antarctic environment.[50]

Following the proposal by the United Kingdom, the XXIII Consultative Meeting adopted Decision 2 (1999), 'Guidelines for Antarctic Shipping and Related Activities',[51] according to which the following was decided:

1. To give priority to the development of guidelines for Antarctic shipping and related activities pursuant to Article 10 of Annex IV to the Protocol; /.../

3. To convene a Meeting of Experts under the provisions of Recommendation IV-24, with the aim of developing draft guidelines for Antarctic shipping and related activities. /.../

5. Pursuant to paragraph 3 above, to request the Meeting of Experts:

   i. to examine the most recent version of the draft Polar Shipping Guidelines being developed for the Arctic by the IMO, and decide which elements of those draft Arctic guidelines should form the basis of the Antarctic guidelines;

   ii. to consider other aspects of the design, construction, manning and equipment of vessels operating in Antarctic waters that might require elaboration in the Antarctic guidelines;

   iii. to take into account existing international instruments regulating shipping activities in Antarctica, including for example MARPOL, SOLAS, UNCLOS and the Environmental Protocol to the Antarctic Treaty;

   iv. to take into account existing guidelines adopted under the Antarctic Treaty, and in particular those adopted under Resolution 6 (1998);

   v. to ensure the guidelines adequately take account of the nature of Antarctic shipping, the environmental conditions of Antarctica and the system of international governance applying to the Antarctic Treaty area.

---

[50] *Ibid.*

[51] See *Final Report of the Twenty-third Antarctic Treaty Consultative Meeting, Lima, Peru, 24 May – 4 June 1999* (Lima: Peruvian Ministry of Foreign Affairs, 1999), Annex B. Text of the report and annexes also available at the website <www.rree.gob.pe/conaan/meeting1.htm>.

COOPERATION BETWEEN THE CONSULTATIVE PARTIES AND THE IMO

The draft Polar Code was submitted to the IMO and attributed to the IMO a certain role as 'the Organisation',[52] despite the fact that the Antarctic Treaty Consultative Parties did not ask IMO to draft a Polar Code and make it applicable to the Antarctic waters.

While the Consultative Parties have usually played the role of 'legislators' in Antarctic affairs, other fora are not to be wholly excluded from the picture. Due to its responsibilities in the field of marine safety and navigation in general, the IMO is certainly one of these fora. One of the purposes of the IMO is 'to encourage the general adoption of the highest practicable standards in matters concerning maritime safety, efficiency of navigation and the prevention and control of marine pollution from ships' (Article 1(a) of the Convention on the IMO)[53]. The LOS Convention provides that states, 'acting through the competent international organization or general diplomatic conference', shall establish international rules and standards to prevent, reduce and control pollution of the marine environment by vessels (Article 212(1)).[54] The IMO is generally understood as the 'competent international organisation' in this respect.

Some coordination has already been established between the IMO and the ATS regarding MARPOL 73/78. In 1990, Annex I (Regulation for the prevention of pollution by oil) and optional Annex V (Regulations for the prevention of pollution by garbage from ships) of MARPOL 73/78 were amended in order to designate the 'Antarctic area', meaning 'the sea area south of 60° South Latitude', as a special area under the said annexes. The same was done in 1992 with respect to Annex II of MARPOL 73/78 (Regulations for the control of pollution by noxious liquid substances in bulk).

Within the ATS, several rules relating to the protection of the marine environment have already been developed by the parties. In addition to a number of recommendations and other measures adopted during the Consultative Meetings, Annex IV to the Environmental Protocol ('Prevention of Marine Pollution') sets forth a series of measures related to the prevention of marine pollution from ships (e.g., measures related to the discharge of oil, noxious liquid substances or sewage, disposal of garbage, ship retention capacity and reception facilities, preventive measures, emergency preparedness and response). Annex IV also provides that 'with respect to those Parties which are also Parties to MARPOL 73/78, nothing in this

---

[52] Para. 3.16 of the 'Guide to the Code'. For example, under paras. 1(2)(1) and 1(2)(2) of the Polar Code, flag states should communicate certain information to the IMO.

[53] Convention on the International Maritime Organisation, Geneva, 6 March 1948, published in UNTS, Vol. 289, pp. 3ff.

[54] It is evident that provisions relating to matters such as design, construction, manning, and equipment of ships should be adopted at a uniform and international level, in order to avoid that differences in domestic legislation prevent the navigation of ships.

Annex shall derogate from the specific rights and obligations thereunder' (Article 14). Annex IV to the Environmental Protocol may thus be seen as a tool that provides for a minimum protection of the marine environment with respect to those parties to the Environmental Protocol, if any,[55] which have not yet agreed to become parties to MARPOL 73/78. The latter is seen by the parties to the Environmental Protocol as an appropriate instrument for regulating pollution from ships.[56]

This should not, however, prevent the establishment of adequate minimum standards specifically applicable to vessels operating in Antarctic waters. It has already been commented that on the design, construction, manning and equipment of ships, which is the main subject of the Polar Code or Guidelines, Annex IV to the Environmental Protocol contains a rather brief and vague provision (Article 10)[57] that needs to be supplemented by a more specific regime.

In 1999 the relationship between the Consultative Parties and the IMO was fully discussed both within MSC and the Twenty-third Consultative Meeting. The already-mentioned document by the United Kingdom[58] identifies the IMO as the most appropriate organisation in which to develop adequate minimum standards, given that

- the IMO has the relevant technical expertise which the Antarctic Treaty System does not, and

- guidelines developed through the IMO would apply to *all* IMO Member States and not solely to those States that are Party to the Antarctic Treaty. Again this is an important consideration, particularly in relation to tourist vessels, a significant proportion of which are flagged with non-Treaty Parties.

Decision 2 (1999) of the Twenty-third Consultative Meeting consequently provided that the Consultative Parties shall 'seek subsequent adoption of these guidelines [i.e., the Guidelines for Antarctic Shipping and Related Activities] by the International Maritime Organisation (IMO) as a means of extending their applicability to members of the IMO that are not Antarctic Treaty Consultative Parties'.[59] It was further agreed that the Meeting of Experts to be convened under Decision 2 (1999) should be hosted in London by the government of the United Kingdom and 'should, as far as possible, be held in conjunction with a meeting of the appropriate IMO expert body'. While the drafting of the guidelines is not

---

[55] As at 29 December 1999, the parties of MARPOL 73/78 comprised 108 states, the combined merchant fleets of which constitute approximately 94 per cent of the gross tonnage of the world's merchant fleet.

[56] See also discussion by Orrego Vicuña, Chapter 3 in this book.

[57] 'In the design, construction, manning and equipment of ships engaged in or supporting Antarctic operations, each Party shall take into account the objectives of this Annex'.

[58] United Kingdom, 'Polar Shipping Code'.

[59] Para. 2 of Decision 2 (1999). How could the applicability of guidelines, which are not binding by their nature, be extended to any state is another, yet rather intriguing, question.

envisaged as a formally IMO process, the basis for cooperation and coordination between the ATS and the IMO was established.

CONCLUDING REMARKS: FROM A POLAR CODE TO GUIDELINES FOR ANTARCTIC SHIPPING AND RELATED ACTIVITIES

Only time can judge the prospects for the former Polar Code, which has now, in the Antarctic context, been transformed into a proposal for consideration in the form of Guidelines for Antarctic Shipping and Related Activities. Substantial progress is only likely to be made if the discussion is limited to the purely technical profile of shipping safety without entering into the political sensitivities and legal intricacies which are peculiar to the Antarctic.

At the Twenty-second Consultative Meeting, held in Tromsø, Norway, in 1998, the Antarctic Treaty Consultative Parties, also on the basis of two papers submitted by Norway,[60] 'agreed that a draft Polar Code, whether adopted as a mandatory or non-mandatory document within the IMO, will have a significant influence on future shipping activity in Antarctica. Consequently, Parties should be actively involved in the development of the Code in order to ensure that Antarctic issues are adequately represented'.[61]

Today, the direction of the active involvement of the Consultative Parties has become clearer. The instrument as a whole, far from having a mandatory application, will be developed as an instrument having a legally non-binding character.[62] The point was very clearly made that this future instrument could relate to the Antarctic waters only because, and in so far as, the Consultative Parties have so agreed. The development of an instrument, perhaps under the title of guidelines, will be facilitated by a Meeting of Experts to be convened under the ATS ambit.[63]

---

[60] Norway, 'The International Code of Safety for Ships in Polar Waters (Polar Code)', doc. XXII ATCM/WP 17, 1998; and Norway, 'The International Code of Safety for Ships in Polar Waters – The Antarctic Issues', doc. XXII ATCM/WP 18, 1998.

[61] See agenda item 10 of the Twenty-second Consultative Meeting, in *Final Report of the Twenty-second Antarctic Treaty Consultative Meeting, Tromsø, Norway, 25 May – 5 June 1998* (Oslo: Royal Norwegian Ministry of Foreign Affairs, 1998), pp. 5 and 15–17. In this spirit, Resolution 3 (1998), 'International Code of Safety for Ships in Polar Waters', adopted at the Twenty-second Consultative Meeting recommended 'that Consultative Parties provide input to IMO, via their national maritime authorities, on the draft Polar Shipping Code as it relates to shipping operations within the Antarctic Treaty area'; text in *ibid.*, Annex C, pp. 124.

[62] This should not exclude later reconsideration, as explained by the United States: 'Therefore, instead of the proposed code, the IMO should, as a first step, develop guidelines to be disseminated under cover of an MSC circular, and then reconsider, in the light of experience gained with their application, whether such a status should be altered'; see IMO, 'Outcome of Discussion at the 71st Session of the Maritime Safety Committee'. Furthermore, the non-binding nature of the guidelines does not prevent States from giving them a mandatory character in their domestic legislation.

[63] More precisely, a meeting convened under Recommendation IV–24, providing that 'Meetings of experts be convened from time to time as the need arises to discuss practical problems relating to Antarctic activities' (para. 1).

The impression is that the former Polar Code, far from having been frozen (as was the fate of other delicate Antarctic issues), is now watered down.[64]

However, this impression is not sufficient to draw negative conclusions from the present situation. The Meeting of Experts convened by the Twenty-third Consultative Meeting is also open to experts of non-Consultative Parties and a number of international organisations, both governmental and non-governmental (IHO, IMO, WMO, IACS, P&I Club, IAATO and ASOC). Close coordination with relevant IMO bodies has been achieved. The need to draft minimum standards for vessels operating in Antarctic waters, especially those engaged in tourism activities, is a serious concern for all Consultative Parties. The Parties have always addressed crucial Antarctic issues in an effective and responsible way, and there is no reason to doubt whether they will do the same in this case also.

POST SCRIPTUM

The Antarctic Treaty Meeting of Experts on Guidelines for Antarctic Shipping and Related Activities was held in London on 17–19 April 2000. It recommended, *inter alia*, that the Guidelines should be non-mandatory and developed in a style of a handbook of information for all vessels operating south of 60° south, that consistency be maintained, where appropriate, between the Arctic and the Antarctic guidelines, and that the next Antarctic Treaty Consultative Meeting continues the development of the Antarctic shipping guidelines using the work undertaken by the Meeting of Experts as its basis.

---

[64] Another very critical point, which is particularly relevant for North Polar waters, is the assumption of the United States that there were conflicts between the draft code 'and the UN Law of the Sea Convention, such as the requirement to give prior notification to the coastal State when entering its EEZ'; see IMO, 'Outcome of Discussion at the 71st Session of the Maritime Safety Committee'. This assumption was bluntly followed by the IMO Maritime Safety Committee, which decided that 'any provisions in the current guidelines which are inconsistent with international law, including the provision for prior notification, should be removed' (*ibid.*). However, the point deserves to be carefully scrutinised; for example, an obligation to give prior notification to the coastal state could be in certain cases consistent with Art. 234 of the UN Law of the Sea Convention.

14

The Antarctic Continental Shelf Beyond 200 Miles: A Juridical Rubik's Cube

*Davor Vidas**

The difficulties that arise when the global law of the sea solutions are applied in the specific legal and political, and indeed natural, context of the Antarctic region have so far been the subject of several analyses.[1] However, there is one emerging and particularly intricate law of the sea issue for states cooperating in management of the Antarctic region that has seldom been explored, that of what is to be done with the requirement contained in the LOS Convention that relates to the submission of information on the outer limit of the continental shelf beyond 200 miles[2] to the Commission on the Limits of the Continental Shelf. This issue can be compared to the Rubik's cube in its juridical form, but which also has the intrinsic nature of a 'time-bomb' because, according to the letter of the United Nations Convention on

---

* This chapter is an updated and revised version of a part of material published by the author in *Ocean Development and International Law*, Vol. 31, 2000. Important developments which occurred soon after the submission of that article for publication in October 1999 have prompted the publication of a part of the material here, in an abridged and updated version.
[1] For an overall study of the law of the sea issues and the Antarctic, see C.C. Joyner, *Antarctica and the Law of the Sea* (Dordrecht: Martinus Nijhoff, 1992), which remains the only available monograph on the subject. Among several shorter studies see especially: F. Orrego Vicuña, 'The Law of the Sea and the Antarctic Treaty System: New Approaches to Offshore Jurisdiction', in C.C. Joyner and S.K. Chopra (eds.), *The Antarctic Legal Regime* (Dordrecht: Martinus Nijhoff, 1988), pp. 97–127; C.C. Joyner, 'The Antarctic Treaty System and the Law of the Sea – Competing Regimes in the Southern Ocean?', *The International Journal of Marine and Coastal Law*, Vol. 10, 1995, pp. 301–331; D. Vidas, 'The Antarctic Treaty System and the Law of the Sea: A New Dimension Introduced by the Protocol', in O.S. Stokke and D. Vidas (eds.), *Governing the Antarctic: The Effectiveness and Legitimacy of the Antarctic Treaty System* (Cambridge University Press, 1996), pp. 61–90; T. Scovazzi, 'The Antarctic Treaty System and the New Law of the Sea: Selected Questions', in F. Francioni and T. Scovazzi (eds.), *International Law for Antarctica*, 2nd edition (The Hague: Kluwer Law International, 1996), pp. 377–394; and P. Gautier, 'The Maritime Area of the Antarctic and the New Law of the Sea', in J. Verhoeven, P. Sands and M. Bruce (eds.), *The Antarctic Environment and International Law* (London: Graham & Trotman, 1992).
[2] All references to miles herein indicate nautical miles.

the Law of the Sea (LOS Convention),[3] the issue awaits the countries claiming sovereignty over portions of territory in the Antarctic 10 years after the entry into force of the Convention for each.

For some time, this issue and its possible implications were only whispered about, first in the literature[4] and then at informal symposia and meetings, such as the March 1995 Symposium on 'The Future of the Antarctic Treaty System' held in Ushuaia, Argentina, and co-sponsored by the Argentine Council for International Relations and the Australian Antarctic Foundation.[5] Although the controversial issue has been defined, its implications had been forewarned and alternative solutions had been suggested by some of the participants in the Symposium,[6] others thought the question to be 'academic and need not be pursued'.[7] As added by some then, '[we are] very happy that we have ten years to present the results of whatever we decide to the Commission ... that is plenty of time'[8]. But, as it often happens, time tends to pass quickly, not least when deadlines are in question.

On 2 December 1999, the Minister for Foreign Affairs and the Minister for the Environment and Heritage of Australia issued a joint press release entitled 'Move to Claim Extended Antarctic Continental Shelf'.[9] By this move, the Australian government had announced that it would take action enabling her to define the limits of the continental shelf off the Australian Antarctic Territory (AAT), by obtaining the necessary data before 16 November 2004. This is when the deadline for Australia, as the first among the seven Antarctic claimant countries, expires for submission of information required to the Continental Shelf Commission. The Government also announced that it would provide approximately 30 million Australian dollars over a 5-year period for the survey work required to establish baselines, determine the foot of the continental slope, and delineate the outer limit of the continental shelf.

This public announcement of the Australian government was, essentially, no more than a statement that it will provide a sizeable funding for the survey work on the continental shelf off the AAT, indeed a decision fully within its own competence. While any later submission of the information so obtained by Australia to the Continental Shelf Commission may still be understood as conditional, pending

---

[3] UN doc. A/CONF.62/122, 10 December 1982; text reprinted in ILM, Vol. 21, 1982, pp. 1,261ff. The LOS Convention entered in force on 16 November 1994.

[4] The first mention of this issue in the literature on Antarctic affairs is found in an article published in 1988, thus twelve years ago: Orrego Vicuña, 'The Law of the Sea and the Antarctic Treaty System', p. 116.

[5] See A. Jackson (ed.), *On the Antarctic Horizon. Proceedings of the International Symposium on the Future of the Antarctic Treaty System* (Hobart: Australian Antarctic Division,1996).

[6] See *ibid.*, pp. 49–50.

[7] *Ibid.*, p. 50.

[8] *Ibid.*, p. 52.

[9] See 'Joint Media Release by Minister for Foreign Affairs, Alexander Downer, and Minister for the Environment and Heritage, Robert Hill', Foreign Affairs and Trade, FA 132, of 2 December 1999, and the accompanying Fact Sheet.

the failure of any other collectively agreed solution among the Antarctic Treaty Consultative Parties, Australia clearly indicated its intention not to forego its rights provided for under the LOS Convention. The effect of this action was actually one of increasing public awareness of the sensitive issue over the Antarctic continental shelf beyond 200 miles,[10] with all the consequences that such a publication was bound to carry.

BACKGROUND: THE LAW OF THE SEA IN THE ANTARCTIC SETTING

In order to provide a brief background, it may be useful to outline what has been in the core of legal, and equally political, controversy regarding the situation of the Antarctic in international law in general, and the law of the sea in particular.

In the first half of the 20th century, seven states – Argentina, Australia, Chile, France, New Zealand, Norway and the United Kingdom – put forward territorial claims to parts of the Antarctic. None of the seven claims has been expressly recognised by any other country apart from fellow claimants,[11] and even then only partially. Moreover, there is a large area south of 60° south that remains unclaimed. However, three claimed sectors – those of Argentina, Chile and UK – all partially overlap in the Antarctic Peninsula area, between 53° and 74° west.[12] This has led to various incidents as well as to the instituting of proceedings before the International Court of Justice by the UK in 1955 – an attempt that proved unsuccessful because of non-acceptance of the Court's jurisdiction, in this case by Argentina and Chile.[13] Eventually, all seven claimant countries and other parties to the 1959 Antarctic Treaty[14] agreed to put aside their competing positions on territorial claims in the Treaty area, and in the interests of establishing a unique form of international governance for the Antarctic, they achieved an 'agreement to disagree' on the sovereignty issue (Article IV).[15]

---

[10] The governmental announcement quickly made the newspaper headlines; see, for instance, '$30m Quest for a Slice of Antarctica', *Sydney Morning Herald*, 4 December 1999, p. 3.

[11] The single possible exception being South Africa's implicit recognition of the Norwegian claim in 1959, in relation to use of an old Norwegian base in Dronning Maud Land; see W.M. Bush (ed.), *Antarctica and International Law: A Collection of Inter-State and National Documents*, Vol. III (London: Oceana Publications, 1988), pp. 168–171 and 195.

[12] The claims of Chile and UK overlap between 53° and 80° W; those of Argentina and UK overlap between 25° and 74° W; and those of Argentina and Chile overlap between 53° and 74° W.

[13] See: *Antarctica case (United Kingdom v. Argentina)*, Order of March 16th, 1956, ICJ Reports 1956, pp. 12–14; *Antarctica case (United Kingdom v. Chile)*, Order of March 16th, 1956, ICJ Reports 1956, pp. 15–17.

[14] UNTS, Vol. 402, pp. 71ff.

[15] For comprehensive studies see Sir Arthur Watts, *International Law and the Antarctic Treaty System* (Cambridge: Grotius Publications, 1992); and P.J. Beck, *The International Politics of Antarctica* (London: Croom Helm, 1986).

This important consensus has enabled the adoption of the Antarctic Treaty, which in turn has provided a foundation for a unique form of international governance of the Antarctic, the Antarctic Treaty System (ATS), which is a regional network of international instruments and decision-making structures for Antarctic affairs.[16] The entire ATS, after being under continuous development for almost four decades, has been based on unresolved sovereignty issues concerning the Antarctic. The essential requirement in the development of the ATS was to build a regime through various means of agreed cooperation, so as not to prejudice the position of any countries that claimed sovereignty in the Antarctic – or of countries that did not recognise these claims.[17]

The *balancing* of approaches to the sovereignty issue remains basic to the ATS and to successful international collaboration between the already numerous parties to the Antarctic Treaty[18]. The ability to achieve balance in the controversial matter over sovereignty has been one of the major factors that has enabled the ATS to develop into the comprehensive international regime that it is today.

However, various challenges for the ATS cooperation remain, or may easily emerge in the foreseeable future, many of which are related to the Antarctic maritime area and its resources. These challenges characteristically involve the need to *reconcile* solutions found in the international law of the sea (*rationae materiae* applicable globally) with Antarctic-regionally applicable rules adopted by a more limited number of states within the ambit of the ATS.[19]

The Antarctic Treaty offers two provisions that are directly relevant to the status of the Antarctic maritime area. Firstly, by stipulating in Article VI that its provisions shall apply to the area south of 60° south latitude, including all ice shelves, the Treaty defines the legal boundary for the Antarctic maritime area within the wider Southern Ocean area. However, although mentioning the high seas, this provision fails to determine *which part* of the area in question actually comprises high seas. Secondly, by 'freezing' the problem of sovereignty claims in Article IV, the Treaty offers no more than an 'agreement to disagree' on the sovereignty issue, thus leading to the ambivalent interpretations with respect to the legal status of the Antarctic

---

[16] For a legal definition of the Antarctic Treaty System, see Art. 1(e) of the Protocol on Environmental Protection to the Antarctic Treaty, text reprinted in ILM, Vol. 30, 1991, pp. 1,461ff. See also discussion by D. Vidas, 'The Antarctic Treaty System in the International Community: An Overview', in Stokke and Vidas (eds.), *Governing the Antarctic*, pp. 36–48.

[17] Note the interplay between various provisions of the Antarctic Treaty, especially Arts. IV and IX.

[18] As of 29 December 1999, there were 44 states parties to the Antarctic Treaty.

[19] For a recent analysis of other emerging law of the sea issues in the Antarctic, including the illegal, unregulated and unreported fishing of the Patagonian toothfish and the controversy over the legal status of the Antarctic seabed, see D. Vidas, 'Emerging Law of the Sea Issues in the Antarctic Maritime Area: A Heritage for the New Century?', *Ocean Development and International Law*, Vol. 31, 2000, pp. 197–222; and D. Vidas, 'Southern Ocean Seabed: Arena for Conflicting Regimes?', in D. Vidas and W. Østreng (eds.), *Order for the Oceans at the Turn of the Century* (The Hague: Kluwer Law International, 1999), pp. 291–314.

maritime area. Both the fact of overlapping claims and the existence of an unclaimed part of Antarctica contribute to the ambivalence of such interpretations.

When the Antarctic Treaty was adopted in 1959, its scope was rather restricted; not least due to the issue of sovereignty claims, it left in ambiguity the legal status of the Antarctic maritime area. One difficulty is related to the *physical properties* of the area. The existing rules of the law of the sea have been conceived for waters adjacent to coasts not possessing such specific properties as those caused by the Antarctic ice-structures.[20] The second difficulty is of a *conceptual nature*. The law of the sea, as embodied both in customary and in conventional rules, relies on two basic concepts – that of the coastal state and, derived from this, that of the baselines of the territorial sea. It was because of the different approaches of the states concerned with respect to the issue of sovereignty claims to Antarctica that the law of the sea was initially approached from divergent 'positions of principle'.[21] These positions of principle were based on the view of either the existence or the non-existence of state sovereignty in Antarctica. They led to a purely legal-logical exposition of the two essentially different lines of reasoning.

Does a coastal state exist in Antarctica? For the claimant states, the answer will be affirmative, with all the consequences resulting from the rules of the law of the sea. Non-claimant states will answer in the negative, again with all the consequences resulting from the rules of the law of the sea – which in particular leads to the conclusion that *all* the sea south of 60° south latitude clearly comprises high seas.

Are those two different lines of reasoning based on false logic? No. Is the question wrongly posed? Perhaps.

The vital need for an Antarctic resources policy has led to a gradual revision of the initial 'positions of principle', or better, to the reconciliation of otherwise irreconcilable positions. The initially restricted scope of the Antarctic Treaty has been widened as the Consultative Parties developed rules for the management of the Antarctic marine resources. Solutions to the legal problems have begun to emerge within the context of various resource regimes: the Seals Convention,[22] CCAMLR and CRAMRA[23] (although the latter is a story in its own right)[24]. These

---

[20] The exception is Art. 234 of the LOS Convention. However, this was adopted with the Arctic rather than Antarctic waters in mind. See discussion by B. Vukas, 'United Nations Convention on the Law of the Sea and the polar marine environment', in D. Vidas (ed.), *Protecting the Polar Marine Environment: Law and Policy of Pollution Prevention* (Cambridge University Press, 2000), pp. 34–56, at pp. 36–37 and 52.
[21] See Orrego Vicuña, 'The Law of the Sea and the Antarctic Treaty System', pp. 98–99.
[22] Convention for the Conservation of Antarctic Seals, adopted in 1972, entered into force in 1978; published in UNTS, Vol. 1080, pp. 175ff, text reprinted in ILM, Vol. 11, 1972, pp. 251ff.
[23] Convention on the Regulation of Antarctic Mineral Resource Activities, adopted in 1988, not in force; text reprinted in ILM, Vol. 27, 1988, pp. 868ff.
[24] See especially W.M. Bush, 'The 1988 Wellington Convention: How Much Environmental Protection?', in Verhoeven et al (eds.), *The Antarctic Environment and International Law*, pp. 69–83; R. Wolfrum, *The Convention on the Regulation of Antarctic Mineral Resource Activities: An Attempt to Break New Ground* (Berlin: Springer–Verlag, 1991); and C.C. Joyner, 'CRAMRA: The Ugly

arrangements managed to circumnavigate the resolution of the Antarctic sovereignty problem.

Significantly, the resource regimes do not differentiate between claimed and unclaimed sectors in the Antarctic Treaty area. Neither does the 1991 Environmental Protocol introduce such a differentiation; for instance, the Protocol prohibits mining activity in both the claimed and unclaimed sectors, and equally for claimants and those who do not recognise claims.[25]

For all juridical solutions in the Antarctic, securing *balance* among the positions on sovereignty remains one of the essential requirements with respect to supporting the functioning of the ATS.

THE ANTARCTIC CONTINENTAL SHELF BEYOND 200 MILES: A DILEMMA FOR THE ANTARCTIC CLAIMANTS

All seven claimant countries are parties to the LOS Convention.[26] In claiming parts of the Antarctic territory, the claimant countries can imply that their rights extend to the continental shelf, as an integral part of their claims, since no express proclamation is required for the continental shelf.[27] Thus, in the spirit of Article IV of the Antarctic Treaty, whereas claimant states can maintain their view on sovereignty and sovereign rights, others can retain a differing view, and this can continue as long as no express proclamation is required. At one point, however, claimant countries may be required to *specify* the extent of their continental shelf around Antarctica. According to the letter of the LOS Convention, the time for doing this seems to be approaching.

In accordance with the LOS Convention, the continental shelf as a legal regime may extend beyond 200 miles from the baselines from which the breadth of the territorial sea is measured, when this is possible due to the configuration of the continental margin in question.[28] Information (meaning co-ordinates, charts and geodetic data) on the limits of the continental shelf with such extent shall be

---

Duckling of the Antarctic Treaty System?', in A. Jørgensen–Dahl and W. Østreng (eds.), *The Antarctic Treaty System in World Politics* (London: Macmillan, 1991), pp. 161–185.

[25] See Art. 7 of the Protocol.

[26] In chronological order, instruments of ratification of the LOS Convention were deposited by: Australia, on 5 October 1994; Argentina, on 1 December 1995; France, on 11 April 1996; Norway, on 24 June 1996; New Zealand, on 19 July 1996; United Kingdom, on 25 July 1997; and by Chile, on 25 August 1997.

[27] See Art. 77(3) of the LOS Convention; and Art. 2(3) of the 1958 Convention on the Continental Shelf. For a detailed discussion, see Orrego Vicuña, 'The Law of the Sea and the Antarctic Treaty System', pp. 113–117.

[28] Art. 76(4)–(6) of the LOS Convention, which provides a lengthy and complex description of the conditions to be satisfied. See discussion by G. Francalanci, 'Technical Problems for the Commission on the Limits of the Continental Shelf', in Vidas and Østreng (eds.), *Order for the Oceans at the Turn of the Century*, pp. 123–132.

submitted by the state concerned to the Commission on the Limits of the Continental Shelf.

This Commission, which became operative in 1997, was set up under Annex II to the LOS Convention. Annex II not only provides the legal basis for establishing the Commission but it also contains a provision that specifies the *deadline* for submission of the particulars on the shelf beyond 200 miles:

> Where a coastal State intends to establish, in accordance with article 76, the outer limits of its continental shelf beyond 200 nautical miles, it shall submit particulars of such limits to the Commission along with supporting scientific and technical data as soon as possible but in any case *within 10 years* of the entry into force of this Convention for that State.[29]

The contents of this rather technical requirement poses both a difficult policy and a legal question of substance for the claimants (and also for other Consultative Parties and the ATS as a whole). If the claimant states individually present limits of their Antarctic continental shelves beyond 200 miles to the Commission, thus entering into discussions with non-claimants, the sensitive balance on sovereignty issue in the Antarctic could as a consequence be disturbed. But if a claimant chooses rather to act jointly, either in the claimant group or together with the other Consultative Parties, how should this be done? Another option is to remain passive and let the 10-year period expire, but this would possibly weaken the claims or allow others to impute renunciation – thus again disturbing the balance on the Antarctic sovereignty issue. It must also be taken into account that part of the Antarctic continental shelf beyond 200 miles may still be within the area of application of the Protocol, and in such case subject to Article 7 of the Protocol, which provides that '[A]ny activity relating to mineral resources, other than scientific research, shall be prohibited'.

CONSEQUENCES OF PASSIVITY

Firstly, it must be known whether there are any consequences for a state not submitting the particulars about the limits of its continental shelf beyond 200 miles to the Commission within the time-limit provided. As noted by one expert, states are not obliged to submit information to the Commission; the coastal state can do this if it so intends.[30] While passivity of a coastal state in this respect will result in no breach of obligation, certain consequences will follow. In its recently adopted Scientific and Technical Guidelines, the Commission specifies that:

> If ... a State does not demonstrate to the Commission that the natural prolongation of its submerged land territory to the outer edge of its continental margin extends beyond the 200-nautical-mile distance criterion, the outer limit of its continental shelf is automatically delineated up to that distance as described in paragraph 1 [of Article 76 the LOS

---

[29] Art. 4, Annex II to the LOS Convention (emphasis added).
[30] Francalanci, 'Technical Problems for the Commission', p. 127.

Convention]. In this case, coastal States do not have an obligation to submit information on the limits of the continental shelf to the Commission /.../.[31]

If a state party to the LOS Convention does not submit the relevant information on its continental shelf beyond 200 miles to the Commission within the time-limit at disposal, that state will have no continental shelf extending beyond 200 miles. This conclusion is also supported by Article 76(8) of the LOS Convention. The Article provides that the coastal state shall establish the outer limit of its continental shelf beyond 200 miles 'on the basis of' recommendations made by the Commission and that the limits so established by the coastal state shall be 'final and binding'. In the absence of such procedure, and with the expiry of the 10-year deadline, the part of the continental shelf (or margin) extending beyond 200 miles would therefore remain, in the LOS Convention terminology, seabed 'beyond the limits of national jurisdiction'. As such, it would fall under the competence of the International Seabed Authority rather than a coastal state.[32] As commented by Nandan and Rosenne:

> The outer limit of the continental shelf constitutes the maximum extent of coastal State jurisdiction. Read in conjunction with article 134, paragraph 3, the outer limit line establishes the boundary between the continental shelf and the Area (i.e., the limits of national jurisdiction).[33]

Moreover, the coastal state shall deposit a copy of charts showing the outer lines of the continental shelf with the Secretary-General of the International Seabed Authority.[34]

It therefore remains for the claimants (or, for that matter, the Consultative Parties) to decide *when* to act in response to the requirement contained in Article 4 of Annex II to the LOS Convention as well as *how* to act.

According to the letter of the LOS Convention, the time available appears limited by the date when the 10-year span from entry into force of the Convention expires for *any* one of the claimants, since this will affect the position of all the remaining claimants. The critical date for claimant countries would thus be 16 November 2004, which is 10 years from the date when the Convention entered in force for Australia. However, a possibility of relaxing this time-limit has subsequently been introduced. In accordance with the LOS Convention, the first election of the members of the Commission was to have been held within 18 months after the entry into force of the Convention (i.e., by 16 May 1996).[35] However, the

---

[31] Para. 2(2)(4) of the Scientific and Technical Guidelines of the Commission on the Limits of the Continental Shelf, UN doc. CLCS/11, of 13 May 1999.

[32] See Art. 1(1) and Part XI of the LOS Convention, especially Art. 134(3).

[33] M.N. Nordquist (ed.-in-chief), with S.N. Nandan and S. Rosenne (eds.), *United Nations Convention on the Law of the Sea: A Commentary*, Vol. II, (Dordrecht: Martinus Nijhoff Publishers, 1993), p. 1,017.

[34] Art. 84(2) of the LOS Convention.

[35] Art. 2(2) of Annex II to the LOS Convention.

Meeting of States Parties to the Convention held in late 1995 agreed that the election be postponed until March 1997,[36] almost a year later than originally envisaged by the Convention. That Meeting also added the proviso that should any state that was already part to the Convention by 16 May 1996

> [be] affected adversely in respect of its obligations under this provision as a consequence of the change in the date of the election, States Parties, as the request of such a State, would review the situation with a view to ameliorating the difficulty in respect of that obligation.[37]

OUTSTANDING QUESTIONS FOR AN ANTARCTIC CONTINENTAL SHELF BEYOND 200 MILES

The following crucial question remains: Regardless of whether claimants act individually, jointly, or with the entire Consultative Party group, *how do they proceed*? Two important aspects of interrelationship between the LOS Convention and the ATS instruments, and subsequent practice based on these, must be taken into consideration.

Where is the 'Baseline' in the Antarctic?

Notwithstanding special political and legal circumstances, the *natural features* of the Antarctic region make it difficult to submit precise information on the outer limit of the continental shelf; such information is dependent on the 'baselines from which the breadth of the territorial sea is measured'.[38] Here, the Antarctic ice structures create a special problem not addressed by the LOS Convention.[39] Churchill and Lowe, while commenting on special geographic conditions for determining baselines, noted the following:

> One special geographic condition for which the Law of the Sea (and Territorial Sea) Conventions make no provision is permanent ice shelves ... may be many miles in width. It is uncertain whether the baseline should be at the outer edge of the ice shelf or the edge of the land. This issue was deliberately not discussed at UNCLOS III for fear of re-opening the question of the legal status of Antarctica.[40]

---

[36] See UN doc. SPLOS/5, para. 20, of 1 December 1995.

[37] *Ibid*. Also para. 82 of the annual 'Law of the Sea – Report of the Secretary General', UN doc. A/51/645, of 1 November 1996.

[38] See Art. 76(1) of the LOS Convention.

[39] See discussion by Joyner, *Antarctica and the Law of the Sea*, pp. 79–87.

[40] R.R. Churchill and V.A. Lowe, *The Law of the Sea*, 3rd edition (Manchester University Press, 1999), p. 33, note 5. However, Orrego Vicuña demonstrates that, while negotiating provisions on the outer limits of the continental shelf, UNCLOS III did address the subject of Antarctica; see Orrego Vicuña, 'The Law of the Sea and the Antarctic Treaty System', pp. 114–115.

When submitting charts and coordinates to the Continental Shelf Commission that indicate the proposed outer limit of the continental shelf, the state shall also submit charts and coordinates that indicate the relevant *territorial sea baselines*.[41] This requirement may create a serious problem for Antarctic claimant countries; moreover, it may also disturb the balance established in Article IV of the Antarctic Treaty, thus becoming a matter of concern for all parties to this Treaty.

Possible Impact on Dormant Disputes over Sovereignty

A second special circumstance concerning the Antarctic is even more directly related to *Article IV* of the Antarctic Treaty and the *dispute over sovereignty* in the Antarctic, which the Treaty encapsulated in that Article. With regard to establishing outer limits of the continental shelf beyond 200 miles, it is stated in the LOS Convention that neither the provisions nor the actions of the Commission shall in any way prejudice matters relating to the delimitation of boundaries between states with opposite or adjacent coasts.[42] In commenting these provisions, Nandan and Rosenne observed the following:

> The phrase 'matters relating to delimitation of boundaries' emphasizes that the Commission is not to function in determining, or to influence negotiations on, the continental shelf boundary between States with overlapping claims ... the Commission is not to be involved in any matters regarding the determination of the outer limits of a coastal State's continental shelf where there is a dispute with another State over that limit. The Commission's role is to make recommendations ... not to be involved in matters relating to delimitation of the continental shelf between States.[43]

In preparing its Rules of Procedure, the Commission – on the basis of the relevant provisions of the LOS Convention – discussed how it should treat 'possible submissions containing areas under *actual* or *potential* delimitation dispute'.[44] In its Rules of Procedure adopted on 4 September 1998, the Commission added an important new phrase; whereas the provisions of the LOS Convention refer solely to cases of maritime delimitation (of the continental shelf) between states with opposite or adjacent coasts, the Commission's Rules of Procedure add 'other cases of unresolved *land or* maritime disputes'.[45]

Moreover, while the LOS Convention provided that the actions of the Commission shall not prejudice matters relating to the delimitation of boundaries

---

[41] See para. 3(a) of 'Modus Operandi of the Commission', UN doc. CLCS/L.3, of 12 September 1997.
[42] See Art. 76(10) of the LOS Convention and Art. 9 of its Annex II.
[43] Nandan and Rosenne, *A Commentary*, p. 1,017.
[44] See para. 5 of 'Statement by the Chairman of the Commission on the Limits of the Continental Shelf on the Progress of Work in the Commission', UN doc. CLCS/7, of 15 May 1998 (emphasis added).
[45] See Rule 44(1) of Rules of Procedure of the Commission on the Limits of the Continental Shelf, UN doc. CLCS/3/Rev.2, of 4 September 1998 (emphasis added).

between states 'with opposite and adjacent coasts', this phrase is omitted in the Commission's Rules of Procedure, which thus refer more broadly to *any* delimitation of boundaries between states.[46] In such cases as referred to in its Rule 44, the Commission 'recognizes that the competence with respect to matters regarding disputes which may arise in connection with the establishment of the outer limits of the continental shelf *rests with States*'.[47] As a consequence, the Commission's Rules provide the following:

> A submission may be made by a coastal State for a portion of its continental shelf in order not to prejudice questions relating to the delimitation of boundaries between States in any other portion or portions of the continental shelf for which a submission may be made *later, notwithstanding the provisions regarding ten-year period* established by article 4 of Annex II to the Convention.[48]

No specific time limit is indicated for the word 'later'. This means that any submission that may prejudice questions relating to the delimitation of boundaries, in respect of areas under actual or potential delimitation dispute, can be postponed for an indefinite period. It is worth noting that such submissions may also be made as a joint submission by two or more states.[49]

LAW OF THE SEA CONVENTION AND RULES OF THE COMMISSION: NEW OPTIONS?

The current regulation, as subsequently specified and adopted by the Continental Shelf Commission, seems to offer various courses of action to the Antarctic claimant states, in contrast to the letter of the LOS Convention alone. This applies whether the claimant states choose to act individually, jointly, or in a group with all other Consultative Parties.

There has been an 'actual' dispute between Argentina, Chile and the United Kingdom regarding the area of their overlapping claims; and though brought before the International Court of Justice, the substance of this dispute was never decided upon. Moreover, various scenarios are possible for 'potential' disputes, not least because there are other states, in addition to the present-day seven claimants, who have reserved their right to claim any part of the Antarctic. Submission of details regarding the outer limits of the Antarctic continental shelf to the Commission contains the potential of re-opening dormant disputes or provoking disputes which are at present only potential. All of these actual disputes, or potentials for new disputes, are currently 'frozen' in Article IV of the Antarctic Treaty.

---

[46] See Rule 44(2) of Rules of Procedure of the Commission; compare with Art. 9, Annex II of the LOS Convention.
[47] Rules of Procedure of the Commission, Annex I, para. 1 (emphasis added).
[48] *Ibid.*, para. 3 (emphasis added).
[49] *Ibid.*, para. 4.

Based on the rules of the Continental Shelf Commission that are analysed above, various options on how to address the requirements contained in the LOS Convention arise. When claimant countries individually make submissions for continental shelves that extend beyond 200 miles adjacent to the areas where their sovereignty over land is undisputed (i.e., their mainland territory), it may be possible for them to make the same or similarly phrased *reservation* in respect to the Antarctic continental shelf, with reference to Article IV of the Antarctic Treaty. It may also be possible for the claimants to act jointly, in the claimant country group or with all the other Consultative Parties, and submit to the Continental Shelf Commission a reservation with reference to Article IV of the Antarctic Treaty.

Other options may exist for the claimants, and for that matter for the Consultative Parties, on how to address the issue and present it to the Continental Shelf Commission. The detrimental solutions appear to be the choice of any of the two extreme options, not only for the claimants' claims but also for the balance established under Article IV of the Antarctic Treaty – and thus for the overall success of the Antarctic cooperation. One of these extreme options would be the decision of a claimant to unilaterally make the submission for the Antarctic portion of its continental shelf extending beyond 200 miles and to fail to qualify this submission by a reservation distinguishing it from a submission made for the continental shelf off its mainland territory. The other extreme option, equally detrimental, would be if the claimants remained passive and allowed the lapse of (a possibly relaxed) ten-year period without any statement submitted to the Continental Shelf Commission.

15

CCAMLR and the Environmental Protocol: Relationships and Interactions

Richard A. Herr

It is more than passing significance that this chapter addressing the role of the Convention on the Conservation of Antarctic Marine Living Resources[1] (CCAMLR) in the implementation of the 1991 Protocol on Environmental Protection to the Antarctic Treaty[2] (Protocol) has been placed here in Part IV rather than in Part II of this book. Despite its own path-breaking origins in the Antarctic Treaty System (ATS) in environmental protection, CCAMLR, together with the institutions established under it, has not been perceived generally by the members of the ATS as providing 'institutional support for the implementation of the Protocol'. This is not to say that all elements of the ATS, or indeed even of CCAMLR, are in agreement that this is, or should be, the case. Since the initial negotiation and including the more latterly implementation of the Protocol, some contributors to CCAMLR processes have objected privately to the institutional compartmentalisation of the ATS's only formal inter-governmental organisation.

There may be a number of explanations for this perception of CCAMLR. The members of the ATS decision-making community tend not to distinguish between the CCAMLR processes and those of the core policy elements of the ATS, principally the Antarctic Treaty Consultative Meetings (Consultative Meetings). Thus, the need to make special provision for CCAMLR in the negotiation or implementation of the Protocol seemed redundant. A second possible explanation derives from political prudence. With the exception of the Protocol and its supplementing the Antarctic Treaty,[3] other ATS instruments – the Convention for

---

[1] ILM, Vol. 19, 1980, pp. 837ff.
[2] ILM, Vol. 30, 1991, pp. 1,461ff.
[3] UNTS, Vol. 402, pp. 71ff. On the Protocol as supplementing the Antarctic Treaty, see Art. 4(1) of the Protocol.

the Conservation of Antarctic Seals[4] (Seals Convention), CCAMLR and the ill-fated Convention on the Regulation of Antarctic Mineral Resource Activities[5] (CRAMRA) – were constructed as free-standing treaties not only to meet fairly limited specific objectives under the ATS banner but also to protect pre-existing agreements from the risks of the new venture. A third factor may be the perception that the Protocol represents the new generation of environmental concerns that has gone beyond CCAMLR.

Whatever the broader reasons, the Final Act of the Eleventh Special Antarctic Treaty Consultative Meeting, at which the Protocol was negotiated, probably stands as both a substantial contributor to CCAMLR's marginalisation from the Protocol and the definitive monument to its exclusion from the implementation process.[6] The Final Act refers to CCAMLR both with respect to preserving the rights and obligations of parties to CCAMLR (paragraph 7) and to exempting CCAMLR activities from Article 8 of the Protocol, which set out the environmental impacts assessment requirements. These provisions also apply to another Antarctic resource-related agreement – the Seals Convention – whereas the paragraph 7 reservation also applies to the International Convention for the Regulation of Whaling.[7] Thus, it appeared to many that commercial activities that were already regulated were to be 'grandfathered' out of the Protocol and so insulated in their core activities from the application of the Protocol.[8]

Whether the holding of CCAMLR apart from the Protocol's implementation was intended to separate the Convention in all respects, as some may think, or rather to benefit CCAMLR, as others might hope, is open to question. This chapter accepts that the separation has incurred benefits and costs for both the Protocol and CCAMLR. It will be argued that the benefits of industry stability and the avoidance of risk for CCAMLR in the Protocol's emergence have come at some cost to both instruments. In pursuit of fulfilling the aims of the Protocol, the opportunity of relating optimally to CCAMLR has been compromised. Perhaps more importantly, CCAMLR has been exposed to additional institutional stress at a time when CCAMLR finds itself pressured by a variety of other factors including the post-Cold War restructuring of international priorities, changes to global legal arrangements and diminishing fish stocks. All in all, it may well have been better for both CCAMLR and the ATS if CCAMLR had had a more significant and direct contribution to make in implementing the Protocol.

---

[4] UNTS, Vol. 1080, pp. 175ff; and ILM, Vol. 11, 1972, pp. 251ff.

[5] ILM, Vol. 27, 1988, pp. 868ff.

[6] The text of the Final Act reprinted in J.A. Heap (ed.), *Handbook of the Antarctic Treaty System*, 8th edition (Washington, DC: US Department of State, 1994), pp. 2,016–2,018.

[7] UNTS, Vol. 161, pp. 74ff.

[8] This same approach clearly did not apply to tourism because this area of commercial activity was not subject to pre-existing international regulation recognised directly by the ATS.

CCAMLR AS A CONSTITUENT OF THE ATS

The multifaceted objectives of CCAMLR are pursued primarily through its Commission (currently with 23 members)[9] that meets annually in the Tasmanian capital Hobart in Australia, which is also the location for the organisation's permanent headquarters. The decisions of the Commission are reported formally as Conservation Measures and Resolutions, the former being the principal mechanism by which the Commission regulates the access of its members to the marine living resources of the Southern Ocean within the CCAMLR area of application. Conservation Measures are used to set total allowable catch limits, season limits, gear type restrictions and the like and may be in force on an annual basis or for an indefinite period. Member states generally enjoy the freedoms of the high seas for access to living resources of the Southern Ocean in the absence of such CCAMLR restrictions (or other treaty obligations). Unlike other arrangements, CCAMLR mandates an ecological approach to fisheries management, and in order to evaluate claims on marine living resources in keeping with this nearly unique environmental constraint, the CCAMLR Commission is assisted in its determinations by a Scientific Committee. This Committee has extraordinary influence within the regional body through the high level of expertise made available to the staff and because of the relative autonomy within the organisational structure of CCAMLR itself. Two specialist 'working groups' within the Scientific Committee undertake detailed assessments of fish stocks and ecosystem impacts, which provide the basis for the Committee's management advice to the Commission.

When the Commission acts on the advice of the Scientific Committee by passing a Conservation Measure, the effect of such a regulation in the form of compliance is not entirely unambiguous regardless of how precisely the Measure is framed.[10] If a member of the Commission feels that it should not be subject to the application of a Conservation Measure, it can object or seek an exemption.[11] Even when there is no objection, a measure may not have the desired impact. Member state compliance with a Conservation Measure in the event depends not just on the political will of the member but very much also on such factors as the domestic legal

---

[9] The 23 members of CCAMLR's Commission are: Argentina, Australia, Belgium, Brazil, Chile, European Union, France, Germany, India, Italy, Japan, Korea (Republic of), New Zealand, Norway, Poland, Russia, South Africa, Spain, Sweden, Ukraine, UK, US and Uruguay. Bulgaria, Canada, Finland, Greece, Netherlands and Peru are parties to the CCAMLR Convention but not members of the CCAMLR Commission.

[10] For a general discussion of compliance under CCAMLR, see O.S. Stokke, 'The Effectiveness of CCAMLR', in O.S. Stokke and D. Vidas (eds.), *Governing the Antarctic: The Effectiveness and Legitimacy of the Antarctic Treaty System* (Cambridge University Press, 1996), pp. 238–239.

[11] This consensual aspect of the Commission's decision-making is regarded by some of its members as a particularly useful mechanism for circumventing the tendency of consensus politics to produce lowest common denominator outcomes. The opt-out clause enables states to allow policies to be carried by the majority that they themselves will not apply to their own vessels. On the other hand, states that do not object have presumably less grounds for subsequent complaint because they could have used the opt-out clause if they were not willing to accept the measure.

status of treaty obligations, internal judicial processes, administrative and political infrastructures and related factors. Thus, not all CCAMLR members are able to guarantee the same level of domestic enforcement.[12] Partially in recognition of the important national differences in members' capacity to secure compliance, all members of CCAMLR are required under Article XXI(2) of the Convention to provide information on national steps to ensure compliance with CCAMLR regulations. Consequently, 'Compliance with Conservation Measures in Force' is a standard agenda item for the annual meetings of the CCAMLR Commission and includes reports from members on both actions that deal with violations by nationals and that promote governmental enforcement of new measures.

CCAMLR's capacity to achieve the specific objectives of its Conservation Measures is constrained by two important general limitations. Firstly, as CCAMLR is essentially a regional organisation, it has a relatively narrow membership. Indeed, not all parties to the Antarctic Treaty are members of CCAMLR. The importance of this constraint has been vividly confirmed in recent years. States that are not parties to the CCAMLR Convention and are technically not bound by its decisions have been implicated as major sources of unregulated fishing. Nevertheless, they are encouraged to voluntarily follow CCAMLR guidelines.[13] Secondly, CCAMLR's geographic scope is essentially defined by a physical boundary, the Antarctic Convergence, and this has not proven to be a secure frontier. Many significant organisms in the Southern Ocean ecology, including seals, whales, seabirds and fish, habitually cross the Convergence; in respect to the species protected by CCAMLR regulations, this can undermine the effectiveness of the regulations, as in the case of seabirds, or create serious regulatory ambiguities, as in the case of albatross.

THE PROTOCOL AND CCAMLR

By all appearances, the Protocol set the broader ATS seal on the innovative ecological thrust to environmental protection that underlay the creation of CCAMLR. Yet, because the Protocol sought to incorporate into the core of the Antarctic Treaty an environmental objective that was a decade more modern than the ideas that are reflected in the CCAMLR Convention,[14] there seemed to be no

---

[12] For an example see Chile's explanation for its own internal difficulties in this regard, in CCAMLR, *Report of the Thirteenth Meeting of the Commission, 26 October – 4 November 1994* (Hobart: CCAMLR, 1995), pp. 15–16. On the Chilean legal system for implementation of international commitments in general, and the Protocol in particular, see Carvallo and Julio, Chapter 18 in this book.

[13] CCAMLR routinely seeks non-member information on fishing in the CCAMLR area and/or support of its regulations. For early examples, see: CCAMLR, *Report of the Tenth Meeting of the Commission, 21 October – 1 November 1991* (Hobart: CCAMLR, 1992), pp. 35–36, and CCAMLR, *Report of the Fourteenth Meeting of the Commission, 24 October – 3 November 1995* (Hobart: CCAMLR, 1996), p. 8. This approach bore exciting new fruit from 1998 when two non-member states, Mauritius and Namibia, accepted CCAMLR's invitation to attend CCAMLR Meetings as observers.

[14] Art. 3 of the Protocol acknowledges this indebtedness to the path-breaking influence when it refers

haste to include CCAMLR in the negotiation of the Protocol. There may well have been several, possibly concurrent and parallel, reasons for the CCAMLR's virtually non-existent involvement in the negotiation of the Protocol. The highly politicised atmosphere surrounding the debate over CRAMRA made it appropriate for the Consultative Parties to deal at the highest levels to negotiate urgently an alternative arrangement. It may have been a consideration that CCAMLR's established environmental mandate should not be complicated or put at risk by the uncertainties that prevailed at time in the Protocol negotiating process. Perhaps there also was a belief that the interests of CCAMLR could be easily compartmentalised from other interests and as such be incorporated into the Protocol package. Whatever the reasons, CCAMLR's institutional absence from the negotiating process for the Protocol proved not to be without implications.

This apparent oversight was more than a mere insignificant slight to CCAMLR by the core decision-making component of the ATS. It bespoke a willingness to relegate CCAMLR to a minor role and to render the Convention ineffective in other serious environmental issues whenever the Consultative Parties deemed it appropriate – or perhaps simply forgot to include CCAMLR in the matter. From discussions on this issue over the years among representatives to CCAMLR and the Consultative Meetings, it has been suggested that CCAMLR's direct involvement in such matters was irrelevant given the substantial overlap in the personnel representing member countries at both CCAMLR Commission and the Consultative Meetings. However, this interpretation tends to regard CCAMLR as a subordinate, rather dependent element of the Consultative Meeting process rather than an element with substantial and institutionally autonomous interests within the ATS. This tendency was evidenced when New Zealand found the Patagonian toothfish issue so urgent and politically compelling that it attempted to make the management of this fishery a central focus for the XXI Antarctic Treaty Consultative Meeting, held in 1997 in Christchurch, New Zealand, despite concern amongst some CCAMLR members for the marginalisation of CCAMLR and its processes.[15] The eventual failure to include this item on the agenda did not prevent similar, albeit less public, efforts at the Tromsø (1998) and Lima (1999) Consultative Meetings.[16]

THE CCAMLR AND PROTOCOL 'MANDATES'

The formal objectives of CCAMLR and the Protocol display significant points of commonality. While these shared objectives are not identities that threaten to make

---

to the protection of the Antarctic environment and 'dependent and associated ecosystems'.

[15] See F. Brenchley, 'Curb to Antarctic Fishing', *Financial Review*, 9 May 1997, p. 23 and A. Darby, 'NZ Warns of Illegal Fishing 'Free-For-All'', *Sydney Morning Herald*, 27 May 1997, p. 10.

[16] The Antarctic and Southern Ocean Coalition (ASOC) report of the 1999 Lima Consultative Meeting makes particular note of these 'corridor' discussions as being significant. See *ASOC Report on the XXIII Antarctic Treaty Consultative Meeting, Lima, Peru 1999*; available at ASOC website <www.asoc.org>.

one or another agreement redundant, there are important areas of overlap and potentially concurrent claims for competence. Undoubtedly the key area of overlap in the mandates of the two regional instruments derives from their mutual concern for the conservation of Antarctica and the Southern Ocean. CCAMLR sums this up in its preamble with the recognition of 'the importance of safeguarding the environment and protecting the integrity of the ecosystem of the seas surrounding Antarctica'. The Protocol, in the wording of its preamble, evidences the benefits of a decade of understanding and developing the ecosystem approach by referring to 'the need to enhance the protection of the Antarctic environment and dependent and associated ecosystems'. This broad and generally shared area of concerns does not in itself create a complication of concurrent and redundant competencies. The shared concerns arise from the detailed development of the fundamental objectives of these two regional instruments.

The problem of overlapping areas of competence, has perhaps been confected from the earliest anticipation of the Protocol. Early conservationist critics of CCAMLR approved of its innovative ecosystem approach to the administration of Antarctica while at the same time complaining that some modifications could be 'necessary to ensure that CCAMLR's management concept worked as the convention intended'.[17] A major issue for environmentalists with regard to CCAMLR has been the orientation of Article II, which includes 'rational use' as part of its definition of the concept of conservation. Unlike the Protocol, which devotes a lengthy provision (Article 3) to identifying and defining its 'environmental principles', the CCAMLR Convention is dominated by the rational use aspects of conserving the Southern Ocean marine living resources. Thus, the Protocol's mandate predictably has enjoyed a greater degree of support amongst environmentally motivated groups than has that of CCAMLR, despite CCAMLR's trailblazing environmental role in the ATS. Indeed, this suspicion regarding CCAMLR's commitment to conservation relative to that of the Protocol persists even today.[18]

As the drafters of the Protocol were naturally aware of the complications that could arise from overlapping interests with CCAMLR, steps were taken to ensure that appropriate lines of demarcation were drawn between the two. The grandfathering of the CCAMLR rights and obligations into the Protocol (Article 4(2)) was one mechanism for avoiding competitive or ambiguous jurisdictions. The requirement under the Protocol's Article 5 added force to this aim:

> The Parties shall consult and co-operate with the Contracting Parties to the other international instruments in force within the Antarctic Treaty system and their respective

---

[17] J. May, *The Greenpeace Book of Antarctica: A New View of the Seventh Continent* (London: Dorling Kindersley, 1988), p. 159. It is significant that this assessment was made on the eve of the collapse of the Consultative Parties' support for CRAMRA and its supplanting by the Protocol.

[18] See, for instance, 'A Question of Values: Can Antarctica Resist Economic Globalisation,' *ECO*, No. 3, 25 May – 6 June 1998, Tromsø, Norway.

institutions with a view to ensuring the achievement of the objectives and principles of this Protocol and avoiding any interference with the achievement of the objectives and principles of those instruments or any inconsistency between the implementation of those instruments and of this Protocol.

As Article 5 anticipated, the apportionment of responsibilities between CCAMLR and the Protocol cannot be achieved by formal demarcations alone; continuing cooperation among the parties to both agreements is also required.

The expectation of interagency cooperation, however, creates a problem in terms of the fundamental objectives of the two instruments. As intimated previously, the CCAMLR agreement attempts to straddle two interests that are often seen as mutually incompatible. On the one hand, CCAMLR has established the machinery to manage the commercial activity of fishing; on the other, its fundamental *raison d'etre* is an ecological mandate to preserve the Antarctic marine environment. This issue of conflicting interests has been the basis of criticism expressed virtually annually by the environmental NGOs of CCAMLR as well as the failure, from the perspective of the NGOs, of CCAMLR to achieve a balance between the commercial and conservation aims of the Convention.[19] The members of CCAMLR do not view the problem in precisely the same way, at least formally, although there have been occasional internal clashes as to whether the body is primarily a fisheries management arrangement or an environmental protection institution.[20]

This internal tension within CCAMLR's membership may contribute to a significant implementation issue between CCAMLR and the Protocol. The Final Act of the Eleventh Special Antarctic Treaty Consultative Meeting exempts activities undertaken pursuant to CCAMLR from the application of the Article 8 (and thus Annex I) requirement to conduct environmental impact assessments. The exemption of activities undertaken pursuant to CCAMLR from the Annex I requirements clearly has the effect of not extending similar exemptions to the other annexes of the Protocol. This can be concluded from, for example, the advice of the CCAMLR Commission to members to adhere to the MARPOL73/78 in support of the Protocol.[21] Nevertheless, the perception that the Final Act exemption was based on an appreciation for the commercial interests inherent in CCAMLR may have helped to foster an expectation that the responsibility for securing compliance with the Protocol rests elsewhere. Otherwise, it is difficult to understand how the Patagonian toothfish long-liners have escaped censure over their use of non-indigenous bait in the Southern Ocean in contravention of the Protocol's Annex II (Article 4) prohibition on the introduction 'of non-native species, parasites and diseases' into the Antarctic Treaty area. This issue is all the more salient given the attribution of

---

[19] For the flavour of this conservationist critique, see the various issues of *ECO*, an ASOC newsletter produced during each session of CCAMLR.

[20] See, for example, Argentina's demarche on this issue in CCAMLR, *Report of the Fourteenth Meeting of the Commission, 24 October – 3 November 1995* (Hobart: CCAMLR, 1996), p. 23.

[21] Information based on the author's personal communication with the CCAMLR Secretariat.

substantial pilchard die-offs in recent years in the Great Australian Bight to lethal parasites introduced by fishing boats using imported bait.

Annex V of the Protocol, which deals with protected areas, is a different matter for CCAMLR in two ways. First, as this annex was not part of the original Protocol package, it was not adopted, nor did it enter into force, simultaneously with the Protocol and Annexes I–IV.[22] Yet, secondly, while Annex V is not yet in force, it is of considerable interest to CCAMLR members because of its provisions for marine Antarctic Specially Protected Areas (ASPAs) and Antarctic Specially Managed Areas (ASMAs). As Annex V asserts that no *marine* area can be designated as an ASPA or ASMA 'without the prior approval' of the CCAMLR Commission (Article 6(2)), this Annex allows for a implementing role for CCAMLR both in proposing management plans for such areas (Article 5(1)) and in vetoing such proposals.

INSTITUTIONAL COOPERATION: FROM THE TEWG TO THE CEP

The Protocol mandated the establishment of the CEP to assist in implementing the objectives of the Protocol. So important was its work perceived to be by the Consultative Parties that a Transitional Environmental Working Group (TEWG) was created to pursue preparatory work even before the Protocol came into force.[23] However, despite the apparent urgency in implementing the Protocol, there has been a somewhat curious and uneven development regarding the establishment of the CEP. Relations with other bodies that have environmental responsibilities have only been partially and incompletely advanced despite recognised commonalities of interest and some formal obligations to harmonise these interests.

The 'expectation' that CCAMLR's Scientific Committee (through its Chair) would be included in the work of the TEWG did not materialise.[24] Perhaps this was an oversight based on the need to establish the machinery of the CEP rather than process proposals for marine ASPAs or assess management plans for areas that would not be forthcoming until Annex V was in force. However, some persons involved with CCAMLR felt there was inconsistency on this point because some aspects of the concept of marine protected areas apparently involved discussions with SCAR but not CCAMLR, notwithstanding the wording of the Annex V on the need for cooperative arrangements on marine ASPAs. It is not necessarily significant that the transitional mechanisms were under-utilised, but at the minimum this 'failure' helped to reinforce the impression that CCAMLR was not to be an important ATS element in the implementation of the Protocol.

The limited experience of the CEP seems to be more encouraging than that the lack of experience of the TEWG. The Chair of CCAMLR's Scientific Committee

---

[22] See discussion by Vidas, Chapter 1 in this book.

[23] On the TEWG see Orheim, Chapter 6 in this book.

[24] Information based on the author's personal communication with the Chair of the CCAMLR Scientific Committee.

has been included in CEP meetings. A proposal to prepare a Management Plan for a marine ASPA around the Balleny Islands, which was submitted at the Lima Consultative Meeting in 1999 by New Zealand, was forwarded to the meeting of the CCAMLR Commission later the same year for consideration as required by Annex V. The proposed plan has been independently forwarded to both the Commission and to the Commission's Scientific Committee, which is a consultative body to the Commission.[25]

POSSIBLE REGIONAL FACTORS AFFECTING FUTURE IMPLEMENTATION

Several developments currently in train within the ATS are likely to have an impact on the future role of CCAMLR and its capacity to contribute to the implementation of the Protocol. These include the continuing development of the Protocol through additional annexes and the work of the CEP. The advent of a 'sixth annex', on the liability for environmental damage in the Antarctic,[26] has the potential to create a new point of friction between the commercial and conservation aims of CCAMLR, especially if this annex (or annexes) were to provide a further basis for regulation of the Southern Ocean fishing industry. On the one hand, it seems likely that the fishing industry would expect continued exemption from the regulation under the Protocol; on the other, exempting CCAMLR activities, which perhaps comprise the largest commercial activity in the Antarctic Treaty area, could significantly complicate negotiations that are already delayed because of their complexity.

Current circumstances do not favour great new investments in the ATS, but the growth and development of the CEP is scarcely beyond reasonable expectations. This could have an adverse impact on CCAMLR, particularly if the CEP were to acquire an independent 'work programme' – a goal to which elements of CCAMLR have long aspired but thus far have failed to achieve. This potential problem is merely speculative for the moment. Currently, the principal source of concern from the perspective of the CEP is SCAR, but CCAMLR could also potentially find itself facing the competitive challenge for an autonomous research programme (funding and responsibilities). Nonetheless, while the chances of conflict are probably very remote in this area, the growth and elaboration of the CEP will overshadow CCAMLR's development, and this may tend to discourage its growth even if the CEP is not a beneficiary.

The question of the need for CCAMLR to adapt its own roles to changing circumstances is increasingly a real prospect. Declining fish stocks have raised the

---

[25] It was partially felt that the two CCAMLR bodies may disagree over the plan because the intention of this proposed plan is based on a 200-nautical-mile-zone rather than on the ecological rationales suggested by Annex V.

[26] See a comprehensive discussion on liability by Skåre and Lefeber, Chapters 9, 10 and 11 in this book.

question of 'CCAMLR Beyond or Without Fisheries?'. Fundamentally, there is an issue of international institutional efficiency for CCAMLR in what its *raison d'etre* would be without a significant fisheries role. Before the CEP, CCAMLR might have been assumed to have expanded and developed its role in monitoring and protecting the environment and ecosystem of the Southern Ocean. However, CCAMLR may now find itself so squeezed by the CEP on environmental issues that, having so little to do, its functions could be consigned to the category similar to that of the Seals Convention, and possibly be being performed through informal mechanisms.

GLOBAL FACTORS IMPACTING CCAMLR AND IMPLEMENTATION

As brilliantly as the ATS regional system has performed over the decades, it is increasingly clear that the contemporary ATS, including both the Protocol and CCAMLR, is affected by the attractions and influences of a growing galaxy of other regional and global treaties, as well as substantial changes in the international system itself. The interactions with the 1982 United Nations Convention on the Law of the Sea become more visible yearly.[27] In addition, the UN agencies, particularly the Food and Agricultural Organisation (FAO), which have been attributed certain responsibilities under the LOS Convention, have both primary and secondary impacts on CCAMLR and on its evolving relations with the Protocol. These centripetal and centrifugal forces would be significant in themselves but they have been intensified by such additions as the Agreement for the Implementation of the Provisions of the United Nations Convention on the Law of the Sea of 10 December 1982 Relating to the Conservation and Management of Straddling Fish Stocks and Highly Migratory Fish Stocks[28] (Fish Stocks Agreement). Finally the post-Cold War international system has had an even more evident impact on the management of Antarctica and the Southern Ocean.

Not only must the interactions between CCAMLR and the Protocol be understood in the context of these dynamic international developments, but these in turn affect the way the elements of the ATS are perceived. In the case of CCAMLR, the development of new mechanisms for reaching the goal of sustainable fisheries has put pressure on CCAMLR to make more use of these measures to ensure greater compliance with its own regulations. CCAMLR has endorsed and supported such FAO initiatives as the Code of Responsible Fishing and measures extending enforcement responsibilities to port states. Although other global instruments, such as the Fish Stocks Agreement, are less supported within CCAMLR, they do add pressure on CCAMLR to stop illegal and unregulated fishing of the Patagonian toothfish. Whatever the CCAMLR response to the global instruments, the global

---

[27] For a recent assessment, see D. Vidas, 'Emerging Law of the Sea Issues in the Antarctic Maritime Area: A Heritage for the New Century?', *Ocean Development and International Law*, Vol. 31, 2000, pp. 197–222.
[28] ILM, Vol. 34, 1995, pp. 1,547ff.

system has put pressure on CCAMLR to better meet environmentalist expectations by creating alternative (non-regional) means for meeting some of the conservation objectives of CCAMLR. In the minds of CCAMLR critics this situation has intensified the perception that CCAMLR is failing in its conservation role. The significance of the global instruments is that, to a certain extent, the critics may amplify the impression that CCAMLR is less necessary than it may once have been perceived to be. This pressure, which hinders CCAMLR from meeting its own objectives, certainly cannot contribute toward improving CCAMLR's chances of promoting the aims of the Protocol.

Another line of international restructuring that has served as a more ambiguous agent of change for the internal adjustments in the ATS is the post-Cold War restructuring of the international system. The reaping of the post-Cold War 'peace dividend' has resulted in new priorities in the funding of science and technology research. In the Antarctic, the replacement of old state actors by new ones, especially that of the role performed earlier by the USSR with the current involvement of Russia and the Ukraine, has significantly altered the capacity and willingness of these players to pay for previous levels of involvement. Related to this development is the reaction of other 'old' actors – Western states such as US and UK – to the absence of a focus on rivalry (previously based on ideological advantage) which encouraged these states to 'shadow' the activities of their rivals. Thus, Western states no longer feel a need to pursue competitive activities in Antarctica merely to offset the apparent interests of a non-existent Soviet Union; and perhaps more important, as these other changes have transpired, the economic rationalism which has been a feature of the more pragmatic international politics has inhibited 'unnecessary' state expenditure.

During the 1990s, the tighter international financial climate contributed to a very cautious approach to international organisations and to the value of these bodies in achieving the national interests of its members. In this respect, some specific factors have inhibited the work of CCAMLR over the decade of the 1990s: a basically static budget, slow member payment of annual dues, and an unwillingness by members to fund innovative programmes. This has worked against CCAMLR as an established inter-governmental organisation. Being at the same time hard-pressed to meet its own mandated commitments, CCAMLR could scarcely contribute substantially to implementing the Protocol. However, the same factors have probably more adversely affected the Protocol and its instruments in attempting to develop its new programmes from a non-established institutional base.[29]

---

[29] For the CEP see Orheim, Chapter 6 in this book.

CONCLUSION

While the Protocol undoubtedly has helped promote and advance many of the aims of the Consultative Parties in establishing CCAMLR, the Protocol has not been without risk for the single, fully formed inter-governmental organisation of the ATS. Many of CCAMLR's current concerns would have emerged even without the possible redundancies and demarcation issues posed by the Protocol. Global and regional changes as well as CCAMLR's diminishing breadth of management responsibilities would have ensured that CCAMLR's routine would have to be changed even if the Protocol had never been adopted. However, as the energy of CCAMLR has appeared to dim, the Protocol has offered a new and possibly brighter source of illumination, at least on environmental/ecological issues, and one which may be more politically sustainable. The collapse of so many Antarctic fisheries and the continuing controversy over the Patagonian toothfish pose dangers on the other primary axis of CCAMLR's twin mandates. Perhaps the one substantial advantage CCAMLR has in maintaining its position in the ATS is an established course with its own interests and influence to which others have to adapt.

This chapter has focused more on the impact of the Protocol on CCAMLR than the reverse because of the limited role assigned to CCAMLR in implementing the Protocol. The compartmentalisation of CCAMLR from the Protocol and its implementation has probably been more complete than might have been expected by all involved in the drafting of the Convention. Whatever the anticipated benefits for either the Protocol or CCAMLR, there have been costs or at least missed opportunities. At a minimum, CCAMLR could have contributed to ensuring a more active role by its members in implementing those Protocol annexes not exempted by the Final Act of the Eleventh Special Consultative Meeting. Given the final stringencies facing the entire ATS in the first decade after the Cold War, there may well have been some benefits to be achieved by more active cooperation. In conclusion, it might have benefited both instruments if some of the recent institutional stress on CCAMLR had been relieved by the perception of close and mutually supportive engagement between the two important contributors to the ATS.

Part V

IMPLEMENTING THE PROTOCOL DOMESTICALLY: THE CONSULTATIVE PARTIES' LEGISLATION AND PRACTICE
(selected case studies)

16

Implementing the Environmental Protocol Domestically: An Overview

Kees Bastmeijer

Although the issue of domestic implementation is important for all international agreements, the uncertain legal status of the Antarctic provides an additional reason for emphasising this issue. Most international agreements on the protection of natural areas, such as the 1971 Ramsar Convention on Wetlands of International Importance Especially as Waterfowl Habitat,[1] require parties to take measures to protect special natural areas or values within their territory. Based on the territorial principle, all human activities within these areas may be subjected to the national legislation of that particular state, regardless of the nationality of the persons conducting these activities.[2] As the sovereignty claims made on parts of the Antarctic by seven states are not recognised by most other states,[3] including states that are not parties to the Antarctic Treaty, the approach based on the principle of territoriality could not be applied for the protection of the Antarctic environment.[4] The legal and practical protection of the Antarctic environment therefore depends on the collective efforts of all the states parties to the Protocol on Environmental Protection to the Antarctic Treaty[5] (Protocol). Each party must undertake steps to ensure that activities undertaken by persons under its jurisdiction in the area south of 60 degrees south are carried out in accordance with the provisions of the Protocol.[6]

---

[1] UNTS, Vol. 996, pp. 245ff.
[2] On the strong connection between the protection of the environment and the territorial sovereignty of states, see, e.g., Art. 4 of the 1972 Convention for the Protection of the World Cultural and Natural Heritage (ILM, Vol. 11, 1972, pp. 1,358ff).
[3] See Art. IV of the Antarctic Treaty (UNTS, Vol. 402, pp. 71ff.).
[4] See discussion by Orrego, Chapter 3 in this book.
[5] ILM, Vol. 30, 1990, pp. 1,461ff. As of 29 December 1999, there were 28 parties to the Protocol (all the 27 Consultative Parties and one non-Consultative Party, namely Greece). See further Vidas, Chapter 1 in this book.
[6] See also Richardson, Chapter 4 in this book.

Recognising that the chapters in Part V all place emphasis on certain aspects that are particularly relevant for the state party being studied, this overview chapter will focus on general aspects of the implementation of the Protocol at the domestic level.

THE NOTION OF 'IMPLEMENTATION' OF THE PROTOCOL AT THE NATIONAL LEVEL

The notion of 'implementation' refers here to all the measures that have been taken by a state party to fulfil the objectives of the Protocol.[7] This complex of measures is the central matter of Article 13 of the Protocol, according to which

> each Party shall take appropriate measures within its competence, including the adoption of laws and regulations, administrative actions and enforcement measures, to ensure compliance with this Protocol.

As this provision implies, an adequate implementation of the Protocol should lead to 'compliance' with the Protocol and hopefully to – depending on the completeness and effectiveness of the Protocol – adequate protection of the Antarctic environment. An adequate implementation of the Protocol at the level of the state parties requires, firstly, incorporation of the relevant provisions of the Protocol into the national legal system, and secondly, practical measures and facilities to implement the Protocol, including the application and enforcement[8] of the national law.

These two elements, referred to under the terms 'legal implementation' and 'practical implementation', will be discussed in more detail later in this chapter.

INTERNATIONAL ATTENTION FOR THE NATIONAL IMPLEMENTATION PROCESS

Soon after the adoption of the Protocol in 1991, the Consultative Parties acknowledged that the implementation process at the national level raised difficult questions. Already in 1992, the Final Report of the Consultative Meeting held that year stated the following:

---

[7] For a general discussion on the meaning of implementation in relation to the UN Convention on the Law of the Sea, see R.R. Churchill, 'Levels of Implementation of the Law of the Sea Convention: An Overview', in D. Vidas and W. Østreng (eds.), *Order for the Oceans at the Turn of the Century* (The Hague: Kluwer Law International, 1999), pp. 317–325.

[8] The term 'implementation' of international or national legislation generally includes the issues of legal incorporation and practical application and not enforcement. For the purpose of this chapter, the issue of enforcement will be discussed in the context of legal incorporation and practical implementation; see also Bush, Chapter 2 in this book.

one Contracting Party stressed the desirability of ensuring uniformity of interpretation of those provisions of the Protocol and its Annexes which required national legislation or other measures by the Parties for their implementation. It was agreed that this aspect was of considerable importance and would require direct contacts between Parties through, for instance, diplomatic channels, as would seem useful.[9]

Not mentioned in the final report from that meeting, but interesting for the subject under discussion here, was the initiative at the same Consultative Meeting to organise an informal discussion group to review the progress of the implementation of the Protocol.

Since the 1992 Meeting in Venice, the issue of the implementation of the Protocol has been on the agenda of the Antarctic Treaty Consultative Meetings. This attention to the implementation process, well ahead of the actual entry into force of the Protocol in January 1998, was a logical consequence of the understanding reached between the parties at the Eleventh Antarctic Treaty Special Consultative Meeting in 1991; the Final Act from that meeting stated:

> Pending the entry into force of the Protocol it was agreed that it was desirable for all Contracting Parties to the Antarctic Treaty to apply Annexes I–IV, in accordance with their legal systems and to the extent practicable, and to take individually such steps to enable it to occur as soon as possible.[10]

However, at most Consultative Meetings held prior to the entry into force of the Protocol, the political and legal aspects which were related to the national implementation of the Protocol were only briefly discussed by the signatories. Little attention was given to items such as the legal status of Article 3 of the Protocol ('Environmental Principles'), the relation between Article 7 of the Protocol on the prohibition of mineral activities in the Antarctic and the provisions of the UN Convention on the Law of the Sea, and the exact jurisdictional scope of the national legislation. These issues have perhaps been considered too sensitive for discussion at the Consultative Meeting. A more indirect means of exchanging views on these difficult issues was through the distribution of copies of the national implementation legislation. The Nineteenth Consultative Meeting, held in Seoul in 1995, 'reaffirmed the importance of making domestic legislation available among the Consultative Parties in order to assist other Parties in drafting implementation legislation'.[11]

---

[9] See *Final Report of the Seventeenth Antarctic Treaty Consultative Meeting, Venice, Italy, 11–20 November 1992* (Rome: Italian Ministry of Foreign Affairs, 1992), para. 32.

[10] Para. 14 of the Final Act of the Eleventh Antarctic Treaty Special Consultative Meeting (ILM, Vol. 30, 1991, pp. 1,460–1,461). The Consultative Parties reaffirmed this at several occasions; see, for example, the Declaration by Contracting Parties on the 30th anniversary of the entry into force of the Antarctic Treaty, in *Final Report of the Sixteenth Antarctic Treaty Consultative Meeting, Bonn, Germany, 7–18 October 1991* (Bonn: German Federal Ministry of Foreign Affairs, 1991), p. 138. In fact, this went beyond the obligation laid down in Art. 18 of the 1969 Vienna Convention on the Law of Treaties (UNTS, Vol. 1155, pp. 331ff) to 'refrain from acts which would defeat the object and purpose of a treaty' prior to its entry into force.

[11] *Final Report of the Nineteenth Antarctic Treaty Consultative Meeting, Seoul, 8–19 May 1995*

Since that Meeting, many Consultative Parties have indeed distributed their legislation to implement the Protocol or have explained in information papers circulated at Consultative Meetings how they incorporated the Protocol into the national legal system. Effective with the entry into force of the Protocol and in accordance with its Article 17(1), the parties are obliged to 'report annually on the steps taken to implement the Protocol'.[12]

Although the implementation issues having a political and legal character have not been discussed in detail at the Consultative Meetings, certain more specific elements of the Protocol, including technical items, have since 1992 received attention during the Meetings. For example, the implementation of the Protocol's provisions on environmental impact assessments[13] (EIAs) has constituted a special agenda item at the Consultative Meetings.[14] The establishment of the Transitional Environmental Working Group (TEWG), the predecessor of the Committe for Environmental Protection (CEP), at the Kyoto Consultative Meeting in 1994 is particularly significant here. From 1995 to 1998, the TEWG discussed

> those items on the agenda of the ATCM ... which, under Article 12 of the Protocol, would be dealt with by the Committee for Environmental Protection.[15]

Non-governmental organisations (NGOs), in particular the environmental pressure groups that cooperate under the umbrella of the Antarctic and Southern Ocean Coalition (ASOC), have also contributed to the discussions on the ratification and implementation of the Protocol. This is not surprising, given the substantial energy invested in the process of creating political will for developing a comprehensive regime for the protection of the Antarctic. First of all, the NGOs made use of many opportunities to speed up the ratification process in the signatory states.[16] Furthermore, environmental NGOs have made a substantial effort in influencing the ideas on the minimum criteria for the legal implementation of the Protocol.[17] While many states parties were still working on the legal implementation

---

(Seoul: Ministry of Foreign Affairs of the Republic of Korea, 1995), para. 37.

[12] Art. 17 of the Protocol.

[13] *Ibid.*, Art. 8 and Annex I.

[14] See, for example, *Final Report of the XIX ATCM*, paras. 91–98.

[15] See *Final Report of the Eighteenth Antarctic Treaty Consultative Meeting, Kyoto, Japan, 11–22 April 1994* (Tokyo: Japanese Ministry of Foreign Affairs, 1994), paras. 39–43. See also *Final Report of the XIX ATCM*, paras. 41–47.

[16] Various articles published in their newspaper ECO, which was distributed at the Consultative Meetings, were among many examples of their advocacy; see, for instance, 'Two and a Half Years On, Ratification Limps Along', ECO, 12 April 1994, p. 1.

[17] See especially ASOC, 'Green Legislation for Antarctica', doc. XIX ATCM/IP 67, 1995. See also: 'Making it Law', ECO, 18 April 1994. For a general overview of the role of NGOs in the ratification and implementation process of the Protocol, see especially J. Barnes and C. Webb, 'Implementing the Protocol: State Practice and the Role of non-Governmental Organizations', in F. Francioni and T. Scovazzi (eds.), *International Law for Antarctica* (The Hague: Kluwer Law International, 1996), pp. 475–503.

of the Protocol, ASOC presented, at the Lima Consultative Meeting in 1999, a detailed implementation checklist[18] as well as a paper on 'good practice' in implementing legislation[19]. In addition to these activities of a more political and legal character, Greenpeace continues to undertake expeditions in the Antarctic in order to monitor the compliance with the Protocol in practice.

In the coming years, the discussions at the Consultative Meetings on the implementation of the Protocol are likely to have an increased focus on the practical application of and compliance with the implementation legislation.

FACTORS INFLUENCING THE NATIONAL IMPLEMENTATION OF THE PROTOCOL

Many factors influence the legal and practical implementation of the Protocol.[20] Some are relevant for the implementation process of any international agreement, while others are of specific relevance for the Antarctic legal and political situation. Although the situation is more complex in practice, the factors influencing the implementation of the Protocol can be grouped into three categories: 1) factors which are directly related to the particular state party, 2) factors related to the ATS and the Protocol, and 3) external factors.

Factors Directly Related to a State Party

Political will or intent of a state to implement is identified as one of the most important factors directly related to a state party.[21] As observed by Miles, 'what is particularly important in tracing the line from international commitment to national performance is not only the administrative capacity of the state but the 'political energy' it can bring to bear on the issue to overcome the inevitable resistance that one can expect'.[22] In this respect it must be acknowledged that the simple fact that a state signed and ratified an agreement does not by itself guarantee a high degree of political will to actually implement that agreement. A state may have a very specific interest in signing an agreement even though it does not necessarily agree with most of the content. A state may have even become party because this seemed a

---

[18] ASOC, 'Protocol Implementation Checklist', doc. XXIII ATCM/IP15, 1999.

[19] ASOC, 'Good Practice in Implementing Legislation', doc. XXIII ATCM/IP 127, 1999.

[20] See, for example, G. Rippingale, 'Effectiveness of an International Environmental Regime: the Protocol on Environmental Protection to the Antarctic Treaty' (unpublished paper, presented in partial fulfilment of the requirements for a Master of Science degree at Lincoln University, Lincoln, New Zealand, 1992, pp. 14–16.

[21] See, for example, E. Brown Weiss, 'Strengthening National Compliance with International Environmental Agreements', *Environmental Policy and Law*, Vol. 27, 1997.

[22] E.L. Miles, 'Implementation of International Regimes: A Typology', in Vidas and Østreng (eds.), *Order for the Oceans at the Turn of the Century*, p. 329.

favourable action at that particular time for the state's international relations; or a state may 'accept' the agreement just because the state does not want to block the adoption or entry into force of the agreement. Moreover, 'political will' in itself is difficult to measure and will be further subject to many more specific factors, depending on particular situation.

'Political intent' will strongly depend on the relationship that the state has with the Antarctic continent. In this respect, the *geographic position* of the state party in relation to Antarctica can be important. 'Gateway' countries to Antarctica, such as Argentina, Australia, Chile, New Zealand and South Africa, have a special interest in safeguarding the peaceful uses of the Antarctic, and they may therefore have an additional reason for emphasising the well functioning of the ATS instruments, including the Protocol. Furthermore, the 'gateway' countries may be confronted with initiatives such as new projects for airborne tourism, and in the future they might have a special responsibility founded on the basis of port state control.[23] Such developments may influence the process of implementing the Protocol by these state parties. The chapter by Klaus Dodds in this book indicates that the *historic involvement* of a state in an international system and the recognition of that state in that system may also have positive effects on the efforts put into the implementation of the Protocol.[24] Moreover, whether a state party is an Antarctic *claimant or non-claimant* may influence that state party's modes of implementation of the Protocol. Non-claimant states, for example, generally base their jurisdiction on the nationality of the organiser or participants of an Antarctic activity and the place where such an activity is organised, while the legislation of most claimant states also applies to foreign persons conducting an activity within the respective claimed Antarctic territories. Furthermore, the *nature, scale, and number of activities* should be distinguished as potential factors of influence. And, in relation to the aspects named above, it will be of importance whether activities are organised by (or on behalf of) the *government or private operators.*[25]

As to the incorporation of the relevant provisions of the Protocol into the national legal systems, the case studies in Part V of this book illustrate that this process depends to a large extent on the *legal system, the existing complex of domestic legislation, and the legislative policy* of the state party involved.[26] The

---

[23] See discussion by Orrego Vicuña, Chapter 3 in this book. See also Bush, Chapter 2, and Richardson, Chapter 4, in this book.

[24] See Dodds, Chapter 21 in this book.

[25] See, for example, Bush, Chapter 2 in this book, observing that one of the factors that has had most influence in transforming the shape and official attitude to domestic environmental law is the prospect of private interests challenging regulatory decisions. See also Richardson, Chapter 4 in this book, discussing Antarctic tourism.

[26] The contributions in this book by Bush (Chapter 17), Njåstad (Chapter 20) and Joyner (Chapter 22) make it clear that in Australia, Norway and the United States, respectively, the incorporation of the Protocol in the national legal order has been influenced by the complex of existing national legislation. For example, with regard to the stricter wording of the general duty of care provision in the Norwegian legislation (para. 4), Birgit Njåstad explains that this stricter wording is to ensure that Antarctic

different aspects of the *practical implementation* of the Protocol may have an impact on the legal implementation process as well. For example, the way in which the practical application of the legislation is organised[27] and the use of other policy instruments may have consequences for the content of the legislation.

Factors Related to the ATS and the Protocol

Besides the 'national factors' described above, the special character of the ATS must be acknowledged as a significant factor in the implementation of the Protocol. Because of the close cooperation required among the states in conducting science and logistic operations, international political tensions that may exist elsewhere in the world need not necessarily be automatically reflected in the Antarctic where states share common interests. On the one hand, this 'family character' of the ATS has served as a special stimulant for an adequate implementation of the Protocol. (Who wants to risk becoming the 'black sheep' of the ATS family?) On the other hand, there is always the risk that the Consultative Parties may be reluctant to criticise each other with regard to failures to comply with the Protocol. In this respect, the CEP will play an important role. The way in which the CEP gives effect to Article 12 of the Protocol, specifying the Committee's functions,[28] and the way in which the Consultative Meeting deals with the advice provided by the CEP will also be important factors in influencing the implementation of the Protocol.

Also the *Protocol* itself may give rise to differences in the ways in which states implement the Protocol in the national legal order or in practice. One problem is that the Protocol, just like many other international (environmental) conventions, contains some *vague and unclear* provisions,[29] either as a result of political compromises or because more specific provisions were not feasible in view of differences in legal systems or the sensitivity of the subject. Moreover, state parties might consider the *Protocol as incomplete* and might feel the need to include additional provisions in the national legislation. Provisions on enforcement of the implementation legislation constitute the most obvious examples.

External Factors

The national implementation process may also be influenced by external factors, in particular by third states, international organisations and NGOs. Third states may raise the issue of an adequate implementation of the Protocol. It is likely that the

---

environmental regulation is in conformity with the general national environmental policy in Norway.

[27] See, for example, the case of Australia, discussed by Bush, Chapter 17 in this book.

[28] See further Orheim, Chapter 6 in this book.

[29] As observed by Kiss, 'the poor drafting of quite a few environmental instruments is an obstacle to sound implementation'; A. Kiss, 'Compliance with International and European Environmental Obligations', Inauguration Lecture at the Erasmus University Rotterdam, 15 November 1996, p. 4.

signing of the Protocol in 1991 played a role in the acceptance of the ATS by third states like Malaysia.[30] Poor implementation of the Protocol by the state parties may, however, raise new doubts as to whether the ATS is the best mechanism to manage the Antarctic. As already discussed in this chapter, environmental NGOs, in particular ASOC, will continue to follow the adequate implementation of the Protocol by the parties. The involvement of other, national NGOs may also have an influence on the implementation process.[31]

LEGAL IMPLEMENTATION

For a better understanding of the scope of the 'legal implementation process', it is useful to distinguish between the different types of provisions that are included in the Protocol:

a) provisions containing obligations for governments of state parties, which primarily play a role in the relation between the state parties, not directly touching on the rights and obligations of individuals;

b) provisions directly referring to rights and obligations of individuals;

c) provisions containing definitions and proclaiming purposes and objective and which should be considered when interpreting and applying the provisions referred to above;

d) other provisions, including those that relate to the position of the Protocol in the broader context of international law.

Incorporation of the Key Elements into the National Legal Order

Not all the types of provisions listed above need to be incorporated into the national law. Generally speaking, 'implementation' of the provisions referred to under (a) will require the states parties to directly undertake the relevant action, without incorporation of these elements into national law. Examples of such provisions in the Protocol are the obligation for 'the Parties to undertake to elaborate rules and procedures relating to liability for damage arising from activities taking place in the Antarctic Treaty area' (Article 16), the obligation to annually send a report to other state parties 'on the steps taken to implement [the] Protocol' (Article 17), and the obligation to cooperate 'in the planning and conduct of activities in the Antarctic

---

[30] See P. Beck, 'The United Nations and Antarctica, 1993: Continuing Controversy about the UN's Role in Antarctica', *Polar Record*, Vol. 30, 1994, pp. 257–264.

[31] The legislative process in the United States, as discussed by Joyner, Chapter 22 in this book, is an interesting example.

Treaty area' (Article 6(1)). Also the provisions included in category (d), for example those on the relation with other international agreements, institutional provisions and provisions on dispute settlement, will not need to be incorporated into national law.

The provisions described under (b) and (c) relate directly to human activities in the Antarctic Treaty area and require incorporation into the national legal order. In fact, two sub-categories may be distinguished:

1. Provisions aimed at the creation of rights for persons, not necessarily limited to persons under the jurisdiction of a state party. The most prominent examples are provisions ensuring the public availability of documents and public participation during the comprehensive environmental evaluation (CEE) process;

2. Provisions aimed at the regulation of activities of persons subjected to the jurisdiction of states parties and which should therefore become legally binding on these persons.[32]

The 'technique' for ensuring that these provisions become part of the national law depends on the legal system of each individual state. For example, the Protocol has been incorporated into Italian law by the 'implementing order procedure', which implies the adoption of a brief legislative act which provides for 'the full and complete' implementation of the treaty in question.[33] The technique followed in the Republic of South Africa is comparable.[34] According to South Africa's Constitution of 1996, 'any international agreement becomes law in the Republic when it is enacted into law by national legislation' (Article 231(4)). In line with this provision, South Africa's 1996 Antarctic Treaties Act provides that the treaties mentioned in Schedule 1 to the Act (including the Protocol) 'shall form part of the law of the Republic'. However, in addition to this simple transposition clause, the implementation legislation of South Africa includes explicit provisions to enable inspection and enforcement.

In many other states, including Australia, Finland, Germany, the Netherlands, New Zealand, Sweden, the United Kingdom and the United States, the relevant provisions of the Protocol have explicitly been 'translated' into domestic legal language in the form of national legislation.

Taking into account these differences and the other factors identified above in this chapter, it is understandable that the legal and practical implementation of the Protocol reflects considerable diversity amongst the parties. As such, these differences are not undesirable, since careful incorporation of the Protocol into

---

[32] It should be noted that these provisions also address activities planned and conducted by the government or governmental institutions of the Contracting Parties (with the exception of Art. 11 of Annex IV to the Protocol). This scope is very important in the Antarctic context because in many states parties most or all scientific research activities are initiated and carried out by or under the auspices of the government.

[33] See the discussion by Pineschi, Chapter 19 in this book.

[34] See the discussion by Dodds, Chapter 21 in this book.

national law contributes to an effective application and enforcement. Like other international conventions, including the case of the Protocol, the existence of these different factors should be respected and room for a national 'transformation and application process' should be provided for.[35] According to the Final Act of the Eleventh Antarctic Treaty Special Consultative Meeting,

> The Meeting acknowledged that, while reservations to the Protocol would not be permitted, this did not preclude a State, when signing, ratifying, accepting or approving the Protocol, or when acceding to it, from making declarations or statements, however phrased or named, with a view, inter alia, to the harmonisation of its laws and regulations with the Protocol, provided that such declarations or statements do not purport to exclude or to modify the legal effect of the Protocol in its application to that State.[36]

As the last part of this statement indicates, the flexibility with regard to the legal implementation of the Protocol is somewhat limited. When adopting the Protocol, the parties were led by the purpose of establishing a minimum level of protection of the Antarctic, and this level of protection must be respected during the implementation process conducted by each state party. In other words, the different factors, like those identified in this chapter, may not be used to justify an inadequate implementation of the Protocol. However, because of certain vague formulations and lacunas in the Protocol, it is not easy to formulate the 'minimum requirements' with regard to all key elements of the Protocol. Questions arise with respect to the exact meaning and scope of certain obligations and prohibitions, the relationship with other international instruments is not always very clear, and different views may be advocated with regard to the question: Which persons or activities should be subjected to the provisions of the national law?

An Overview of Selected Questions for Domestic Legislation

To illustrate the complexity of questions that have to be dealt with by the states parties during the domestic legislative process, some of the most important questions are addressed in this section, with examples of how some of the states parties have dealt with these in their implementing legislation.

What are the values that the legislation aims to protect? For example, is the 'wilderness value' of Antarctica, which is specially recognised in Article 3 of the Protocol, to be explicitly included in domestic legislation?

---

[35] This consideration is not only included in the general provision with regard to implementation, i.e. Art. 13(1) of the Protocol; more specific provisions of the Protocol also show that Contracting Parties have the levy to give the provisions of the Protocol a logical place in the context of national law. See, for example, Art. 1(1) of Annex I of the Protocol with regard to EIA ('The environmental impacts of proposed activities referred to in Art. 8 of the Protocol shall, before their commencement, be considered in accordance with appropriate national procedures').

[36] Para. 12 of the Final Act.

Implementing the Protocol Domestically

In the legislation of some countries, for example that of Norway, the protection of wilderness value is contained within the description of the purpose of the legislative act.[37] Moreover, in the domestic legislation of some other countries, such as New Zealand and the Netherlands,[38] the wilderness value has explicitly been included in the definition of the Antarctic environment. However, the wilderness value of Antarctica is not mentioned at all in the legislation of some of the other parties, including the United Kingdom and Sweden. This of course does not necessarily mean that these countries deny the importance of the wilderness value of Antarctica.[39] Other questions, similar to that of wilderness value, may also be raised with regard to other notions found in the Protocol, for example the notion of 'dependent and associated ecosystems'.

Should the principles contained in Article 3 of the Protocol be incorporated into domestic legislation? The legislation of several countries, including Australia, the United Kingdom and the Netherlands,[40] require that their competent national institutions respect the principles of Article 3 of the Protocol during the decision-making process. In other countries, for example South Africa and Norway,[41] these principles are formulated by legislation as directly binding on the operators in the Antarctic Treaty area. In the legislation of the United States, the principles contained in Article 3 of the Protocol have no legal status at all.[42]

Which persons should be addressed by the national legislation? Most countries, including the United Kingdom and the Netherlands,[43] seem to base their jurisdiction on Article VII(5) of the Antarctic Treaty. Some of these countries seem to focus on the organisation of activities within, or activities proceeding from, their territory, while other countries, for example Sweden and Japan,[44] also apply the permit requirement for Antarctic activities to nationals who organise an activity

---

[37] Para. 1 of the Norwegian Regulations Relating to Protection of the Environment in Antarctica, Royal Decree of 5 May 1995; see Njåstad, Chapter 20 in this book.

[38] See Art. 7(1) of the New Zealand Antarctic (Environmental Protection) Act 1994, No. 119; also Art. 1(2)(b) of the Antarctica Protection Act, *(the Netherlands) Bulletin of Acts and Decrees*, No. 220, 1998.

[39] For a discussion on the wilderness value of Antarctica, see, e.g., United Kingdom, 'Wilderness and Aesthetic Values in Antarctica', doc. XXII ATCM/IP 2, 1998.

[40] For Australia see Art. 7A of the Antarctic Treaty (Environmental Protection) Act 1980, also discussed by Bush, Chapter 17 in this book. For the United Kingdom, see Art. 15 of the Antarctic Act 1994. For the Netherlands, see Art. 4(g) of the Antarctica Protection Act 1998.

[41] For South Africa, see Art. 3 in relation to Schedule 2 of the Antarctic Treaties Act 1996. For Norway see para. 4 of the Regulations Relating to Protection of the Environment in Antarctica; and see also discussion by Njåstad, Chapter 20 in this book.

[42] See discussion by Joyner, Chapter 22 in this book.

[43] See Arts. 3–11 of the United Kingdom Antarctic Act 1994; and Art. 1(e) in relation to Art. 8 and Art. 32 of the Dutch Antarctica Protection Act 1998.

[44] See Art. 1 in relation to Art. 16 of the Swedish Antarctica Act; and Art. 2 in relation to Art. 5(1) of the Japanese Law Relating to Protection of the Environment in Antarctica, 1997.

outside their territory. Most claimant states also apply their legislation to foreign persons who visit the part of the Antarctic that is subject to a territorial claim of that state. For example, the New Zealand Antarctica (Environmental Protection) Act applies to 'any person in the Ross Dependency' (Art. 2(a)), while the Norwegian Regulations Relating to Protection of the Environment in Antarctica apply to 'anyone staying, or responsible for activity, in Dronning Maud Land and on Peter I Island' (para. 2(3))[45]. The legislation of the United Kingdom does not have an explicit provision that states that the legislation applies to all persons within their claimed area.

Because of the severe overlap of jurisdictional scope of application, most domestic laws explicitly state that the legislation will not apply to persons that already have obtained approval by another state party to the Protocol.

What exactly are the minimum requirements for the preliminary assessment (PA) of a proposed activity in the Antarctic? According to Article 3(2)(c) of the Protocol, activities in the Antarctic Treaty area shall be planned and conducted on the basis of information sufficient to allow prior assessment of, and informed judgements about, the possible impacts of the activities on the Antarctic environment and dependent and associated ecosystems and on the value of Antarctica for the conduct of scientific research. Further to this, Article 8(1) of the Protocol states that the proposed activities shall be subject to procedures set out in Annex I for prior assessment of the impacts of those activities. Have these requirements of Articles 3(2)(c) and 8(1) under the Protocol been incorporated in domestic legislation? In the United Kingdom, the preliminary assessment has the form of an application for a permit and the competent authority may require information to be attached to that application.[46] In the United States, the preliminary assessment has the form of a memorandum to be sent to the competent authority.[47] In New Zealand, the requirements for a preliminary environmental evaluation have been laid down in a separate provision of the legislation.[48] In the legislation of some other states parties, including Norway[49] and the Netherlands,[50] no distinction has been made between the preliminary assessment, on the one hand, and the initial environmental evaluation (IEE) procedures,[51] on the other, so the requirements for the IEE procedure form the minimum level of environmental impact assessment for all Antarctic activities.

---

[45] For the further discussion of Norwegian legislation, see Njåstad, Chapter 20 in this book.
[46] See Art. 4 of the UK Antarctic Regulations 1995.
[47] See Joyner, Chapter 22 in this book.
[48] See Art. 17 of the New Zealand Antarctica (Environmental Protection) Act 1994.
[49] See para. 10 of the Norwegian Regulations Relating to Protection of the Environment in Antarctica; see also discussion by Njåstad, Chapter 20 in this book.
[50] See Arts. 8 and 10 of the Antarctic Protection Act 1996 in relation to the Antarctic Protection Decree 1998.
[51] See Art. 2 of Annex I to the Protocol.

How should the phrase 'minor or transitory impact' on the Antarctic environment be interpreted? This question, of course, is directly related to the requirement under the Protocol regarding the development of a comprehensive environmental evaluation (CEE) for an activity which has been determined likely to have 'more than a minor or transitory impact'.[52] Despite the many initiatives the parties have taken to date,[53] the parties have failed to provide detailed criteria that would specify the formulation of 'minor or transitory impact'. At the first CEP meeting in 1998, 'It was recognised, however, that it may not be possible to make precise definitions and that the concepts will evolve from practical experience'.[54] This view was confirmed by the Consultative Meeting with the adoption of the EIA guidelines in 1999.[55] According to these guidelines, 'the interpretation of this term will need to be made on a case by case site specific basis'. This explains why most of the parties to the Protocol simply incorporate the unclear terminology found in the Protocol and its annexes into their own national legislation. As a consequence, the national competent authorities will make their own decisions as far as the required level of EIA is concerned, although openness with regard to this decision-making (e.g. distribution of IEEs) enables an exchange of views on this issue.

Does the Protocol prohibition on mineral activities also apply in the Antarctic deep seabed?[56] The mining ban in the United Kingdom Antarctic Act 1994 (Section 6), for example, does not apply to the deep seabed because the definition of 'Antarctica' as laid down in Article 1(1) of the Act does not include the deep seabed. Section 12 (in conjunction with Section 11) of the New Zealand Antarctica (Environmental Protection) Act 1994 explicitly prohibits New Zealand citizens and corporations from undertaking mineral resource activities in the deep seabed south of 60 degrees south. South Africa,[57] as well as many other parties to the Protocol,[58] applies the mining ban to the 'Antarctic Treaty area' without making specific references to the deep seabed area.

---

[52] See Art. 8(1) of the Protocol and Annex I, especially Art. 3 of Annex I.

[53] See, for example, doc. XIX ATCM/INFO 14, 1995, and doc. XXII ATCM /WP 14, 1998.

[54] See para. 30 of the 'Report of the Meeting of the Committee for Environmental Protection, Tromsø, 25–29 May 1998, in *Final Report of the Twenty-second Antarctic Treaty Consultative Meeting, Tromsø, Norway, 25 May – 5 June 1998* (Oslo: Royal Norwegian Ministry of Foreign Affairs, 1998), Annex E.

[55] See Resolution 1 (1999), in *Final Report of the Twenty-third Antarctic Treaty Consultative Meeting, Lima, Peru, 24 May – 4 June 1999* (Lima: Peruvian Ministry of Foreign Affairs, 1999), Annex C.

[56] For a detailed analysis of this issue, see D. Vidas, 'The Southern Ocean Seabed: Arena for Conflicting Regimes?', in Vidas and Østreng (eds.), *Order for the Oceans at the Turn of the Century*, pp. 291–314, especially pp. 303–307.

[57] See Art. 3 in conjunction with Schedule 2 of the South African Antarctic Treaties Act; see also discussion by Dodds, Chapter 21 in this book.

[58] See, for example, para. 5 of the Norwegian Regulations and Art. 6(1)(h) of the Netherlands Antarctica Protection Act 1998.

Further questions. Domestic legislation for implementation of the Protocol raises a number of other complex questions, only some which are named here. For instance, should monitoring also be required for IEE activities and what are the concrete requirements for the monitoring of IEE and CEE activities? Can ship-borne tourism without any landings south of 60 degrees south be regulated, taking into consideration that the Protocol 'shall supplement the Antarctic Treaty and shall neither modify nor amend that Treaty', which in turn safeguards the high seas rights in the Antarctic Treaty area?[59] Apart from Article 3 of the Protocol, what are the criteria for determining whether a proposed activity in Antarctica may proceed? What should the consequence be of the lacunas in knowledge regarding the effects of these activities on the Antarctic environment? Should the precautionary principle be taken into account in the decision-making process?[60] Is it possible to prohibit an activity, although the impact is not 'significant' or 'more than minor or transitory'? And finally, what legal sanctions may be imposed in case of violation of the implementation legislation and/or the Protocol? For example, the types of sanctions for one and the same violation, as well as the maximum penalties envisaged, differ among parties. For instance, the maximum period for imprisonment for the illegal taking of a penguin is 6 months under New Zealand law,[61] one year under Norwegian law,[62] and up to two years under the law of the United Kingdom.[63]

Assessment and International Discussion

With an eye to the questions above and the differences that exist in domestic implementation legislation, it is surprising that the legal implementation of the Protocol has received so little attention during the Consultative Meetings.[64] Although copies or summaries of the national implementing legislation have been distributed in the past years among the parties, in all the years since the adoption of the Protocol in 1991 there has not been any in-depth discussion on the content of the implementing legislation of a state party. Of course, with regard to certain issues, like the issue of territorial jurisdiction over a claimed area, this may easily be explained by the existing sensitivities among the parties and more specifically by the 'agreement to disagree' on the Antarctic sovereignty issue, laid down in Article IV of the Antarctic Treaty. Furthermore, it is possible that an in-depth discussion during

---

[59] See Art. 4(1) of the Protocol in conjunction with Art. VI of the Antarctic Treaty.
[60] As Bush, Chapter 2 in this book, notes, the environmental regime of the Protocol is far from being a model example of the application of the precautionary principle.
[61] See sections 28 and 33 of the New Zealand Antarctica (Environmental Protection) Act 1994.
[62] See paras. 14 and 32 of the Norwegian Regulations.
[63] See sections 7 and 20 of the UK Antarctic Act 1994.
[64] During the Twenty-third Consultative Meeting in Lima in 1999, which was the second meeting since the entry into force of the Protocol, the issue was discussed in Working Group I for about ten minutes altogether.

the first years after the adoption of the Protocol might have raised too many uncertainties or even sensitivities among the Consultative Parties and thus possibly hampering the ratification process. However, now that the Protocol has entered into force, parties should become willing to exchange views with regard to the existing differences between the legal implementation of the Protocol and questions that have arisen during the process of implementation. Hopefully, the suggestions tabled at the XXIII Consultative Meeting for the development of a standard for the reports under Article 17 of the Protocol, the initiatives of CEP participants, and the continuing advocacy work of organisations such as ASOC will stimulate the commencement of these discussions.

PRACTICAL IMPLEMENTATION OF THE PROTOCOL

As Article 13 of the Protocol reflects, the protection of the Antarctic environment requires more than merely an incorporation of the relevant provisions of the Protocol into the national legal order.[65] As Robin Churchill observed in the context of implementation of the UN Convention on the Law of the Sea, it is necessary to look not only at the relevant legislation, but also at relevant administrative practices ... and at how the legislative and administrative framework actually operates in practice'.[66] The practical implementation of the Protocol requires that the parties make all the arrangements necessary for an *effective application* of their national legislation. For example, an organisational structure will have to be developed for the well functioning of the EIA provisions and the permit requirements.[67] Of particular importance is the designation of the competent national authority and the measures to ensure that this authority will have sufficient knowledge and capacity to fulfil its tasks. An interesting question in this context is whether the competent authority may directly be involved in the organisation and conduct of Antarctic activities. In some countries, including Norway[68] and the United States,[69] the organisations that have been designated as the competent authority with regard to scientific research activities are the same that are directly involved in conducting scientific research in the Antarctic. On the one hand, this is a logical choice from the perspective of available knowledge and the network that these organisations possess.

---

[65] For instance, in addition to 'adoption of laws and regulations', Art. 13(1) requires each party to the Protocol to take appropriate measures within its competence to ensure compliance with the Protocol, including 'administrative actions' and 'enforcement measures'.

[66] Churchill, 'Levels of Implementation', p. 323.

[67] See, for instance, an overview of the many agencies and institutions that are involved in the implementation of the Protocol in Chile and Norway, respectively, by Carvallo and Julio, Chapter 18, and Njåstad, Chapter 20, in this book.

[68] According to the Norwegian Regulations Relating to Protection of the Environment in Antarctica, the Norwegian Polar Institute is the competent national authority. For the arguments for this choice and a detailed description of the tasks of the Institute, see Njåstad, Chapter 20 in this book.

[69] See Joyner, Chapter 22 in this book.

On the other hand, this choice may give rise to questions regarding the accountability of the competent authority.

Furthermore, the practical implementation of the Protocol includes the *use of other instruments* in addition to the typical legal instruments such as specific prohibitions, EIAs and permit requirements. In particular, the education and training of persons who intend to visit Antarctica are important. As observed by Lee Kimball, 'it is almost pointless to strengthen legal requirements or to develop detailed guidelines for implementing waste disposal, impact assessment, or other environmental protection measures in Antarctica if the countries involved do not widely publicise the measures in a form easily comprehensible to all those active in the region'.[70] In this association, however, it can be added that the distinction between legislation and other instruments is not always very clear. An interesting illustration of this is para. 21 of the Norwegian Regulations Relating to Protection of the Environment in Antarctica, which includes a concrete obligation to undergo training: 'Every activity in Antarctica must have a person responsible for waste management. The person responsible shall undergo training approved by the Norwegian Polar Research Institute'.

In addition to educating and training those who intend to actually visit the Antarctic Treaty area, educating the general public on the special character and values of the Antarctic is important. For this purpose, many countries put forth a great effort to develop information material (such as brochures and comprehensive websites), and to produce special programmes for teachers to be used in primary and secondary schools and visitor centres.[71]

Other instruments, that are not explicitly stated in the Protocol, but may nevertheless strengthen the protection of the Antarctic environment, include the following:

– the possibility to require a minimum level of environmental care as a condition for securing financial and logistic support;
– the obligation for the operator to provide financial security to cover certain costs (for example, the removal of certain facilities, products, or wastes) or to cover liability claims;[72]
– the obligation for the operator to take part in certain 'self-regulating' initiatives (e.g. IAATO);[73]

---

[70] See L.A. Kimball, *Southern Exposure: Deciding Antarctica's Future* (Washington, D.C.: World Resources Institute, 1990), p. 22.

[71] See, for example, websites of the British Antarctic Survey (www.antarctica.ac.uk/), Antarctica New Zealand (www.antarcticanz.govt.nz/), and the Australian Antarctic Division (www.antdiv.gov.au/). See also Australia, 'Antarctic Treaty Introductory Booklet', doc. XXIII ATCM/WP 14, 1999. This aspect of information is also addressed by Vidas and Njåstad, Chapter 8 in this book.

[72] See, for example, para. 31 of the Norwegian Regulations Relating to Protection of the Environment in Antarctica and Art. 16(f) of the Dutch Antarctica Protection Act.

[73] See Richardson, Chapter 4 in this book.

- the obligation for the operator to produce a public report after having completed an activity, giving information on, for example, the environmental impact, measures taken to prevent or mitigate adverse impacts, and the way in which waste was disposed.[74]

In addition to the organisational arrangements necessary to enable the application of the national legislation and the development of other instruments, the 'practical implementation' of the Protocol also includes the fulfilment of the provisions in the Protocol that are primarily addressed to the *governments* themselves and therefore have not been incorporated by all states parties into national law. Examples are

- the development of contingency plans as required by Article 15 of the Protocol,
- the development of waste management plans as referred to in Article 8 of Annex III to the Protocol,[75]
- arrangements that are necessary to fulfil the different reporting obligations that are laid down in Article 17 of the Protocol and Article 6 of Annex I to the Protocol, and
- initiatives to stimulate the international cooperation as required under Article 6 of the Protocol.[76]

Finally, arrangements to enable effective *enforcement* of the Protocol at two levels are necessary: Firstly, enforcement of the obligations which are established in the relation among states, and secondly, enforcement of the obligations (often laid down in implementation legislation) in the relation between the government of a state party and legal subjects under its legislation.

Three Major Concerns

Considering the differences in the legal implementation of the Protocol and the increase of non-governmental activities, in particular *three issues* discussed below will require special attention in the near future.

Informal Network of Competent Authorities. The international discussion on the implementation of the Protocol should not be limited to the annual CEP meeting and Consultative Meeting. The available time at these meetings must be divided over many subjects and discussions are often influenced by political considerations.

---

[74] See, for example, para. 8 of the Norwegian Regulations Relating to Protection of the Environment in Antarctica and Art. 33 of the Dutch Antarctica Protection Act.
[75] For an example of national legislation in which this obligation has been incorporated, see Art. 28 of the Finnish Act on the Environmental Protection of Antarctica (October 1996).
[76] At the XXIII Consultative Meeting, Peru stressed the importance of Art. 6 of the Protocol and proposed to give this issue more specific attention in the coming years.

Having in mind the special legal and political status of Antarctica, informal cooperation and the exchange of information between the persons directly involved in the application of the national law would be advantageous, in particular regarding the assessments of EIAs and applications for permits. Because of the many different views among the states parties with regard to the issue of jurisdiction, and because the participants of many Antarctic activities are nationals of different states, close cooperation between the competent authorities of all states parties is necessary to prevent private operators from misusing the gaps in the various national legislative systems. For example, many Contracting Parties do not require their nationals to conduct an EIA or to apply for a permit if the relevant activity has been assessed and authorised by another state party to the Protocol.[77] If a person in state X invokes such a provision, although the person involved in fact only receives logistic support from contracting party Y, direct contact between the relevant national authorities will clarify this. Furthermore, through the exchange of experience, this network of competent authorities could identify issues that should be discussed by the CEP, which could prepare some of these discussions.

Cumulative Impacts. Another aspect of implementing the Protocol that will deserve special attention in the next few years is the difficulty of assessing cumulative impacts. One of the most important consequences of the legal status of the Antarctic, and the way in which this region is managed, is that the activities in the area are assessed and approved by competent national authorities in (currently) 28 states parties to the Protocol.[78] In particular, the increase of tourist visits and the concentration of tourism in certain areas of Antarctica (such as the Peninsula region) require an approach that is not limited to the assessment of the likely impacts of individual activities. The aspect of assessing cumulative impacts was discussed at the 1999 CEP meeting in Lima. Although most ship-born tourist activities are subject to EIA procedures in accordance with the Protocol, the accumulation of tourist activities at certain Antarctic sites is of concern; consequently, the CEP, and the Consultative Meeting, welcomed the suggestion of ASOC to think about a more strategic approach with regard to EIA for tourist activities.[79] It is to be hoped that this discussion will be continued at the next CEP meeting, which will take place in the Hague, 11–15 September 2000.

Enforcement. Adequate supervision and enforcement of legal provisions concerning activities in the Antarctic area are generally considered as problematic. As the proper assessment of activities must be ensured in 28 states, which are

---

[77] See, for example, Art. 6 of the Finnish Act on the Environmental Protection of Antarctica, and Art. 3(4) of the UK Antarctic Act 1994.

[78] See also Bush, Chapter 2 in this book. Bush points to the absence of a method to take account of cumulative impacts of issuing permits under Annex II of the Protocol.

[79] See *Final Report of the XXIII ATCM*, paras. 120 and 121. See also ASOC, 'Large-scale Tourism in

currently parties to the Protocol, the same problem is manifested with regard to the enforcement of the national legislation and permit conditions. Furthermore, the possibilities for supervision and enforcement are limited because of the geographic remoteness, the climate conditions, the costs involved in sending governmental officers to the area, questions with regard to jurisdiction,[80] and possible conflicts of interest in cases of government-sponsored activities. In view of these difficulties, the establishment of an international, independent inspectorate for the Antarctic has often been advocated. However, this option appeared to be unacceptable for several states, as the idea directly touches upon the sovereignty issue and sensitivities with regard to sovereign immunity of government-sponsored activities. In the absence of such an international inspectorate, the questions that arise with regard to the enforcement of the national law are numerous. How has a minimum level of supervision in the area been arranged? What sorts of sanctions are available (civil, administrative, punitive)? Does the legislation provide for defences? Who is the enforcing authority? Who may initiate an enforcement action? Are mechanisms developed to improve compliance (e.g., systems of self-regulation, education and training and reporting obligations)?[81] On the basis of national reports and discussion papers,[82] these and related questions should be given attention by the CEP and at the Consultative Meeting. The title of an agenda item that is regularly accepted for the Consultative Meetings – 'Compliance with the Protocol, General Matters and Implementation by the Consultative Parties' – provides a suitable basis for these discussions. Another option might be to change the agenda item 'Inspections under the Antarctic Treaty' to 'Supervision and Enforcement', in order to broaden the scope of this item to include aspects other than the international inspections under the Treaty.

CONCLUSIONS AND FUTURE DEVELOPMENTS

In the absence of undisputed sovereignty claims on the Antarctic continent, the legal and practical protection of Antarctica depends heavily on the collective efforts of all the 28 parties to the Protocol. The chapters in Part V of this book, each of which focuses on a selected Consultative Party, provide an overview of the most important steps that have been taken by these countries to implement the Protocol.

---

the Antarctic', doc. XXIII ATCM/IP 121, 1999.

[80] See, e.g., S. Blay and J. Green, 'The Practicabilities of Domestic Legislation to Prohibit Mining Activity in Antarctica: A Comment on the Australian Perspective', *Polar Record*, Vol. 30, 1994, pp. 23–32. Amongst other issues, Blay and Green discuss the possibilities and difficulties of enforcing Australian law with regard to violations committed by foreign nationals. They distinguish and discuss difficulties of a legal, practical and political nature.

[81] For a more detailed discussion of 'elements of an effective system of enforcement', see Bush, Chapter 2 in this book.

[82] On enforcement issues related to the Protocol, see, e.g., Chile, doc. XXII ATCM/IP 43.

The modes of incorporation of the Protocol in the national law of these countries demonstrate important differences. In view of the complexity of the factors that influence the legal implementation process, the existence of differences as such is easy to understand and should not cause concern. In fact, the effectiveness of the Protocol will be improved if the relevant provisions are carefully incorporated into the existing national legal systems and if due account is taken of the specific situation and circumstances of each state party. However, this flexibility still has certain limits because the minimum level of protection that the Protocol aims to establish should be respected. The problem, however, is that – largely due to vague formulations in, and incompleteness of, the Protocol – the precise minimum requirements with regard to several key elements of the Protocol are not perfectly clear. This chapter has reviewed a selection of important questions that are to be dealt with during the legal implementation process. In that context, the differences in the legal systems and approached employed in the national legal implementation of the relevant provisions of the Protocol are likely to affect the effectiveness of the Protocol

It is therefore remarkable that the legal implementation of the Protocol has received so little attention during the Consultative Meetings. Now that the Protocol has entered into force, its parties should give priority to an exchange of views with regard to the existing differences between the legal implementation of the Protocol and questions that have arisen during the legal implementation process. Hopefully, the suggestions put forth at the Lima Consultative Meeting in 1999 to develop a standard for reports under Article 17 of the Protocol, the initiatives of CEP participants, and the continuing work of non-governmental organisations such as ASOC, IAATO and others, will stimulate these discussions. In this respect, it is important to note that Article 12 of the Protocol does not limit the advisory role of the CEP to technical aspects, as discussions at the CEP meetings and the Consultative Meeting sometimes seem to suggest.

As indicated by Article 13 of the Protocol, the legal implementation is only one part of the whole implementation process. On the one hand, the effects of an inadequate legal implementation of the Protocol may be compensated by an effective practical implementation. On the other hand, a perfect set of national statutes may have little effect if there are no arrangements for adequate application and enforcement. The parties should therefore take practical measures to implement the Protocol, including arrangements that are necessary for an effective application of the national law (assessments of EIAs and permit applications), the fulfilment of the provisions of the Protocol that are primarily addressed to the governments (e.g., contingency and waste management plans, reporting activities), the use of other instruments (e.g., education and training) and arrangements that are necessary for an effective supervision and enforcement of the national law (e.g., designation of inspectors and organising inspections in cooperation with other states parties).

Although many of these issues are the subjects of fruitful discussions at the CEP meetings, it would be desirable to have an informal network of competent

Implementing the Protocol Domestically

authorities of all states parties. Within such a network, in addition to CEP meetings, special attention should be paid to the issues of jurisdiction, cumulative impacts, and supervision and enforcement.

Finally, one should also look forward and be ready for timely responses to possible new developments.[83] As Mr. Upton, Environmental Minister of New Zealand, stated at the Ministerial on Ice meeting in January 1999, 'Antarctica[s] greatest defence was isolation but that isolation has evaporated rapidly'. The last decade of the twentieth century was characterised by the increase of Antarctic tourism, which has increased almost 800 per cent within less than 10 years.[84] What will the next decades bring? An increase of airborne tourism and permanent tourist facilities on the continent? Exploitation of icebergs?[85] Increased scientific research for commercial purposes? Or even mineral activities under the guise of scientific research?

It has been stated many times that the Protocol should be recognised as a first step – not as a final step – towards a comprehensive protection of the Antarctic environment.[86] Although many problems may be resolved by a creative application of the Protocol by its parties and cooperation among states,[87] a further strengthening of the Protocol may prove to be necessary.

---

[83] See, however, Bush, Chapter 2 in this book. Bush argues that the ability of the ATS to respond to new developments is limited, in particular because of the fact that the Consultative Meetings take place only once a year. Another characteristic aspect of the ATS that underlines his opinion is the consensus rule.

[84] See Richardson, Chapter 4 in this book.

[85] Note that para. 6 of the Final Act of the Eleventh Special Antarctic Treaty Consultative Meeting explicitly excluded ice from the mining ban of Art. 7 of the Protocol. For a detailed discussion, see F. Trombetta-Panigadi, 'The Exploitation of Antarctic Icebergs in International Law', in F. Francioni and T. Scovazzi (eds.), *International Law for Antarctica*, pp. 225–257.

[86] John Blincoe, New Zealand Member of Parliament, during the parliamentary discussions on the New Zealand implementation legislation, described the Protocol as 'a crucial first step towards world park status for Antarctica', see Weekly Hansard 25, Report of Foreign Affairs and Defence Select Committee, 17 November 1994, Antarctic (Environmental Protection) Bill, p. 5,019. Similar wording was used by Greenpeace in their 1994 brochure 'State of the Ice', p. 22.

[87] See Bush, Chapter 2, and Richardson, Chapter 4, in this book.

17

Australian Implementation of the Environmental Protocol

*William Bush**

The Australian legislative scheme for the implementation of the Protocol on Environmental Protection to the Antarctic Treaty[1] (Protocol) is straightforward. Within little more than a year of the finalisation of the Protocol in October 1991, Australia had enacted two complex pieces of legislation to give effect to the Protocol in Australian law.[2] The first dealt with Annex IV on the prevention of marine pollution. The legislation amended the Protection of the Sea (Prevention of Pollution from Ships) Act 1983 (the PS(PPS) Act),[3] the legislation by which Australia implements obligations that it has accepted under the International Convention for the Prevention of Pollution from Ships of 1973, as modified by the 1978 Protocol (MARPOL 73/78). The second amending act gave effect to the Environmental Protocol itself and all other annexes to the Protocol – including Annex V on area protection and management. This legislation amended (transformed would be a better description) legislation passed initially in 1980 to implement the 1964 Agreed Measures for the Conservation of Antarctic Fauna and Flora, i.e., the Antarctic Treaty (Environment Protection) Act 1980 (ATEP Act).[4] As amended, the ATEP Act is the principal environmental legislation applicable to Australian Antarctic

---

* The author gratefully acknowledges the assistance of Mr. Andrew Jackson and Mr. Martin Betts of the Antarctic Antarctic Division but emphasises that the assessments are, of course, his own.

[1] ILM, Vol. 30, 1991, pp. 1,461ff.

[2] Transport and Communications Legislation Amendment Act (No. 2) 1992 (No. 71 of 1992) and Antarctic (Environment Protection) Legislation Amendment Act 1992 (No. 156 of 1992) in W.M. Bush (ed.), *Antarctica and International Law: A Collection of Inter-State and National Document* (Dobbs Ferry, NY: Oceana Publications, 1992–), Booklet AU92B, pp. 42–61.

[3] The 1983 act as amended is reproduced in Bush (ed.), *Antarctica and International Law* (1992–), Booklet AU92A, pp. 4–48.

[4] The 1980 Act as amended is reproduced in Bush (ed.), *Antarctica and International Law* (1992–), Booklet AU/Rev. May–June 1980, pp. 26–77.

activities. Regulations made by the Governor-General under the ATEP Act supplement the Act. So far, these regulations concern environmental impact assessment,[5] waste management[6] and seals.[7]

The amendments to the PS(PPS) Act have yet to 'commence' (i.e., come into operation). In the meantime, Annex IV of the Protocol is being given effect to administratively. All the amendments to the ATEP Act have commenced with the exception of those that give effect to the protected area regime under Annex V which itself has yet to become effective.

This chapter will look at some aspects of this implementing Australian legislation on the assumption that it has all commenced.

ISSUES OF SUBSTANCE

The scope of the Australian legislation can be described in terms of the geographic reach, the jurisdictional limits recognised within that reach and the range of activities covered. The various aspects of the scope of this legislation can be illustrated by enquiring whether a ship voyaging on the Southern Ocean would be subject to the environmental impact assessment of the ATEP Act. The activity might be covered if the act is intended to apply to the entire area south of 60° south latitude. Nevertheless the activity would be excluded if Australia did not assert jurisdiction over the vessel. Even if the vessel were within Australian jurisdiction, for example by being an Australian flagged vessel, it might be excluded on the grounds that the activity in which it was engaged – for example, transiting the Southern Ocean to a point outside the Antarctic – was one that the legislation did not purport to control.

Geographic Scope of Domestic Legislation

By all appearances, the geographic scope of the Australian legislation is broad. Subject to the jurisdictional limits on the entities covered (an issue discussed below in this chapter), most of the obligations of Australian legislation apply to the 'Antarctic', which is defined to mean 'the area south of 60° south latitude, including all ice shelves in the area' (ATEP Act, Section 3(1)). Are there any geographic limitations, horizontal or vertical, on this apparently broad scope?

---

[5] Antarctic Treaty (Environment Protection) (Environmental Impact Assessment) Regulations (Statutory Rules 1993, No. 115) in Bush (ed.), *Antarctica and International Law* (1992–), Booklet AU93, pp. 1–8.

[6] Antarctic Treaty (Environment Protection) (Waste Management) Regulations (Statutory Rules 1994, No. 36) in Bush (ed.), *Antarctica and International Law* (1992–), Booklet AU94, pp. 17–23.

[7] Antarctic Seals Conservation Regulations (Amendment) (Statutory Rules 1993, No. 289) that amended Statutory Rules 1986, No. 398 (Australia, *Statutory Rules*, 1986, pp. 1,432–1,440). See Bush (ed.), *Antarctica and International Law* (1992–), Booklet AU85–86, pp. 54–67.

Horizontal geographic limits. The qualified geographic scope of the Antarctic Treaty and the Agreed Measures, imported by the definition of 'Antarctic', creates doubts whether the ATEP Act covers the entire sea area south of 60° south latitude. The suggestion is that the high seas may be excluded by virtue of the saving of high seas rights found in both those international instruments.[8] Whatever the high seas exception signified, most if not all parties accepted that at least for the 'preservation and conservation of living resources in Antarctica' the Antarctic Treaty[9] and Agreed Measures did not apply to the high seas.[10] That was perhaps the main reason for the negotiation in 1972 of the Convention for the Conservation of Antarctic Seals[11] (Seals Convention).

Such a limitation might be taken to affect the ATEP Act for several reasons. The definition of 'Antarctic' introduces a doubt by referring to ice shelves. Why refer to ice shelves and not also to the high seas? Moreover the reference to ice shelves echoes the similar formulation in the Agreed Measures that the ATEP Act was intended to implement. Sections 4(2) and 3(5) may reinforce this suggestion. Section 4(2) makes the act subject to the 'obligations of Australia under international law, including obligations under any international agreement binding on Australia' and 'any law of the Commonwealth giving effect to such an agreement'. Section 3(5) of the ATEP Act in its original form specifies that 'an expression that is used in both this Act and either the [Antarctic] Treaty or the Agreed Measures (whether or not a particular meaning is assigned to it by the Treaty or those Measures) has, in this Act, the same meaning as in the Treaty or those Measures, as the case may be'.

In spite of these indications, there is no doubt that the executive government believed, when the 1992 amendments were passed and even before, that the ATEP Act extended to the high seas. This is clear from regulations made under the act[12] and the explanatory memorandum that accompanied the 1992 amendments.[13] Doubt lingers, though, because the 1992 amendments did not alter the definition of 'Antarctic', which is the same as appeared in the 1980 legislation. Even so, the better view is that 'Antarctic' under the ATEP Act as amended will be interpreted to cover the high seas. Should the point ever be litigated, it could become an interesting example of recognition by an Australian court that the meaning of a treaty may be varied by subsequent agreement or practice in accordance with a rule of

---

[8] Art. VI of the Antarctic Treaty and Art. I of the Agreed Measures.

[9] UNTS, Vol. 402, pp. 71ff.

[10] See discussion in, for example, F. Orrego Vicuña, *Antarctic Mineral Exploitation: The Emerging Legal Framework* (Cambridge University Press, 1988), pp. 133–134 and 138–139; W.M. Bush (ed.), *Antarctica and International Law: A Collection of Inter-State and National Document* (London: Oceana Publications, 1982–88), Vol. I, pp. 64ff and 147.

[11] UNTS, Vol. 1080, pp. 175ff; and ILM, Vol. 11, 1972, pp. 251ff.

[12] Reg. 11(3) read in conjunction with Reg. 2(1) (definition of 'convention area') Antarctic Seals Conservation Regulations, Statutory Rules 1993, No. 289, as amended.

[13] Australia, Parliament, Senate, *Antarctic (Environment Protection) Legislation Amendment Bill 1992, Explanatory Memorandum* (Canberra, 1992), p. 23.

interpretation articulated in Article 31(3) of the Vienna Convention on the Law of Treaties 1969.[14]

Vertical geographic limits. The ATEP Act explicitly recognises that at least in some respects the Act embraces airspace and zones below the earth's surface. It refers to the use of aircraft[15] and goes into greater detail than the Protocol in specifying the vertical limits of Antarctic specially protected areas (ASPAs) and Antarctic specially managed areas (ASMAs). These areas may extend to the subsoil to a specified depth and the seabed including subsoil of the seabed (Section 8(4)).

The vertical extent is particularly relevant to the ban on mining 'in or on an area of the Antarctic' found in Section 19B. The explanatory memorandum of the 1992 legislation that amended the ATEP Act leaves no doubt that it was the intention of the drafters that the mining ban in Section 19B should embrace the whole of the continental shelf and the high seas (and thus deep seabed) that were not already covered by the ban in Section 19A relating to the Australian Antarctic Territory (AAT).[16]

Regulation of Jurisdiction, Control and Enforcement

As one of the seven claimants, Australia in the ATEP Act asserts both territorial jurisdiction on the basis of its claims and jurisdiction elsewhere in the Antarctic on the basis of jurisdictional links to persons and things such as ships and aircraft. In so doing, Australia is laying down, with the precision expected of domestic law, jurisdictional rules that the Protocol deliberately left undefined. Typically of the Antarctic Treaty and instruments associated with it, Article 13 on implementation of the Protocol stipulates no more than that the parties are to take 'appropriate measures'. The Protocol thus leaves unstated those for whom each party to the Protocol is responsible with regard to ensuring compliance; national legal systems cannot afford to allow such vagueness. This section will consider first assertions that are based on territorial reach over land, ice and maritime zones, then assertions defined in terms of persons both natural and juridical and persons as groups, and finally assertions defined in terms of things such as ships and aircraft.[17]

---

[14] For the application by Australian courts of the rules of interpretation of the Vienna Convention on the Law of Treaties 1969, though it seems not as yet any instances of a change in the original meaning of a treaty, see the following High Court decisions: *De L* v. *Director-General, NSW Department of Community Services* (10 October 1996) <www.austlii.edu.au/au/cases/cth/high_ct>; *'Applicant A' and Anor v. Minister for Immigration and Ethnic Affairs* (24 February 1997) *ibid.*; *Victrawl Pty Ltd v. Telstra Corporation Ltd* CLR, Vol. 183, 1995, p. 595ff, at pp. 621–622. See also D.C. Pearce, *Statutory Interpretation in Australia*, 4th edition (Sydney: Butterworths, 1996), pp. 45–47.

[15] Sections 29(2)(j) and 29(2)(a). Similarly see Reg. 11(4)(b) of Antarctic Seals Conservation Regulations as amended (Statutory Rules 1993, No. 289, as amended).

[16] Australia, Parliament, Senate (1992), p. 23.

[17] For a recent general account of Australian practice in the assertion of jurisdiction in other than

Territorial reach. With only limited exceptions, the ATEP Act has unambiguous territorial application to all persons in the Australian Antarctic Territory. The ATEP Act states that 'this Act applies in the Territory in relation to any persons and property, including foreign persons and property' (Section 4(1)(a)).[18] In spite of some more recent equivocation,[19] Australia probably still accepts the position adopted at the 1926 Imperial Conference that the land should include 'ice-barriers which are to all intents and purposes a permanent extension of the land proper'.[20] The ATEP Act makes clear that ice and even snow can be land. Section 3(1) defines 'land' to include 'ice' and 'ice' to include 'snow'. Even so, ice-covered sea would probably still be regarded as sea.

The status of ice-covered areas blurs the boundary of the major juridical significance between the land and the sea. This has not prevented Australia from asserting or declaring that the Australian Antarctic Territory generates maritime zones[21], namely internal waters, a territorial sea, an exclusive economic zone,[22] a continental shelf[23] and the potential for a fishing zone.[24] The inner limit of the territorial sea is incorporated in the measurement of one or both of the inner and outer limits of all the other zones. The currently proclaimed territorial sea of 12 nautical miles is 'measured from the baseline established under international law or as otherwise determined by Proclamation under section 7 of the Seas and

---

Antarctic contexts, see I. Shearer, 'Jurisdiction', in S. Blay, Piotrowicz and M. Tsamenyi (eds.), *Public International Law: An Australian Perspective* (Melbourne: Oxford University Press, 1997), pp. 161–192.

[18] The AAT is defined in Section 4 of Australian Antarctic Territory Act 1954 and Section 2 of Australian Antarctic Territory Acceptance Act 1933.

[19] Australia, *Parliamentary Debates*, House of Representatives, 30th Parlt., Vol. 107, 8 November 1977, p. 3,124.

[20] UK, Imperial Conference 1926, *Report of Committee on British Policy in the Antarctic*; *Memorandum E. 130 (revise)* (London, 19 November 1926) in PRO file: CAB32/47, pp. 264–266; Bush (ed.), *Antarctica and International Law* (1982–88), Vol. II, p. 103, para. 31. Similarly Orrego Vicuña, *Antarctic Mineral Exploitation: The Emerging Legal Framework* (Cambridge University Press, 1988), p. 205 and pp. 160–161. See also discussion in C.C. Joyner, *Antarctica and the Law of the Sea* (Dordrecht: Martinus Nijhoff, 1992), pp. 80–87 and 198–202 and S. Kaye, *Australia's Maritime Boundaries* (Wollongong: Centre for Maritime Policy, University of Wollongong, 1995), pp. 199–204.

[21] For doubts about the legitimacy of such zones, see Joyner, *Antarctica and the Law of the Sea*, p. 96 and also pp. 124, 140 and 253 and D.R. Rothwell, *The Polar Regions and the Development of International Law* (Cambridge University Press, 1996), pp. 275–281.

[22] Australia, *Commonwealth of Australia Gazette*, No. S 290 (29 July 1994) in Bush (ed.), *Antarctica and International Law* (1992–), Booklet AU94, pp. 55–58.

[23] Proclamation 10 September 1953 (Australia, *Commonwealth of Australia Gazette*, No. 56, p. 2,563 (11 September 1953); Bush (ed.), *Antarctica and International Law* (1982–88), Vol. 2, pp. 172–173).

[24] Proclamation under the Fisheries Management Act 1991 excepting the waters around AAT from the AFZ (Australia, *Commonwealth of Australia Gazette*, No. S 52 (14 February 1992) and Bush (ed.), *Antarctica and International Law* (1992–), Booklet AU92A, pp. 1–2) and Reg. 4(1) of Fisheries Management Regulations as amended by Statutory Rules 1998, No. 24 (Bush (ed.), *Antarctica and International Law* (1992–), Booklet AU98A, p. 85).

Submerged Lands Act 1973 from time to time'.[25] Where an ice barrier exists, such as the Shackleton and Amery Ice Shelves, the territorial sea would probably be measured from its edge as the United Kingdom asserted long ago in the case of the Ross Ice Shelf.[26] On 2 December 1999, Australia announced it would undertake a survey programme to define the limits of the physical continental shelf beyond 200 nautical miles off the Australian Antarctic Territory for submission by 2004 to the Commission on the Limits of the Continental Shelf.

The territorial sea, being part of the Australian Antarctic Territory under Australian law, the ATEP Act would apply to the activities of foreigners in that maritime area. Thus it would be an offence under Section 19(1)(a) for a foreigner there to 'interfere with a native bird' or, under Regulation 11(4)(a) of the Antarctic Seals Conservation Regulations as amended, 'intentionally or recklessly [to] injure a seal'.[27] The ATEP Act asserts a right of an inspector under the Act to search 'a foreign vessel' or a 'foreign aircraft' while it is in the Territory' (Section 17(1)–(3)).

In addition, the PS(PPS) Act, as amended in 1992 to implement Annex IV to the Protocol in conjunction with related MARPOL obligations, applies to the territorial sea off the Australian Antarctic Territory. Section 6 states that the Act 'extends to every external Territory'. The drafting scheme adopted in the PS(PPS) Act describes offences such as the following specific Antarctic offence dealing with sewage:

> ... if any discharge of untreated sewage occurs from a ship (other than a ship certified to carry not more than 10 persons) into the sea in the Antarctic Area, the master of the ship and the owner of the ship are each guilty of an offence punishable, upon conviction, by a fine not exceeding $200,000 (Section 26BC(1)).

Each such description is followed by a provision limiting the application of the offence in the case of foreign ships. The specific Antarctic offence quoted above does not apply to the discharge of sewage from a foreign ship unless the discharge occurs in the territorial sea or internal waters (Sections 26BC(2) and 3(1B)). In this way most offences under the Act apply to foreign vessels in both the territorial sea and internal waters of the Australian Antarctic Territory.

The prohibition in Section 19A of the ATEP Act on mining activities applies on a territorial basis to the continental shelf of the Australian Antarctic Territory. The section similarly applies to that part of the continental shelf of the territory of Heard Island and the McDonald Islands that extends south of the 60th parallel, but curiously not to a similar extension of the continental shelf around Macquarie Island. The Antarctic Mining Prohibition Act 1991 that the 1992 amendment to the ATEP

---

[25] Proclamation dated 9 November 1990 (Australia, *Commonwealth of Australia Gazette*, No. S 297, p. 1 (13 November 1990) in Bush (ed.), *Antarctica and International Law* (1992–), Booklet AU90, pp. 73–74 in accordance with Section 4 of the Seas and Submerged Lands Act 1973 in *ibid.*, Booklet AU94, pp. 1–7).

[26] Foreign Office file: W1985/87/50/1927 in Public Record Office file: FO371/12644, p. 280; Bush (ed.), *Antarctica and International Law* (1982–88), Vol. 3, p. 56.

[27] Statutory Rules 1993, No. 289, as amended.

Act replaced had not prohibited mining on any of the continental shelf of the territory of Heard Island and the McDonald Islands or the continental shelf of Macquarie Island.[28]

The ATEP Act is not expressed to apply to the exclusive economic zone off the Australian Antarctic Territory, although the area within that zone will be covered as a place 'outside Australia' to the extent that the Act applies to the entire Antarctic south of 60° south latitude. On the other hand, the PS(PPS) Act was amended to apply to the exclusive economic zone of the Australian Antarctic Territory that was proclaimed in July 1994.[29] For external territories such as the AAT, the outer limits of the zone are 'the lines that are 200 international nautical miles seaward of the baselines established under international law'.[30] Article 57 of the United Nations Convention on the Law of the Sea prescribes that these are 'the baselines from which the breadth of the territorial sea is measured'. The inner limit of the exclusive economic zone is described as 'beyond and adjacent to the territorial sea'.[31]

The extension of the PS(PPS) Act to the controversial Australian exclusive economic zone off the Australian Antarctic Territory is a significant assertion of Australian jurisdiction. Australian authorities are now empowered to exercise substantial control over foreign fishing and whaling vessels, re-supply ships and tourist ships within the territory's exclusive economic zone and territorial sea as well as internal waters. Thus, in the context of a suspected offence such as the discharge of sewage contrary to Section 26BC(1) quoted on page 314, Australian authorities may require information of 'a foreign ship [that] is navigating in the territorial sea or the exclusive economic zone' (Section 26G(1)). The authorities may inspect the ship (Section 27(1A)) and even detain it if 'there is clear objective evidence' of a pollution breach threatening major damage (Section 27A). When action is taken, the act requires Australian authorities to notify the flag state in accordance with Article 231 of the United Nations Convention on the Law of the Sea (Section 27B).

The ATEP Act excepts some activities of foreigners in the Australian Antarctic Territory. The application of the Act is made 'subject to subsection 4(1) of the *Antarctic Treaty Act* 1960' (Section 4(1)). This is the Australian law that exempts observers and exchanged personnel as required by Article VIII(1) of the Antarctic Treaty. A person defined as such is 'not subject to the laws in force in the Territory in respect of any act or omission occurring while he is in Antarctica for the purpose

---

[28] Act No. 43 of 1991 (Bush (ed.), *Antarctica and International Law* (1992–), Booklet AU91, pp. 6–8.
[29] Transport and Communications Legislation Amendment Act 1994 (No. 64 of 1994). Amendments made by this act are noted in footnotes to Bush (ed.), *Antarctica and International Law* (1992–), Booklet AU92A, pp. 4–55.
[30] Australia, *Commonwealth of Australia Gazette*, No. S 290 (29 July 1994) in Bush (ed.), *Antarctica and International Law* (1992–), Booklet AU94, pp. 55–58.
[31] Art. 55 United Nations Convention on the Law of the Sea (LOS Convention). Under Section 10B of the Seas and Submerged Lands Act 1973, the limits of the exclusive economic zone must be consistent with Arts. 55 and 57 of the LOS Convention.

of exercising his functions'.[32] Curiously, the PS(PPS) Act that implements Annex IV to the Protocol and the Antarctic-specific obligations of MARPOL does not include a similar saving for subsection 4(1) of the Antarctic Treaty Act 1960. In addition, the 1960 statute has effect on only 'laws in force in the Territory' (i.e., the Australian Antarctic Territory) and, in the absence of a statutory provision to the contrary, the statute would have no effect on Australian law applicable beyond the territorial sea of the AAT. This is because areas beyond the territorial sea are regarded normally as outside the scope of Commonwealth statutes applicable to a state or territory.[33] Thus, Commonwealth statutes applicable to the continental shelf of the AAT would not be laws in force *in* the AAT that are referred to in Section 4(1) of the Antarctic Treaty Act 1960.[34]

According to the explanatory memorandum associated with the 1992 amendment to the ATEP Act, the 1960 Act would not shield observers or exchanged personnel engaged 'in a mining activity' from prosecution under Australian law:

> [T]heir activities cannot legitimately extend to mining. Under Article VIII.1 such persons are outside Australia's jurisdiction for the purpose of exercising their Treaty functions. Any mining activity that was not scientific research in accordance with the Treaty would thus bring them within the operation of this legislation.[35]

This Australian interpretation of Article VIII(1) is not beyond controversy. It could just as well be said that observers or exchanged scientists who commit murder or indeed other criminal offences would be acting outside their treaty functions. The test for immunity under Article VIII(1) and Section 4(1) of the 1960 statute is whether the 'acts or omissions' occurred while the observers or exchanged scientists were 'in Antarctica for the purpose of exercising their functions'. The acts or omissions do not need to have been committed as part of their functions. Indeed acts or omissions constituting an offence are rarely likely to be part of the functions of observers or exchanged personnel. Moreover, given the close similarity between legitimate geological research and illegal prospecting,[36] it is easy to conceive how the activities of an exchanged scientist could cross the boundary to illegal mining activity. The collection of data about mineral resources with no intention of making this public or the collection of geological samples for the purpose of sale would probably both fall within the definition of forbidden mining activity in Section 3(1)

---

[32] Section 4(1) of Antarctic Treaty Act 1960 (Bush (ed.), *Antarctica and International Law* (1982–88), Vol. 1, pp. 192–193).

[33] Section 17 (definitions of 'Australia' and 'external territory') and Section 15B (application of law to coastal sea) Acts Interpretation Act 1901.

[34] Note, for example, that the application by the Sea Installations Act 1987 of the laws of the Australian Antarctic Territory to its continental shelf does not include the Antarctic Treaty Act 1960 (see sections 5(6), 45, 47 and schedule of the former).

[35] Australia, Parliament, Senate (1992), p. 23.

[36] See, for example, discussion in the context of Art. 7 of the Protocol in Bush (ed.), *Antarctica and International Law* (1992–), Booklet AU91C, pp. 23–24 and in the context of the definition of 'mining activity' in the ATEP Act in *ibid.*, Booklet AU/Rev. May–June 1980, pp. 35–36.

of the ATEP Act. In such circumstances, it is hard to understand why the government of nationality of the scientist should not be the one to exercise jurisdiction under Article VIII(1).

The Australian position is that those on private tourist and other ventures and even members of foreign official expeditions are not entitled to exemption from Australian jurisdiction under Article VIII(1) of the Antarctic Treaty.[37] Even so, the ATEP Act effectively excludes such persons to the extent that they (or anyone else) are acting under an authority issued by another contracting party. Thus, Section 7(1) states the following:

'... no action or proceeding lies against any person for or in relation to anything done by that person to the extent that it is authorized by a permit or by a recognised foreign authority.'

This is by no means a general immunity of persons on foreign expeditions from the Australian law. Any exemption applies only for activities covered by a specific authority 'issued, given or made by a Party (other than Australia) to the Madrid Protocol that has accepted under that Protocol the same obligations as Australia in relation to the carrying on of that activity in the Antarctic' (Section 3(1)). Moreover, the ATEP Act does not recognise the legitimacy of any authority to carry on activities such as keeping a dog in Antarctica, bringing into the Antarctic unsterile soil or packing material such as polystyrene beads, or permitting the escape of a bacterium or fungus that would be contrary to the Protocol and its annexes.[38] To permit any of these actions in the Australian Antarctic Territory constitutes an offence against the Act. In another respect, the ATEP Act places obligations on foreign expeditions. In spite of the ATEP Act's recognised legitimacy of authorities issued by foreign governments, Section 21(1) requires holders of such an authority to provide information on activities carried out under it such as entering ASPAs and details of birds and mammals killed or otherwise interfered with. This provision is discussed further below.

In addition to these legislatively recognised jurisdictional limits, general international law imposes varying limits on the extent to which it is permissible for any state to bind foreign nationals in each of the maritime zones off that territory. For example, Australia acknowledges the obligation to respect the right of innocent passage of vessels in the territorial sea around the Australian Antarctic Territory. On the continental shelf to which Section 19A of the ATEP Act applies and the exclusive economic zone to which the PS(PPS) Act applies, the power of a coastal

---

[37] Paras. 6–14 of opinion dated 16 September 1991 of the Attorney-General's Department in Australia, Parliament, House of Representatives, Standing Committee on Legal and Constitutional Affairs, *Reference on the Legal Regimes of Australia's External Territories and the Jervis Bay Territory; phase II—the Australian Antarctic Territory and the Territory of Heard Island and McDonald Islands; Submissions Authorised for Publication*, Vol. 3 (Canberra, 1991), pp. S391–393; Bush (ed.), *Antarctica and International Law* (1992–), Booklet AU91, pp. 84–90 at pp. 84–86.
[38] Section 19(3)(c) ATEP Act.

state is limited to what is necessary to protect the sovereign rights.[39] Any such universally recognised limitations of customary international law would probably be applied as part of the common law. Whereas rules of customary international law may be excluded by statutes, this is not the case with the ATEP Act, the application of which is subject to 'the obligations of Australia under international law' (Section 4(2)).[40] It is possible for parties to a treaty to waive their rights under customary international law and so permit the coastal state to detain their vessels in circumstances beyond those permitted by customary international law. Tentative as it is on the question of jurisdiction, the Protocol contains no such waiver.

Persons and things beyond territorial reach. Legislation extends Australian jurisdiction beyond the territorial boundaries that Australia in some way claims. In company with other claimants, Australia must assert jurisdiction extraterritorially if only to give effect to the various treaties and measures under the Antarctic Treaty system that apply to the Antarctic region that not only encompasses but also extends beyond the boundaries of any one claim. Indeed, some obligations, including reporting obligations, involve the parties' taking action well outside the Antarctic region, however defined.[41]

Any extraterritorial reach depends on the establishment of a link between Australia and the person or things over which jurisdiction is asserted. International law judges the sufficiency of that link just as it judges the legitimacy of the territorial reach but, in like manner, a link invalid under international law would still be valid under Australian domestic law if the legislative intent was sufficiently clear. Legislation can focus on individual natural or legal entities, on collective groups of people with something in common apart from their link to Australia, and finally on things such as ships and aircraft. Because in any particular fact situation a substantial overlap between the heads of jurisdiction often exists, Australian jurisdiction may well be established on several grounds.

The strongest link to justify the exercise of extraterritorial jurisdiction is, of course, the nationality of a human being. Thus the ATEP Act applies 'outside Australia in relation to ... Australian citizens' (Section 4(b)(i)). On this basis it would be possible for an Australian tourist or scientist to be prosecuted for an offence committed on the Antarctic Peninsula. The act also applies extraterritorially to legal entities. Two formulations are used that seem to have the same meaning. The formulation of general application refers to 'a corporation that is incorporated in

---

[39] An official Australian statement of these limitations is set out in paras. 5–8 of an opinion of the Attorney-General's Department of 31 May 1991 in Standing Committee on Legal and Constitutional Affairs, *Submissions Authorised for Publication* (1991), Vol. 3, pp. S361–368 and Bush (ed.), *Antarctica and International Law* (1992–), Booklet AU91, pp. 12–17.

[40] The relationship between international and Australian domestic law and the effect of Section 4(2) of ATEP is discussed below in this chapter.

[41] Discussed in Bush (ed.), *Antarctica and International Law* (1982–88), Vol. 1, p. 64.

Australia or whose activities are carried on principally in Australia'.[42] The formulation concerning the extraterritorial prohibition on mining in Section 19B applies to an 'Australian national' which is defined to include 'a body corporate that is incorporated in Australia or carries on its activities mainly in Australia' (Section 3(1)).

The ATEP Act also applies extraterritorially to a collection of natural or legal persons. Thus, in addition to a corporation, the term 'Australian organisation' is defined to include an 'unincorporated body or association the majority of whose members are Australian citizens or domiciled in Australia' (Sections 3(1) and 4(1)(b)). The same act also asserts jurisdiction over 'Australian expeditions and members of Australian expeditions' (Section 4(1)(b)). It defines an 'Australian expedition' as 'an expedition organized by one or more of any of the following:

(a) an Australian organisation;
(b) an Australian citizen;
(c) a person resident or domiciled in Australia (Section 3(1)).

This definition is wide enough to cover unofficial as well as official expeditions and would probably embrace tourist ventures, though one suspects that tourist overflights are not covered because they do not touch down in the Antarctic.

A common practice is to assert jurisdiction over people who are in some way associated with means of transport. The ATEP Act applies generally to the crew (including the person in charge) of Australian ships, aircraft, hovercraft and other vehicles (Section 4(1)(b)(iv)). A result of limiting jurisdiction to crew is that passengers (including tourists) would probably not be embraced by this head of jurisdiction, although passengers who are Australian citizens would, of course, be covered by virtue of their nationality. This is discussed further below in this chapter.

Both the ATEP Act and the PS(PPS) Act of 1983 identify 'things' as the subject of extraterritorial jurisdiction. Both acts apply to vessels registered in Australia and also, in the case of the PS(PPS) Act, to 'an unregistered ship having Australian nationality'.[43] The ATEP Act applies to aircraft registered as Australian and to both vessels and aircraft 'in Australian control'. Something is 'in Australian control' if it is . . . 'in the control or possession of one or more of' the Australian Commonwealth (including an arm of the Defence Force) or a State or Territory, a corporation established for a public purpose, a body corporate in which the Commonwealth has a controlling interest or, finally, a person such as an Australian citizen, an Australian corporation or a member of an Australian expedition.[44] With the notable exception of the Australian flagged *Aurora Australis* commissioned in 1990, it has been

---

[42] Section 3(1), definition of 'Australian organization', and use of that term in Section 4(1).
[43] ATEP Act Section 3(1), definition of 'Australian property', and PS(PPS) Act Section 3(1), definition of 'Australian ship'.
[44] Sections 3(1) (definition of 'in Australian control') and 4(1)(b) ATEP Act.

Australia's practice to charter foreign re-supply vessels. These have been under the control of the Commonwealth.

The application of the PS(PPS) Act to Australian vessels is consistent with one of the few clear statements of jurisdictional scope in Antarctic instruments: Article 2 of Annex IV on prevention of marine pollution of the Protocol acknowledges that 'with respect to each Party,' the Annex applies not only to 'ships entitled to fly its flag' but also to 'any other ship engaged in or supporting its Antarctic operations, while operating in the Antarctic Treaty area'.

Beyond ships and aircraft, the ATEP Act asserts extraterritorial jurisdiction over property other than vessels and aircraft that 'is in Australian control'.[45] Property is defined widely to mean 'property of any description' and can include vehicles among which hovercraft are counted.[46]

Assertion of jurisdiction over a thing suggests an asserted competence to affect property rights invested in the thing or otherwise to control it. In addition, assertion over a thing can be associated with assertion of jurisdiction over persons associated with it. Traditionally, a state's jurisdiction over vessels flying that state's flag extends to the exercise of jurisdiction over all persons on board (whether nationals of the flag state or not). Whether similar consequences flow from a statutory assertion of jurisdiction over a thing will depend on an interpretation of the statute. A statute such as Section 4(1)(b)(iv) of the ATEP Act that expressly gives jurisdiction over a vessel in some way linked to Australia and its crew would probably not be interpreted as an assertion of jurisdiction over passengers; in the absence of a contrary indication, the statute would be a case for the application of the rule of thumb *expressio unius exclusio alteris.*

Enforcement. The remoteness and inhospitality of the Antarctic pose problems for enforcement of Australian environmental legislation; this predicament is compounded by the lack of general agreement on acceptable grounds upon which parties to the Protocol may exercise jurisdiction.[47] Steps such as constituting station leaders as inspectors under Section 13 of the ATEP Act and special constables under Section 4(1) of the Criminal Procedure Ordinance 1993 overcome only some of the practical difficulties in enforcement. The lack of agreement on jurisdiction seriously impedes the development of cooperative arrangements with other parties on matters such as taking evidence and joint policing of protection measures that are necessary to give the legislation teeth.

Presently, Australia is able to exercise effective control over persons taking part in Australian official expeditions because the ATEP Act applies to the activities of those expeditions and as official expeditions still constitute the greatest bulk of

---

[45] Sections 3(1) (definition of 'Australian property') and 4(1)(b)(v) ATEP Act.
[46] Sections 3(1) (definitions of 'property' and 'vehicle') and 4(1)(b)(iv) of *ibid.*
[47] D.R. Rothwell, 'Australian and Canadian Initiatives in Polar Marine Environmental Protection: A Comparative Review', *Polar Record*, Vol. 34, 1998, pp. 305–316 at p. 311.

Australian Antarctic activity, it can be said that the act is effective. Ironically, other means of control exist for those expeditions that would have allowed Australia to give effect to the Protocol without the need of the ATEP Act. These means of control include powers of direction and disciplinary action against persons who may be public servants. For persons who are not public servants, sanctions may be applied for breach of contractual commitments whereby expeditioners[48] may be returned to Australia. There is also the likelihood that the Antarctic Division would not support the future research of persons who had previously failed to follow environmental guidelines. The ability of Australia to manage without legislation that specifically implements the Protocol is illustrated by the delay in commencement of the PS(PPS) Act to implement Annex IV to the Protocol. In the meantime, the Australian authorities secure compliance administratively by the few vessels concerned, for example by securing the inclusion of appropriate terms in the charter party. With respect to the prevention of marine pollution, the Antarctic Division has reported that

> The ships that Australia charters to support its Antarctic program already meet the majority of the requirements arising from Annex IV and modifications are being made where necessary. Marine pollution contingency plans have been developed for both ships and are being finalised for each of Australia's Antarctic stations. As part of this planning, oil pollution response equipment has been placed on the ships and at the stations. Study of further equipment requirements is continuing.[49]

Substantial private activity is what will put Australian legislation to the test. At present Australia has more influence on tourism than would be expected considering the country's relatively small population. Recent figures show that Australia ranks third with the United Kingdom as a source of tourists, with some ten per cent of all Antarctic tourists being Australians. Of even more significance, it is estimated that currently between 15 and 17 per cent of sea-based tourism travels on non-Australian ships either chartered or sub-chartered by Australian operators.[50] In addition, there are Australian companies who fill berths booked on voyages operated by non-Australian companies. This industry pattern means that the ATEP Act under which the Australian operators are authorised has application to much tourism operating from South American ports to the Antarctic Peninsula area. Despite Australia's significant influence on tourism, operators could rearrange their affairs so as to sever this link to Australia if the Antarctic Division came to be perceived as too zealous in the enforcement of environmental regulations. On the other hand, the Antarctic Division has no formal control over the few tourist voyages to the Australian Antarctic Territory by operators such as Quark Expeditions authorised by other treaty parties. Even fewer voyages call at Australian ports. It is probably the

---

[48] See ANARE code of personal behaviour (Bush (ed.), *Antarctica and International Law* (1992–), Booklet AU87, pp. 22–29.

[49] See <www.antdiv.gov.au/environment/> as at 29 December 1999.

[50] Information from Mr. Martin Betts, Australian Antarctic Division, 17 December 1999.

goodwill and cooperation that presently exists between the Antarctic Division and private operators rather than the risk of prosecution that is most effective in securing compliance with Australian environmental regulation. The continuing growth in tourism and other private ventures is likely to strain this goodwill and cooperation.

Type of Activities Covered

The range of activities covered by the ATEP Act is broad. Within the Act's reach, those proposing 'to carry on an activity' are obliged to follow the environmental impact assessment procedures of part 3.[51] The ATEP Act draws no distinction between private and official activities in that the legislation is expressed to bind 'the Crown in right of' the Australian Commonwealth, each of the states and the self-governing territories such as the Australian Capital Territory (Section 6). Accordingly, members of expeditions, whether private or official, would commit an offence under the act if, for example, they carry on an activity without an authorisation or permit (Sections 9, 19(3)(c) and 21A), breach the condition of a permit (Section 20), keep a dog in the Antarctic (Section 19(1)(caa)), or carry on mining activity (Sections 19A and 19B).

Only a few activities are excluded from this broad net. The minister is empowered to declare the environmental impact assessment procedures inapplicable to classes of activity where, 'because of their nature, those activities are likely to have no more than a negligible impact on the Antarctic environment' (Section 12C(1)). As discussed above, activities authorised by other parties to the Protocol are largely exempt from proceedings under the ATEP Act (Section 7(1)).

It is arguable that the ATEP Act excludes other activities. These exclusions could derive from a number of key provisions of the Protocol (though not explicitly from Annex II) being confined to 'activities undertaken in the Antarctic Treaty area pursuant to scientific research programmes, tourism and all other governmental and non-governmental activities in the Antarctic Treaty area for which advance notice is required in accordance with Article VII(5) of the Antarctic Treaty, including associated logistic support activities'.[52] An Australian flagged vessel transiting the Southern Ocean or even an Australian aircraft following a polar route over the continent between points outside the Antarctic are examples of activities for which the Antarctic Treaty does not require advance notice. In contrast, the Treaty requires advance notice for a vessel or aircraft heading to or from the Antarctic continent. For this reason Australia has subjected tourist overflights in Qantas aircraft to assessment procedures. The Treaty does not require advance notice of commercial fishing in the Southern Ocean. On the other hand, it is Australian practice to give

---

[51] See, for example, Section 12D.
[52] Arts. 3(4), 8(2) and 15(1)(a) of the Protocol and Annex III, Art. 1(1).

advance notice under the Antarctic Treaty of marine research in the Southern Ocean.[53]

Unlike implementing legislation of some other Protocol parties,[54] the scope of Australian legislation does not expressly exclude transiting or commercial fishing, but it is arguable that such exclusions may be implied. The provisions of the ATEP Act to which appeal could be made are Sections 4(2) and 3(5). These provisions were discussed above in the context of whether there was an implication that the high seas exception of Article VI of the Antarctic Treaty limited the geographic scope of the ATEP Act. It may also be possible to appeal to the intention recited in the preamble that the legislation was 'to make provision for giving effect to' the Environmental Protocol. Section 4(2) makes the act subject to the 'obligations of Australia under international law, including obligations under any international agreement binding on Australia' and 'any law of the Commonwealth giving effect to such an agreement'. Section 3(5) of the ATEP Act specifies that 'an expression that is used in both this Act and either the [Antarctic] Treaty or the Madrid Protocol (whether or not a particular meaning is assigned to it by the Treaty or the Protocol) has, in this Act, the same meaning as in the Treaty or the Protocol, as the case may be'.

The point under discussion could arise in, for example, defending a prosecution of a vessel or aircraft transiting the Southern Ocean or the Antarctic continent for failing to follow the environmental assessment procedures set out in part 3 of the ATEP Act. It would be arguable that part 3 was limited to implementing the Protocol requirements on environmental impact assessment without the introduction of an additional element. In support of this proposition, reference could be made to the explanatory memorandum that states that part 3 was intended to 'embody the general principles of Article 8 of the Protocol and Annex I'. This would be evidence that part 3 was not among those elements of the legislation designed, in the words of the preamble of the Act, 'to make other provision relating to the protection of the environment in the Antarctic' and thus not be intended to go beyond the Protocol. Relying on Section 3(5), the next step would be to argue that the 'activities' that part 3 focused upon should be read in the same sense of the term 'activities' used in the Protocol, which in Article 8(3) limits the term to those activities for which advance notice must be given under Article VII(5) of the Antarctic Treaty. Were a court to uphold such an argument, it would be limited to those obligations of the Protocol that are limited by reference to Article VII(5). Non-interference with birds and mammals laid down by Annex II to the Protocol is an example of an obligation that does not appear to be limited by reference to Article VII(5). Whatever the force of

---

[53] This is consistent with Recommendations VI–13 and VIII–6, Annex, para. XV.

[54] E.g. Section 5(2) Antarctic Act 1994 (1994 c. 15) (UK) and Art. 3(2) Gesetz zur Ausführung des Umweltschutzprotokolls zum Antarktis-Vertrag (Germany). The German legislation excludes 'research on or use of' marine living resources and seals associated with legislation enacting the seals and marine living resources conventions.

these arguments, the conclusion must be that the scope of the activities embraced by the ATEP Act is not as clear as it should be.

In contrast to this uncertainty of the ATEP Act, the PS(PPS) Act clearly applies to all activities of any vessel within the geographic and jurisdictional scope of the Act. Thus fishing, marine research, Antarctic re-supply vessels, tourist vessels and transiting container ships are all bound by its provisions implementing Annex IV. The Act contains dispensations that appear to reflect dispensations found in the Protocol, such as the dispensation in Section 26BC in favour of ships certified to carry not more than 10 persons in relation to discharge of sewage.[55]

PROCEDURAL AND INSTITUTIONAL ISSUES

National Legislative Process and Institutions Involved

The Commonwealth of Australia is a federation of six states with legislative competence divided between the federal and state legislatures. Under the Commonwealth constitution the federal parliament may make laws for external affairs and for territories.[56] These heads of power are the ones most relevant to legislation of the federal parliament on Antarctic matters. The AAT is one of Australia's external territories that comprise every territory of the Commonwealth other than one of the three 'internal' territories, namely the Australian Capital Territory, the Jervis Bay Territory and the Northern Territory.[57] The AAT was placed under the authority of the Commonwealth by a British Order-in-Council of 1933[58] and accepted by the Commonwealth in the Australian Antarctic Territory Acceptance Act 1933.[59]

The ATEP Act is an example of a statute of the federal parliament that is applicable both to the AAT as an Australian territory and extraterritorially to Australian activities elsewhere in the Antarctic. The regulations made by the executive government under the ATEP Act are a form of subsidiary legislation. Ordinances for the territory made by the Governor-General under the Australian Antarctic Territory Act 1954 are another form.[60] Relatively few ordinances have

---

[55] See Art. 6 of Annex IV. Note also reference in Art. 3 of the same Annex to permitted discharge of oil or oily mixture under Annex I of MARPOL 73/78.

[56] Sections 51(xxix) and 122 of the Australian Constitution.

[57] Section 17 of the Acts Interpretation Act 1901.

[58] Australia, *Commonwealth of Australia gazette*, 1933, No. 15, p. 365 (16 March 1933); Bush (ed.), *Antarctica and International Law* (1982–88), Vol. 2, pp. 142–143.

[59] Act No. 8 of 1933 in Australia, *Commonwealth Acts 1901–1950*, Vol. 1, p. 227; Australia, *Commonwealth Acts, 1933*, p. 12; Bush (ed.), *Antarctica and International Law* (1982–88), Vol. 2, pp. 146–147.

[60] Section 11 of the Australian Antarctic Territory Act 1954, reprinted as at 30 September 1993; Bush (ed.), *Antarctica and International Law* (1992–), Booklet AU/Rev 1954, pp. 13–17.

been made. The one of most relevance to the enforcement of the ATEP Act is the Criminal Procedure Ordinance 1993.[61]

Since about 1980, much federal environmental and related legislation of general application to Australia has been expressed to apply to Australian external territories and thus to the AAT. Examples are the Wildlife Protection (Regulation of Exports and Imports) Act 1982[62] the Sea Installations Act 1987[63] and, notably, the omnibus Environment Protection and Biodiversity Conservation Act 1999.[64] The 1999 Act consolidates and rationalises earlier environmental legislation implementing treaties such as the Convention on Biological Diversity adopted at Rio de Janeiro on 5 June 1992, the Convention on the Conservation of Migratory Species of Wild Animals adopted at Bonn on 23 June 1979 and the Convention for the Protection of the World Cultural and Natural Heritage adopted at Paris on 23 November 1972. Within this constellation of legislation, the ATEP Act, specific as it is to the Antarctic, is the Australian environmental legislation that is of most practical application to the region.

Legislation Implementing the Protocol and Australian Implementation of Treaties Generally

The ATEP and PS(PPS) Acts are typical examples of the manner in which Australia implements treaties in domestic law. Consistent with the Westminster system, the executive arm of the federal government has authority to commit Australia to treaties but, with few exceptions, this action will not alter domestic law. Parliament has no constitutional role in the treaty-making process. To avoid the embarrassment of being in breach of a treaty obligation when domestic law does not permit effect to be given to it, the practice is for the executive arm of government to defer indicating Australian consent to be bound to a treaty until the domestic law has been brought into conformity with the treaty obligation.[65] Australian courts adopt a different approach to the relationship between customary international law and domestic law. Unlike the position adopted by English courts, customary international law is not automatically received as part of domestic law. Instead, Australian courts are prepared to refer to customary international law as a 'source' of or 'influence' on the

---

[61] Ordinance No. 2 of 1993; Bush (ed.), *Antarctica and International Law* (1992–), Booklet AU93, pp. 75–106.

[62] Act No. 149 of 1982; Bush (ed.), *Antarctica and International Law* (1992–), Booklet AU/REV 1982–86, pp. 2–13.

[63] Bush (ed.), *Antarctica and International Law* (1992–), Booklet AU87, p. 34–37.

[64] Act No. 91, 1999. It is unlikely to commence until July 2000.

[65] For a brief description of Australian treaty-making practice and of the initiation of procedures to give Parliament some oversight of the process, see J. Linehan, 'The Law of Treaties,' in Blay, Piotrowicz and Tsamenyi (eds.), *Public International Law: An Australian Perspective*, pp. 95–118, at pp. 111–114.

law of Australia and, it seems, the courts regard themselves as bound to apply universally recognised principles of customary international law.[66]

Courts recognise that Parliament may legislate contrary to international law although, wherever possible, courts will interpret a statute so as to be consistent with established rules of international law.[67] The ATEP Act is unusual in that Parliament has stated in Section 4(2) that the Act is subject to international law, thus going well beyond that rule of statutory interpretation. In doing so, Section 4(2) recognises that court should regard treaties differently from customary international law:

> 'This Act has effect subject to:
> (a) the obligations of Australia under international law, including obligations under any international agreement binding on Australia; and
> (b) any law of the Commonwealth giving effect to such an agreement'.

The ATEP Act is thereby made subject to treaties that are implemented by any earlier or later statute. The Act is also subject to other international obligations (being customary rules) whether or not a statute has given effect to them. The section thus forestalls any argument that Parliament may have intended to enact a provision in the ATEP Act contrary to international law. For example, Section 4(2) would ensure that the ATEP Act is not given a jurisdictional reach in relation to maritime zones that exceeds Australia's competence under customary international law as perceived by Australian courts.

The ATEP and PS(PPS) Acts largely translate the Protocol obligations into statutory language. A less common method of treaty implementation is to enact that the language of the treaty itself will have the force of domestic law. To a small extent this is done in both the statutes in question by stating that the meaning of terms should generally correspond with the meaning of similar terms used in the treaties. For example, Section 3(5) of the ATEP Act states:

> Except so far as the contrary intention appears, an expression that is used in both this Act and either the [Antarctic] Treaty or the Madrid Protocol (whether or not a particular meaning is assigned to it by the Treaty or the Protocol) has, in this Act, the same meaning as in the Treaty or the Protocol, as the case may be.[68]

This method of treaty implementation allows Australian courts to refer to the treaty texts to interpret the statute as canvassed above.

The ATEP Act is also notable in requiring the Minister, in the exercise of discretions to grant permits and the like under that statute, to decide 'in a manner that is consistent with the basic environmental principles' which are those set out in

---

[66] R. Balkin, 'International Law and Domestic Law', in Blay, Piotrowicz and Tsamenyi (eds.), *Public International Law: An Australian Perspective*, pp. 119–145, at pp. 121–123.
[67] *Ibid.*, pp. 132–33.
[68] Similarly see Section 26BA PS(PPS) Act.

Article 3 of the Protocol.[69] This direction in the statute goes beyond the regard that the Australian High Court has ruled that a decision maker should have for a treaty in the absence of that language.[70] A result of this language in the statute is to leave the way open for private litigants to challenge decisions under administrative law on the ground that the decisions were not made in accordance with the correct interpretation of complex Article 3 of the Protocol.

The Domestic Institutions Involved in Implementation of the Legislation

The agency of government in Australia that has long had prime authority for Antarctic affairs is the Antarctic Division. Having started its life in the 1940s in the then Department of External Affairs, the Division, after various moves, was transferred in 1987 to its present location in the department responsible for the environment. The administration of the ATEP Act and other Antarctic specific legislation such as the Antarctic Treaty Act 1960 is allocated to the minister responsible for the department.[71] The government rejected a bid from the then National Parks and Wildlife Service to administer the ATEP Act.[72] This would have left the Antarctic Division with its role as manager of the Australian Antarctic programme and administrator in other respects of the AAT. The Antarctic Division has put in place a number of measures such as arranging for external scrutiny of environmental assessments to reduce the possibility of conflicts of interest between its programme management and environmental regulatory functions.[73]

Although the National Parks and Wildlife Service was not given general responsibility for Antarctic environmental management, it was made responsible for the protection of cetacea – animals of particular relevance to the Southern Ocean – under the Whale Protection Act 1980. Cetacea are currently administered within the 'biodiversity group' of the Department of the Environment and Heritage, and thus within the same department as the Antarctic Division. Similar arrangements are expected to continue when replacement legislation, the Environment Protection and Biodiversity Conservation Act 1999, commences.

---

[69] Sections 7A, 9(2), 12J(2), 12L(2) and (3), 12N and 3(1) (definition of 'basic environmental principles') ATEP Act.

[70] *Minister for Immigration and Ethnic Affairs v. Teoh* (1995) 128 ALR 353 discussed in Balkin, 'International Law and Domestic Law', pp. 136–138.

[71] Details administrative arrangements orders are given in Bush (ed.), *Antarctica and International Law* (1992–), Booklet AU/Rev 1909–48, pp. 3–5.

[72] Paras. 4.28–4.36 of Australia, Parliament, House of Representatives, Standing Committee on Legal and Constitutional Affairs, *Australian Law in Antarctica; The Report of the Second Phase of the Inquiry Into the Legal Regimes of Australia's External Territories and the Jervis Bay Territory, November 1992* (Canberra: Australian Government Publishing Service, 1992); Bush (ed.), *Antarctica and International Law* (1992–), Booklet AU92B, pp. 1–39.

[73] For a discussion of these measures see W.M. Bush, 'Australian Environmental Legislation and the Antarctic: The Meeting of International and Domestic Law and Politics', *Antarctic Project Reports*, No. 8 (Lysaker: Fridtjof Nansen Institute, 1999), section 3, pp. 21-29.

The growth in recent years of activity and interest in the Antarctic has led to an increasingly complex legal and administrative environment. For reasons of efficiency, the government has allocated principal authority for two important issues, namely fisheries and pollution from ships, to government agencies other than the Antarctic Division. The growth of Australian interest in commercial fishing in the Southern Ocean has led to the Australian Fisheries Management Authority (AFMA), which works under the Fisheries Management Act 1991, to assume an important role for Australia in the commission for the conservation of Antarctic marine living resources. Australian commercial activity has so far been confined to the area around the territory of Heard Island and the McDonald Islands and around Macquarie Island. All these islands lie north of the area of application of the ATEP Act. The role of the fisheries authority leaves the responsibility of the Antarctic Division for the ATEP Act unaffected beyond the sort of interactions that exist between the Protocol and the Antarctic marine living resources regimes on matters such as CEMP sites (provided for in Section 9(2B) of the ATEP Act) and the overlap of responsibility implicit in the wide scope of both treaty regimes that is reflected in, for example, the broad definition of marine living resources in the convention (Article 1(2) of CCAMLR) and the obligation on the commission to take an ecosystem approach. The coordination of Australian policy is facilitated by close cooperation between the Division and AFMA, by the Division's continuing responsibility for marine scientific research (including research into marine living resources regulated by the Antarctic Marine Living Resources Conservation Act 1981) and by the Division's leadership of Australian delegations to the commission.

Under the PS(PPS) Act, considerations of efficiency also led the government to entrust the administration of Annex IV of the Madrid Protocol to the Australian Maritime Safety Authority (AMSA) rather than give it to the Antarctic Division under the ATEP Act. Australian expertise in the regulation of shipping resides in the authority. Through the PS(PPS) Act, the Maritime Authority is also charged with implementation of Australian obligations under MARPOL instruments of the International Maritime Organization.

ASSESSMENT

Effectiveness of Domestic Legislation and Institutions in Implementing the Protocol

In the absence of a general audit, it is difficult to assess the effectiveness of Australia's efforts to implement the Protocol, but the indications are that within the scope of Australian official activities its efforts are indeed effective. The changes in legislation and subsidiary legislation listed at the beginning of this chapter were only one element of the changes required to bring Australian practice into concordance with the Protocol.[74] We have seen that the Antarctic Division administers all of this legislation other than the PS(PPS) Act. The division has developed or revised a large array of manuals, guidelines and management plans to facilitate compliance with both the legislation, provisions of the Protocol and related guidelines developed by COMNAP. Examples of this effort include management plans for the three Australian bases of Mawson,[75] Davis[76] and Casey;[77] guidelines for the preparation of initial and comprehensive environmental evaluations,[78] a waste management strategy[79] and an environmental policy statement on quarrying in the Antarctic.[80] Australia has not yet found cause to require anyone to prepare a comprehensive environmental evaluation under the Protocol. In contrast, each year proponents prepare 80 to 90 preliminary assessments of environmental impacts[81] and up to the beginning of 1999 about 13 initial environmental evaluations.[82]

Documents such as manuals play an important role in disseminating information about what may be termed static obligations of the Protocol and its annexes, such as dogs not being allowed in the Antarctic. These documents also inform about dynamic obligations to move to higher environmental standards and describe the bounds within which decision-makers under the legislation should exercise their discretion. The management plans of bases and their environs are relevant to dynamic obligations in providing a framework for the approval and conduct of activities. In the case of higher or evolving standards, there is implicit interim tolerance of activities or situations that do not meet those higher or evolving standards. Examples are obligations to 'clean up' pre-existing waste disposal sites and to 'develop' as well as monitor waste management plans and to meet evolving

---

[74] See <www.antdiv.gov.au/environment/> as at 29 December 1999.

[75] Australian Antarctic Division, *Mawson Station, Mac. Robertson Land – Antarctica, Management Plan* (Kingston: Australian Antarctic Division, 1993); see <www.antdiv.gov.au/­aad/p&p/ems/mawson/mawson_manag.htm>.

[76] Australian Antarctic Division, *Davis Station, Ingrid Christensen Coast – Antarctica, Management Plan* (Kingston: Australian Antarctic Division, 1993); Bush (ed.), *Antarctica and International Law* (1992–), Booklet AU93, pp. 8–48.

[77] See <www.antdiv.gov.au/aad/p&p/ems/casey.html>; Bush (ed.), *Antarctica and International Law* (1992–), Booklet AU97A, pp. 67–108.

[78] See <www.antdiv.gov.au/environment/index.html>; Bush (ed.), *Antarctica and International Law* (1992–), Booklet AU93, pp. 49–63.

[79] M. Arens, *A Waste Management Strategy for Australia's Antarctic Operations Prepared* (Kingston: Australian Antarctic Division, 1994); Bush (ed.), *Antarctica and International Law* (1992–), Booklet AU90, pp. 26–55.

[80] Australian Antarctic Division, *Quarrying in the Antarctic: An Environmental Policy Statement* (1996) in Bush (ed.), *Antarctica and International Law* (1992–), Booklet AU/Rev. 1995–96, pp. 1–14.

[81] Australian Antarctic Division, 'Impact Assessments of Australian Activities in Antarctica –1989 to present', at <www.antdiv.gov.au/aad/p&p/ems/aust_iees.html>, viewed 17 June 1998.

[82] Bush (ed.), *Antarctica and International Law* (1992–), Booklet AU/Rev May–June 1990, p. 57; see <www.antdiv.gov.au/aad/p&p/ems/aust_iees.html>. Examples are reproduced in D.R. Rothwell and R. Davis, *Antarctic Environmental Protection: A Collection of Australian and International Instruments* (Leichardt, NSW: Federation Press, 1997), pp. 279–295 and Bush (ed.), *Antarctica and International Law* (1992–), Booklet AU90, pp. 49–75.

emission standards for incineration (Annex III Articles 1(5), 3(1) and 10(a)). Administrative guidelines on these matters supplement related obligations in the Antarctic Treaty (Environment Protection) (Waste Management) Regulations.

In the light of the evolving nature of many Protocol obligations, actual practice is always likely to lag. The Antarctic Division recognises this lag and asserts that it has made 'significant progress' in addressing areas of Australian Antarctic practice that have required action and that it has identified areas where further action is required.[83] The identification of 'gaps' is not necessarily the same as identification of breaches. The approach is reflected in the laudatory report of Davis station conducted in 1995 by a United States inspection team. Comments that might be taken as negative fell into category of 'gaps'. They included lack of monitoring, treatment or filtering of air emissions from the power plant, no use of alternative energy sources, the presence of a halon fire suppression system and the existence of 'numerous roads ... extending for an undetermined distance into the Vestfold Hills'.[84] The continuing evidence, such as the roads, of earlier unremediated unsatisfactory environmental activity was also referred to in an interim report of Belgian and French inspectors who visited Mawson, Davis and Casey stations at the beginning of 1999:

> '... the Australian operators of these facilities are paying special attention to measures aiming at the protection of the Antarctic environment and its ecosystems. However, the situation of the abandoned station of Wilkes and adjacent old tip site raises the question of the cleanup of contaminated sites and removal of old facilities from the Antarctica Treaty area as provided for by the Madrid Protocol'.[85]

It is clear from the jurisdictional reach of Australian legislation that Australian effectiveness in implementing the Protocol cannot be judged only by the conduct of official expeditions. The pretence of application to virtually everyone in the AAT means that foreign official expeditions as well as private tour operators come within the ambit of the Australian legislation. Even so, as a matter of policy Australia does not seek to enforce its law against members of a national expedition of another party to the Antarctic Treaty.[86] As inspection reports reveal, foreign bases within the Australian Antarctic Territory are maintained to different standards from the standards maintained at Australian bases.[87] Australia has had some experience of

---

[83] From <www.antdiv.gov.au/environment/> as at 29 December 1999.

[84] US, Department of State and Arms Control and Disarmament Agency, *Report of the United States Antarctic Inspection February 9 to March 11, 1995* (1995), p. 32, appended to doc. XX ATCM/INF 129, 1996.

[85] Belgium and France, 'Joint Inspection in Eastern Antarctica Conducted in 1999 by Belgium and France under Article VII of the Antarctic Treaty', doc. XXIII ATCM/ IP 42, 1999, para. 5.

[86] Para. 9 of opinion dated 9 September 1991 of the Department of Foreign Affairs and Trade in Standing Committee on Legal and Constitutional Affairs, *Submissions Authorised for Publication*, 1991, Vol. 3, pp. S391–393, in Bush (ed.), *Antarctica and International Law* (1992–), Booklet AU91, pp. 80–82.

[87] See comments on Mirniy and Zhongshan in *Report of the United States Antarctic Inspection 1995*.

placing official observers on tourist voyages[88] but, as already discussed earlier in this chapter, Australia may have few practical means of enforcing legal controls over an uncooperative operator.

In sum it can be said that Australia is effective in implementing the Protocol for its official activity in the Antarctic. Beyond that, the ATEP Act is probably ineffective. Australia cannot take credit for compliant behaviour by other foreign official expeditions within the AAT even though the ATEP Act extends to these; Australia does not seek to regulate environmental compliance by those expeditions. The ATEP Act is thus in practice irrelevant to such expeditions. Multinational tourist ventures visiting the AAT would generally have approvals of other parties. There are presently a substantial number of Australian private operators who are called on to comply with Australian legislation. If an operator regards Australia as too zealous, it can easily arrange its affairs to escape much of the rigour of Australian regulation and enforcement action. Even in the case of Australian official expeditions where the ATEP Act is applied, the Act is probably unnecessary in that Australia, through the Antarctic Division, could implement the Protocol by other means. As mentioned above, the Antarctic Division has other powers and leverage by which to encourage and enforce compliance with its guidelines and other administrative directions. Small-scale Australian private ventures are few and likely to be cooperative. The consequence of all this is that the ATEP Act is more a show pony than a work horse in the stable of the Protocol.

Contributions to Antarctic Law and Policy: Consistency and Additions

This final section will describe some elements of the Australian legislative scheme for implementation of the Protocol and the annexes to the Protocol that are not explicitly enjoined by those instruments. Whereas no aspect of Australian implementation seems contrary to these international instruments,[89] there are many examples of where Australia has exercised judgement permitted by the instruments to implement obligations in a manner not specified in them. There are also Protocol obligations that are not embodied in domestic legislation on the grounds, presumably, that the effect can be given to the international instruments without the need for legislation.[90] In other respects Australia has more precisely defined obligations contained in the instruments.

---

[88] Observer report for Voyage 1 *Kapitan Khlebnikov* to Ross Sea area, 29 December 1994 until 20 January 1995, in Australia, 'Monitoring of Tourist Activities by a Shipboard Observer', doc. XIX ATCM/INF 33, 1995.
[89] Note, though, Australia's questionable assertion that foreign exchanged personnel who breach the ban on mineral resource activities would be subject to Australian jurisdiction. This relates to the interpretation of Art. VIII(1) and the scope of Antarctic Treaty Act 1960 rather than to an interpretation of the Protocol.
[90] Emergency response action under Art. 15 is an example.

Article 13 of the Protocol gives parties to the Protocol immense discretion to determine themselves the 'appropriate measures' they will take to implement the Protocol. The first paragraph of Article 13 states that 'Each Party shall take appropriate measures within its competence, including the adoption of laws and regulations, administrative actions and enforcement measures, to ensure compliance with this Protocol'.

The above-described jurisdictional reach that Australia has set as a claimant is a key area on which Article 13 of the Protocol is mute. That jurisdictional reach is challengeable on familiar grounds by non-claimants. Australia has chosen the criminal law to underpin many of the Protocol obligations (Sections 19–23). Article 13 has provided for an enforcement system that regularly appoints inspectors, such as station and voyage leaders, who are endowed with coercive powers (Sections 13–18).

Australia has amplified in a number of ways the single, most significant prohibition of the Protocol, that on '[a]ny activity relating to mineral resources, other than scientific research . . .' (Article 7). The ATEP Act defines both 'mineral' and 'mining activity', key terms used in Sections 19A and 19B to give effect to Article 7; this is not done in the Protocol. The ATEP defines 'mining activity' to mean 'an activity carried on for, or in connection with, the recovery or exploitation of minerals (including prospecting and exploring for minerals); but the definition does not include an activity that is necessary for scientific investigation or scientific research within the meaning of the Treaty' (Section 3(1)). The intended scope of this definition is further amplified in the explanatory memorandum of the 1992 amendments.[91] More recently, Australia seems to have supported a broad interpretation of the scientific research exception. A 1997 report to the Government on the future of Australia's Antarctic programme referred to the need 'to determine the mineral and petroleum resource potential of Antarctica' for the purpose of 'strategic management and decision-making'.[92] The government did not dissent from the report on this point.[93]

It was concluded above in this chapter that the mining ban in the ATEP Act most probably includes the deep seabed as well as other areas south of 60° south latitude. This seems the scope intended by the executive government and as such reveals an Australian interpretation of the Protocol that is not shared by all its parties. Australia's view that, in spite of Article VIII(1) of the Antarctic Treaty,

---

[91] Australia, Parliament, Senate (1992), p. 6, quoted in Bush (ed.), *Antarctica and International Law* (1992–), Booklet AU/Rev. May–June 1980, p. 35.

[92] Australia, Antarctic Science Advisory Committee, *Australia's Antarctic Program Beyond 2000: A Framework for the Future; a Report to the Parliamentary Secretary for the Antarctic by the Antarctic Science Advisory Committee* (Hobart: Department of the Environment, 1997) para. 5(11); see <www.antdiv.gov.au/foresight/>; Bush (ed.), *Antarctica and International Law* (1992–), Booklet AU97B, pp. 13–48.

[93] Bush (ed.), *Antarctica and International Law* (1992–), Booklet AU98A, pp. 88–92; see <www.antdiv.gov.au/asac/asac_resp.html>.

exchanged scientists, observers and their staff would be subject to the Australian mining ban within the AAT is likely to be even more controversial (see page 316 above).

The role described at page 326 that the ATEP Act accords to the environmental principles of Article 3 of the Protocol reveals an Australian view of that complex article. Australia rejected the alternative courses of omitting those principles from the legislation or, at the other extreme, making it an offence to carry on an activity that fails to comply with them. Instead, the ATEP Act requires decision makers to exercise their many discretions under the ATEP Act in a manner consistent with the principles (Section 7A). This will allow those aggrieved by a decision to challenge the decision on administrative legal grounds that the decision did not have regard to or was otherwise inconsistent with the principles.

The scheme of environmental impact assessment established in the Australian legislation includes a number of interesting extensions and amplifications of the Protocol and Annex I. Part 3 of the ATEP Act, together with the related regulations[94] and guidelines,[95] describe not only the procedures for comprehensive environmental evaluations that Annex I also specifies in some detail, but it also describes procedures for the initial environmental evaluations that Annex I only lightly touches on and preliminary assessments that Article 1 of Annex II leaves entirely to 'appropriate national procedures'. Australian legislation and practice is based on the view that the environmental impact assessment requirements of the Protocol do not apply to pre-existing activities unless there is an increase or decrease in intensity of them (Article 8(3)). This is reflected in Section 12D that requires preliminary assessments only if 'after the commencement of' the provisions, a person 'proposes to carry on an activity' or 'a change is proposed or occurs'. For example, the Antarctic Division has issued the following environmental policy statement on quarrying:

> 'The Madrid Protocol does not envisage, nor does the AT(EP) Act require, environmental assessment of pre-existing activities. Article 8.3 of the Protocol, however, requires that an increase or decrease in the intensity of an existing activity be subject to assessment. Accordingly, there is no legal requirement or international obligation to undertake environmental evaluation of quarrying activities that were ongoing on the date the Protocol was signed, or at the time the legislation commenced, and are continuing at that level'.[96]

---

[94] Statutory Rules 1993, No. 115.

[95] Australian Antarctic Division, 'Australian Guidelines for Preparation of Initial and Comprehensive Environmental Evaluations' (1993), in Bush (ed.), *Antarctica and International Law* (1992–), Booklet AU93, pp. 49–63; see <www.antdiv.gov.au/environment/ index.html>.

[96] Australian Antarctic Division, *Quarrying in the Antarctic: An Environmental Policy Statement* (1996), para. 2(7).

In the case of long-term Antarctic programmes, it is a fine judgement whether what is being carried on is a new or a continuing activity and when a continuing activity changes in intensity.

As they presently stand, Australian regulations on environmental impact assessment require the circulation of draft comprehensive environmental evaluations and comments 'to each signatory to the Madrid Protocol' (Regulations 9(a) and 11) rather than to all parties as Annex I Article 3(3) requires.[97] The regulations thus could enjoin circulation to states that have not become a party, which was the intention while the regulations had effect before the Protocol entered into force; but at the same time the regulations do not require circulation to parties to the Protocol that become bound to it by accession and thus without signature.

Of the matters dealt with by Annex II, the ATEP Act defines a native bird and native mammal to include dead specimens (Section 3(1)). The Antarctic Seals Conservation Regulations make similar provision for dead seals (Regulation 1).[98] In contrast, Annex II does not specifically refer to dead specimens in its definition of native mammals and native birds (Article 1). In addition to specified precautions regarding poultry, living birds and non-sterile soil, Article 4(6) of Annex II required only that each Party take 'precautions ... to prevent the introduction of micro-organisms ... not present in the native fauna and flora'. Section 10(4) of the Australian legislation implements this loose obligation by requiring permit holders to keep micro-organisms under 'such control as is specified in the permit' and to destroy the micro-organisms promptly or remove them from the Antarctic after they have served their purpose. The act thus applies the same requirements to micro-organisms as to non-indigenous animals and plants, which in Article 4 of Annex II are treated differently.

As already mentioned, Section 21(1)(a) requires the holders of authorities issued by other parties as well as holders of Australian permits to provide Australian authorities with information on matters such as plants collected or native birds, mammals or invertebrates killed or taken. This obligation applies to everyone in the AAT and Australian nationals elsewhere in the Antarctic. The main reason for insisting on this information is to enable Australia to provide returns of information required by Article 6 of Annex II. It is unimaginable that Australia would be charged with breach of the provision if it failed to include information from persons authorised by other parties, whether or not the authorised persons are Australian nationals. If the requirement is intended to be more than a further hollow assertion of sovereignty, it is likely to be a fallible attempt to obtain as early as possible for environmental planning purposes information directly from all those active in the area of the Antarctic (the AAT) of particular interest to the Antarctic Division.

The ATEP Act provides for protected areas to include sea as well as land (Section 8(1)), the latter being defined to include ice and ice includes snow (Section

---

[97] Statutory Rules 1993, No. 115.
[98] Statutory Rules 1993, No. 289, as amended.

3(1). Designated marine areas can thus be protected by the Act. On the other hand, the Act goes beyond the Annex V in making explicit that a protected area can include subsoil to a specified depth, waters and seabed and subsoil of the seabed (Section 8(4) mentioned earlier in this chapter). At the Twentieth Consultative Meeting there was an unresolved discussion in which the United Kingdom argued that management plans for ASMAs should not include mandatory provisions. Australia took the lead in arguing that such provisions may be included.[99] The ATEP Act makes it an offence under Australian law to carry on an activity in an ASMA otherwise than as authorised by its management plan (Section 19(1)(e)). However, the Act does make different arrangements for ASPAs and ASMAs. A permit is required to enter an ASPA (Section 19(1)(d)) but not to enter the latter. The permit for an ASPA includes conditions which must be consistent with the area's management plan (Section 10(7)) and breach of those conditions is made an offence (19(3)(c)).

The ATEP Act seeks to coordinate the administration of protected areas under the Protocol with CCAMLR CEMP sites for the commission's ecosystem monitoring program. The implementation of the latter is provided for by the Antarctic Marine Living Resources Conservation Regulations[100] made under the Antarctic Marine Living Resources Conservation Act 1981.[101] Section 9(2B) of the ATEP Act prohibits the issue under the ATEP Act of a permit to carry on an activity in a CEMP site in the absence of a permit issued under the regulations of the 1981 Act or by another contracting party to the marine living resources convention. A parallel limitation is included in the Antarctic Seals Conservation Regulations (Regulation 4(2D)).[102]

---

[99] *Final Report of the Twentieth Antarctic Treaty Consultative Meeting, Utrecht, 29 April – 10 May 1996* (The Hague: Netherlands Ministry of Foreign Affairs, 1997), paras. 154–156, and United Kingdom, 'The Status of Protected Area Designations under Annex V to the Environmental Protocol', doc. XX ATCM/WP 16, 1996.

[100] Statutory Rules 1994, No. 345; Australia, *Statutory Rules, 1994*, Vol. 3, pp. 3,171–3,192; Bush (ed.), *Antarctica and International Law* (1992–), Booklet AU94, pp. 60–66.

[101] Bush (ed.), *Antarctica and International Law* (1982–88), Vol. 2, pp. 251–261.

[102] Statutory Rules 1993, No. 289, as amended.

18

Implementation of the Antarctic Environmental Protocol by Chile: History, Legislation and Practice

María Luisa Carvallo and Paulina Julio[*]

Several unique circumstances have influenced the national history of Chile as related to the Antarctic. Situated closest to Antarctica, Chile has been stressed by chroniclers and poets of the like of Alonso de Ercilla, who rendered the country as 'a notable and fertile province of the famous Antarctic region' ('fértil provincia y señalada de la región antártica famosa', *La Araucana*, 1556). The governors of Chile during Colonial times were concurrently the governors of *Terra Australis*. The first real Antarctic navigation by Gabriel de Castilla in 1603 and the discovery of the Shetland Islands by William Smith in 1819 were achieved in ships sailing from the Chilean port of Valparaíso; and it was in the neighbourhood of that port that John Miers drew the first Antarctic chart.

This geographical consciousness is also evident in the 1884 map of Chilean territory, with the Andean Range extending to the then scarcely known orography of the Austral Lands, a map that the Government printed for use in all the schools of the Republic.

From the beginning of the twentieth century, Chile was linked to the polar continent through its diplomatic, political and administrative activities as well as through other activities such as fishing. In 1902 Chile began granting hunting and fishing franchises, and these extended indefinitely south into the Southern Ocean. In 1906, the first concession of Antarctic islands and lands was granted by Chile. That same year the Chilean government organised an expedition that was to be thwarted by the dreadful earthquake of 1906, and the first Antarctic Commission was established as an advisory agency to the Chilean Minister of Foreign Affairs.

---

[*] The authors gratefully acknowledge the assistance and advice of Ambassador Jorge Berguño, Deputy Director of the Chilean Antarctic Institute, in the preparation of this chapter.

Various other activities at that time witness on the incorporation of Antarctic activity into the national legal system. Examples are 1) the publication in a report of the Ministry of Foreign Affairs of the advancement of the diplomatic negotiations with Argentina on a Complementary Treaty on Boundaries Demarcation in Antarctica; 2) the granting of legal rights to whaling companies to operate from South American territory and from Antarctica itself; and 3) the legal organisation of those companies under the laws of the Republic.[1]

Because of the Norwegian invitation to a Polar Congress and the US initiative to undertake the exploration and development of South American Antarctica, the Chilean government, through Decree No. 1541 of 7 September 1939, appointed Professor Julio Escudero to study the political and legal problems of the Antarctic. Through Supreme Decree No. 1723 of 2 November 1940, the government stated its intention 'to fully incorporate the polar Antarctic region, over which Chile has sovereignty, to the life of the Nation' and decided to centralise in one agency only, the Ministry of Foreign Affairs, 'the knowledge and resolution of all affairs concerning Chilean Antarctica, or the Chilean Antarctic Region, whatever their nature'. Through Supreme Decree No. 1747 of 6 November 1940, Chile's obligation of accurately setting the territorial boundaries was complied with. Although these territorial boundaries had not been determined 'in the part which extends towards the polar region named American Antarctica', the initiation of the studies of the relevant boundary settings had been placed on record in 1906. The contents of the decree, which sets the Chilean Antarctic Territory between meridians 53° and 90° west longitude west, is of significance in the definition of the geographical scope of Chilean Antarctic legislation.

Law No. 11,846 of 17 June 1955 stipulated that the Chilean Antarctic territory, 'in view of [its] special nature', should be managed by means of a 'special regime' that would be provided for through the by-laws of the Statute of the Antarctic Territory. A year later, Decree Law No. 298 of 17 July 1956, the subject matter of which comprised the said by-laws, was issued. This decree, which is still in force, regulates special situations that might occur in Chilean Antarctica. It should be pointed out that this legal instrument was adopted prior to the 1959 Antarctic Treaty[2] and is currently under review for possible up-dating. By giving the Governor of Magallanes (currently the Governor of the XII Region) competence to hear and resolve all management affairs pertaining to the Chilean Antarctic Territory, the Decree Law No. 298 of 17 July 1956 fully inserted that territory into the administrative, political and juridical structure of the Nation.[3]

---

[1] O. Pinochet de la Barra, *La Antártica Chilena*, 2nd edition (Santiago: Del Pacifico Publisher, 1948), pp. 93–119; and *ibid*., 'Antecedentes históricos de la política internacional de Chile en la Antártica. Negociaciones chileno-argentinas 1906, 1907 y 1908', in F. Orrego Vicuña *et al.* (ed.) *Política Antártica de Chile* (Santiago: Instituto de Estudios Internacionales de la Universidad de Chile, 1984), pp. 67–80.

[2] UNTS, Vol. 402, pp. 71ff.

[3] Laws and decrees are published in the *Diario Oficial* (Official Gazette of Chile). English speaking

CHILE AND THE ANTARCTIC TREATY SYSTEM

Chile was one of the twelve signatory countries of the Antarctic Treaty. Together with Argentina and Australia it was one of the last three countries to simultaneously ratify the Antarctic Treaty on 23 June 1961, on which date the Treaty entered into force. The Antarctic Treaty was put into force in Chile through the Supreme Decree No. 361 of 24 June 1961.[4] Chile contributed decisively in the negotiations of the Antarctic Treaty, in particular through the ideas of 'freezing' not the claims but the Antarctic litigation (reflected in Article IV of the Treaty) and of protecting the marine living resources (reflected in Article IX(1)(f) of the Treaty).[5]

The Protocol on Environmental Protection to the Antarctic Treaty[6] was put into force in Chile through Supreme Decree No. 396 of the Ministry of Foreign Affairs on 3 April 1995. The decree was signed by President Eduardo Frei Ruiz-Tagle in Punta Arenas.[7] The ratification of the Protocol by Congress did not create much public debate and the instrument of ratification was sent to the Depository Government on 11 January 1995.

It should be pointed out that Chile actively participated in the development of the Protocol. In view of the fact that the Convention on the Regulation of Antarctic Mineral Resource Activities (CRAMRA) was not going to be ratified by all the Consultative Parties required for its entry into force, so the problem of a possible mining activities in Antarctica would remain open, Chile, at the XV Consultative Meeting in Paris, October 1989, put forth the proposal for 'Comprehensive Measures for the Protection of the Antarctic Environment and of its Associated and Dependant Ecosystems'.[8] Because of the open issue on mining and in accordance with the proposal put forth by Chile, it was agreed that the XI Special Antarctic Treaty Consultative Meeting would be convened in Viña del Mar, Chile, from 19 November to 6 December 1990. In his opening speech at the Meeting, the acting Minister of Foreign Affairs, Edmundo Vargas, stated that Chile desired 'a clean Antarctica, but [that] we also want an Antarctica that is useful for man. We are faced with the challenge of conceiving an Antarctica free of pollution and, at the same time, open to human activity'.[9] These words have foreseen the basic principles of the Protocol that was adopted the in the following year.

---

readers will find it useful to consult W.M. Bush (ed.), *Antarctica and International Law: A Collection of Inter-State and National Documents* (London: Oceana Publications, 1982), Vol. II, pp. 275–467.

[4] Supreme Decree No. 361 of 24 June, published in the *Diario Oficial* (Official Gazette of Chile), of 14 July 1961.

[5] At the opening of the Antarctic Conference on 15 October 1959, the leader of the Chilean Delegation, Senator Marcial Mora Miranda, stated: 'My Government would suggest ... that it would be useful to supplement these aims with an agreement on the protection of antarctic marine resources'.

[6] ILM, Vol. 30, 1991, pp. 1,461ff.

[7] Supreme Decree No. 396 of 3 April 1995, published in the Official Gazette of 18 February 1998, upon the entry into force of the Protocol on 14 January 1998.

[8] See doc. XV ATCM/WP 7, 1989.

[9] Chile, 'Intervención del Ministro Subrogante de Relaciones Exteriores Señor Edmundo Vargas

In the course of three weeks of the first session of the XI Special Consultative Meeting, the body of the Protocol and its annexes took form. After further negotiation, the Protocol was adopted on 4 October 1991 in Madrid, Spain, by all the (then) 26 Consultative Parties and 12 non-Consultative Parties – less than a year after negotiations started.[10] It is important to underscore the political will of the various countries to negotiate and adopt an international document of such extent and significance in such a short time with the purport of enabling Antarctica to remain the cleanest continent on the Earth and a 'natural reserve dedicated to peace and science'.

The entry into force of the Protocol was very gratifying for the Chilean government. Not only had Chile contributed to the development of the concept of systematic protection of the Antarctic environment and the dependent and associated ecosystems, but the Protocol, by placing environmental protection at the centre of the Antarctic Treaty System (ATS) and by becoming a supplement to the Antarctic Treaty, rectified the tendency to autonomy and fragmentation of individual conventions and provided a significant step toward the integration of the ATS, which has always been a goal supported by Chile.

GEOGRAPHICAL SCOPE OF THE NATIONAL LEGISLATION

The basic principle of Chilean legislation is that the Chilean Antarctic Territory is considered as part of the national territory, with Chilean law applying to the inhabitants of this territory under the same conditions as at all the mainland territory of the Republic. Nevertheless, as mentioned above, considering its particular nature, the Chilean Antarctic Territory is governed by a special regime, specified in the Antarctic Statute. This distinction is a first level of speciality of the law; because of the nature of the physical environment of the Antarctic territory, national standards that apply to Chilean Antarctica differ in certain aspects from the laws that apply to the Chilean mainland territory.

The second level of speciality derives from the norms of the ATS. The Antarctic Statute is compatible with the obligations stipulated in the Antarctic Treaty and is also in harmony with the designation of Antarctica, including the Chilean Antarctica, as a 'natural reserve devoted to peace and science' (Article 2 of the Protocol).

Here, the principles of interpretation of legal instruments are of importance.[11] The principle of specialisation, *lex specialis*, applies namely to the special rules'

Carreño en la sesión inaugural de la XI Reunión Consultiva Especial del Tratado Antártico' (Speech by the acting Minister of Foreign Affairs, Edmundo Vargas Carreño, at the opening session of the XI Special Antarctic Treaty Consultative Meeting), November 1990, p. 3.

[10] The Protocol and Annexes I to IV were adopted in Madrid on 4 October 1991. Annex V was adopted as an attachment to Recommendation XVI–10 at the XVI Consultative Meeting held in Bonn, Germany, 7–18 October 1991.

[11] These rules of interpretation come from the domain of the law of treaties, but are equally applicable

taking precedence over the general rule. In other words, the rules of the Antarctic Treaty, the Protocol and other complementary instruments that have the rank of treaty have precedence over national legislation to the extent that these instruments do not oppose the Political Constitution of the State. Within the national legislation there also exists a *lex specialis*, the Antarctic Statute, the articles of which make reference to applicable international agreements, to which it adapts. The general Chilean legislation applies in every aspect that does not contradict the Antarctic Statute nor the rules of the ATS.

The Chilean legal system, *mutatis mutandis*, also applies the principle of *lex posterior*, that is, the prevalence of 'the new' rules over 'the older' when applied to the same matter. In the internal legislation this means that the clauses of the Antarctic Statute are modified as a consequence of institutional and legislative developments, even though the Statute remains as a framework of a special regime to administer the Chilean Antarctic Territory. In the implementation of rules of the ATS, for instance, Article 7 of the Protocol, which prohibits mineral activities in the Antarctic, supersedes CRAMRA (which Chile signed but has not ratified).

Finally, the Chilean legal system also applies the principle of *integration* and of *complementarity*. The general laws and regulations currently in force in the Republic apply in all matters not explicitly considered by the Antarctic Statute or regulated by the rules of the ATS or by other international conventions in force for Chile (Articles 10 and 16 of the Antarctic Statute). Nevertheless, when such laws, decrees or rulings are not fully consistent with the special regime of the Antarctic Statute or with the international obligations of Chile, they will apply as long as no inconsistency occurs.

An interesting example of the above is the pre-eminence of the Protocol, as it applies to the Chilean Antarctic Territory, over Law 19.300 on General Bases of the Environment Law, which applies within Chile as from 9 March 1994. By virtue of this law, a Regulation on the Environmental Impact Evaluation System was adopted and applies to the whole country, with the exception of the Chilean Antarctic Territory where special legislation takes precedence over the general. Article 8 of the Protocol and its Annex I comprise such *lex specialis*. Nevertheless, Law 19.300 may be applied in a complementary and subsidiary manner to the rules of the Protocol and its Annexes. In practice, the above-mentioned law contains principles of environmental protection, definitions of environmental damage, a definition of protected areas, environmental conduct, and, above all, the obligation of state operators to carry out environmental monitoring, the latter which reinforces the obligations stipulated in the Protocol.

National legislation integrated the Antarctic Territory with the regime of regions, provinces and communes that exist in Chile. But the special regime of the

---

to domestic conflict of laws. See I. Brownlie, *Principles of Public International Law*, 3rd edition (Oxford University Press, 1985); and Sir Ian Sinclair, *The Vienna Convention on the Law of Treaties*, 2nd edition (Oxford University Press, 1984).

Antarctic Statute as well as the regime of international obligations of the ATS apply exclusively to that part of the XII Region of Magallanes and Chilean Antarctica that corresponds with the Antarctic commune, which is identified as the Chilean Antarctic Territory – a sector claimed by Chile between the 53° and 90° west. Moreover, this geographical definition of the scope of application requires a more accurate definition in relation to the maritime areas.[12]

In the Chilean Antarctic Territory, all maritime limits referred to by Articles 593 and 596 of the Civil Code that do not affect the maritime limits of the Republic apply: the territorial sea, the contiguous zone, the exclusive economic zone and the continental shelf. Decree 1.747, approved in 1940, stated that the Chilean Antarctic Territory is composed of 'all land, islands, islets, reefs, glaciers (pack-ice) and others, already known or to be known, and their respective territorial sea, existing within the limits of the segment formed by meridians 53° longitude West of Greenwich and 90° longitude West of Greenwich'. Application of the sector theory appears to be based on the expression 'known or to be known', but in the maritime area the references of interest are the 'territorial sea' – the outer limits of which extended to three nautical miles in 1940 when the decree was adopted – with 'pack-ice', which tends to assimilate the ice of ice floes, still not considered permanent to the land.

On 23 June 1947, the President of Chile, Gabriel González Videla, proclaimed national sovereignty over the entire continental shelf adjacent to island and continental coasts of the national territory and over all seas, whatever their depth, adjacent to these coasts, and in addition sovereignty over all extensions necessary to protect Chile's natural resources. All Chilean official acts reflected the conviction that these claims regarding the continental shelf and the column of water extending to 200 nautical miles applied without exception to both American and Antarctic Chile. This was also the interpretation of the whaling company INDUS, whose lawyers had prompted the preparation of the 1947 declarations when requesting a concession lease in the Chilean Antarctic Territory. The concession was granted through Supreme Decree No. 166 on 17 February 1948, which was symbolically signed by President González Videla in Puerto Soberanía Base in Antarctica (currently Prat Base).

The 1940 decree only traced the limits by means of the meridians that converged on the South Pole, but similar to the Norwegian delimitation instrument, the decree did not contain any definition of its northward extent. In accordance with the Antarctic Treaty and the Protocol, such jurisdictional limit would correspond unequivocally to 60° south latitude. Nevertheless, on the occasion of signing (and

---

[12] On 27 February 1948 President González Videla announced in Punta Arenas that the 'Chilean Antarctic' would be incorporated into the Magallanes Province and would have its administrative capital on Navarino Island, off Tierra del Fuego. Although this decision was confirmed in 1975, the Chilean Antarctic Territory consists only of the Commune of Antarctica and does not include the Commune of Navarino that is situated in the continental Chile. *Keesing's Contemporary Archives (1946–48)*, pp. 9 and 133–134.

later on ratifying) the Protocol, Chile made an interpretative statement regarding the scope of application of the prohibition on mineral activities, as established by Article 7 of the Protocol; the Chilean interpretative statement stated that mining on the continental shelf of the Antarctic Peninsula and of the South Shetland Islands would also be forbidden, even where it extends *beyond* 60° south latitude. Such interpretation was apparently based on the fact that the Consultative Parties, while regulating mining activity in CRAMRA, had excluded island platforms which projected toward the application zone of the Antarctic Treaty but had included those platforms generated by Antarctic continental or insular masses which overflowed the 60° south latitude. Indeed, a more difficult question was: Are *all* areas *south* of 60° south latitude included in the prohibition under Article 7 of the Protocol?[13]

The territorial application of Chilean law in the Chilean Antarctica has numerous exceptions. To facilitate the carrying out of the functions and rights stipulated in the Antarctic Treaty, and with consideration for the diversity of juridical positions of the parties to the treaty regarding jurisdiction,[14] national jurisdiction regarding actions or omissions in the Chilean Antarctic Territory is not applied to the following categories of foreigners:

a) observers designated in accordance with Article VII of the Antarctic Treaty, Article 14 of the Protocol and Article XXIV of the Convention for the Conservation of Antarctic Marine Living Resources[15] (CCAMLR);

b) scientific staff exchanged under Article III(1)(b) of the Antarctic Treaty;

c) scientific and logistic staff associated with the foreign bases and stations located in the Chilean Antarctic Territory, under Article III of the Antarctic Treaty and any other personnel under the authority of the commander of a base or station;

d) foreign ship and aircraft personnel in the Chilean Antarctic Territory, as per Articles I(2), II and VII(4) and (5) of the Antarctic Treaty; and of ships flying the flag of CCAMLR parties and performing marine research or fishing activities in the zone of application of CCAMLR;

e) foreign tourists and visitors on trips organised by foreign tour operators without organisational logistic support in Chile's American territory. The supply of services, ship arrival in port and landing in national strips or airports are not necessarily considered to be organisational logistic support.

---

[13] The complexity of the subject is discussed by Davor Vidas; see D. Vidas, 'The Southern Ocean Seabed: Arena for Conflicting Regimes?', in D. Vidas and W. Østreng (eds.), *Order for the Oceans at the Turn of the Century* (The Hague: Kluwer Law International, 1999), pp. 291–314. See also Chile, 'Relationship between the Protocol on Environmental Protection to the Antarctic Treaty and Other International Environmental Protection Treaties', doc. XX ATCM/ WP 30, 1996.

[14] See discussion by Orrego Vicuña, Chapter 3 in this book.

[15] ILM, Vol. 19, 1980, pp. 837ff.

These are traditional exceptions in accordance with the Antarctic Treaty and its complementary instruments. Nevertheless, Article 13 of the Protocol establishes stricter compliance requirements, imposing 'rules and regulations, administrative acts and coercive measures' to ensure the enforcement of the Protocol. This standard of compliance requires the revision and updating of the Chilean Antarctic legislation in order to determine which matters might require validating legislation, on the one hand, and which matters can be applied immediately or only require guidelines for their execution, on the other.[16] The latter is seemingly a case where the main part of the obligations are imposed by the Protocol in terms of preventive conduct and rules that the annexes of the Protocol have codified but which were already implemented through earlier instruments.

Nevertheless, it has been interpreted that, on assuming the obligations of the Antarctic Treaty and its Protocol, Chile becomes responsible for the activities by its nationals within the whole scope of application of the Treaty and the Protocol. As a consequence of the foregoing, national jurisdiction will be applied to Chilean citizens who find themselves in other regions of the Antarctic in the following cases:

a) Chilean citizens who sponsor scientific, tourist or other visits or explorations of areas located outside the Chilean Antarctic Territory, particularly if such expeditions, visits or explorations have as the final starting point the Chilean South American territory;

b) observers designated by the Chilean Government, according to Article VII of the Antarctic Treaty, Article 14 of the Protocol or Article XXIV of CCAMLR;

c) Chilean scientific personnel exchanged in accordance with Article III(1)(b) of the Antarctic Treaty;

d) officers and crew of ships which are, or should be, under Chilean flag; and of Chilean aircraft which are under control, licence or authorisation of the Chilean Air Force or of national aeronautical authorities; and

e) Chilean citizens who enter or remain in the unclaimed area of the Antarctic, adjacent to the Chilean Antarctic Territory, which is south of 60° south latitude, between meridians 90° and 150° longitude west, to the South Pole.

ENFORCEMENT AND PRACTICAL APPLICATION OF THE PROTOCOL

As stated above, the Chilean Government enacted the Protocol through an Executive Decree signed by the President of the Republic and the Minister of Foreign Affairs

---

[16] For distinctions made in this sense, see discussion by Pineschi, Chapter 19 in this book.

in the city of Punta Arenas, on 3 April 1995. The enacting decree contains the usual sentence: 'Be it observed and enforced as an Act and a certified copy thereof be published in the Official Gazette'. Upon publication in the Official Gazette on 18 February 1998, the Protocol entered into force for Chile and must be complied and executed as law in the Republic.

Recommendation XVI-10, on Annex V to the Protocol, was enacted by Decree No. 583 of 24 April 1997 and published in the Official Gazette on 9 August 1997. Upon enactment, the Recommendation, as well as the contents of Annex V, were adopted pursuant to Article IX of the Antarctic Treaty. Both are accordingly subject to such enacting modalities as may govern all instruments adopted at the Antarctic Treaty Consultative Meetings pursuant to Article IX.

The Protocol and its Annexes can be considered only to a certain extent as a 'self-executing treaty' in Chilean legislation.[17] The Antarctic Treaty was itself considered such a self-executing instrument, being a framework treaty that advanced toward its objectives through the means of recommendations. Once adopted, those recommendations acquired legal force in the Chilean domestic legislation through their enactment by decree and publication in the Official Gazette. Nevertheless, as has been stated above, the requirement for coercive enforcement contained in Article 13 of the Protocol entails a higher standard of domestic implementation because 'enforcement measures' cannot be applied if no punishments have been stipulated previously for violations to or infringement of the regulations of the Protocol.

A tentative solution to this problem can, only in principle, be found in the prospective application of the legal body designated as Law on the General Basis of the Environment (Law No. 19.300). Under the integrative principle contained in the Statute of the Chilean Antarctic Territory, this law can be applied in all aspects that are consistent with Chile's international obligations and with the special regime established by the Statute. Title III of Law No. 19.300 includes the concept of criminal liability for environmental damage and a presumption of guilt of any party that infringes 'regulations on environmental protection, preservation or conservation contained in this law or *any other legal or statutory provisions*' (emphasis added).

Pursuant to Law No. 19.300, 'protection, preservation or conservation regulations' means 'any regulation that may be enacted on the matter'. However, the fact that an all-embracing umbrella legislation exists to justify the treatment of violations and infringements of the Protocol by activities that should be penalised does not fully respond to all the requirements of Article 13 of the Protocol. Consequently, it is not sufficient that the Law on the General Basis of the Environment considers a conduct as criminal and deserving punishment. Specific violations or infringements require a scale of appropriate penalties, which can only be established by special legislation to that effect.

There is one further aspect where enforcement cannot be carried out under the present Chilean legislation, since Law No. 19.300, Section 51(2) suggests that

---

[17] Compare the analysis by Pineschi, Chapter 19 in this book.

'regulations on liability for environmental damage contained in special laws shall prevail over those herein stated'. To a certain extent, this implies that Law No. 19.300 is irrelevant in matters of liability for damage to the Antarctic environment or to dependent and associated ecosystems. Although this conclusion is not to be construed as a *lacunae* in the process of complying with Article 13 of the Protocol; all such rules that are in the future agreed upon by the parties to the Protocol pursuant to Article 16 of the Protocol should apply as *lex specialis*, as is the case with any other aspect related to liability for damage to the Antarctic environment and dependent and associated ecosystems.

In 1998, Chile submitted an information document to the XXII Consultative Meeting that contained a description of its approach toward the enforcement of the Protocol. The document provided guidance on the current level of compliance and on the actions envisaged to further advance the process of attaining the standards introduced by Article 13.[18] At the XXIII Consultative Meeting the following year, in respect of the Arbitral Tribunal and in conformity with the requirements of Article 2 of the Schedule to the Protocol ('Arbitration'), Chile designated three arbitrators: Ambassadors José Miguel Barros, Fernando Zegers and María Teresa Infante.[19]

Discussion of enforcement issues should be completed with an overview of the legal and practical steps taken to comply with all the annexes to the Protocol, indicating in each case the agency or agencies responsible for compliance with the specific domain regulated by the respective annex. Moreover, the sections below will consider the work being done in interagency meetings, the participation of non-governmental organisations, and the programme for future work.

Annex I

During 1999, Regulations for the application of Annex I to the Protocol were drawn as a matter of high priority. This instrument was approved by the Chilean authorities and put into provisional application without waiting for its formal enactment through a decree. This was done to foster compliance by the Chilean operators and establish some uniform method for the various environmental impact assessments established by the Protocol and Annex I to the Protocol.

The provisional Regulation creates a National Committee for Environmental Impact Assessments (CONAEIA) under the authority of the Council for Antarctic Policy (CPA). The Committee is chaired by the Executive Director of the National Environmental Commission (CONAMA) whose staff constitutes the technical support secretariat for the Committee. All the state operators, including the Chilean Antarctic Institute (INACH), the Chairman of the Chilean SCAR Committee (CNIA), the Chilean Representative to the CEP and other organisations, are

---

[18] Chile, 'Enforcement of the Protocol on Environmental Protection to the Antarctic Treaty', doc. XXII ATCM/IP 43, 1998.
[19] Chile, 'Designation of Arbitrators', doc. XXIII ATCM/IP 70, 1999, para. 2.

represented. All categories of impact assessment recognised by the Protocol and Annex I fall under the competence of CONAEIA.

Once the requirements established in the regulation are properly satisfied, CONAEIA issues a resolution certifying that the impact assessment provisions of the Protocol have been complied with at the level of a preliminary, initial or comprehensive environmental assessment. All submissions are required to conform to the format of the Guidelines prepared by COMNAP and approved by the XXIII Consultative Meeting held in Lima, Peru, in 1999. It is important to highlight that the Regulation recognises that procedures to apply and monitor compliance with the requirements of the Protocol for planning and assessing possible adverse impacts belong to an international process where standards must be equivalent and the driving force must be international cooperation.

It is interesting to note in this context that an impact assessment prepared by the tour operator Adventure Network International was approved by the British authorities pursuant to submission. When the same impact assessment was afterward presented by the operator's representative to CONAEIA, it was accepted as valid only after some debate. Nevertheless, a discussion arose with regard to the disposal of wastes from that operator and others, in the city of Punta Arenas. A majority in the National Committee stated that those matters were beyond CONAEIA's competence and should be taken up by other governmental agencies.

INACH requires that all proponents of scientific projects submit a preliminary assessment and consider whether their prospective activities may require either an initial or a comprehensive environmental evaluation. CONAEIA certified not only the assessments accompanying scientific projects but also those required for all associated logistic activities. CONAEIA has also been considering the merits of monitoring, auditing and other follow-up procedures devised to insure that the certified activity proceeds after its certification along the lines drawn when it was authorised.

Annex II

Chile informed SCAR in 1960 that 'Chilean national legislation protected wildlife throughout the country, including areas in Antarctica'.[20] The Government of Chile approved the Agreed Measures for the Conservation of Antarctic Fauna and Flora and ratified the Convention for the Conservation of Antarctic Seals[21] (Seals Convention). Since 1964, the INACH has been responsible for a system of permits, including permits for the taking of specially protected species, which may be issued only if there is a compelling scientific purpose for doing so and provided that the activities undertaken pursuant to the issuance of such a permit do no jeopardise the

---

[20] SCAR Bulletin No. 8, in *Polar Record*, Vol. 10, 1961, p. 536.
[21] UNTS, Vol. 1080, pp. 175ff; and ILM, Vol. 11, 1972, pp. 251ff.

natural ecological system or endanger the survival of the species concerned. Taking or 'harmful interference' with Antarctic mammals and birds is only allowed for the purposes of scientific research and the information so obtained for the exclusive use of scientific institutions.

While the taking of marine mammals in the Antarctic Treaty area requires permits that are issued jointly by the Under-Secretary for Fisheries and INACH, the Under-Secretary, in conformity with Law No. 18.892 (General Law on Fisheries, Aquaculture and the regulatory decree of the Hunting Law), has the authority to allow for the taking of fishes, molluscs and other marine living organisms for purposes of providing specimens for museums, aquariums, educational institutions and other non-commercial legitimate scientific uses. On the other hand, INACH issues any permit that may be allowed under the Seals Convention. In full compliance with the Protocol, the Under-Secretary for Fisheries has ruled that no aquaculture is allowed to take place in the Chilean Antarctic Territory.

Annex III

The Protocol contains in this Annex a rather elaborate scheme for the handling and disposal of various types of wastes. However, the fundamental principle of prohibiting the disposal of any waste in Antarctica and of making the generator of the waste liable for its disposal out of the Antarctic Treaty area was accepted and applied by Chilean Operators, including INACH and each of the Armed Forces in their respective bases, even before Chile signed the Protocol. INACH applies a Manual, both in its bases and for use by its expeditions, that stipulates the relevant prohibitions and regulations applicable to the burning of waste. The combustion of waste in incinerators must meet the emission standards drawn up by the COMNAP and other bodies of the ATS. At present there is a high degree of compliance with the principles and provisions of Annex III, but a recent Anglo-German Inspection Report detected a lack of homogeneity in the manner in which these principles and norms were being enforced in the area of the Antarctic Peninsula and the South Shetland Islands by Antarctic bases of various nations.[22]

Annex IV

Obligations stated in Annex IV differ from other matters where the main source for national obligation stems directly from the Protocol and its respective Annexes and indirectly from principles contained in domestic legislation, namely Law 19.300 on the General Basis of the Environment. A distinct *corpus iuris* originates in

---

[22] Chile, 'Report on the Protection of the Environment', doc. XXI ATCM/IP 6, 1997. United Kingdom and Germany, 'Report of a Joint Inspection under Article VII of the Antarctic Treaty by United Kingdom and German Observers' (London: Foreign and Commonwealth Office/Bonn: German Federal Ministry of Foreign Affairs, 1999).

instruments adopted by the International Maritime Organisation (IMO), whose institutional spheres of competence in the domestic field lie with the Maritime Territory and Merchant Marine Directorate (DIRECTEMAR).

DIRECTEMAR is divided into 16 maritime administrations, one of them located at the Prat Base on Greenwich Island, which oversees operations of the port authorities at Fildes Bay, Sovereignty Bay, Covadonga Anchorage and Paradise Bay. These maritime authorities act as co-ordinators for search and rescue operations, oil spills control and containment, the monitoring of maritime traffic, the prevention of accidents and damage to vessels, and the protection of the Antarctic ecosystem. They also operate under the National Coordinating Centre for Pollution Control in Punta Arenas, to relay meteorological, sea ice and other related reports, and monitor terrestrial sources of marine pollution.[23]

The Maritime Training Centre (CIMAR), the 'Piloto Pardo' Antarctic Navigation School, and the Search and Rescue Services contribute to Antarctic maritime safety. But the Naval Antarctic Patrol, which has now coordinated its activity with the Argentinean Navy, can inspect at sea vessels and detect non-compliance with Annex IV provisions. The Naval Antarctic Patrol may also require information from a foreign ship navigating in the territorial sea or the exclusive economic zone of the Chilean Antarctic Territory and report to the flag state any infringement, pollution breach, or presumption of illegal fishing. At Chilean ports, a corps of specialised inspectors check the operational conditions of ships and their compliance with domestic and international maritime standards, including standards that enable the ship to navigate safely in the Antarctic.

Annex V

Permits authorising entry into an Antarctic Specially Protected Area are issued by the INACH only in a manner consistent with the approved management plan or in the case of a compelling scientific purpose for such entry that cannot be fulfilled in another comparable area not subject to protective restriction. As with the taking of species, action allowed under permits of entry must not jeopardise the ecological system in the protected area. Under the Protocol, exception or derogation to the prohibitions of the various annexes, including Annex V, are made if the forbidden acts are committed 'under emergency circumstances involving the safety of human life or of ships, aircrafts, or equipment or facilities of high value, or the protection of the environment'. No permits are issued which carry the implication that unlawful or unauthorised acts are included in the permit. Nonetheless, if the described emergency circumstances arise and imperatively require the commission of an unlawful act, submission of a report explaining that invocation of the 'emergency clause' was necessary.

---

[23] Chile, 'Chile's Contribution to Maritime Safety in the Antarctic', doc. XX ATCM/IP 24, 1996.

Chile has been actively involved in the two workshops organised immediately before the CEP Meetings held in conjunction with the Tromsø and the Lima Consultative Meetings in 1998 and 1999, respectively. Chile was likewise involved in the process of developing guidelines for protected areas and of dealing with the issue concerning a systemic environmental geographic framework for Antarctica referred to in Article 3(2) of Annex V.

Other significant areas of Chilean interest can be found in the promotion of the concepts of 'dependent and associated ecosystems', which were enshrined both in the Protocol, pursuant to a number of recommendations produced by the XV Consultative Meeting, and in various proposals for the expansion of the existing protection to include historical and cultural sites, artefacts and events of Antarctic history.

INSTITUTIONS, NORMS, AND POLICIES: AN ASSESSMENT

Compliance with the Environmental Protocol is one of the declared objectives of the document on National Antarctic Policy; it was signed by President Ricardo Lagos in the Chilean Antarctic Territory on 1 April 2000 – thus 40 years and four months after the signature of the Antarctic Treaty. All eleven objectives pursued by the Antarctic policy relate, in one way or another, to the Protocol, but some of them, including 'the preservation of the Peace Zone, scientific activities and the Natural Reserve', 'international cooperation', 'linkage of Antarctic science to global trends', 'controlling tourism' and 'land-use planning', bear a direct relationship to the Environmental Protocol.[24]

Moreover, there are two fundamental areas where either implementing legislation or the broad application of executive regulatory powers, in the form of by-laws, must advance if the existing fragmented, piecemeal and sometimes contradictory Antarctic domestic legal structure is going to be trimmed and improved. The demand for enforcement in Article 13 of the Protocol is certainly one area that needs improvement, while the updating of the institutional framework of the national Antarctic policy is another significant priority.

Ideally, both tasks could be undertaken in one statute, and a comprehensive law project would be the best way to address the objective of full integration of the domestic and the international components of Antarctic obligations. This would also be an effective way to appropriately deal with the need of achieving a measure of equilibrium between the general Antarctic policies and institutions, on the one hand, and the geographic component that also demands attention at a time when so many pressures, risks and opportunities arise for the so-called 'gateway states' to the Antarctic, on the other.

---

[24] The text of decree No. 429, of 28 March 2000, and the attached document on the National Antarctic Policy has been published in *Boletín Antártico Chileno*, Vol. 19, No. 1, May 2000; available also at INACH website <www.inach.cl>.

This global approach encounters significant, practical difficulties. Taking into account considerations of time and convenience, national as well as international, it would be more expeditious to solve all issues and address all matters that could be regulated through executive powers, leaving for the legislative stage substantive problems of jurisdiction, enforcement in penal matters and, eventually, liability and state responsibility. The government of Chile has chosen this approach and has enacted some by-laws through decrees. It is presently engaged in the study and analysis of other decrees for the implementation of the Protocol, such as the establishment of a national committee entrusted with the follow-up and support of decisions taken by the Committee for Environmental Protection,[25] which is established under the Protocol.

As the previous section of this chapter has demonstrated, there are a number of domestic institutions directly or indirectly involved in the implementation of the Protocol. The Ministry of Foreign Affairs maintains a tuition on all Antarctic affairs; however, in 1978, the need to integrate the input of many actors from the public as well as private sectors led to the establishment of the Antarctic Policy Council in the organic statute of the Ministry. More recently, Decree No. 495 of 7 April 1998 regulated the functions of the Council and, in particular, the functions of the Council's subsidiary bodies and working committees. As described above, one such subsidiary body is CONAEIA, which is entrusted with the certification of environmental impact assessments required by Article 8 and Annex I to the Protocol.

Within the framework of the Council, the various ministries, operators, the National Environmental Commission, the INACH, and specialised agencies, may interact. The Council is a consultative body for the Minister of Foreign Affairs, who also chairs the Council and is entrusted with the determination of the political, legal, scientific and economic basis for Chilean presence in the Antarctic.[26]

Parallel with the Antarctic Policy Council, a network of more informal relations exists among the main operators, the scientific and environmental bodies, the universities, non-governmental organisations and private groups interested in Antarctica. Historically, Chilean participation in Antarctic affairs has been associated with the Antarctic Commissions but is now superseded by the Council. Other landmarks can be named: the National Commission for the International Geophysical Year, created by Decree No. 76 of 17 June 1955; the Commission for the Study and Planning of Scientific, Technical Oceanographic and Antarctic Activities created by supreme Decree No. 1305 of 4 May 1960; the National Committee for the Scientific Committee for Antarctic Research (CNIA) created by Decree No. 363 of 13 September 1962; and Decree No. 103 of 10 September 1963,

---

[25] On the CEP see Orheim, Chapter 6 in this book.

[26] Decree with force of law No. 161 of 3 March 1978, reorganising the Ministry of Foreign Affairs and establishing the Antarctic Policy Council; text also reprinted in Bush (ed), *Antarctica and International Law*, Vol. II, pp. 463–465.

approving the organic regulations of the Chilean Antarctic Institute (INACH). At present, in addition to INACH, CNIA, and the CONAMA, which acts as technical Secretariat for CONAEIA, other institutions such as the Natural History Museum, the Antarctic Circle (a club of Antarctic 'old hands') and regional organisations from the XII Region of Magallanes and Chilean Antarctica, led by the University of Magallanes, are keenly interested and involved in various aspects of the implementation of the Protocol.

The main lesson to be learned from the processes of implementation concerns the significant roles played by policy at the various levels, and perhaps even more important, by the *behaviour* of the various actors, including social constraints facilitating a degree of enforcement of the Protocol. Education programmes, training, publication of manuals, visitors' guidelines, and other steps taken to foster compliance, have all a role in advancing the principles and objectives of the Protocol.

The domestic institutional framework and the more informal network of pro-Antarctic institutions coalesce in two of the permanent working committees of the Antarctic Policy Council: the Committee for General Affairs and the Committee for Financial Affairs, entrusted with the elaboration of proposals for the Council's budget. Recently, the Committee for General Affairs has drawn an agenda that includes as a priority the modernisation of domestic legislation and the establishment of a National Committee for the CEP.

Entry into force of the Protocol has generated a greater awareness of the relative strength and weaknesses of the current system of implementation of Antarctic decisions and measures. The fact that the Protocol is, to a great extent, a codification of previous recommendations, seemed to indicate that most annexes could be complied without further implementing legislation. The outstanding exception was Annex I, and CONAEIA was provisionally set up for that reason so that the formal promulgation of its rules of procedure could take place soon.

However, the interaction of institutions, norms and policies seems to point to further complexities. As stated above, legislation will be required only for matters related to jurisdictional, criminal and perhaps liability issues. It is likely that those issues may expand. Leaving aside the more technical aspects of sanctions for transgressions of the annexes, and issues of liability that must be addressed collectively by the ATS, the major test for implementing legislation is, undoubtedly, the issue of jurisdiction.

By virtue of the Antarctic Treaty, Chile has jurisdiction over activities undertaken by its nationals in areas of Antarctica outside the Chilean Antarctic Territory, in particular in the unclaimed sector. By analogy, that official position was maintained by the government with regard to activities by its flag vessels in the CCAMLR area and on the high seas. Nevertheless, owners of those vessels challenged this analogy by requesting the Chilean courts to state that they were unable to pass judgement on offences committed beyond the maritime jurisdictions of the Republic. Although the courts supported the governmental interpretation, which was sustained by the Supreme Court, cumbersome litigation forced the authorities to approve a modification of the Fisheries Law. This modification clarified the authority of Chilean courts to adjudicate cases involving violations of international maritime agreements and conservation measures adopted pursuant to those agreements. This exercise may need to be repeated if the situation arises where a Chilean national accused of transgressions to the Protocol in areas outside the Chilean Antarctic Territory were to challenge the jurisdiction vested in the Courts of Punta Arenas.

Issues concerning the exercise of rights of jurisdiction may put the whole ATS in a delicate position. So far, only a few cases have been handled either among the countries concerned or through a constructive expansion of the jurisdiction of the base commanders. For purposes of application of the Protocol, a precedent arises from the fatal accident involving three mountaineers who met their deaths in parachute-jumping at Patriot Hills, a mountainous area of Antarctica claimed only by Chile. The accident was examined by the Criminal Court in Punta Arenas. More than a territorial jurisdiction, the jurisdiction in this case was perhaps a 'port state jurisdiction'[27] because the expedition had departed from and returned to Punta Arenas. The issue was not, but could have been, an environmental issue. As suggested by Chile to the Consultative Meeting:

> True to the spirit of Article IV, it is essential not to prejudice or affect the legal position of the Parties in the implementation of this principle. Practical consideration should also be born in mind as regards the efficacy of the jurisdiction claimed and the reality of its presumed applicability. Finally, it is critical in the light of Article 13 itself, that no action or omission infringing the Protocol should be unpunished.[28]

---

[27] On this concept see Orrego Vicuña, Chapter 3 in this book. For possible application to activities such as tourism, see Richardson, Chapter 4 in this book.

[28] Chile, 'Enforcement of the Protocol on Environmental Protection to the Antarctic Treaty', XXII ATCM/IP 43, 1998.

19

A Self-Executing Treaty?
Italian Legislation and Practice in
Implementing the Environmental Protocol

Laura Pineschi

The provisions of the Protocol on Environmental Protection to the Antarctic Treaty[1] (Protocol) have been incorporated into the Italian legal system by means of a legislative act (Law No. 54 of 15 February 1995),[2] according to the so-called *implementing order* procedure.

In Italy, an international treaty can be incorporated into the domestic legal system according to two different procedures.[3] The first, the *ordinary procedure* ('procedimento ordinario'), consists of the promulgation of a legislative act that provides for all the changes in the domestic legislation that a proper fulfilment of the treaty provisions in the national legal order involves. In other words, the international rules are reformulated at the domestic level to ensure the existence in the internal legal order of provisions that comply with the international ones. This technique has some drawbacks. In particular, the mechanism is burdensome for the

---

[1] ILM, Vol. 30, 1991, pp. 1,461ff.

[2] *Ord. Suppl.* to *Gazzetta Ufficiale della Repubblica Italiana* (hereinafter: *G.U.*) No. 48 of 27 February 1995. The instrument of ratification of the Protocol was deposited by Italy on 31 March 1995.

[3] For a survey of the Italian method of incorporation of international treaties into the national legal order, see: G. Gaja, 'Italy', in F.G. Jacobs and S. Roberts (eds.), *The Effect of Treaties in Domestic Law* (London: Sweet and Maxwell, 1987), pp. 87–108; A. D'Atena, 'Adattamento del diritto interno al diritto internazionale', *Enciclopedia Giuridica*, Vol. I, pp. 1–8; M. Giuliano, T. Scovazzi and T. Treves, *Diritto Internazionale. Parte generale* (Milan: Giuffré, 1991), pp. 578–583; E. Cannizzaro, 'Trattato internazionale (adattamento al)', *Enciclopedia del Diritto*, Vol. XLIV, 1992, pp. 1,394–1,425; T. Treves and M. Frigessi di Rattalma, 'Italie', in P.M. Eisemann (ed. in chief), *L'intégration du Droit international et communautaire dans l'ordre juridique national* (The Hague: Kluwer Law International, 1996), pp. 365–406; G. Cataldi, 'Rapporti tra norme internazionali e norme interne', *Digesto delle discipline pubblicistiche*, Vol. XII, 1997, pp. 391–411; B. Conforti, *Diritto Internazionale*, 5th edition (Napoli: Editoriale Scientifica, 1997), pp. 295–297.

legislator who must foresee all the consequences that a proper fulfilment of the international provisions will imply at the domestic level. Moreover, an incidental misinterpretation of an international rule by the national legislator will inevitably affect the subsequent application of the domestic rules.

The second procedure for incorporating an international treaty is the *implementing order* ('ordine di esecuzione') procedure (or *special procedure* – 'procedimento speciale'). This procedure implies the adoption of a legislative act that provides for 'the full and complete' implementation of the treaty in question, whose text is usually reproduced in an annex in one of the official languages and in a non-official Italian translation. The same law authorises the conclusion, and orders the application of the treaty.

The decision of the Italian legislator to give effect to a treaty by means of the implementing order procedure is discretionary but subject to an important condition: the provisions included in the international instrument are to be considered to be 'self-executing'. In principle, a treaty can be deemed self-executing when it already contains all the elements that allow for its implementation in the domestic legal system and no further integration needs to be added by state organs.

It is worth noting, however, that this assumption implies a corollary: the specific wording and content of a treaty alone are not sufficient to assess its direct applicability. Whether a treaty is or is not self-executing is a question that must be appreciated also in the light of the national legal system. Treaty provisions can be thus considered directly applicable if the existing domestic legislation already contains all the elements that are lacking at the international level to give proper effect to the relevant obligations.[4]

The assumption that the Protocol is a self-executing treaty is questionable. It is also difficult, however, to maintain that all the elements that are necessary to give effect to the Protocol already exist in the Italian legislation or can be easily subsumed from certain domestic provisions.

A comparative analysis of some selected Protocol provisions and existing Italian legislation can be useful in determining whether and to what extent the implementing order contained in Law No. 54 of 15 February 1995 really is sufficient to give proper effect to Protocol obligations in the Italian legal system.

---

[4] See L. Condorelli, *Il giudice italiano e i trattati internazionali* (Padova: Cedam, 1974), p. 53, and, more generally on the distinction between self-executing and non-self-executing treaties: J.J. Paust, 'Self-Executing Treaties', *American Journal of International Law*, Vol. 82, 1988, pp. 760–783; T. Buergenthal, 'Self-Executing and Non-Self-Executing Treaties in National and International Law', in Académie de Droit International, *Recueil de Courses*, 1992, Vol. IV, pp. 305–400; J.H. Jackson, 'Status of Treaties in Domestic Legal Systems: A Policy Analysis', *American Journal of International Law*, Vol. 86, 1992, pp. 310–340; C.M. Vázquez, 'The Four Doctrines of Self-Executing Treaties', *American Journal of International Law*, Vol. 89, 1995, pp. 695–723.

THE PROVISIONS OF THE PROTOCOL

It is well accepted that the direct applicability of a treaty must be assessed in the light of its specific provisions. In other words, a treaty is not a *unicum*: certain obligations may be self-executing while others are not. In order to know whether a treaty can be considered directly applicable, it is therefore necessary to analyse each provision on a case-by-case basis.

As far as the Antarctic Environmental Protocol is concerned, many of its provisions are so detailed that no margin of appreciation is left to the contracting parties. Moreover, other provisions do not require any action within the domestic legal system of the states parties as far as they express a prohibition. Very clear *non facere* obligations are for example contained in Article 7 of the Protocol ('Any activity relating to mineral resources, other than scientific research, shall be prohibited'), Article 4(2) of Annex II ('Dogs shall not be introduced onto land or ice shelves') and Article 4(1) of Annex III ('Wastes not removed or disposed of in accordance with Articles 2 and 3 shall not be disposed of onto ice-free areas or into fresh water systems').

However, not all the obligations of the Protocol and its Annexes can be considered self-operative. First of all, there are provisions that are vague or that are not sufficiently detailed. The parties are therefore left a certain freedom of interpretation when they draw up their domestic legislations.

There are also rules of the Protocol that are clearly non self-executing; these leave the parties a margin of appreciation in the choice with respect to the most appropriate national means by which to give effect to these. This is the case, for example, of Article 1 of Annex I to the Protocol, according to which 'the environmental impacts of proposed activities referred to in Article 8 of the Protocol shall, before their commencement, be considered *in accordance with appropriate national procedures*' (emphasis added).

Provisions that produce their full effects only when an adequate organisational regime has been previously established at the domestic level cannot be considered self-executing. This is the case, for example, of the Protocol provisions that envisage duties of cooperation among the contracting parties. Correct enforcement of these provisions involves at least specification of the national organs responsible for the collection and dissemination of information as well as coordination with foreign governments for the running of joint activities.

Finally, there are important issues that are typically left to regulation by the national legislator in the Antarctic as well as in other international regimes. Suffice here to mention the exercise of national jurisdiction over public or private bodies that operate in the Antarctic Treaty area, the specification of penalties that may be imposed for deliberate or inadvertent violation of the rules of law, and the allocation of national funds for the implementation of the obligations arising out of the provisions of the Protocol. All these issues are completely remitted to the domestic regulation of the contracting parties.

Beyond these preliminary remarks, a closer look at the scope of the Protocol and its content is indeed required. The purpose of the analysis which follows in this chapter is threefold: 1) to single out the matters in which the intervention of the national legislator is necessary; 2) to assess the degree of discretion that is left to the contracting parties in the establishment of domestic procedures; and 3) to check whether the Italian legislation already contains all the elements necessary for giving proper effect to the Protocol obligations.

THE REGULATION OF THE ENVIRONMENTAL IMPACT ASSESSMENT PROCEDURE

The Provisions of the Protocol

Several provisions on environmental impact assessment (EIA) have been enacted within the Antarctic Treaty System (ATS) since 1975. It was however with the conclusion of the Protocol that the Consultative Parties recorded noticeable progress in the regulation of EIA procedures. After the improvements specifically devoted to EIA were introduced by the Protocol and its Annex I, the ATS could be considered the most developed and detailed system of legally binding rules so far on EIA elaborated at the international level[5]. It is also true, however, that many rules are far from being self-operative.

Determination of the person in charge of the environmental assessment procedure. One of the points not specifically addressed by the Protocol is designation of the person who should be in charge of undertaking the EIA. This important question is left to the domestic jurisdiction of every contracting party with the consequence that different solutions are found by different national legislations. In certain cases the EIA is to be made by a state organisation, in others by the developer (the applicant for authorisation for a project), and still in others by an independent firm of environmental consultants on behalf of the developer. Any of these solutions seems to be compatible with the Protocol.[6]

On this point, a specific enactment by contracting parties seems hardly avoidable. Legislation already adopted for projects to be carried out in the national territory cannot be easily presumed to meet the particular conditions of the Antarctic and its environment.

---

[5] Suffice here to mention EC Council Directive 85/337, concerning the assessment of the effects of certain public and private projects on the environment of 27 June 1985 (*Official Journal of European Communities* L 175 of 5 July 1985), as amended by Directive 97/11 of 3 March 1997 (*Official Journal of European Communities* L 73 of 14 March 1997), and the UN/ECE Convention on Environmental Impact Assessment in a Transboundary Context (Espoo, 25 February 1991), ILM, Vol. 30, 1991, pp. 802ff.

[6] As a matter of fact, it could be difficult to find an independent consultant firm with adequate experience of the Antarctic environment.

The threshold of the adverse impact. The Protocol does not contain a definition of 'adverse impacts on the Antarctic environment and its dependent and associated ecosystems' (Article 3(2)(a)). The environmental principles that according to Article 3 must be 'fundamental considerations in the planning and conduct of all activities in the Antarctic Treaty area' are very wide in scope. The Protocol does not supply a list of mandatory and non-mandatory projects[7] to ensure that the most dangerous activities will always be subject to an initial environmental evaluation (IEE) or a comprehensive environmental evaluation (CEE). Nor does the Protocol provide guidance for a uniform interpretation of the terms 'minor' and 'transitory' impact contained in Articles 1, 2 and 3 of Annex I.[8]

The latter omission already raised discussions and proposals during the most recent Consultative Meetings,[9] and very general Guidelines for EIA in Antarctica were adopted in 1999, during the Twenty-third Consultative Meeting[10]. However, until an agreement is concluded among the parties on the interpretation or the application of the Protocol or a practice is asserted in the application of this instrument 'which establishes the agreement of the parties regarding its interpretation'[11], a wide range of discretion will be left to the parties in the choice of the methodology to be applied in identifying the appropriate level of EIA for different planned activities. Under the Protocol, the final decision on whether to proceed with an activity is remitted to the interested state that, despite criticisms or contrary advice, is not barred from proceeding with the controversial activity[12].

There is, however, at least one limitation that national legislations cannot ignore. The wording of the Protocol and Annex I clearly suggests that the effect of the proposed activities must be assessed on a case-by-case basis. This implies that a domestic legislation providing for a pre-established list of activities which are *ipso jure* excluded from the duty of prior IEE or CEE cannot be considered consistent with the letter and the spirit of the Protocol. On the contrary, in the absence of an agreement or an accepted practice of the parties, nothing could prevent the national legislator from providing guidance to public and private bodies with the indication

---

[7] See, for example, the Espoo Convention, Appendix I, and EC Council Directive 85/337, Annexes I and II.

[8] See further discussion by Bastmeijer, Chapter 16 in this book.

[9] See, e.g., New Zealand, 'Developing an Understanding of "Minor" and "Transitory"', doc. XX ATCM/INF 2, 1996; Russian Federation, 'Contribution to Further Understanding of the Terms "Minor" or "Transitory" Impacts. Russian Viewpoint: Brief Version', doc. XXI ATCM/IP 80, 1997; and Australia, 'Environmental Impact Assessment – The Role of EIA Guidelines in Understanding "Minor" and "Transitory"', doc. XXII ATCM/WP19, 1998.

[10] Resolution 1 (1999), *Final Report of the Twenty-third Antarctic Treaty Consultative Meeting, Lima, Peru, 24 May – 4 June 1999* (Lima: Peruvian Ministry of Foreign Affairs, 1999), Annex C.

[11] Art. 31(3)(b) of the 1969 Vienna Convention on the Law of Treaties, UNTS, Vol. 1155, pp. 331ff.

[12] According to the Antarctic and Southern Ocean Coalition (ASOC), 'the sole option remaining to halt an activity is binding dispute settlement; (...) an option that is not likely to be used', ASOC, 'A Critique of the Protocol to the Antarctic Treaty on Environmental Protection', doc. XVI ATCM/INFO 21, 1991, p. 6.

of methodologies for measuring the duration and the magnitude of the potential impact of different planned activities.

Public participation. From Article 3(3) of Annex I to the Protocol[13] it should be inferred that the right for public comment and scrutiny extends to all individuals, whatever their nationality.

Unlike the UN Convention on Access to Information, Public Participation in Decision-Making and Access to Justice in Environmental Matters (Århus, 25 June 1998)[14], the Protocol does not contain a definition of 'public'[15] or 'public concerned'[16]. However, in Antarctica, where there are no native human inhabitants, any restrictive definition of 'public participation' does not make sense. In the case of Antarctica, where the protection of the environment is, as stated in the preamble of the Protocol, 'in the interest of mankind as a whole', the interested public is every individual and every non-governmental organisation. When a general interest is in question and nobody is directly affected, everybody is affected.

The only procedural limit explicitly set forth in Annex I of the Protocol is the period of time allowed for the receipt of comments: 90 days (Article 3(3)). The domestic authorities are entitled to choose the most appropriate ways for providing the public with the information gathered for the preparation of draft CEEs and for the collecting of expressed opinions. However, the duty established by the Protocol would not be considered properly complied with if the national legislator assumed a mere formalistic approach that failed to give the public a clear, concrete and adequate opportunity for participation.

Under Article 6(2) of Annex I, the duty to provide information in case of IEE is extremely vague: 'Any Initial Environmental Evaluation prepared in accordance with Article 2 shall be made available *on request*' (emphasis added). From the wording of this provision, it is impossible to infer whether information on request is to be provided to the other parties only or also to the public. The most restrictive approach, however undesirable, does not seem to be in conflict with Article 6(2).

No clear duty of public participation is expressed for the EIA in the preliminary stage. At this level, Annex I leaves the parties completely free to assess the environmental impacts 'in accordance with appropriate national procedures' (Article 1(1)).

---

[13] 'The draft Comprehensive Environmental Evaluation shall be made publicly available and shall be circulated to all Parties, which shall also make it publicly available for comment'.

[14] Doc. ECE/CEP/43, 21 April 1998. As of 29 December 1999, the Convention has not entered into force. For a comment see K. Brady, 'New Convention on Access to Information and Public Participation in Environmental Matters', *Environmental Policy and Law*, Vol. 28, 1998, pp. 69–75.

[15] See Art. 2(4) of the Århus Convention.

[16] See *ibid.*, Art. 2(5).

International cooperation. A number of recent international instruments provide for the requirement of international cooperation (information, consultation and negotiation) with other states during the EIA process.[17]

The practice of states in the implementation of duties of international cooperation in the case of projects which are likely to have significant effects on the environment of other states is rather disappointing. For instance, the first EC Commission's five-year review on the implementation of EC Directive 85/337 within member states[18] clearly shows that inter-governmental cooperation is one of the fields where the implementation of the Directive has been most defective[19].

In Antarctica, where the extensive experience of fruitful cooperation among Consultative Parties is a matter of fact, the willingness of the parties to give effect to the Protocol provisions on exchange of information on EIA is evident. In 1995, a Resolution on the EIA circulation of information among Consultative Parties was adopted at the end of the Nineteenth Consultative Meeting.[20] The appeal for a prompt transmission of information on IEEs and CEEs prepared by contracting parties has been constantly reiterated during recent Consultative Meetings.

Besides the requirement for transparency, there are also other important obligations of international cooperation established by the Protocol. For instance, Article 6(1)(b) provides for 1) duties of assistance to other parties in the preparation of EIA; 2) sharing of information that may be helpful to other parties in planning and conducting their activity in the Antarctic Treaty area 'with a view to the protection of the Antarctic environment and dependent and associated ecosystems' (para. 2); and 3) cooperation with parties that exercise jurisdiction in areas adjacent to the Antarctic Treaty area 'with a view to ensuring that activities in the Antarctic Treaty area do not have adverse environmental impacts on those areas' (para. 3).

The effective implementation of these duties depends largely on the capacity and willingness of the institutions in charge of foreign relations of each state party to cooperate with other contracting parties. It cannot be ignored, however, that the dialogue at the international level is easier if national legislations are expressly

---

[17] In addition to EC Directive 85/337 as amended by EC Directive 97/11 (Art. 7) and the 1991 Espoo Convention (Arts. 3 and 5) provisions on exchange of information and consultations among the parties concerned may be found in the ECE Convention on the Protection and Use of Transboundary Watercourses and International Lakes (Helsinki, 17 March 1992), ILM, Vol. 31, 1992, pp. 1,313ff (Arts. 3(1)(h) and 9(2)(j)), and in the UN Convention on the Law of the Non-Navigational Uses of International Watercourses (New York, 21 May 1997), ILM, Vol. 36, 1997, pp. 700ff (Art. 12).

[18] Doc. COM(93) 28, of 2 April 1993.

[19] In 1991, only few member states (Denmark, Germany, Greece, Ireland and Spain) had enacted rules on consultation of other member states in case of significant transboundary impacts. See doc. COM(93) 28, p. 27A.

[20] Under Resolution 6 (1995), the Consultative Parties should provide a list of IEEs and CEEs that were submitted to them during the preceding calendar year to the host country of the subsequent Consultative Meeting, no later than 1 March. For a summary of information on EIA procedures prepared by Consultative Parties, which were circulated during the most recent Consultative Meetings in compliance with Resolution 6 (1995), see docs. XXI ATCM/IP 46 and XXI ATCM/IP 57, Rev. 1, 1997; docs. XXII ATCM/IP 24 and XXII ATCM/IP 25, 1998 and doc. XXIII ATCM/IP 10, 1999.

drafted in order to facilitate international cooperation and specific provisions have been introduced in order to better achieve this aim.[21] In other words, the duty to cooperate, which is clearly established by the Protocol, is directly binding for contracting parties. These are therefore under the obligation to behave in good faith. What is left to the national legislator is only the organisation of human and financial resources that are necessary for the correct implementation of the international duty to cooperate at the national level.

Italian Legislation and Practice

Italian general legislation on EIA is modelled on EC Directive 85/337. The application of the Protocol as *lex specialis*, instead of any other general and incompatible rule, is beyond any discussion. The Protocol aims to enhance the protection of the very special Antarctic ecosystem. Its provisions apply to human activities that take place prevalently in the Antarctic Treaty area and can produce adverse impacts on the Antarctic environment and its dependent and associated ecosystems.

It could be added that not only the aim but also the spirit of the two instruments, the EC Directive and the Protocol, are different. Suffice it to say here that the EC Directive makes a distinction between mandatory and non-mandatory projects,[22] whereas under the Protocol the effect of the proposed activities must be assessed on a case by case basis. The EC Directive provides for exemptions that do not even appear in the text of the Protocol or its Annex I.

Last but not least, the system of rules that implements the EC Directive in Italy is rather complicated and transitory. The system is complicated, because several provisions have been subsequently adopted since the enactment by the EC Council Directive 85/337. This causes evident problems of coordination among different sources of law.

The present regulation is also transitory because a bill containing a general and organic regime on EIA has been under discussion in the Italian Parliament for many years.[23] The objective of the bill is to guarantee the certitude of law through a

---

[21] Interesting provisions are, for example, contained in the German implementing legislation of the Protocol, of 22 September 1994. According to Art. 8 (permit procedure accompanied by CEE), the applicant must submit the results of an assessment of the activity and its environmental impact in the English and German languages. The assessment must include, *inter alia*, a description of the proposed activity, including the likely area of impact. The latter requirement, which is not mentioned in Art. 3 of Annex I to the Protocol (literally transposed in Art. 8 of the German law) should clearly facilitate the detection of adverse environmental impacts upon areas adjacent to the Antarctic Treaty area and subject to the jurisdiction of other state parties.

[22] It is clear that this expression means that mandatory is the EIA procedure (and not the project).

[23] A text of 19 articles and 4 annexes, approved by the Senate in July 1998, is currently under discussion at the Chamber of Deputies. On the bills on EIA, tabled and discussed in the Italian Parliament after 1994, see G. Francescon,'L'evoluzione legislativa in tema di valutazione di impatto ambientale', *Rivista Giuridica dell'Ambiente*, Vol. 10, 1995, pp. 769–788.

clarification and simplification of the national decision-making process. It is also appreciable that the bill is not limited to remedying the deficiencies of the past, as it contains some rules aimed at complying with EC Council Directive 97/11, which amends EC Directive 85/337. Some of the draft provisions specify the national procedures that are to be followed to comply with the Espoo Convention, another treaty that has been incorporated into Italian domestic legislation by means of a simple implementing order.[24] However, whether and when a definitive law will be approved by the Parliament cannot be predicted.

All these remarks lead to some obvious conclusions. Under the pressure of the EC Commission control and the evolution of international treaty law or EC instruments on EIA, Italian legislation is presently under a process of reorganisation and adaptation to more advanced standards of environmental protection. The results of this process may still be uncertain with regard to the consistency and effectiveness of the standards that evolve. What is certain, however, is that unfortunately there are no clear signals of parallel efforts aimed at incorporating into the domestic legislation all the specific rules required to give not only formal but also substantive and effective implementation to the Protocol and its Annex I. At present there is only one statement from which it can be inferred that the Italian Government is aware of the need to adopt specific national regulations for giving effect to the Protocol properly. In the report annexed to the bill containing the ratification and the implementing order of the Protocol, it is clearly stated that the Protocol is only a framework agreement whose further specification and evolution is remitted to national measures of compliance.[25]

Both the laws in force that regulate the EIA procedures for certain public and private activities in Italy and the bill which is presently under discussion at the Italian Parliament can hardly be useful to clarify the ambiguities of the Protocol or compensate for the lack of precision of some of its provisions. From the Protocol, combined with the domestic provisions presently in force in Italy, it is difficult to detect 1) *who* is responsible for certifying and monitoring that a certain planned activity can have less than a minor or transitory impact, a minor or transitory impact, or more than a minor or transitory impact on the Antarctic environment and its dependent and associated ecosystems; 2) *how* and *where* effective public participation in the EIA process is ensured;[26] and 3) *what* happens if private or public bodies fail to comply with provisions on EIA for a planned Antarctic activity.

---

[24] See Law No. 640 of 3 November 1994, *Ord. Suppl.* to *G.U.* No. 273 of 22 November 1994.

[25] *Atti Parlamentari*, Camera dei Deputati, XII Legislatura, Act No. 1458–A, p. 6.

[26] See e.g., the German implementing legislation of the Protocol, of 22 September 1994: Art. 7(4), regulating information to the public in case of IEE; Art. 9, containing measures on public inspection and objections in case of CEE; and Art. 16, concerning EIA of other Parties. Under this Article, documents on EIA circulated by other Parties to the Protocol 'should be laid out publicly at the headquarters of the Federal Environmental Agency for a period of three weeks' (para. 2) and 'comments delivered within the prescribed period should be transmitted to the Parties concerned' (para. 3).

But even supposing that certain national provisions in force could apply through analogy to Antarctic activities, evident requirements of time and cost-effectiveness and, above all, the certainty of law impose the adoption of provisions specifically tailored for Antarctic activities. A few examples may be sufficient to illustrate this assumption.

Only one central, well-identified authority should be responsible for application, granting, revocation or suspension of a permit and the monitoring of permitted activities.[27] The involvement of different bodies in the decision-making process can only complicate and slow down the procedure. The principle of subsidiarity has no reason to be invoked in the case of Antarctic activities.

Finally, sufficiently dissuasive sanctions should be conceived considering the particularly serious (or even irreversible) danger the Antarctic environment is exposed to by incautious action. Article 18(2) of the bill that is presently under discussion in the Italian Parliament provides for administrative penalties against those who disregard EIA procedures. The penalties consist of payments of fines from fifty to one thousand million lira, or a sum of money equivalent to twenty per cent of the total cost of the work if immediately quantifiable. Additional measures would also be appropriate, such as the withdrawal of the permits and coercive removal of the work at the organiser's own expense. A guarantee fund could also be established to recover liabilities the Italian Government could incur at the international level (as in the case of transboundary movements of hazardous wastes)[28].

THE CONSERVATION OF ANTARCTIC FAUNA AND FLORA

The provisions of Annex II to the Protocol update and strengthen the 1964 Agreed Measures for the Conservation of Antarctic Fauna and Flora (Agreed Measures)[29]. Formally, the issues concerning the protection of Antarctic species within their habitat and the conservation of areas worthy of special consideration are dealt with separately, by two different instruments: Annex II (Conservation of Antarctic Fauna and Flora) and Annex V (Area Protection and Management). Annex V is embodied in a Consultative Meeting recommendation (Recommendation XVI-10 of 17

---

[27] A number of national legislations follow this approach. See, e.g., the Finland Act on the Environmental Protection of Antarctica of 18 October 1996, doc. XXI ATCM/IP 101, 1997 (Ministry of the Environment); the 1997 Japanese Law Relating to Protection of the Environment in Antarctica, doc. XXI ATCM/IP 112, 1997 (the Director General of the Environmental Agency); the New Zealand Antarctica (Environmental Protection) Act, 1994, No. 119 (Minister of Foreign Affairs and Trade); the Norwegian Regulations Relating to Protection of the Environment in Antarctica of 5 May 1995 (Norwegian Polar Institute; for a detailed discussion see Njåstad, Chapter 20 in this book); the United Kingdom Antarctic Regulations, 1995, No. 490 (Secretary of State).

[28] See Law No. 475 of 9 November 1988 (Art. 9 *bis*(3)), *G.U.* No. 264 of 10 November 1988; DPCM No. 457 of 22 October 1988 (Art. 6(2)), *G.U.* No. 256 of 31 October 1988; Decree of the Ministry of the Environment of 26 April 1989, *G.U.* No. 128 of 3 June 1989.

[29] Recommendation III–VIII of 1964.

October 1991) that was accepted by Italy on 31 March 1995. Because Recommendation XVI-10 has not yet become effective at the international level,[30] attention will be focused on the implementation of Annex II to the Protocol.

The Provisions of the Protocol

Annex II contains several prohibitions[31] and some stringent *facere* obligations[32] for the conservation of the Antarctic flora and fauna prevailingly, but not exclusively, *in situ*.

In principle, almost all the provisions of the former category can be derogated in accordance with a permit.[33] This, however, does not affect the self-executing character of these rules. The issue of national permits is subject to several restrictions that are clearly expressed by the Protocol. Furthermore, the use of vague expressions such as '*unavoidable* consequences of scientific activities',[34] '*compelling* scientific purpose',[35] or use of non lethal techniques '*where appropriate*'[36] does not affect the content of the obligation. The scope of these rules is to leave reasonable flexibility to administrative bodies who are responsible for the issuance of permits and who decide on a case by case basis, in the light of different variables, whether certain activities are consistent with the Protocol.

Similarly, certain provisions of Annex II to the Protocol (e.g., Article 1(h)(ii) and Article 4(1) erode the freedoms of the high seas, which in principle are not affected by the Protocol (Article 1(b)). Also in this case it will be necessary to determine case by case whether the exercise of the traditional freedoms of the high seas conflicts with the scope of the protection of Antarctic fauna and flora.

On the other hand, the intervention of the national legislator is absolutely necessary at the organisational level. Without the adoption of clear domestic provisions, specifically regulating procedures for the issue of permits, control over scientific and tourist activities and appropriate penalties for unlawful behaviour, it is hardly possible to ensure effective compliance with Annex II to the Protocol.

Italian Legislation and Practice

An appraisal of Italian legislation to ascertain whether the domestic rules on the conservation of flora and fauna already contain all the elements necessary to give

---

[30] See discussion by Vidas, Chapter 1 in this book.

[31] See, e.g., Art. 3(1) and Art. 4(1) and (2).

[32] See, e.g., Art. 4(4) providing for a duty of removal or elimination of non-native plants and animals that have been imported to Antarctica in accordance with a permit.

[33] An exception is Art. 4(2) of Annex II, prohibiting the introduction of dogs into Antarctica.

[34] Art. 3(2)(c) of Annex II to the Protocol.

[35] *Ibid.*, Art. 3(5)(a).

[36] *Ibid.*, Art. 3(5)(c).

proper effect to Annex II to the Protocol in Italy is a complicated and frustrating process.

The first step is obviously to refer to the provisions that Italy should have enacted to give effect to the Agreed Measures. Nothing is forthcoming, however, since these provisions do not exist. The Agreed Measures were accepted by Italy in 1987, along with all the Consultative Meeting recommendations in force then, when the country acquired the status of Consultative Party to the Antarctic Treaty. However, no specific implementing measures have been adopted by the Italian legislator since then.

On the other hand, it is rather difficult to extrapolate from the Italian legislation in force all the elements necessary to give proper effect to Annex II to the Protocol. In Italy, a general law for the protection of fauna and flora does not exist;[37] as has happened in several other countries, the sectorial approach (protection of certain species and areas and regulation of certain activities, such as hunting and fishing) has been preferred. Furthermore, some of these provisions give effect to international treaties[38] and EC Directives. Consequently, all these rules must be carefully integrated with the international provisions to which they refer.[39]

In principle, however, the national provisions in force are inadequate to cover the peculiar scope of the Antarctic Environmental Protocol. Law No. 157 of 11 February 1992, concerning the protection of homeothermic wild fauna and hunting activities[40], evidences broad scope in purporting to protect 'wild fauna in the interest of the national and international community' (Article 1(1)). Furthermore, in addition to species that are present on the national territory, law provides for 'particular protection' to all species that EC directives or international conventions consider to be under threat of extinction (Article 2(1)(c)). However, this law has shortcomings with respect to incorporating sufficient elements that would give proper effect to

---

[37] Art. 9 of the Constitution generally provides that the Italian Republic protects the landscape ('paesaggio').

[38] See e.g., Law No. 874 of 19 December 1975, giving effect to the Convention on International Trade in Endangered Species of Wild Fauna and Flora (CITES) (Washington, 3 March 1973) (*G.U.* No. 49 of 24 February 1976), supplemented by Law No. 150 of 7 February 1992 (*G.U.* No. 44 of 22 February 1992); Law No. 42 of 25 January 1983 (*Ord. Suppl.* to *G.U.* No. 48 of 18 February 1983), giving effect to the Convention on the Conservation of Migratory Species of Wild Animals (Bonn, 23 June 1979); or Law No. 124 of 14 February 1994 (*Ord. Suppl.* to *G.U.* No. 33 of 23 February 1994), giving effect to the 1992 Convention on the Biological Diversity (Nairobi, 22 May 1992).

[39] As far as the protection of the Antarctic environment is concerned, suffice here to recall, by way of example, that none of the species that are listed in the Appendices to the Bonn Convention are present in the Antarctic Treaty area. Furthermore, Italian Presidential Decree No. 357 of 8 September 1997, providing for the implementation of EC Directive 92/43 of 21 May 1992 on the protection of natural habitats (*G.U.* No. 248 of 23 October 1997), is irrelevant because the sphere of application of the EC directive is limited to the protection of natural habitats and wild flora and fauna that are present in the European territory of EC member states (Art. 2(1)).

[40] *Ord. Suppl.* No. 41 to *Gazz. Uff.* No. 46 of 25 February 1992. For an interesting comment that focuses on the consistency of this law with international treaties and EC provisions on the protection of wild fauna, see M.C. Maffei, 'Aspetti internazionalistici della nuova legge italiana sulla caccia (L. 11 febbraio 1992, n. 157)', *Rivista Giuridica dell'Ambiente*, Vol. 7, 1992, pp. 939–944.

Annex II to the Protocol in Italy. Law No. 157 of 1992 has, as a primary objective, the protection of wild fauna (only) within the national territory. Concerning the importation to Italy of certain specimens, the law adopted in 1992 contains provisions that are more restrictive than the Protocol. The latter admits the 'taking', a term that includes, *inter alia*, the killing or capture,[41] of a native mammal or bird (which in principle is prohibited)[42] only under a specific permit, with the aim of providing specimens for scientific study and information, as well as for museums, zoological gardens and other educational or cultural institutions or uses[43]. On the contrary, Article 20 of the Italian law admits the importation of living specimens of wild fauna, providing that the specimens belong to indigenous species and will only serve the purposes of repopulation and genetic improvement. To this end, authorisations are issued by the Ministry of Forests and Agriculture 'in observance of international conventions'. Nothing is said, however, about the importation of dead specimens.

Italian jurisprudence has shown sensitivity toward the idea of a broad protection of wildlife. A recent decision, rendered by the Court of Cassation in 1994, clearly states that the notion of 'fauna' contained in the domestic legislation must not be associated with a certain territorial dimension, but with the specific interest of the international community to a general and objective protection of species.[44] In contrast with previous decisions of other domestic tribunals, which restricted the protection provided by Italian legislation to animals actually living in the state territory,[45] the Court of Cassation has recognised the existence of a general principle. This principle prohibits the possession and trade of specimens belonging to species that cannot be hunted under Italian law, whether the specimen in question (dead or alive) belongs to species living in Italy or to species imported from other parts of the world.

It is also true, however, that a restrictive interpretation of the notion of 'fauna' was preferred by the same Court in another decision rendered at the end of the same year.[46]

In this scenario, the adoption of domestic provisions specifically conceived for the protection of Antarctic flora and fauna is absolutely necessary. In particular clear 'organisational' provisions should be enacted that provide, for instance, pre-

---

[41] Art. 1(g) of Annex II to the Protocol.

[42] *Ibid.*, Art. 3(1).

[43] *Ibid.*, Art. 3(1) and (2).

[44] See Court of Cassation, III Penal Section, 18 February 1994 (8 March 1994), in *Rivista Giuridica dell'Ambiente*, Vol. 10, 1995, pp. 693–700 (see in particular pp. 696 and 700), with a comment by A. Maestroni, 'Il concetto di specie e il principio di extraterritorialità nella tutela della fauna selvatica', *ibid.*, pp. 700–702.

[45] See, e.g., Court of Cassation, III Penal Section, 17 August 1993, No. 1013, *Rivista penale*, 1994–I, p. 569.

[46] See, e.g., Penal Court of Cassation, Joint Sections, 28 December 1994, No. 25, *Cassazione penale*, Vol. 35, 1995, p. 892.

established standard forms[47] or specification of the basic information required from the applicant for a permit[48]. Such provisions would benefit both applicants for activities in the Antarctic Treaty area and the national authorities that are competent for the collection and dissemination of records and statistics to other contracting parties (Article 6 of Annex II). Clarification should also be given regarding who is responsible at the national level for the production, revocation and suspension of permits and for the organisation of certain preventive inspection and control activities.[49] Finally, the national legislator should enact adequate rules to provide for 1) the cessation of activities that cause harmful interference with Antarctic flora and fauna, 2) the removal of animals, plants, soil, substances or products that have been brought into the Antarctic without a permit, 3) the conditions on which permits are revoked or suspended, and 4) the determining of adequate penalties for deliberate or negligent violations of the rules of law.

WASTE DISPOSAL AND WASTE MANAGEMENT

The Provisions of the Protocol

Some of the basic considerations that have been expressed with regard to Annex II to the Protocol can be extended to Annex III, concerning waste disposal and waste management. This Annex lays down several stringent commands (see, e.g., Article 2) and expresses prohibitions (see, e.g., Article 7). In both cases, no further legislative activity is required because all the necessary and sufficient elements for the correct implementation of the Protocol can be inferred directly from its own rules. There are also many provisions of the same Annex that are expressed in terms of soft-law.[50] This, however, does not exclude the self-executing nature of such provisions. On the contrary, it is evident that Annex III also contains provisions that can become effective if a specific initiative is undertaken by the domestic legislator to regulate some procedural aspects concerning the issuance of permits, control of activities and determination of administrative and criminal penalties.

Unlike Annex II, however, where the adoption of appropriate domestic rules can be only implicitly inferred from the *ratio* of its provisions, Annex III provides also for some obligations that are openly placed upon the domestic legislator. Article 8, for instance, requires contracting parties active in Antarctica to prepare and annually review and update their waste management plans; these must contain specific information and conditions for fixed areas, field camps and certain ships. Needless

---

[47] Questionnaires have been set up, for instance, by Argentina (see doc. XX ATCM/INF 28, 1996).

[48] See, e.g., Arts. 4–7 of the Draft Decree on the Protection of Antarctica, submitted by the Netherlands during the XXII Consultative Meeting (doc. XXII ATCM/IP 33, 1998).

[49] For instance, inspection of dressed poultry that is brought to the Antarctic is required by Annex II, Appendix C, to prevent the spread of certain diseases in the Antarctic Treaty area.

[50] See, e.g., Art. 1(2), Art. 4(2) and Art. 5(1)(a).

to say, such a provision is not applicable without interposing a domestic act providing for the designation of the competent authority at the national level to collect data, draw up Antarctic waste management plans and control their implementation. Even supposing that this were not sufficiently evident from the *ratio* of the rule, Article 10 of the same Annex reminds contracting parties that a waste management official must be designated at the national level 'to develop and monitor waste management plans'.

Italian Legislation and Practice

Italian legislation on waste has recently been revised and updated by Legislative Decree No. 22 of 5 February 1997, which contains measures for the implementation of EC Directives 91/156 on waste, 91/689 on harmful waste and 94/62 on packaging and packaging waste (the so called 'Ronchi Decree', from the name of its proposer)[51]. No rules of the decree contain useful elements to integrate the Protocol provisions on wastes generated in Antarctica, the principal concern of the Protocol. The Ronchi Decree is primarily concerned with the management of wastes produced in the national territory. Chapter III contains provisions regulating the establishment of management plans of wastes, which must be set up by regions. This solution is the most suitable for a wise management of wastes at the local level; it is also clear, however, that the waste management planning activity provided for under Article 8 of Annex III to the Protocol must be organised at the central level if an efficient mechanism for the organisation and control of the production and management of wastes in Antarctica is to be established. Article 26 of the Ronchi Decree provides for the establishment of a National Observatory on Wastes within the Ministry of the Environment, but no provisions specifically address the aims pursued by Article 10 of Annex III to the Protocol.

As far as the transboundary movement of hazardous wastes is concerned, the Ronchi Decree refers to EC Council Regulation No. 259/93 of 1 February 1993 on the Supervision and Control of Shipments of Waste within, into and out of the European Community[52]. This Regulation prohibits all exportation of waste for disposal and recovery, with some exceptions. These, however, have no consequences for the protection of the Antarctic environment because a general ban to export hazardous wastes or other wastes for disposal within the area south of 60° south ('whether or not such wastes are subject to transboundary movements'[53]) is contained in Article 4(6), of the Convention on the Control of Transboundary

---

[51] For a comment, see *Rivista Giuridica dell'Ambiente* (special issue devoted to the Ronchi Decree), Vol. 12, 1997, pp. 387–622.

[52] *Official Journal of European Communities* L 30 of 6 February 1993.

[53] The wording of this provision is appreciable, since it excludes any doubts on the existence of a ban also on shipping movements between a claimant state and its claimed Antarctic sector.

Movements of Hazardous Wastes and their Disposal (Basel, 22 March 1989)[54]. Italy, which is party both to the Basel Convention and the Antarctic Environmental Protocol, will therefore be bound to the general prohibition contained in the EC Council Regulation, even though Article 4(6) of the Basel Convention does not explicitly prohibit the import of hazardous wastes into Antarctica.

With regard to the importation of wastes from Antarctica to Italy, the EC Council Regulation is ineffective because Article 1(2)(e) explicitly excludes the application of its provisions to the shipments of waste into EC territory, which fall under the application of the Protocol.[55]

Despite the limits of its domestic legislation, the Italian Antarctic National Research Programme (PNRA) has undertaken appreciable initiatives to ensure environmentally sound waste management and waste disposal since its first Antarctic season. In an information paper submitted by Italy to the XXII Consultative Meeting, held in Tromsø in 1998, it is said that Waste Management Plans have been prepared since April 1989 and when the Protocol was signed in 1991 'it was found that the Waste Management Plan that had already been applied was in line with the Protocol'.[56] All wastes from Antarctic activities (Terra Nova station, its field camps, logistic and research vessels) are removed from the Antarctic Treaty area; 'the only waste that is disposed of in Antarctica is the treated effluent from the sewage treatment plant operating at the base'[57]. Terra Nova station is equipped with a double chamber incinerator whose 'gases pass through a scrubber system before discharge into the atmosphere'.[58] A completely new sewage plant has been operating at the same station since 1998 and 'the sludge resulting from the treatment is retrograded to Italy'.[59] For the final disposal, contracts have been signed with waste treatment and disposal organisations that operate in Italy and elsewhere, but rigorously outside the Antarctic Treaty area.

In principle, Italian practice seems to be consistent with the precepts of the Protocol. The Italian Antarctic National Research Programme (PNRA) has also shown, in many respects, that it is inclined to go beyond a strictly literal interpretation of the environmental obligations established at the international level.

---

[54] ILM, Vol. 28, 1989, pp. 652ff. The Convention, which entered into force at the international level on 5 May 1992, has been binding for the European Community and Italy since 7 February 1994.

[55] Incidentally, it can be interesting to recall that in its first report on the implementation of the EC Council Regulation 259/93 (doc. COM(1998)475 def., 28 July 1998), the EC Commission clearly states that no member state has submitted information on wastes to which the EC Council Regulation does not apply according to Art. 1(2). As far as other information is concerned, the general situation is disappointing: the information that has been supplied to the Commission by most member states is inadequate, while the information provided by others (including Italy) is 'non-existent'.

[56] Italy, 'Waste Management at the Italian Terra Nova Bay Station', doc. XXII ATCM/IP 35, 1998.

[57] 'The only waste which is disposed of in Antarctica is the treated effluent from the sewage treatment plant operating at the base', *ibid*.

[58] *Ibid*.

[59] *Ibid*.

There are some limits, however, which are very evident at the present state of affairs. Suffice here to say, with regard to the effective implementation of Article 8 of Annex III to the Protocol, the practice undertaken by the Italian National Programme is commendable; however, it is a simple practice that can be modified, in the absence of general criteria established by law, at the discretion of those responsible for the Italian National Programme. Furthermore, it cannot be ignored that no subjects other than those from the National Agency for Atomic Energy (ENEA), the agency directly involved in the organisation of Italian activities in the Antarctic, supervise PNRA's waste management and waste disposal activities in the Antarctic Treaty area. These are sufficient arguments to conclude that Italy has not fully implemented some fundamental obligations that are incumbent on it as a contracting party to the Protocol.

PREVENTION OF MARINE POLLUTION

The Provisions of the Protocol

Any discussion on the correct implementation of Annex IV to the Protocol on the prevention of marine pollution at the domestic level must begin from fundamental considerations. Most provisions of this Annex closely track provisions of the International Convention for the Prevention of Pollution from Ships of 1973, as amended by the Protocol of 1978 (MARPOL 73/78)[60]. Parties to the Protocol that are also parties to MARPOL 73/78 can therefore apply to the Antarctic Treaty area the provisions they have already enacted at the domestic level to give effect to the MARPOL 73/78. Rules of Annex IV that go beyond MARPOL 73/78 (such as rules regulating the disposal into the sea of garbage other than plastics in Article 5(2) and (3)) can be effectively implemented either by supplementing the domestic rules aimed at giving effect to MARPOL 73/78[61] or by treating them as conditions for the granting of a permit[62].

Various problems arise with respect to the correct implementation of other provisions of Annex IV, as for instance with Article 12 of Annex IV, which concerns marine pollution emergency issues. This Article places upon the parties the burden to prepare 'contingency plans for marine pollution response in the Antarctic Treaty area'. These plans cover emergencies where ships ('other than small boats that are part of the operations of fixed sites or of ships') or coastal installations can cause, in particular, oil pollution of the marine environment. The obligation is

---

[60] *Marpol 73/78. Consolidated Edition, 1991* (London: International Maritime Organisation, 1992).
[61] This is the approach followed by the Dutch Government; see the Netherlands, 'Information on Dutch Legislation Implementing the Protocol', doc. XXII ATCM/IP 31, 1998, p. 1.
[62] This is the United Kingdom approach. It was clearly explained in a letter of 22 October 1996 by Ms. Karen Miller (Foreign and Commonwealth Office, Polar Regions Section, South Atlantic and Antarctic Department) who kindly replied to some of the author's queries.

expressed in mandatory terms ('the Parties ... shall develop') and requires that an *ad hoc* initiative (in other words, an initiative expressly conceived with regard to the specific environmental conditions of the Antarctic) is undertaken by the national authorities of each contracting Party. The same Article provides for a duty of cooperation with other contracting parties, as well as with the CEP and the IMO (Article 12(1)(a) and (b)), which can produce its full effects provided that an adequate organisational regime has been previously established at the domestic level. These two considerations are sufficient to lead to the obvious conclusion that Article 12 cannot be considered self-executing.

Italian Legislation and Practice

MARPOL 73/78 was integrated into Italian legislation by Law No. 662 of 29 September 1980[63] and Law No. 438 of 4 June 1982, which contain the implementing order of the amending Protocol of 1978[64]. The 'material obligations' of MARPOL 73/78 (in other words, the provisions of the Convention which regulate the conduct of maritime operators through uniform technical parameters) are the law of the state and as such they must be observed by Italian ships both within and outside Italian territorial waters.

There are, however, some problems connected with the concrete implementation of MARPOL 73/78 in Italy that are worth mentioning. One issue relates to the lack of both an adequate system of port facilities for the reception and treatment of ship-generated waste and cargo-residues and an efficient system of control over compliance. The EC Commission has recently focused its attention on these problems and the drafting of a Community directive is at present under discussion[65]. It is to be hoped that under the pressure of EC initiatives, the Italian legislator will be induced to review domestic legislation through the adoption of specific measures.

A second issue arises from the existing contrast between MARPOL 73/78 and Italian Law No. 979 of 31 December 1982 regarding rules for the protection of the sea[66]. According to Article 16 of this law, ships flying the Italian flag are bound to stricter obligations than those required by MARPOL 73/78 because Italian ships are forbidden to discharge into the high seas hydrocarbons, mixtures of hydrocarbons and chemical substances listed in Annex A to that law. On the conflict between Article 16 of the Italian law and MARPOL 73/78, divergent decisions have been rendered by Italian courts[67] and divergent solutions have been suggested by

---

[63] *Ord. Suppl.* to *G.U.* No. 292 of 23 October 1980.
[64] *Ord. Suppl.* to *G.U.* No. 193 of 15 July 1982.
[65] See doc. COM(1998)452 def., 17 July 1998.
[66] *Ord. Suppl.* to *G.U.* No. 16 of 18 January 1983.
[67] See. e.g., Pretura of Genoa, Penal Section, 12 May 1995, No. 3989, *Rivista Giuridica dell'Ambiente*, Vol. 11, 1996, pp. 736–744; Court of Cassation, III Penal Section, 22 September 1995 (15 November 1995), *Rivista Giuridica dell'Ambiente*, Vol. 12, 1997, pp. 683–684 (with a comment by A. Merialdi, 'Il

academic writers[68]. The issue has also been raised before the EC Court of Justice; however, the court excluded its competence to pronounce on the consistency of a domestic provision adopted by a member state with MARPOL 73/78[69].

Concerning the protection of the Antarctic environment, the two issues must be considered in light of the provisions of both Annex IV to the Protocol and the 1990 amendments to MARPOL 73/78. The amendments to MARPOL were adopted by the Marine Environment Protection Committee (MEPC) of the IMO on 16 November 1990 and proclaim the Antarctic area south of 60° south as a special area under Annexes I and V of MARPOL 73/78 (hereinafter 1990 MEPC Resolution)[70]. Under this instrument, which entered into force on 17 March 1992, *any discharge* into the sea of oil or oily mixture from *any ship* shall be prohibited. For states that are parties to both the Protocol and Annex V to MARPOL 73/78, this unambiguous preclusion prevails over Article 3 of Annex IV to the Protocol (providing that any discharge into the sea of oil or oily mixture are prohibited, except in cases permitted under Annex I of MARPOL 73/78). Therefore, no conflict between the Italian legislation and MARPOL 73/78 can be envisaged with regard to the protection of the Antarctic high seas.

Ship retention capacity and reception facilities are also regulated by both Annex IV to the Protocol and MARPOL 73/78, as amended by the 1990 MEPC Resolution. In this case, however, the obligations that are set forth by the two instruments are substantially coincident. States parties to the Protocol (and MARPOL 73/78) ensure that ships flying their flags (and any other ship engaged in or supporting their operations, under the Protocol) before entering the Antarctic Treaty area meet certain technical requirements that guarantee sufficient capacity for the retention of wastes on board while operating in the Antarctic Treaty area.[71]

---

contrasto tra norme interne e Convenzione MARPOL: diritto internazionale del mare, codice penale e possibili sviluppi normativi', *ibid.*, pp. 684–693; Pretura of Venezia, Penal Section, 28 February 1996, No. 1400, *ibid.*, pp. 750–754 (with a comment by T. Scovazzi, 'L'illuminazione in un contesto indigesto', *ibid.*, pp. 754–756); Court of Cassation, III Penal Section, 19 November 1996 (29 January 1997), *ibid.*, pp. 917–923 (with a comment by A. Merialdi, 'Legge 979/1982 e Convenzione MARPOL: la Cassazione cambia indirizzo', *ibid.*, pp. 923–924); Court of Cassation, Joint Penal Sections, 24 June 1998 (24 July 1998), *Rivista Giuridica dell'Ambiente*, Vol. 14, 1999, pp. 249–261 (with a comment by A. Merialdi, 'Legge 979/1982 e Convenzione MARPOL: la Cassazione Penale si pronuncia a Sezioni Unite', *ibid.*, pp. 261–265).

[68] See, e.g., A. Merialdi, 'Il contrasto tra norme interne e Convenzione MARPOL: diritto internazionale del mare, codice penale e possibili sviluppi normativi', pp. 684ff. and T. Scovazzi, 'L'illuminazione in un contesto indigesto', pp. 754ff.

[69] In the opinion of the Court, since the European Community is not a contracting party to the London Convention, this cannot produce effects on Community law. EC Court of Justice, decision of 14 July 1994 (*re* C-379/92). For a critical stand on the decision of the Court, see P. Di Leo, 'La tutela del mare: la legge n. 979/82 e la convenzione Marpol', *Il Diritto Marittimo*, Vol. 97, 1995, pp. 505–512.

[70] Resolution MEPC 42(30).

[71] Arts. 9(1) and 10 of Annex IV to the Protocol and Reg. 10 of Annex I to MARPOL 73/78. The expressed mention of noxious liquid substances in Art. 9(1) of the Protocol is the only relevant difference between the two instruments.

Contracting Parties whose ports are used by ships departing for or arriving from the Antarctic Treaty area also undertake to establish 'as soon as practicable' adequate reception facilities for the reception of wastes (sludge, dirty ballast, tank washing water, and other oily residues and mixtures from all ships) produced in the Antarctic.[72] Unlike other 'special areas' proclaimed under MARPOL 73/78, where the coastal states of the areas undertake to install adequate reception facilities at their ports, in the Antarctic these facilities must be established outside the area. The Consultative Parties have in fact agreed that the establishment of waste reception facilities for vessels operating in the Antarctic is not an adequate solution, because there is the risk of transferring the waste disposal problem from vessels to Antarctic stations and facilities.[73] The Protocol therefore provides for a duty of cooperation among contracting parties so that the establishment of port reception facilities does not place a burden only upon parties adjacent to the Antarctic Treaty area[74].

At present, there are only few ships that sail in the Antarctic Treaty area under the Italian flag. Nevertheless, this fact does not exonerate the Italian government from the duty to adopt all the necessary measures – at both the legislative or administrative levels – to ensure that vessels flying the Italian flag comply with the international requirements before entering the Antarctic Treaty Area. This implies, *inter alia*, that Italy must establish whether its vessels en route to Antarctica comply with international requirements at the points of issuance of certificates, and the same vessels must be regularly checked to ensure that they continue to satisfy the requirements laid down in Annex IV to the Protocol and MARPOL 73/78. A responsible step toward comprehensive marine environmental protection of the Antarctic Treaty area could also be taken through the adoption of national measures that bring all vessels owned or operated by Italy under the regulations and restrictions of Annex IV to the Protocol, despite the sovereign immunity accorded by Article 11(1) of the same Annex.[75]

Adequate measures should also be taken to ensure appropriate means of control over ships engaged in or supporting Italian Antarctic activities (Article 9(1) of the Protocol).

The establishment of appropriate port facilities in Italy for the recovery of wastes produced by ships engaged in Antarctic operations is not a problem to be faced in the short term. Furthermore, although the subjects to which Article 9(3) of this provision refers are defined very broadly, it would be difficult to infer from the

---

[72] Art. 9(2) of Annex IV to the Protocol and Reg. 10(8) of Annex I and Reg. 5(5) of Annex V to MARPOL 73/78.

[73] *Final Report of the Fifteenth Antarctic Treaty Consultative Meeting, Paris, 9–20 October 1989* (Paris: French Ministry of Foreign Affairs, 1990), para. 92.

[74] Art. 9(3) of Annex IV to the Protocol.

[75] According to Art. 11(1) of Annex IV 'shall not apply to any warships, naval auxiliary or other ship owned or operated by a state and used, for the time being, only on government non-commercial service'. See discussion by C.C. Joyner, 'Protection of the Antarctic Environment Against Marine Pollution under the 1991 Protocol', in D. Vidas (ed.), *Protecting the Polar Marine Environment: Law*

literal wording of the Article ('Parties *operating ships*') that the duty to undertake consultations is placed not only upon the flag states of the ships navigating in the Antarctic Treaty area, but also upon states (as Italy could be) which are active in the area with the support of ships flying the flag of other states. When the parties decided to extend a certain obligation both to states that are active in the Antarctic Treaty area with their own ships and to states that operate in the Antarctic with the support of ships flying the flag of other states, they stated this explicitly (see Article 9(1)).

Nevertheless, it would be appreciable if Italy gave a broad interpretation to its undertakings, paying more attention to the spirit and the *ratio* of the rule than to the literal wording. It would be desirable if, under domestic legislation, Italian expeditions to the Antarctic could be allowed to make use of foreign vessels provided that the flag state had established adequate reception facilities for ships engaged in or supporting Antarctic operations in its ports.

CONCLUSIONS

The mere fact that in Italy certain treaties acquire the status of domestic law upon the enactment of a simple *implementing order* does not necessarily mean that the provisions of such treaties create directly enforceable legal rights or obligations. In other words, by means of the *implementing order* procedure a treaty is validly incorporated into the Italian legal system and its provisions must be observed as a law of the state.[76] This procedure is not sufficient, however, to make the provisions of treaties which are non self-executing (as certain provisions of the Protocol are) concretely applicable.

It is true that even non self-executing treaties can produce legal effects. Although such treaties cannot operate directly without specific implementing legislation, they can be used indirectly for interpretative purposes, in other words, as a means for interpreting relevant domestic legal provisions.[77] Non self-executing treaties can also affect domestic law indirectly; in principle, a rule of national law conflicting with international law would not be applied by domestic courts. Finally, there are relevant consequences from the perspective of international responsibility. When a state is charged with a breach of its international obligations, it cannot claim as a defence that it was unable to fulfil the obligations because its internal law was defective or contained rules in conflict with international law. These consequences are very limited, however, and they are not sufficient to bring Italy into conformity

---

and Policy of Pollution Prevention (Cambridge University Press, 2000), pp. 117–119.

[76] This is the formula that usually appears in the legislative act containing the *implementing order*. As has been correctly observed, it is not a mere stylistic formula at all; see T. Scovazzi, 'Un gorilla, cinque tigri, nove scimpanzé, una convenzione e una costituzione', *Rivista Giuridica dell'Ambiente*, Vol. 11, 1996, pp. 336–340, at p. 339.

[77] Paust, 'Self-Executing Treaties', p. 781.

penalising measures for certain acts or omissions in violation of the Protocol.[78] However, Article 13(1) of the Protocol clearly places a burden upon its contracting parties to take appropriate measures ('including the adoption of laws and regulations, administrative actions and enforcement measures') *to ensure compliance* with the Protocol. It would be contrary to the principle of good faith, which must always be taken into consideration in the correct interpretation of international treaties,[79] to give effective compliance at the national level with international treaties without the definition of adequate penalties for unlawful acts or omissions.

At the national level, there is the risk of a conflict between Law No. 54 of 15 February 1995 (containing the *implementing order* of the Protocol) and the Italian Constitution. Article 25(2) of the Constitution implies that crimes and sanctions must be determined before certain acts can be punished (*nullum crimen sine lege, nulla poena sine lege*). In the specific case, there is in Italy a law that prescribes the observance of certain environmental obligations for Antarctic activities, but there are no laws that provide for adequate penalties if such obligations are infringed. Consequently, Italian judges who apply Law No. 54 of 1995 cannot inflict penalties without infringing at the same time Article 25(2) of the Italian Constitution.[80]

No further measures have been enacted in Italy since the publication on the Official Journal of Law No. 54 of 1995, which orders 'the full and complete' implementation of the Protocol into the domestic legal system. Nothing, however, prevents the national legislator from following the *ordinary* procedure to give full effect to a treaty that has already been incorporated within the domestic legal order according to the *special* procedure. Some recent examples illustrate this practice with specific reference to international environmental agreements.[81]

In conclusion, with the adoption of Law No. 54 of 15 February 1995, the Italian legislator has done the minimum necessary to proceed to the ratification of the Antarctic Environmental Protocol. In principle, the practice shows a responsible environmental attitude of the Italian bodies that are involved in the organisation and

---

[78] For instance, under Art. VIII(1) of CITES: 'The Parties shall take appropriate measures to enforce the provisions of the present Convention and to prohibit trade in specimens in violation thereof. These shall include measures: (a) to penalise trade in, or possession of, such specimens, or both; and (b) to provide for the confiscation or return to the state of export of such specimens'.

[79] See Art. 31(1) of the Vienna Convention on the Law of Treaties.

[80] In this sense, *mutatis mutandis*, with reference to the implementation in Italy of CITES, see Scovazzi, 'Un gorilla, cinque tigri, nove scimpanzé', p. 336 ff.

[81] For instance, CITES has been incorporated within the Italian legal system by means of an *implementing order* (Law No. 874 of 19 December 1975), but in 1992 a specific law was enacted to regulate sanctions for offences deriving from the violation of this Convention (Law No. 150 of 7 February 1992, *G.U.* No. 44 of 22 February 1992). Law No. 549 of 28 December 1993 (*G.U.* No. 305 of 30 December 1993) gives full effect by means of the *ordinary procedure* to some specific provisions of the Convention for the Protection of the Ozone Layer and the Protocol on Substances that Deplete the Ozone Layer for which Italy had already adopted laws containing an *implementing order* in 1988 (see Law No. 277 of 4 July 1988, in *G.U.* No. 170 of 21 July 1988 and Law No. 393 of 23 August 1988, in *G.U.* No. 211 of 8 September 1988).

achievement of the objectives of the Italian National Programme. The intervention of the domestic legislator is still necessary, however, not only to give substantial effect to the Protocol obligations within the domestic legal system, but also to support the activities of Italian operators in the Antarctic with clear terms of reference in the organisation of their undertakings.

20

Norway: Implementing the Protocol on Environmental Protection

Birgit Njåstad

Extensive Norwegian whaling activity and a tradition for exploring, mapping and research were the primary reasons for Norway's claims in the Antarctic region. Claims were put forward not to exclude other nations, but to ensure access to the Antarctic and sub-Antarctic whaling grounds for Norwegian whalers. Norway put forth claims to three distinct areas and established these as her dependencies: the sub-Antarctic island Bouvetøya (54°25' S, 3°21' E), Peter I Øy in the Bellinghausen Sea (68°50' S, 90°35' W) and Dronning Maud Land (its western and eastern boundaries at 20° W and 45° E longitude respectively). As a party to the Antarctic Treaty,[1] Norway has agreed that while the Treaty is in force, no acts or activities taking place shall constitute a basis for asserting, supporting or denying a claim to territorial sovereignty in Antarctica or create any rights of sovereignty in Antarctica[2].

Presently, tourism and scientific research are the primary activities in the Norwegian Antarctic context. Research constitutes the major focus, implemented within the framework of the annual governmental Norwegian Antarctic Research Expedition (NARE). The number of expedition members varies extensively from year to year and can range from 5 to more than 50 participants. In addition, there is a low, but regular, flow of Norwegian tourists to Antarctica. Although many of these join the organised cruises to the continent, a surprisingly large number are participants of private adventure expeditions, such as climbing, skiing, skydiving.

Being one of the original signatories to the Antarctic Treaty, Norway has since placed great importance to the Antarctic Treaty System (ATS) as a means of preserving and enhancing international cooperation in respect of the Antarctic.

---

[1] UNTS, Vol. 402, pp. 71ff.
[2] Antarctic Treaty, Art. IV(2).

When in 1989 it became obvious that the Convention on the Regulation of Antarctic Mineral Resource Activities[3] (CRAMRA) would not enter into force, Norway became actively involved in negotiations that had at that time been initiated on the Antarctic environmental protection regime.

RATIFICATION OF THE PROTOCOL AND EFFECTUATION OF INTERNATIONAL TREATIES IN NORWAY

The Protocol on Environmental Protection to the Antarctic Treaty[4] (Protocol) was signed by Norway on the occasion of its adoption in Madrid, on 4 October 1991. Subsequent to the signing of the Protocol, the national ratification process was initiated. The question of ratification was considered by the Norwegian Parliament (*Storting*) in accordance with Article 26 of the Constitution of the Kingdom of Norway,[5] which requires Parliamentary approval of international treaties that are of special importance or that necessitate the adoption of new legislation. No political issues surfaced during the process, and the Parliament agreed to ratification in their deliberations on 11 February 1993. Ratification was subsequently approved by the cabinet on 30 April the same year, and an instrument of ratification was deposited with the United States government on 16 June 1993.

The signing and ratification of the Protocol entailed obligations for Norway with respect to implementation of the Protocol provisions. In the Norwegian domestic legal system, international treaties are not directly legally binding in their entirety. Provisions relating to the duties of the state are in principle self-executing and are effective once the treaty is in force. Obligations relating to the rights and duties of the citizens must, however, be implemented through Norwegian legislation to be legally binding domestically. It follows that the provisions of the Protocol concerning Norway's responsibilities within the ATS, for instance the exchange of information and annual reporting, consequently became effective for Norway immediately following entry into force of the Protocol, while the provisions regulating the rights and duties of citizens had to be effectuated through national legislation.

NATIONAL LEGAL FRAMEWORK

The primary legal framework for Norwegian Antarctic policy is the Dependencies Act of 1930,[6] which established Bouvetøya, Peter I Øy[7] and Dronning Maud Land[8] as Norwegian Antarctic and sub-Antarctic dependencies.[9]

---

[3] ILM, Vol. 27, 1988, pp. 868ff.
[4] ILM, Vol. 30, 1991, pp. 1,461ff.
[5] 'Kongeriget Norges Grundlov', adopted by the *Rigsforsamlingen*, at Eidsvold, 17 May 1814.
[6] 'Lov av 27. feb. 1930 nr. 3 om Bouvetøya, Peter I's øy og Dronning Maud Land m.m' (Act of 27 February 1930 relating to Bouvetøya, Peter I Øy and Dronning Maud Land, etc.).

In 1960, as a consequence of Norway's ratification of the Antarctic Treaty, the Act was amended to include a new provision (Article 7) that allows the government to draw up further regulations regarding implementation of the provisions of the Act and any international recommendations to which Norway has agreed.[10] A further amendment to this provision in 1990 specifically notes that such provisions can be made applicable to other areas of the Antarctic than the Norwegian dependencies.[11] Article 7 of the Dependencies Act thereby provides a legal framework for the implementation of any instruments under the ATS.

According to Article 2 of the Dependencies Act, Norwegian civil, penal and procedural law apply to the Norwegian Antarctic dependencies unless the contrary has been provided.[12] The cabinet may decide to which extent Norwegian legislation shall apply to the Antarctic dependencies.[13] The geographic scope of several Norwegian acts and regulations has consequently been extended to include the Norwegian dependencies in the Antarctic, including acts pertaining to nuclear energy, gene technology, broadcasting and telecommunications.

In addition, Norwegian domestic legislation regulates activities in the Antarctic that are under Norwegian jurisdiction according to international law. Relevant examples here include the 1951 Sealing Act[14] and the 1939 Whaling Act[15] that, *inter alia*, regulate sealing and whaling activities carried out by Norwegian citizens or companies in international waters. Other relevant examples include the 1983 Saltwater Fisheries Act[16] that, *inter alia*, regulates fishing activity from Norwegian vessels in international water, and the 1903 Seaworthiness Act,[17] which regulates all

---

[7] Amended by legislation on 24 March 1933. See *Odelstingsproposisjon* (Norwegian Government proposition to the *Odelsting*; hereafter Ot.prp.) No. 5 (1933) 'Om forandring i lov om Bouvetøya' ('Regarding changes to the Act relating to Bouvetøya').

[8] Amended by legislation on 21 June 1957. See Ot.prp. No. 64 (1957) 'Om lov om endring i lov av 27 februar 1930 om Bouvetøya og Peter I's Øy' ('Act regarding changes to Act of 27 February 1930 relating to Bouvetøya and Peter I's Øy').

[9] Dronning Maud Land was claimed as a Norwegian dependency by Royal Decree already on 14 January 1939, but due to World War II its incorporation into the Dependencies Act was delayed significantly.

[10] Added by legislation on 2 June 1960. See Ot.prp. No. 72 (1959–60) 'Om lov om endring i lov av 27 februar 1930 om Bouvetøya, Peter I's Øy og Dronning Maud Land' ('Act regarding changes to Act of 27 February 1930 relating to Bouvetøya, Peter I's Øy and Dronning Maud Land').

[11] Amended by legislation on 24 August 1990. See Ot.prp. No. 41 (1989–90) 'Om lov om endringer i lov av 27. feb. 1930 nr. 3 om Bouvetøya, Peter I's øy og Dronning Maud Land m.m.' (Act regarding changes to Act of 27 February 1930 no. 3 related to Bouvetøya, Peter I's Øy and Dronning Maud Land, etc.').

[12] Dependencies Act, Article 2.

[13] *Ibid.*

[14] 'Lov av 14. desember 1951 om fangst av sel'.

[15] 'Lov av 16. juni 1939 om fangst av hval'.

[16] 'Lov av 3. juni 1983 om saltvannsfiske'.

[17] 'Lov av 9. juni 1903 nr. 7 om Statskontrol med Skibes Sjødygtighed'.

aspects pertaining to the safety of operations of vessels owned by Norwegian legal subjects.

KEY FEATURES OF ANTARCTIC ENVIRONMENTAL REGULATIONS

Preparation and Adoption

With authority in Article 7 of the Dependencies Act, and in accordance with the Public Administration Act of 10 February 1967[18] and the Instructions for preparation of official studies and reports[19], the Norwegian government prepared and adopted national legislation to implement the provisions of the Environmental Protocol that pertain to the rights and duties of Norwegian citizens in Antarctica.

The Ministry of the Environment coordinated the extensive task of preparing the legislation. During this process it became clear that in reality there was no opposition to the adoption of legislation for implementing the relevant provisions of the Protocol. Consequently, the Regulations Relating to Protection of the Environment in Antarctica[20] (Antarctic Environmental Regulations – AER) were adopted by Royal Decree on 5 May 1995 and entered into force upon adoption. Expeditions travelling to Antarctica within one year of the entry into force of the AER were required to comply as far as possible with the Regulations.[21]

Purpose and Scope of the AER

The objective of the Regulations is articulated in Article 1 and reflects the text of Article 3(1) of the Protocol:

> The purpose of these Regulations is to protect the environment in Antarctica and the dependent and associated ecosystems, to preserve the intrinsic value of Antarctica, including its wilderness and aesthetic values, and to maintain its value as an area for the conduct of scientific research. These considerations shall be fundamental to the planning and implementation of all activities in Antarctica.

The designation of Antarctica as a natural reserve, as articulated in Article 2 of the Protocol, is not considered in the AER. Article 2 of the Regulations establishes

---

[18] 'Lov av 10. februar 1967 om behandlingsmåten i forvaltningssaker'.
[19] Instructions for preparation of official studies and reports were adopted by Royal Decree on 16 December 1994. The Instructions stipulate the procedures pertaining to the preparation of public documents and legislation, e.g., when economic and administrative consequences must be evaluated, how to proceed with public hearings, and related matters.
[20] 'Forskrift om vern av miljøet i Antarktis' (available as publication from the Norwegian Pollution Control Authority, Oslo).
[21] AER, Art. 33.

that the AER provisions apply to the entire area south of 60° south latitude.[22] The commentaries to the AER state specifically that the area of application has been chosen to correspond with the area of application of the Antarctic Treaty.[23]

Although the importance of 'associated ecosystems' is emphasised in the AER statement of purpose, this should not be interpreted to imply that the AER provisions apply outside the Antarctic as defined by the AER, but rather that associated ecosystems shall always be *considered* when applying the provisions of the Regulations.

The AER provisions apply to all Norwegian nationals, legal persons, ships and aircraft, resident foreigners in Norway, and foreigners who are members of or responsible for expeditions to Antarctica organised in Norway or departing from Norway[24]. The Regulations do not, however, apply to persons staying in Antarctica as participants of an expedition that is organised by another party to the Protocol that has adopted provisions corresponding to those of the Antarctic Regulations. This provision applies, *inter alia*, to Norwegian tourists on cruise vessels run by non-Norwegian tour operators or on a ship under foreign flag where the responsibility for obtaining the necessary permits lies with the respective national authorities.[25]

Norway has also assumed the right to enforce jurisdiction on foreign nationals in the Norwegian dependencies of Dronning Maud Land and Peter I Øy when the activity these partake in is organised from the territory of a non-contracting party.[26]

Types of Activities Covered by the AER

The term 'activity' is not defined in the Regulations, although it has been presupposed that 'activity' refers to all human presence in the Antarctic. No distinction has been made between governmental and non-governmental activities in this context, and it has been clearly stated that the obligation to give advance notice applies to both governmental and non-governmental activities.[27]

No activity is clearly excluded from being covered by the Regulations, with the exception of actions taken in emergency situations.[28] However, considering the area of application of the AER, as articulated in Article VI of the Antarctic Treaty, it can be inferred that nothing in the Regulations shall prejudice or in any way affect the

---

[22] For discussion of the consequences of so defining the scope of application of domestic application for particular issues, see Bastmeijer, Chapter 16, and Dodds, Chapter 21, in this book.

[23] See 'Comments on the Regulations Relating to Protection of the Environment in Antarctica', in *Vedlegg til forskrift om vern av miljøet i Antarktis* (Oslo: Norwegian Pollution Control Authority, 1995) (hereafter Annex to AER), comments to Art. 2. For the area of application of the Antarctic Treaty, see Art. VI of the Treaty.

[24] AER, Art. 2.

[25] Annex to AER, comments to Art. 2

[26] AER, Art. 2.

[27] Annex to AER, comments to Art. 9.

[28] AER, Art. 7.

rights, or the exercise of the rights, under international law with regard to the high seas within the area south of 60° south latitude. It may be deduced then, that the Regulations do not apply to activities that are implemented in accordance with provisions of international agreements for the high seas, i.e., fishery activities covered by the Convention on the Conservation of Antarctic Marine Living Resources[29] (CCAMLR) and sealing activities covered by the Convention for the Conservation of Antarctic Seals[30] (Seals Convention), or whaling activities covered by the International Convention for the Regulation of Whaling[31].

IMPLEMENTING PROTOCOL OBLIGATIONS THROUGH THE ANTARCTIC ENVIRONMENTAL REGULATIONS

The basic position of the AER is that no permit is required to travel to the Antarctic. However, permits are required for certain types of activities, while other types of activities are prohibited all together. For example, in accordance with the Protocol, permits are required for the taking and collection of flora and fauna (Article 14), for the introduction of plant or animal species (Article 15), and for entry into specially protected areas (Article 25). All activities related to mineral resources, except scientific research, are prohibited, reflecting the ban specified in Article 7 of the Protocol.[32]

Although travel to Antarctica in itself does not require a permit, any person planning an activity that is to take place in Antarctica (e.g., a research project or an adventure expedition) is nonetheless obliged to submit an advance notice to the authorities,[33] who subsequently can further regulate the activity.[34] It is the prerogative of the authorities to require that an activity is postponed or cancelled if the potential impacts of the activity are considered to be in contradiction to the purpose of the legislation.

Through the above-described approach, the Regulations emphasise that the primary concern is not so much which activity should be permitted in the Antarctic, but rather the manner in which an activity is carried out.

The environmental principles articulated in Article 3 of the Protocol are implemented as general obligations for any person planning activity in the Antarctic; all activity shall, namely, be planned so as to induce the least possible impact on the

---

[29] ILM, Vol. 19, 1980, pp. 837ff.

[30] UNTS, Vol. 1080, pp. 175ff; and ILM, Vol. 11, 1972, pp. 251ff.

[31] ILM, Vol. 161, pp. 74ff. Norway supports the international agreement on the moratorium on whaling in the Southern Ocean, and does not engage in any whaling activity in this region.

[32] AER, Art. 5. No time frame for the ban has been identified and a re-consideration of the AER is necessary in order to suspend the prohibition.

[33] *Ibid.*, Art. 9.

[34] *Ibid.*, Art. 12.

environment.³⁵ As underscored in the AER, *all* effects on the environment should be avoided, not just 'adverse impact' and 'significant adverse effects' as articulated in Article 3 of the Protocol. This stricter wording is to ensure that the AER is in conformity with the general national environmental policy of Norway.

Environmental Impact Assessment

Articles 9 through 13 of the AER implement the Protocol provisions that relate to environmental impact assessments. In contrast to the Protocol, no differentiation is made in the Regulations between the preliminary evaluation stage and the initial environmental evaluation stage.³⁶ Accordingly, an initial environmental evaluation (IEE) must be submitted along with the advance notice for *all* planned activities subject to the AER, unless it is found that impacts are more than minor or transitory, in which case a comprehensive environmental evaluation (CEE) is required. Separate forms and questionnaires have been developed to aid proponents in preparing IEEs and to assist the authorities in processing such evaluations.³⁷

The Regulations furthermore provide the authorities with an opportunity to require a post-activity report from those responsible for an activity in the Antarctic.³⁸ Currently, reports are required for all activities implemented in the Antarctic, and separate report forms have been developed for this purpose.³⁹ Synthesising information from such reports enables the Norwegian authorities to fulfil the national reporting requirements specified in the Protocol.

Protection of Fauna and Flora

Articles 14 and 15 of the AER implement the Protocol provisions relating to the protection of flora and fauna. It is interesting to note that the Regulations specifically include *harmful traffic* as a prohibited harmful interference.⁴⁰ Many years of experience with managing the fragile environment on the Arctic archipelago of Svalbard have provided the authorities with ample knowledge of the destructive consequences of traffic. Specific mention of the harmful impact of traffic in the AER thus ensures that this issue is sufficiently considered in the planning process.⁴¹

---

³⁵ *Ibid.*, Art. 4.

³⁶ See Arts. 1 and 2 of Annex I to the Protocol.

³⁷ Initial Environmental Impact forms have been developed for yacht tourism, terrestrial tourism and for research projects. The forms are common for Finland, Norway and Sweden, the three Nordic countries that are Antarctic Treaty Consultative Parties and parties to the Protocol.

³⁸ AER, Art. 8.

³⁹ Common report forms have been developed for Finland, Norway and Sweden.

⁴⁰ AER, Art. 14.

⁴¹ Public Consultation Paper related to Draft Regulations relating to protection of the environment in

The Regulations stipulate, as does the Protocol, that the introduction of plants and animals requires a permit. In contrast to the Protocol provision, which applies only to the introduction of species non-native to the Antarctic Treaty area,[42] the somewhat stricter AER provision also applies to the introduction of species from one area to another within the Antarctic.[43]

Waste Disposal and Management

Articles 16 through 21 of the AER implement the Protocol provisions that relate to waste disposal and management. Whereas the Protocol contains an elaborate scheme for the handling and disposal of various specific types and categories of waste, the Regulations prohibit the disposal of any waste in the Antarctic but allow the combustion of waste in incinerators that meet emission standards drawn up under the ATS.[44] This approach reflects two primary concerns: Firstly, other wastes than those listed in Article 2 of Annex III to the Protocol may become undesirable for disposal in the Antarctic in the future; and secondly, 'out of sight, out of mind' is not a suitable disposal strategy, and whoever generates waste should also be responsible for removing it for proper disposal outside the Antarctic.

The authorities can initiate clean-up operations and require that the costs be reimbursed by the person(s) responsible for the activity if clean-up is unsatisfactory after the conclusion of an activity in the Antarctic.[45] This provision may constitute an effective control mechanism and provide a strong incentive for proponents to comply with the general provisions relating to waste management.

Article 18 of the AER prohibits emission of substances or products that can be harmful to the environment. Article 18 has been included in the AER as a reflection of the provisions of Article 7 in Annex III, although a listing of harmful products corresponding to the list of Article 7 has not been included. Such a listing has been considered undesirable for two reasons. Firstly, a list will need to be updated continuously as new environmentally harmful products come into use. Secondly, the Article 7 listing is not a complete list of substances currently known to be harmful and which are prohibited on the Norwegian mainland.[46]

Antarctica, circulated by the Ministry of the Environment, 22 February 1995.
[42] The Protocol, Annex II, Art. 4(1).
[43] AER, Art. 15.
[44] AER, Art. 17. See also Annex to AER, comments to Art. 17.
[45] AER, Art. 20.
[46] Annex to AER, comments to Art. 18.

Pollution by Ships, Specially Protected and Specially Managed Areas, and Other Provisions

International agreements relating to pollution by ships, in particular MARPOL 73/78,[47] are in Norway implemented through the Seaworthiness Act and regulations issued pursuant to this Act. However, as Norway has not acceded to Annex IV of MARPOL 73/78,[48] the discharge of sewage is not regulated through the Seaworthiness Act. Consequently, the only Protocol provisions of Annex IV that are set out in the AER are the provisions relating to the discharge of sewage, which are thereby made legally binding for Norwegian ships while sailing south of 60°south latitude.

Articles 25 through 27 of the AER implement the Protocol provisions that relate to protected areas management. Although, unlike Annexes I–IV to the Protocol, Annex V is yet to become effective,[49] the relevant AER provisions are already binding for Norwegian legal subjects.

The Regulations require that all activity in the Antarctic be covered by insurance for costs potentially incurred in search and rescue operations.[50] Although this provision cannot be considered strictly environmentally related, the Regulations have been found to be the most appropriate legislative framework for this provision.

CONTROL AND ENFORCEMENT THROUGH THE ANTARCTIC ENVIRONMENTAL REGULATIONS

Control

Through Article 29 of the AER, the Norwegian Polar Institute (NP), or any entity authorised by the Institute, has been delegated the responsibility of supervising the implementation of the AER provisions. The responsibility encompasses supervision of pre- and post-activity requirements, such as the submission of environmental impact assessments (EIAs) and final reporting, and supervision of compliance in the field. In carrying out this latter capacity, NP is to have unrestricted access to all installations, means of transport and areas where activities are conducted.

---

[47] The International Convention for the Prevention of Pollution from Ships, 1973, as modified by the Protocol of 1978 (MARPOL 73/78); ILM, Vol. 12, 1973, pp. 1,319ff (Convention) and ILM, Vol. 17, 1978, pp. 546ff (Protocol).
[48] Regulations for the Prevention of Pollution by Sewage from Ships.
[49] See further Vidas, Chapter 1 in this book.
[50] AER, Art. 31.

Enforcement

Deliberate or negligent violation of the Regulations, or other instruments issued pursuant to them, is punishable by fines or imprisonment up to one year, or both.[51]

In instances where the control authorities observe acts that are in obvious contravention to the AER, legal action can be filed. Some guidance to such action is provided through the AER.

Firstly, acts in contravention with the provisions of the AER committed by Norwegian nationals in Dronning Maud Land or on Peter I Øy can be pursued by legal action. This is stated in Article 2 of the AER, and furthermore supported by Article 8 of the Dependency Act, which states that actions that are in contradiction with the Dependencies Act or regulations issued pursuant to it are punishable by fines or imprisonment for up to one year or both.

Secondly, when Norwegian nationals commit acts in contravention with the provisions of the AER in areas outside the Norwegian claims, legal action still can be taken pursuant to Article 2 of the AER, but only with the consent of the Ministry of Foreign Affairs as specified in Article 3(b) of the AER.

Thirdly, legal action against foreign nationals can only be filed if the activity in question was not organised by person(s) from another party to the Protocol. In these cases legal action can furthermore only be filed if the act in contravention of the provisions takes place in Dronning Maud Land or on Peter I Øy. The AER Article 3(a) specifies that the consent of the Ministry of Foreign Affairs is also required in these situations.

The Ministry of Justice is the legal enforcement authority with respect to Norwegian activities in the Antarctic. This follows from a Royal Decree of 19 September 1930, which appoints the Ministry of Justice as Chief of Police on Bouvetøya. Furthermore, it follows from the Dependencies Act and interpretations of this Act that the Ministry of Justice also has such responsibility in the other Norwegian Antarctic dependencies. When breaches are observed, the supervising authority should file charges with the Ministry of Justice, who in turn will determine whether prosecution should be initiated.

PENDING PROTOCOL ISSUES AND THE ANTARCTIC ENVIRONMENTAL REGULATIONS

Liability and Emergency Response

The questions of liability and emergency response are still pending in the ATS, and are unresolved in the Protocol.[52] The Regulations do not pose any provisional

---

[51] *Ibid.*, Art. 32. On sanctions for violations under various domestic legislation see also Bastmeijer, Chapter 16 in this book.

[52] See discussion by Skåre, Chapter 9, and Lefeber, Chapters 10 and 11, in this book.

resolution to these issues. The question of liability is not considered in the AER, and furthermore it is specifically stated in the AER that emergency actions are not covered by the legislation,[53] which includes no requirements regarding response action. Any decision in the ATS with regard to these issues will have to be followed by amendment of the AER in accordance with Article 33(2) of the Regulations, or by introducing new legislation that covers these aspects of the environmental provisions for the Antarctic.

Minor or Transitory Impact

The Protocol introduced the terms 'minor' and 'transitory' impact, which play an important role in the execution of the Protocol provisions. The terms are decisive in determining when a planned activity is to be thoroughly evaluated by initiating a CEE process before an activity is allowed to proceed. The terms 'minor' and 'transitory' have proven difficult to define, and there is no agreement within the ATS as to what these terms entail.[54] In fact, as to the understanding of these terms and when to initiate a CEE, the Consultative Parties have in reality agreed to rely on developing practice.[55]

With respect to deciding at which point a CEE should be prepared, the same trigger mechanism is prescribed in the AER as in the Protocol, namely that a CEE shall be prepared for 'any activity that is likely to have more than a minor or transitory impacts on the environment'[56]. No attempt has been made in the Norwegian legislation to define the terms, and the decision as to when a CEE should be triggered relies on practice as it develops within the ATS.

DOMESTIC INSTITUTIONS TAKING PART IN IMPLEMENTING THE ANTARCTIC ENVIRONMENTAL REGULATIONS

In Norway the main responsibility of achieving the overall national goals pertaining to environmental protection, including cultural heritage, rests with the Ministry of the Environment. The Ministry exercises some of its authority through its directorates, such as the Norwegian Pollution Control Authority for issues relating to

---

[53] AER, Art. 7.
[54] See also discussion by Bastmeijer, Chapter 16 in this book.
[55] See para. 30 of the 'Report of the Meeting of the Committee for Environmental Protection, Tromsø, 25–29 May 1998', Annex 2, in *Final Report of the Twenty-second Antarctic Treaty Consultative Meeting, Tromsø, Norway, 25 May – 5 June 1998* (Oslo: Royal Norwegian Ministry of Foreign Affairs, 1998), Annex E; and para. 45 of the 'Report of the Second Meeting of the Committee for Environmental Protection, Lima, 24–28 May 1999', Annex 1, in *Final Report of the Twenty-third Antarctic Treaty Consultative Meeting, Lima, Peru, 24 May – 4 June 1999* (Lima: Peruvian Ministry of Foreign Affairs, 1999), Annex G.
[56] AER, Art. 11(1).

pollution, the Directorate for Nature Conservation for issues pertaining to nature conservation, the Directorate for Cultural Heritage for cultural heritage issues and the Norwegian Polar Institute for polar-related issues.

The Norwegian Polar Institute

In the context of the Antarctic environmental legislation, it was agreed that the Norwegian Polar Institute should have the practical responsibility for implementing and supervising the Antarctic environmental legislation. This approach of delegating the responsibility to an agency with expertise in a given *geographic* area differs from the more common approach of delegating the responsibility to agencies with expertise related to specific functional issues (e.g., pollution, nature conservation or cultural heritage). The reasoning behind this approach is twofold: Firstly, the Norwegian Polar Institute harbours a broad expertise in issues pertaining to the polar regions.[57] In the event that the authority were to rest with any other agency, NP would nevertheless have to be consulted. Secondly, as Norwegian activity in the Antarctic is relatively limited, the legislative caseload was expected to be so low that it would be unreasonable and inefficient to delegate the authority to a number of directorates or agencies.

Consequently, NP is the principal agency responsible for implementing and overseeing the provisions of the AER. This entails a number of direct and indirect tasks for the Institute:

Firstly, NP is responsible for evaluating and processing advance notifications and EIAs and for ensuring that proposed and planned activities are not in contradiction with the provisions of the AER.[58] If the planned activity is in contradiction with the purpose of the AER, NP may require alterations to the activity, or postpone or prohibit it all together.[59] It is also within NP's mandate to issue permits for taking or collecting of flora and fauna, for introducing plants and animal species, and for permission to enter protected areas. Presently the Institute processes between 3 and 20 notifications, IEEs and permit applications per year, with peaks in years in which the larger national research expeditions take place.

Secondly, NP collects post-activity reports[60] and evaluates these against the environmental evaluations prepared in advance of an activity. The Institute

---

[57] The Norwegian Polar Institute, established in 1928 with a history from 1906, is the principal Norwegian Institution concerned with mapping and scientific and environmental investigations of Norwegian polar regions in the Arctic and Antarctic. The objective of the Institute is to advance national polar research and to contribute towards national and international environmental management in the polar regions. With the exception of the AER, NP has no legislative implementation power, but acts rather as an advisor to the national authorities.

[58] AER, Art. 9 and 12.

[59] *Ibid.*, Art. 12.

[60] *Ibid.*, Art. 8.

consolidates the information from these reports into reports that comply with the provisions for exchange of information and reporting within the ATS.[61]

Thirdly, also vested in NP's responsibility is the authority to issue guidelines that will assist in implementing the provisions of the AER, including reporting procedures,[62] guidelines for the preparation of advance notices,[63] guidelines for the prevention of introduction of micro-organisms[64], and guidelines for incineration[65]. The Institute has initiated extensive cooperation with the Antarctic authorities of Finland and Sweden to develop a common framework for some of these guidelines. This common framework will aid these countries in achieving a level of uniformity in the process of implementing the Protocol nationally.

Fourthly, NP is responsible for supervising compliance with the provisions of the AER, both with respect to pre- and post-activity requirements and with respect to compliance in the field. The supervising role of NP implies that when acts in contravention of the AER are observed, NP is obliged to file a suit so that proper legal action may be taken.

At the Norwegian Polar Institute, the Polar Environmental Management Department (PEMD) is for all practical purposes responsible for the tasks described above. The Department processes advance notices and EIAs, develops further guidelines, and ensures that reports are filed with the appropriate national authorities. NP has an extensive research department with expertise in geology, glaciology, oceanography, biology and polar history, and the PEMD utilises this in-house expertise in implementing its responsibilities. These in-house resources are most often sufficient for the types of issues common within the Norwegian Antarctic framework. However, if an in-house process indicates that external expert evaluation is needed, then out-of-house expertise is commissioned. In fact, in delegating authority to NP, the Ministry of the Environment specifically noted that the Institute should seek advice from relevant agencies when its in-house expertise was insufficient in a particular matter.[66] Consequently, NP relies on a number of outside agencies, institutes and organisations in implementing its responsibilities.

The Norwegian Pollution Control Authority. In evaluating proposals for activities that may have significant consequences with respect to pollution, or in developing further guidelines pertaining to waste management, contingency planning and other pollution related issues, NP is required to initiate a cooperative effort with the Norwegian Pollution Control Authority.

---

[61] Provisions for exchange of information are found in Art. 17 of the Protocol, as well as Art. 6 of Annex I, Art. 6 of Annex II, Art. 9 of Annex III and Art. 10 of Annex V.
[62] AER, Art. 8.
[63] *Ibid.*, Art. 9.
[64] *Ibid.*, Art. 15.
[65] *Ibid.*, Art. 17.
[66] Case brief submitted by the Ministry of the Environment to the cabinet in conjunction with the adoption of Royal Decree of 5 May 1995.

The Directorate for Cultural Heritage. NP collaborates with the Directorate for Cultural Heritage when processing issues pertaining to cultural heritage within the framework of the AER.

Relevant research institutions. When external expertise is needed to evaluate issues related to environmental impacts, research and monitoring, NP puts an inquiry to external research institutes (e.g., the Norwegian Institute for Nature Research and the universities). Advice is often sought directly from scientists known to hold relevant expertise.

The National Animal Research Authority. In Norway, no person may carry out biological research on animals without a licence from the National Animal Research Authority.[67] Although the Animal Welfare Act does not apply to Antarctica, NP requires that a project has been evaluated by the National Animal Research Authority before it issues permits for the taking of animals in accordance with Article 14 of the AER.

Norwegian Telecommunications Authorities. The telecommunication authorities receive permit applications for radio amateurs planning expeditions to the Antarctic.[68] A procedural agreement is in place between NP and the Norwegian Telecommunications Authorities, ensuring that no radio licence is issued until NP has received and processed an advance notice in accordance with Article 9 of the AER.

The Ministry of the Environment

Although freed from the practical responsibilities, the Ministry of the Environment remains the overall authority with respect to the implementation of the AER provisions. In addition, the Ministry is responsible for evaluating EIAs and issuing permits when the Norwegian Polar Institute is responsible for activity that is to take place in the Antarctic.

Norwegian organised activity in the Antarctic is relatively limited; in fact, most of the activity is carried out by NP, the agency delegated with the authority to implement and supervise the national legislation. NP is the operator of the national Norwegian Antarctic Research Expeditions, which are normally conducted on an

---

[67] 'Lov av 20. desember 1974 om dyrevern, Art. 21' ('Act of 20 December 1974 relating to Animal Welfare').

[68] Use of radio equipment in Antarctica requires license in accordance with 'Lov av 23. juni 1995 om telekommunikasjon', Art. 5(3) ('Act of 23 June 1995 relating to telecommunications') and 'Forskrift om stedlig virkeområde for lov 23. juni 1995 nr. 39 om telekommunikasjon vedrørende Svalbard, Jan Mayen, Bilandene og Antarktis', Art. 2 ('Regulations regarding the application of Act of 23 June 1995 no. 39 relating to telecommunications to Svalbard, Jan Mayen, the Dependencies and Antarctica').

annual basis, but with peak activities every third or fourth year. In a peak season, NARE activity will likely constitute nearly all of the total Norwegian organised activity in the Antarctic.

It is obvious that there would be a procedural flaw if NP in these cases were to be the final authority in evaluating its own activities against the provisions of the AER. The Ministry of the Environment has therefore made it quite clear that it is the Ministry itself that will consider the notifications and environmental evaluations, as well as issue any relevant permits when NP is responsible for an activity in the Antarctic.[69] In such cases, NP submits an advance notice and an impact evaluation to the Ministry, in accordance with Article 9 of the AER. NP considers each major expedition as a separate activity and consequently submits an IEE to the Ministry for all its expeditions.

The Ministry of the Environment is also the appeal authority with respect to the AER. When proponents are in disagreement with any decision taken by NP, appeals are lodged with the Ministry which will then process these in accordance with the Public Administration Act[70].

The Ministry of Foreign Affairs

The Ministry of Foreign Affairs has two main roles in the implementation of the environmental legislation for the Antarctic. Firstly, a legal action can be taken towards person(s) committing acts in contravention of the AER; this may in some instances, as described above, require the consent of the Ministry of Foreign Affairs.[71]

Secondly, the Ministry of Foreign Affairs is responsible for ensuring Norwegian compliance with the Protocol provisions that are not articulated in the AER. This relates, *inter alia*, to the provisions regarding exchange of information and annual reporting.

The Ministry of Justice

The Ministry of Justice is an enforcement authority. When charges against breaches have been filed, the Ministry decides whether prosecution shall be initiated. The Ministry of Justice may also appoint inspection teams to carry out inspections as called for in Article 17 of the Protocol and reflected in Article 6 of the AER. This authority is vested in Article 5 of the Dependencies Act, which states that the

---

[69] Case brief submitted by the Ministry of the Environment to the cabinet in conjunction with the adoption of Royal Decree of 5 May 1995.
[70] Public Administration Act, Ch. VI.
[71] AER, Art. 3.

Ministry of Justice may appoint persons to inspect stations, installations, equipment, ships, aircraft, etc., in the Norwegian dependencies and other areas in the Antarctic.

The Ministry of Trade and Industry

As has been described earlier in this chapter, the Regulations do not cover the provisions of Annex IV to the Protocol regarding pollution by ships. This is because the MARPOL 73/78 provisions reflected in Annex IV are already implemented in Norway through the Seaworthiness Act and regulations issued pursuant to this Act. The Ministry of Trade and Industry, through the National Maritime Directorate, is responsible for implementing these regulations.

The Ministry of Fisheries

During the last half century a number of regulations relating to the fishing and harvesting in the Southern Ocean have been adopted, some having their basis in international agreements. For example, pursuant to the Sealing Act,[72] the following have been adopted: Regulations Prohibiting Harvest of Ross Seal[73] and Regulations Prohibiting Harvest of Fur Seal and Elephant Seals on Peter I Øy and Fur Seal on Bouvetøya[74]. The Ministry of Fisheries, through the Fisheries Directorate, is the implementing authority with respect to these regulations.

Although the above regulations are not considered as environmental legislation *stricto sensu* and have not been adopted to implement Protocol provisions, they nevertheless contain some provisions that overlap or are somewhat in contradiction with the AER; these should therefore be considered in the context of the environmental legislation.

The Inter-ministerial Polar Committee

The Norwegian government has established an Inter-ministerial Polar Committee. The Polar Committee is a forum for coordination of ministerial affairs, including legislation, concerning the Arctic and Antarctic. Pursuant to the government's provisions, all draft bills concerning the Antarctic must be submitted to this committee for consideration before being further processed. It follows that the Inter-ministerial Polar Committee also will be consulted with respect to questions relating to the interpretation of the AER or Protocol provisions.

---

[72] 'Lov av 14.12.51 om fangst av sel'.

[73] 'Forskrift av 05.07.68 om forbud mot fangst av Ross-sel'. The Regulations state that harvesting or killing Ross seal on land or on fast-ice south of 60° south latitude is prohibited.

[74] 'Forskrift av 27.02.53 om forbud mot fangst av pelssel og sjøelefanter på Bouvetøya og Peter I.'s øy'.

EVALUATING THE ANTARCTIC ENVIRONMENTAL REGULATIONS AND THE INSTITUTIONS INVOLVED IN IMPLEMENTATION

The Regulations have been adopted to ensure that Norwegian nationals comply with the provisions of the Protocol that relate to the planning and implementation of activities in the Antarctic. Several issues demonstrate that the AER ensures an efficient implementation of the provisions of the Protocol:

Firstly, 'user-friendliness': The Regulations synthesise the provisions of the Protocol to a manageable size and format and state in simple language what the responsibilities of an individual are. It is assumed that this format will contribute to ensuring a clear understanding of the obligations that pertain to the persons addressed, thereby increasing the efficiency of implementation.

Secondly, comprehensiveness: All the Protocol provisions, with the exception of the MARPOL 73/78 regulations, are articulated in one piece of legislation rather than in several separate parts, thereby increasing the awareness of the persons addressed of their responsibilities pertaining to the implementation of the proposed activity.

Thirdly, clear delegation of authority and minimal bureaucracy: The Regulations provide clear guidance in the processing of cases relating to activity in the Antarctic. By delegating the majority of the responsibility to one agency, NP, the level of bureaucracy has been minimised. The efficiency of processing is maximised by delegating the responsibility to an agency that has most of the expertise necessary for proper handling in-house.

Fourthly, a holistic approach: By delegating the responsibility of enforcement to only one agency, it is possible to take a holistic approach in the process of evaluating proposed activities, thereby avoiding contradicting decisions at various levels of the decision-making process.

All in all, the Regulations provide an efficient framework for the implementation of the provisions laid out in the Environmental Protocol. However, a few obstacles to full efficiency still remain, some of the most obvious of which are summarised below.

The Protocol terms *'minor' or 'transitory' impact* have been incorporated into the AER text, although no satisfactory definition of the terms exists. It is therefore quite unclear at what level an IEE is not sufficient, calling for the preparation of a CEE instead. This issue remains unresolved in Norway, as it does in the ATS, and a framework for such decisions will have to evolve with practice within the ATS and Norway itself.

Pre-activity requirements, control and enforcement are dependent on whether an advance notice for a proposed activity is in fact submitted. Proper handling is dependent on NP's awareness of the activity being planned. If a proponent is not aware of the AER requirements (or neglects the requirements), it will only be by chance that NP will become aware of the activity and able to enforce the AER obligations. This is normally not a problem because most persons planning an activity in the Antarctic will first seek NP's advice on other matters pertaining to the Antarctic and thus will also be made aware of their AER responsibilities. Antarctic adventures are also usually highly exposed in the media, and the authorities can become aware of proposed activities in this manner.

Although regularly present in the Antarctic, NP is rarely in the areas where private enterprises take place; NP is therefore rarely able to act in its capacity as a *supervisor*. In order for the Regulations to be truly efficient, there is a need to look further at the establishment of practical procedures that will enable control and supervision in the field.

The advantages of having vested NP with the operational authority are obvious and have been noted above. The obvious flaw with respect to NP's role is that the agency that has the operational authority is the *same agency* that is responsible for most of the Norwegian activity in the Antarctic.[75] The appropriateness of NP's being the sole authority was questioned when the Regulations were being developed. Procedurally, the issue has been solved by placing the final authority with the Ministry of the Environment.

CONCLUSION

In Norway, the ratification of the Environmental Protocol and the adoption of domestic legislation have been uniquely non-controversial processes. For comparison, one needs only consider the complicated and time-consuming process behind the preparation of a draft text for a comprehensive environmental legislation for the Arctic archipelago of Svalbard. Three years were required before a specially appointed committee was able to integrate the views and needs of a number of widely ranging interests into draft legislation. It may be assumed that the ease with which the Antarctic environmental legislation was passed most likely reflects the fact that the provisions of the Protocol did not interfere significantly with Norwegian interests and activities in the Antarctic.

However, it is too early to provide a comprehensive evaluation of the efficiency of the national legislation, considering that the body of practice is still very limited. Because there are a number of unclear terms in the Protocol and the Regulations, it is also still difficult to assess what the consequences of the implementation of the provisions will be. These unclear terms can only be defined based on the basis of national practice and through the advice from the Committee for Environmental Protection[76].

---

[75] See also discussion by Bastmeijer, Chapter 16 in this book.
[76] See also Orheim, Chapter 6 in this book.

For the time being, however, there is no doubt that the Regulations provide a strict legal framework for regulating Norwegian activities in the Antarctic in such a manner that the activities are implemented with the least possible environmental impacts. Time and practice will tell whether this framework is in fact sufficient.

21

South Africa: Implementing the Protocol on Environmental Protection

*Klaus Dodds**

South Africa was among the twelve original parties to the 1959 Antarctic Treaty.[1] As a non-claimant, yet geographically proximate state to the Antarctic, successive political leaders have argued passionately that the Republic has an interest in shaping the future politics of the polar continent. During the *apartheid* era, the meetings within the fora of the Antarctic Treaty System (ATS) and related bodies such as the Scientific Committee on Antarctic Research (SCAR) provided a relatively benign forum for South Africa's participation in segments of international politics. Since the mid-1980s, however, this position started to be questioned by some Third World countries, following up the initiative by Malaysia in the UN General Assembly, which resulted in a debate not only about the future status of the ATS but also whether South Africa should remain a member, given its record of human rights abuses and racist governance.[2] In May 1993, one publication from the then Department of Environment Affairs (DEA) acknowledged the following:

---

* I owe a debt of gratitude to the officials affiliated with the Department of Environmental Affairs and Tourism and the Department of Foreign Affairs in Pretoria for their assistance. Thanks to David Simon and to the participants in the workshop of the project 'Implementing the Environmental Protection Regime for the Antarctic', held at the Fridtjof Nansen Institute, Norway, 3–6 September 1999, for their advice. Special thanks are due to Kees Bastmeijer for his advice on South African law. I would also like to thank the Arts and Humanities Research Board for funding a period of research leave in 1998–99.

[1] UNTS, Vol. 402, pp. 71ff.

[2] See P. Beck, *International Politics of Antarctica* (London: Croom Helm, 1986); and P. Beck and K. Dodds, *Why Study Antarctica?* (CEDAR Discussion Paper Series Number 26, Royal Holloway, University of London, 1998). For details on South Africa's geopolitical and legal interests in Antarctica see, K. Dodds, 'South Africa and the Antarctic, 1920–1960', *Polar Record*, Vol. 32, 1996, pp. 25–42; and K. Dodds, *Geopolitics in Antarctica: Views from the Southern Oceanic Rim* (Chichester: Wiley, 1997), pp. 185–212.

During a time when South Africa found many doors closed in the international political arena, the Antarctic Treaty Council Meetings [sic] remained open. Excellent international relations were established, and South Africa found support from Treaty member countries.[3]

There was considerable fear that South Africa's continuing participation in the ATS could be threatened given that these political tensions coincided with controversies over the Convention on the Regulation of Antarctic Mineral Resource Activities[4] (CRAMRA) negotiations (1982–1988).[5] At the height of the debates over minerals and mining rights, Third World critics in the United Nations were demanding the dismantling of the ATS in favour of a solution imposed by the UN. The dismantling of *apartheid* in the early 1990s, in conjunction with the adoption of the Protocol on Environmental Protection to the Antarctic Treaty[6] (Protocol) in October 1991, helped to suppress these demands for radical change, not least since the Protocol imposed a ban on mineral resource activity in the Antarctic Treaty area. Furthermore, South Africa's continued participation in the ATS was assured when the new administration under President Nelson Mandela promised to sustain the South African National Antarctic Programme (SANAP) in 1994.

This chapter is concerned with exploring the implications for South Africa in the aftermath of *apartheid* and the domestic ratification of the Protocol. SANAP is first considered within the political context of post-*apartheid* South Africa. Thereafter, attention is given to the implementation of the Protocol and some of the outstanding legal and political issues that must be confronted in the near future. Given South Africa's interests in the sub-Antarctic Prince Edward Islands (comprising Marion and Prince Edward), the creation of the Prince Edward Island Management Committee in 1995 will be briefly considered in conjunction with the polar programme. The Management Plan for these remote islands contains provisions that are arguably more robust then the requirements established in the Annexes to the Protocol. Finally, the discussion is concluded with an assessment of the financial and political implications of the Protocol for a country which is struggling to reconstruct after 40 years of unequal and uneven economic and human development.

---

[3] *Draft Comprehensive Environment Evlauation (CEE) of the Proposed New SANE IV Facility at Vesleskarvet, Queen Maud Land, Antarctica* (Pretoria: Department of Environment Affairs, 1993) p. 3.
[4] ILM, Vol. 27, 1988, pp. 868ff.
[5] See D. Vidas, 'The Antarctic Treaty System in the International Community: An Overview', in O. Stokke and D. Vidas (eds.), *Governing the Antarctic: The Effectiveness and Legitimacy of the Antarctic Treaty System* (Cambridge University Press, 1996), pp. 35–60.
[6] ILM, Vol. 30, 1991, pp. 1,461ff.

SOUTH AFRICA AND THE PROTOCOL: POLITICAL CONTEXT

The 1994 general election in South Africa witnessed the formal cessation of white minority rule. South Africa's foreign affairs had been dominated by *apartheid* politics, and the country's involvement in the polar continent was not entirely immune from these political currents.[7] As an original signatory to the 1959 Antarctic Treaty, South Africa played a role in shaping the environmental, political and scientific dimension of the ATS. During the *apartheid* era, successive governments had argued strongly in favour of retaining the demilitarised status of the Antarctic and advocated the negotiation and adoption of the Convention on the Conservation of Antarctic Marine Living Resources[8] (CCAMLR) in 1980.[9] South Africa was a founding member of SCAR, and South African scientists such as Dr. Denzil Miller have contributed significant research on living resources in the Southern Ocean.[10] During the period from the 1950s through the 1970s, South Africa managed to participate in the meetings of SCAR and the Antarctic Treaty Consultative Meetings without any great diplomatic difficulties. However, by the 1980s the issue of *apartheid* had become firmly tied up with the future well being of the ATS, with Third World nations seeking to condemn South African membership. This situation also produced some diplomatically embarrassing moments when Australia, as the convenor of the Consultative Parties' contact group in the UN, defended South Africa's participation in the ATS even though the Australian government had formally condemned *apartheid*.[11]

Subsequently, the new government, led by the African National Congress (ANC), confirmed that the major investment programme for SANAE IV would not be unduly affected by the demands of the Reconstruction and Development Programme (RDP).[12] While in opposition, the ANC had in a strategic working paper acknowledged the significance of the Antarctic:

---

[7] For reviews of South African foreign policy, see P. Vale, 'Zuid-Afrika's zoekocht naar een buitenlands beleid', *Internacionale Spectator*, Vol. 50, 1996, pp. 243–249; and R. Davies, *South Africa's Foreign Policy Options in a Changing Global Context* (Bellville: University of Western Cape, Centre for Southern African Studies, 1995).

[8] ILM, Vol. 19, 1980, pp. 837ff.

[9] See *Special Issue of South African Journal of Antarctic Research*, Vol. 21, 1991, for a detailed review of the history of South African involvement in Antarctica and the PEI from 1948 onwards.

[10] See, for example, D. Miller, 'Conservation of Antarctic Marine Living Resources: A Brief History and a Possible Approach to Managing the Krill Fishery', *South African Journal of Marine Research*, Vol. 10, 1991, pp. 321–329.

[11] See R. Woolcott, 'Speech to the Australian Institute of International Affairs, 10 August 1990', *World Review*, Vol. 30, 1990, pp. 48–56.

[12] For details on the RDP, see A. Lemon (ed.), *The Geography of Change in South Africa* (Chichester: Wiley, 1995).

As good environmental citizens we will strongly support the comprehensive protection of Antarctica. We urge the development of instruments which will enable the continent to become a 'Nature Reserve – Land of Science.[13]

Moreover, a significant factor that contributed to South Africa's commitment to polar research and the ATS has undoubtedly been the enduring presence of key officials in the Department of Environmental Affairs and Tourism (DEA&T). As the lead government agency in South Africa's Antarctic interests, the DEA&T, in conjunction with the Department of Foreign Affairs, argued strongly in 1993–94 for the retention of the SANAP and future investment programmes.[14]

Notwithstanding these new commitments to the SANAP, Antarctica and the Southern Ocean has never been a major feature of South African foreign policy. During the *apartheid* era, the minority white governments were far more preoccupied with the frontline states of Southern Africa, relations with the United States and the general geopolitics of the Cold War. In the 1970s, as a consequence of diplomatic and military relations with the military juntas of Argentina and Chile, government attention was focused on the South Atlantic as a possible zone of military co-operation.[15] While the proposals for a South Atlantic Treaty Organization (SATO) never came to fruition, the new ANC government has ensured that South Africa has remained an active member of the South Atlantic Zone of Peace and Cooperation and the Indian Ocean Rim Initiative, which have sought to promote diplomatic and economic relations amongst the littoral states. South Africa wants to ensure that, within this oceanic framework, the Antarctic remains demilitarised, environmentally protected and uncontaminated by military confrontation.

SOUTH AFRICA AND THE PROTOCOL: INSTITUTIONAL CONTEXT AND IMPLEMENTATION

The Protocol was ratified by the South African government in July 1995, and the domestic legislation concerning implementation was passed in 1996. As a long-time supporter of CRAMRA, South Africa subsequently embraced the Protocol with considerable public vigour. At the 1989 Paris Antarctic Treaty Consultative Meeting, the leader of the South African delegation 'welcomed the new awareness in dealing with environmental matters in Antarctica. The delegation is therefore

---

[13] Cited in G. Mills (ed.), *From Pariah to Participant: South Africa's Evolving Foreign Relations 1990–1994* (Johannesburg: South African Institute of International Affairs, 1994), p. 235.

[14] Based on interviews of public officials carried out by the author in Cape Town, Durban and Pretoria, in September 1995.

[15] On details concerning SATO and the South Atlantic, see A. Hurrell 'The Politics of South Atlantic Security', *International Affairs*, Vol. 59, 1983, pp. 179–193 and D. Venter, 'South Africa, Brazil and South Atlantic Security: Towards a Zone of Peace and Cooperation in the South Atlantic', paper presented to the International Relations Research Institute on Brazil – Africa du Sol, Rio de Janeiro,

committed to a strong stand on environmental matters at this meeting and future deliberations'.[16] After the 1991 Bonn Consultative Meeting, South African delegations to the Consultative Meetings have shown considerable interest in developing particular elements of the Protocol including oil spill contingency plans, environmental evaluation of Antarctic activities, and support for the creation of an Antarctic Treaty Secretariat.[17] While the South African delegation is a relatively minor player compared to Consultative Parties such as the United States, the United Kingdom, Norway and Australia, the South Africa has established important precedents concerning the environmental evaluation of its activities in the Antarctic and the implementation of the Protocol.

Institutional Context

South Africa's political system is based on a parliamentary system, with the main institutions located in Cape Town and Pretoria. The executive branch of government is contained in the administrative capital of Pretoria. Until 1994, democratic elections were the privilege of the white minority in South Africa, despite some electoral concessions to the Indian and coloured population. The black majority was systematically excluded from the mainstream political process. In the last five years, after the ANC-led government under President Mandela swept into power in June 1994, the demise of apartheid and a South African government based on democratic principles have become realities. Recent elections in May 1999 have witnessed the election of the ANC's Thabo Mbeki as the second President of post-*apartheid* South Africa.

Unlike other areas of foreign policy making, the Antarctic affairs comprise a relatively narrow and discrete issue. While select parliamentary committees (such as the Environmental Portfolio Committees) debate, monitor and approve particular spending programmes, the Department of Environmental Affairs and Tourism, rather than the Department of Foreign Affairs, is the lead agency when it comes to the administration and coordination of Antarctic policy. The Pretoria-based Directorate, Antarctica and Islands of the DEA&T, liaises with this Directorate's logistical centre in Cape Town and in the Antarctic with the team leaders at the SANAE IV base and the base on Marion Island.

Within the government, coordination between government departments is facilitated through the Antarctic Management Committee (AMC) responsible for the SANAP, the Prince Edward Islands Management Committee and the South African

---

24–25 September 1995.
[16] Opening Address by Dr. J. Serfonstein, Leader of the South African Delegation to the Fifteenth Consultative Meeting in Paris, 9–19 October 1989.
[17] South Africa's interest in marine affairs and maritime pollution was further cemented by membership of the International Maritime Organisation (IMO) in February 1995 and re-admission to the Intergovernmental Oceanographic Commission in June 1995.

Committee for Antarctic Research (SACAR). Membership of these committees comprises government departments such as the National Public Works Department (NPWD) and the Department of Foreign Affairs, research institutions, and universities such as Cape Town and Pretoria. The Department of Foreign Affairs (Sub-Directorate: Marine and Antarctica) in conjunction with the DEA&T is responsible for representing South Africa at Consultative Meetings and for ensuring that international legal obligations are implemented.

The history of South African involvement in the Antarctic dates from 1960, when the Norwegian government offered the Republic access and subsequently ownership of their base in Dronning Maud Land.[18] Initially, the Department of Transport (with the specific advice of the Weather Bureau) was the lead government agency in terms of providing logistic and administrative support for the SANAP (previously called the South African National Antarctic Expedition). Since 1985, the DEA&T has been, as noted above, the lead agency in terms of administering and implementing the SANAP. The Chief-Directorate of Environmental Management of the DEA&T supplements the work of the Directorate of Antarctica and Islands by preparing additional legislation and procedures for environmental protection in the region. The NDPW organises the maintenance and servicing of the SANAE IV and Marion bases; the supply ship, the *SA Agulhas*, is the responsibility of the DEA&T's Chief Directorate: Marine and Coastal Management; the Department of Land Affairs is responsible for mapping projects relating to South African Antarctic and Island endeavours.

The research component of SANAP is carried out under the auspices of SACAR which is composed of representatives from government departments, research institutions and universities. This Committee acts as an advisory body to the inter-departmental Antarctic Management Committee which is responsible for the final decisions relating to the SANAP. The Antarctic research has four major components: biological sciences, earth sciences, oceanographic (in the southern ocean) and physical sciences. Meteorological observations are undertaken by the Chief Directorate: Weather Bureau of the DEA&T.[19] For the past 40 years, Antarctic research has been complemented by South African scientific studies centred on the Islands of Gough and Marion (part of the Prince Edward Islands) in the South Atlantic and Indian Oceans respectively.

---

[18] See J. Kingwill, 'First Ten Years of South African Antarctic Research', *South African Journal of Antarctic Research*, Vol. 1, 1971, pp. 2–3. There is a longer history of South African whaling and sealing which dates back to the nineteenth century and was based at the ports of Cape Town and Durban.

[19] See J. King, 'Marion Island', *Lantern*, Vol. 3, 1954, pp. 298–304 and 433–436; A. Crawford 'Establishment of the South African Meteorological Station on Marion Island, 1947–1948', *Polar Record*, Vol. 40, 1950, pp. 576–579; and V. Smith, 'Surface Air Temperatures at Marion Island, Sub-Antarctic', *Sud-Afriaanse Tydskrif vir Wetenskap*, Vol. 88, 1992, pp. 575–578. Smith provides a review of weather programme and observations regarding temperature changes since 1949.

Implementation

In November 1996, the Antarctic Treaties Act[20] was signed by the President of South Africa. This legislation contains details of the implementation procedures. Unlike the procedures relating to the United States,[21] the 1996 Act effectively incorporates the Protocol (without modification) into South African domestic legislation. Article 3 of the 1996 Antarctic Treaties Act deals with this transposition:

> (1) Subject to this Act, the treaties mentioned in schedule 1 shall form part of the law of the Republic.
> (2) The Minister shall as soon as practicable after the promulgation of this Act, cause to be published in the *Gazette* the texts of the treaties mentioned in Schedule 1.
> (3) The Minister may by notice in the *Gazette* amend Schedule 1 to reflect the number and date of a *Gazette* in which the treaties were published.[22]

This mode of incorporation follows the South African Constitution which notes the following:

> Any international agreement becomes law when it is enacted into law by national legislation; but a self-executing provision of an agreement that has been approved by Parliament is law in the Republic unless it is inconsistent with the Constitution or an Act of Parliament.[23]

Therefore, Article 3(1) of the 1996 Act incorporates the Protocol (and other international treaties) into national law in a manner consistent with the Constitution of South Africa. Once published in the Government *Gazette*, the Protocol effectively becomes part of South African law.[24]

The 1996 Antarctic Treaties Act also reiterated South Africa's commitment to implementing the legal obligations stemming from the Antarctic Treaty, CCAMLR and the Convention for the Conservation of Antarctic Seals[25]. Articles 4–9 of the Act relate to the enforcement of the Protocol and detail the power of the Minister attached to the Department of Environmental Affairs and Tourism to enforce the Protocol. Article 9 is particularly significant because it relates to enforcement:

> Any person who contravenes a provision of a treaty mentioned in Column 1 of Schedule 2 [i.e. the Antarctic Treaty, the Protocol, CCAMLR and CCAS] shall, subject to the particular treaty, be guilty of an offence and on conviction liable to a fine or to

---

[20] Act No. 60 of 1996.
[21] See Joyner, Chapter 22 in this book.
[22] Art. 3 of the 1996 Antarctic Treaties Act.
[23] Chapter 14, Art. 231(4) of the South African Constitution. The text of the Constitution is available at <http://star.hsrc.ac.za/constitution/rsacontents.html>.
[24] Compare this with the procedures under the legal system of Italy; see discussion by Pineschi, Chapter 19 in this book.
[25] UNTS, Vol. 1080, pp. 175ff; and ILM, Vol. 11, 1972, pp. 251ff.

imprisonment for a period not exceeding the period mentioned in Column 2 of that schedule opposite the number of that provision.[26]

Given the mode of legal incorporation and since there is no qualification of the Protocol and its contents, the effectiveness of the 1996 Antarctic Treaties Act will depend on their practical implementation.

The area of application of the Protocol is assumed to be the area of application of the Antarctic Treaty.[27] While the Act does not define 'Antarctica', the domestic legislation does specify that the Prince Edward Islands are included within the zone of application. Similar to countries such as Germany, the domestic legislation draws no distinction between the continental shelf of Antarctica, the high seas and the deep seabed within the Antarctic Treaty's area of application. It is thus not clear whether South Africa's domestic legislation recognises the legal dilemmas relating to the Protocol and the provisions of the Law of the Sea Convention (LOS Convention).[28] As Davor Vidas recently noted, these two regimes, with a significant overlap of states parties, apply specific yet contradicting rules to the same area of application.[29] Under Part XI of the LOS Convention, mining regulation on the seabed beyond the national jurisdiction is to be administered by the International Seabed Authority (ISBA). In contrast, Article 7 of the Protocol bans all mineral activities south of 60° south latitude. As a consequence, there is a potential for disagreement as to whether the Southern Ocean deep seabed is subject to the authority of the ISBA or is exempt from the ISBA in regard to the application of the Antarctic Treaty. The Protocol clearly prohibits all forms of mineral activities in the Antarctic Treaty area while the LOS Convention contains a regulation of mineral activities in the deep seabed rather than a blanket ban. The domestic legislation of South Africa does not recognise this problematic issue and makes no reference to the deep seabed and possible exemptions from the provisions of the Protocol.

The domestic legislation is more forthcoming, however, on the types of activities covered by South Africa's implementation of the Protocol. The following are included in the South African implementation procedures (with formal judicial authority being delivered within the magisterial district of Cape Town): all South African citizens; all citizens who are members of expeditions which have been organised in South Africa for the purpose of visiting the Antarctic and all

---

[26] Art. 9 of the 1996 Antarctic Treaties Act. Schedule 2 of the 1996 Act also includes the relevant provisions of the Protocol which should be respected when planning and conducting activities in the Antarctic Treaty area. Practical application and supervision will then be critical in ensuring implementation of the 1996 Act.

[27] See Art. VI of the Antarctic Treaty.

[28] Ratified by South Africa on 23 December 1997. South Africa argues that the Protocol should take precedence over the LOS Convention and that Antarctica should be accorded a special status in international law of the sea; information based on e-mail correspondence between the author and officials attached to the DEA&T in May – June 2000.

[29] D. Vidas, 'The Southern Ocean Seabed: Arena for Conflicting Regimes?', in D. Vidas and W. Østreng (eds.), *Order for the Oceans at the Turn of the Century* (The Hague: Kluwer Law

companies, NGOs and corporations which operate and/or use ships, vessels and/or planes registered in South Africa. Exemptions exist for those persons who are members of expeditions organised by other Antarctic Treaty parties and for citizens performing official functions such as scientists, inspectors or observers within the Antarctic Treaty region. The details regarding the application of the 1996 Antarctic Treaties Act do not specifically mention the status of non-contracting parties to the Antarctic Treaty, i.e., third states.[30]

Practical implementation of the Protocol rests primarily with the Minister of Environmental Affairs and Tourism. With the specific advice of the Directorate: Antarctica and Islands, the Minister is charged with ensuring that the obligations under the Protocol are respected. The 1996 Antarctic Treaties Act empowers the Minister to act in instances where the Antarctic environment is damaged or detrimentally affected, for example to curtail particular activities such as expedition trips, to insist that any damage to the environment is repaired (where possible) and to recover any expenditure from a person or organisation who failed or refused to comply with these instructions. In cases where inspectors have produced *prima facie* evidence relating to environmental damage in the Antarctic region, the Act has a specific schedule of offences which can be considered by the Cape Town court. For instance, any person (not having immunity from prosecution) found guilty of damaging a historic site or monument can be imprisoned for up to two years in a South African prison.[31] At present, these disciplinary powers relating to the Protocol have not yet been tested and only relate to South African citizens and activities.

The ratification of the Protocol by South Africa coincided with a major new development in the SANAP. Owing to the poor condition of their permanent base SANAE III, the De Klerk government in power at the time approved funding for a new base called SANAE IV to be sited at Vesleskarvet in Dronning Maud Land[32]. After receipt of the draft documentation of the Protocol in 1991, it was announced that any new construction programme would have to adhere, as closely as possible, to Articles 1 and 2 of Annex I of the Protocol, which dealt with environmental impact assessments (EIAs). With the assistance of the NDPW, the DEA&T began a feasibility study in 1991 that concluded that the new base should be located in Dronning Maud Land. In 1992, armed with 'The Integrated Environmental Management Guideline Series', the DEA&T commissioned a formal investigation

---

International, 1999), pp. 291–314.

[30] The Antarctic Treaties Act declares that the provisions relating to the Protocol are applicable to all citizens visiting Marion Island.

[31] As for variation of severity of sanctions under implementing legislation of different countries, see Bastmeijer, Chapter 16 in this book.

[32] Built in the 1979–80 season, the condition of SANAE III had worsened considerably in the 1987–88 season when the base structure began to collapse due to pressure from the ice. Planning and budgeting for a new base began in 1989–90 and it was anticipated that the new base would be complete in the 1994–95 summer season. After funding and building difficulties, SANAE IV was officially opened in the 1997–98 season.

into the construction and the likely environmental impact of a new research station in this particular locale.

The results of this consultation and planning process were detailed in the draft comprehensive environmental evaluation (CEE) of the proposed new SANAE facility in Vesleskarvet which was published in May 1993. This was an important document because it detailed the South African approach to the implementation of the Protocol (which at that stage had not been formally ratified by the government) through so-called integrated environmental management (IEM). At the heart of this evaluative process was an inspection visit in 1992/3 to the proposed site, coupled with a widespread consultation exercise with interested parties in South Africa. The EIA concluded that the baseline environmental state at Vesleskarvet did not contain any breeding colonies of birds and mammals and that the long-term environmental, health and safety implications of the proposed base were minimal. This conclusion was reached after a number of individuals from South African government departments and universities had studied the site and devised an Environmental, Health and Safety Management System (EHSMS) that related to base maintenance, effluent treatment, research implementation and the impact of base construction. The official report concluded the following:

> Taken that no single potential impact has a major detrimental effect, it can be stated that the key to minimising the overall impact lies in the stringent management of all activities relating to the proposed facility. This is especially important in the light of the nature of the environment at Vesleskarvet, which is not so forgiving as the site of the existing SANAE III facility. [33]

This honest appraisal of the new site was critical, given that South Africa planned to abandon a substantial part of SANAE III to the elements. But the report also acknowledged that little established meteorological or geographical information existed for the proposed base in Dronning Maud Land.[34]

The subsequent review of the decommissioning of SANAE III in 1994–95 recruited expert reviewers from Antarctica New Zealand and the British Antarctic Survey, in contrast to the report of the late 1980s on the Marion Island landing strip.[35] International credibility was considered to be at a premium for South Africa's environmental managers who were concerned with the aftermath of the Marion Island airstrip saga.[36] The initial environmental evaluation (IEE) report on SANAE III published in 1998 details the procedures followed in terms of compliance with

---

[33] Cited in the *Draft Comprehensive Environmental Evaluation of the Proposed New SANAE IV Facility at Vesleskarvet, Queen Maud Land, Antarctica* (Pretoria: DEA&T, May 1993), p. xxix.

[34] See South Africa, 'Comprehensive Environmental Evaluation of SANAE IV', doc. XVIII ATCM/INFO 54,1994.

[35] The Marion Island landing strip proposal in the late 1980s was also not subjected to external scientific review. See Anon, 'An Albatross around South Africa's Neck', *New Scientist*, Vol. 113, 1987, p. 26.

[36] See K. Dodds, 'South Africa, the South Atlantic and the International Politics of Antarctica', *South*

Article 2 of Annex I of the Protocol relating to the IEE and impact on the Antarctic environment.[37] While rejecting the proposal to simply abandon SANAE III, South Africa committed itself to removing various materials from the base, such as hazardous substances prohibited by the Protocol, catwalks, electrical cabling and other supplies left at the base.[38] After consultation with various experts in South Africa, the DEA&T concluded that removing the base structure (now embedded in the ice) was not practical, even though some of the polyurethane panels would be left in contravention of the Protocol (Annex III). As part of the evaluation process, a team from SANAP were sent to SANAE III to carry out an IEE using criteria based on the nature of the activity and the potential impact of the site on human health and safety and on the environment. In addition to advocating the removal of waste, the report notes: 'Since the health and safety of the salvage team will be afforded the top priority in the operation [to decommission SANAE III], a decision has been made to leave the panels [composed of polyurethane] in situ'.[39] Other materials left on site included ventilation ducts, scaffolding and the Armco shell and supporting steel girders of the base.

It would be fair to conclude that the implementation of the Protocol in the South African context is progressing. Contingency plans for possible oil spills (required under Article 15 of the Protocol) in relation to the SANAP and an emergency response plan for the ship *SA Agulhas* and SANAE IV are implemented.[40] The DEA&T's annual report on the Protocol acknowledged that, 'further implementation of the Act are [sic] being developed and are in the process of finalisation'.[41] Some areas of implementation have advanced considerably since the 1980s, such as the Waste Management Plan (prepared in accordance with Article 9 of Annex III of the Protocol) which has been in operation since the 1991–92 season and insists that all waste relating to the SANAE IV base be first separated and then shipped back to Cape Town. Other regulations of the Protocol, such as an inventory of past activities in accordance with Article 8 of Annex III, have yet to be implemented by the DEA&T.

The SANAE III decommissioning process raised important issues relating to the general implementation of the Protocol. While it was known for some time that the SANAE III base was submerging into the ice, the South African government

---

African Journal of International Affairs, Vol. 3, 1995, pp. 60–80.

[37] South African National Antarctic Programme, *Initial Environmental Evaluation for the Decommissioning of the SANAE III and Sarie Marais Bases in Dronning Maud Land, Antarctica* (Pretoria: Department of Environment Affairs and Tourism, 1998).

[38] The Sarie Marias base at Grunehøgna in Dronning Maud Land will be removed in conjunction with the SANAE III decommissioning process.

[39] South African National Antarctic Programme, *Initial Environmental Evaluation for the Decommissioning of the SANAE III*, p. 4.

[40] The Oil Spill Contingency Plan, which has been effective since 23 October 1996, is based on the guidelines devised by the Council of Managers of National Antarctic Programmes (COMNAP).

[41] See Department of Environment Affairs and Tourism, *Annual Report Pursant to the Protocol on Environmental Protection to the Antarctic Treaty* (Pretoria: DEA&T, 1998).

decided that the construction of SANAE IV would take priority over any recovery operation relating to the older research base. In part, this decision was based on a financial judgement that any future investment should be channelled into securing a new scientific base for the SANAP. The uncertain political climate of the early 1990s also made it imperative for officials in the DEA&T to secure the Antarctic science programme before the transfer of power to the black majority government in 1994. However, this review process also demonstrated that the Protocol regulations regarding environmental impact and waste disposal would be partially implemented if there were evidence of only minor and transitory impact to the environment. The official report on the SANAE III decommissioning made it clear that, because the base had been enveloped in the ice, resources would be devoted to the recovery task in the 1998–99 season only.[42] The report also concluded that the long-term impact of SANAE III would be monitored because residual impact of the base is 'considered only minor as long as the base remains entrapped within the ice shelf'.[43] As with other areas of implementation, a trade-off between environmental impact, physical uncertainty and health and safety remains to be negotiated in particular institutional and geographical contexts.[44]

The final point to be made here concerning the South African implementation of the Protocol in the Antarctic is specific to Dronning Maud Land. This area of Antarctica has not yet been subjected to the pressures of tourism and activities of non-governmental organisations. As a non-claimant state, South Africa conducts its research exclusively at the SANAE IV base. Rather than monitoring activities such as tourist landings at popular destinations like Port Lockroy and Paradise Harbour in the Antarctic Peninsula, scientific resources have been focused on the decommissioning of former bases and ensuring that SANAE IV complies, as far as possible, with the requirements of the Protocol. In any case, limited financial and logistical resources prevent any further extensions of SANAP's environmental and scientific commitments.

THE PRINCE EDWARD ISLANDS MANAGEMENT COMMITTEE AND ENVIRONMENTAL PROTECTION IN THE SOUTHERN OCEAN

South Africa's formal involvement with the Prince Edward Islands dates from 1948 when the then ruling British government agreed to transfer sovereignty to the Union

---

[42] The latest information from the DEA&T suggests that the decommissioning of SANAE III will not be completed until the 1999/2000 summer season. An independent Environmental Officer is charged with carrying out an annual audit on all activities at SANAE IV.

[43] South African National Antarctic Programme, *Initial Environmental Evaluation for the Decommissioning of the SANAE III*, p. 6.

[44] The DEA&T is responsible for implementing an annual review process of environmental related issues relating to SANAE IV, ship and helicopter-based operations and over-snow transport.

of South Africa.[45] In 1949, a weather station was created in order not only to monitor weather conditions in the Southern Ocean but also to confirm South Africa's effective occupation of these remote Islands. Since the 1950s, scientific research on Marion Island has concentrated on monitoring the Islands' ecosystems through annual research visits organised formerly by the Department of Transport and subsequently by the DEA&T. According to Cooper and Headland, scientific research in the Prince Edward Islands began to change in the 1970s from an 'intermittent, opportunistic and informal activity to a properly organised and full time activity' under DEA&T supervision.[46] While scientific research on Marion Island has expanded over time, the management of the Prince Edward Islands has been plagued by controversy because South Africa not only sought to construct a landing strip but also proposed shooting the feral cat population in 1986–87.[47] After some considerable debate, the DEA eventually conceded in May 1987 that the landing strip proposal should be abandoned because it was 'not desirable ... [because] the impact it will have on that fragile environment particularly during the construction stage'.[48] The official report on the landing facility proposed that the Prince Edward Islands should be declared a wilderness region.

However, because the Prince Edward Islands are not located in the area of application of the Antarctic Treaty, the measures relating to the implementation of the Protocol *specifically exclude* the Islands. Nevertheless, CCAMLR recognised that the management of living resources in Antarctica and the Southern Ocean would have to include island groups such as Prince Edward in any ecosystem approach to the Southern Ocean. The geographical location of these island groups on the edge of the Antarctic Convergence means that they are an integral part of CCAMLR's management strategies.[49] In the last ten years, the rich fishing grounds around the Prince Edward Islands have attracted increasing numbers of distant water fishing fleets and there are fears that illegal fishing has severely depleted fish stocks

---

[45] For further details on the annexation process in 1948, see J. Marsh, *No Pathway Here* (Cape Town: Howard Timmins, 1948) and in a more critical vein Dodds, *Geopolitics in Antarctica*, pp. 192–196. On the legal implications of this South African claim, see P. Monteiro, *Marion and the Prince Edward Islands: The Legal Regime of the Adjacent Maritime Zones* (unpublished paper, University of Cape Town, Graduate Diploma in Law Research, 1987).

[46] See J. Cooper and R. Headland, 'A History of South African Involvement in Antarctica and the Prince Edward Islands', *South African Journal of Antarctic Research*, Vol. 21, 1991, pp. 77–91.

[47] The proposal to shoot the feral cat population was made after the failure of earlier attempts to cull the population by introducing specific viruses in 1977. The cats were introduced to the Prince Edward Islands in 1948, shortly after South Africa annexed the Islands, in an attempt to control the mice population. See G. Anderson and P. Condy, 'A Note on the Feral House Cat and House Mouse on Marion Island', *Suid-Afrikaanse Tydskrif vir Antarktiese Navorsing*, Vol. 4, 1974, pp. 58–61.

[48] See G. Heymann, *Report to the Minister of Environment Affairs on an Environmental Impact Assessment of a Proposed Emergency Landing Facility on Marion Island* (Pretoria: South African National Scientific Programme Report 140, 1987).

[49] Legislation dealing with marine resources around the Prince Edward Islands include Art. 54 of the Sea Fishery Act of 1988 which provides for the control and conservation of the Islands' territorial waters and the Fishing Zone. See also the *1978 Fishing Industry Development Act*.

such as Patagonian toothfish.[50] Under the evolving law of the sea, South Africa declared a 200-mile exclusive fishing zone around the Prince Edward Islands in 1979.[51] Since that period, the South African Navy has devoted time and resources to protecting this zone from illegal, unreported and unregulated fishing.

In the early 1990s, the DEA&T in consultation with academic experts, announced that the Prince Edward Islands should be a fully protected wilderness area. In November 1995, under Ordinance Number 5592, the DEA&T declared that the Prince Edward Islands would become Special Nature Reserves governed by a Management Committee headed by Professor Steven Chown of the University of Pretoria.[52] The transformation was carried out under the existing legislation of the 1989 Environment Conservation Act and the 1988 Sea Fishery Act. The implication of this designation was more than merely a semantic intervention; as Special Nature Reserves the Prince Edward Islands would be treated according to stringent environmental regulations contained within the Management Plan (MP) for the Prince Edward Islands. As the draft 1995 Management Plan noted:

> the primary aim [of the MP] the conservation and sustained preservation of this unique ecosystem for all the people of South Africa and for the scientific community at large. The PEI are an integral part of South Africa's national heritage and territorial integrity; their rational and strong management will serve to offer a model to the world in keeping with the new and emerging international political order.[53]

This position was in sharp contrast to the published 'guidelines' of the 1970s and 1980s, which governed South African management of the Prince Edward Islands. In 1988, the DEA&T issued a new Code of Conduct which reiterated the demand that all activities follow the Agreed Measures for the Conservation of Fauna and Flora, and CCAMLR.[54]

The formal framework for implementing environmental protection in the Prince Edward Islands follows the obligations expressed in the Annexes to the Protocol. According to the first chair of the Prince Edward Islands Management Committee (PEIMC), Professor Steven Chown, these obligations are at least as stringent as the Protocol. With the advice of the PEIMC, the DEA&T stipulates that all visitors to

---

[50] See Herr, Chapter 15 in this book.

[51] See *Republic of South Africa's Gazette 1979* and the 1989 Environment Conservation Act. The key section is 18 of the 1989 Act, which specifies the obligations connected with 'Special Nature Reserves'.

[52] Department of Environmental Affairs and Tourism, *Prince Edward Islands Management Plan* (Pretoria: DEA&T, 1996). See also C. Hanel and S. Chown, *An Introductory Guide to the Marion and Prince Edward Island Special Nature Reserves* (Pretoria: DEA&T, 1999). My thanks to Carol Jacobs of DEA&T for sending copies of these reports to the author, 25 May 1999.

[53] Department of Environmental Affairs and Tourism, *A Management Plan for the Prince Edward Islands* (Pretoria: Prince Edward Islands Management Plan Working Group, 1995).

[54] See W. Visagie, *Code of Conduct for Prince Edward Islands* (Pretoria: DEA&T, 1988), p. 45. From 1989 onwards, research proposals for Marion Island have been assessed and approved by the DEA&T. Previously the Council for Scientific and Industrial Research (CSIR) had advised on scientific

the Islands must be granted entry permits and that only 64 persons are allowed on Marion Island at one time.[55] Some of the regulations concerning visits to the Prince Edward Islands exceed the requirements of the Protocol. For example, the Prince Edward Islands Management Plan (PEIMP) does not allow fresh produce to be brought onto Marion Island because of the risks of introducing fungal, bacterial and or viral pathogens.[56] All forms of logistical equipment, including camping supplies, must be fumigated before arrival on the Island. The team leader at Marion Island is also responsible for the implementation of the PEIMP and has to report back to the DEA&T in the event of any violations to the regulations governing behaviour on the Island.

As a consequence of its limited geographical area and undisputed sovereignty claim, the PEIMP also introduced a geographical classification for the Prince Edward Islands.[57] While the Protocol specifies particular areas of Special Protection in the Antarctic, Marion Island is divided into four zones of protection and usage:

Zone 1: Service Zone (this is the area occupied by the base near Transvaal Cove).
Zone 2: Natural Zone (a specific area close to the base which allows limited free walking and access for base personnel).
Zone 3: Wilderness Zone (open to research but permits must be issued by the Director-General of DEA&T).
Zone 4: Protected Zone (special entry areas with access governed by the Director-General of DEA&T).

As breeding areas and historical sites on Marion Island and the entire area of Prince Edward Island are classified as Zone 4, access to and research in these areas are strictly limited. No building structures can be constructed in Zone 4 and all field time is strictly limited by the access permits which are only granted to research programmes that have been reviewed by the PEIMC and the DEA&T.[58]

Given South African sovereignty over the Islands, the implementation of the PEIMP has been unchallenged, but the regulation of tourism and tour operators remains an open challenge of the future. Although a number of unauthorised visits by yachts and tourist vessels have occurred since the 1960s, at present no Antarctic tour operators deploy schedules which include visits the Prince Edward Islands.. The

---

programmes.

[55] This figure is derived from the maximum number of persons that can be accommodated at the base on Marion Island.

[56] My thanks to Professor Steven Chown of the University of Pretoria for his comments and advice regarding the operation of the PEIMC.

[57] At present, scientific activity is centred on Marion Island and there are no scientific sites on neighbouring Prince Edward Island.

[58] The 'appropriate authority' as specified in the Agreed Measures for the Protection of Antarctic Fauna and Flora has never been questioned by other parties, not least because all visitors to Marion have been legally associated with the South African government and the DEA&T. Since 1948, the majority of visitors to Marion Island have been South African citizens.

PEIMP acknowledges, however, that tourism and *bona fide* educational visits should be considered. The PEIMC has initiated a study of the potential EIA of tourism limited to Zone 2 of Marion Island, but at the time of writing the results of this analysis have not yet been released to the public. The PEIMP notes that the DEA&T will be advised by the PEIMC of the merits of proposed visits, but does not consider the implications of unplanned and/or unreported landings at Marion Island or Prince Edward Island. South Africa's capacity to monitor movements around the Prince Edward Islands is severely limited and there is little prospect of naval resources being devoted to anything bar periodic monitoring of fishing in the South Indian Ocean.

The jurisdiction of the PEIMP covers any person who is a member of an expedition and any person responsible for organising an expedition that proceeds from South Africa as its final point of departure. Every approved visitor to the Prince Edward Islands is formally contracted to the DEA&T and accepts this application of port state jurisdiction. In doing so, such a movement contributes further positive momentum to what Orrego Vicuña (Chapter 3 in this book) has called 'gateway port' arrangements that recognise that proximate Consultative Party nations, such as South Africa, should play a key role in the implementation of the Protocol.[59] As a non-claimant nation in the Antarctic and a claimant in the sub-Antarctic, South Africa retains a unique position regarding implementation and jurisdiction. The environmental regulations governing the Prince Edward Islands might at the very least function as a model for other Southern Ocean claimant states such as Norway, Australia and France who have witnessed a growth of activity around their islands, mainly in the form of tourism and fishing.

CONCLUSIONS

As a geographically proximate nation, South Africa has been an active player in the international environmental politics of Antarctica and the Southern Ocean. Despite having suffered widespread condemnation for *apartheid*, South Africa was able to maintain close diplomatic and scientific connections with the Consultative Parties. As a non-claimant state, the Republic has also sought to ensure that the territorial freeze on all Antarctic claims has been maintained in conjunction with the demilitarised status of the polar continent. However, as the sovereign claimant to the Prince Edward Islands in the South Indian Ocean, South Africa has also sustained a small but significant research programme concerned with meteorology and other scientific investigations over a wide ecological spectrum.

Significant developments have been made with regard to EIA, contingency plans and the monitoring of environmental protection in the Antarctic. It could be argued that South Africa has been one of the leading exponents of EIA and

---

[59] On the geopolitical significance of proximity, see Dodds, *Geopolitics in Antarctica*.

monitoring, which is demonstrated by their the country's commitment to SANAE IV. As the lead government agency, the DEA&T has a considerable task at hand in implementing the demands of the Protocol in Antarctica and the Prince Edward Islands. Armed with limited resources, the environmental monitoring of previous sites such as SANAE III in combination with the effective management of SANAE IV will continue to dominate South African proceedings. The development of the PEIMP indicates, however, that the spirit of the Protocol will be implemented further in a thorough and consistent manner. In the future, the PEIMP may provide a model for managing fragile island ecosystems in the Southern Ocean, and South Africa can also, through the Committee for Environmental Protection,[60] provide invaluable information on the long-term implementation of the Protocol in Antarctic and Southern Ocean Island environments.

---

[60] On the CEP, see Orheim, Chapter 6 in this book.

22

The United States: Legislation and Practice in Implementing the Protocol

Christopher C. Joyner

On 2 October 1996, President William Clinton signed into law the Antarctic Science, Tourism and Conservation Act of 1996 (ASTCA).[1] This act concluded four years of intense discussions between the administration and environmental organisations over the content and scope of US legislation needed for implementing the 1991 Protocol on Environmental Protection to the Antarctic Treaty.[2] In the process, several critical environmental issues and policy considerations were debated intensely in order to fuse together the legislation necessary to ratify the Protocol.

THE RATIFICATION PROCESS

Though the Bush administration initially viewed negotiation of the Environmental Protocol warily, tremendous public pressure compelled government officials to recognise the benefits of the Protocol for US national interests, and the United States signed the Protocol on 4 October 1991. On 6 February 1992, President Bush transmitted the Protocol and its annexes to the Senate for 'advice and consent' to ratification.[3] On 4 May 1992, the Senate Foreign Relations Committee held a hearing on the Protocol during which both administration and environmental witnesses strongly supported ratification of the Protocol. On 11 June 1992, the Senate Foreign Relations Committee ordered the Protocol and its annexes reported

---

[1] Antarctic Science, Tourism and Conservation Act of 1996, Public Law No. 104–227, *Statutes*, Vol. 110, section 3034.

[2] ILM, Vol. 30, 1991, pp. 1,461ff..

[3] Treaty Document 102–22 (1992).

out favourably to the Senate for its advice and consent. The Senate gave its advice and consent to ratification of the Protocol on 7 October 1992.[4]

The 'advice and consent' procedure, however, was not sufficient to consummate the ratification process for the United States. In the Senate's view, the Environmental Protocol fell short of being self-executing.[5] That is, the Environmental Protocol could not automatically be assimilated into US domestic law. Special implementing legislation by Congress was necessary to make US laws conform to obligations in the Protocol, as well as to link and activate the Protocol's provisions to current US laws relating to activities in the Antarctic.

Implementing legislation for the Environmental Protocol had to abide the US congressional legislative process, a special legislative process that is cumbersome and protracted. First, a bill must be introduced in separate actions by a member to the House of Representatives as well as by a Senator to the Senate. Once introduced in each chamber, the bill is referred to a committee, which then refers it to a subcommittee. For action to be taken on the bill, subcommittee members must be sufficiently interested in doing so. Since subcommittees in both chambers wanted to proceed with various bills implementing the Environmental Protocol, hearings were held in 1992, 1993, 1994 and 1996, during which witnesses testified about respective merits and deficiencies of the Protocol. While these hearings were low-profile events, they provided fora in which interested groups could express opinions about the intent and nature of a bill, and members of Congress could learn about the substantive and political impact of the proposed legislation.

Once hearings were held, a 'mark-up' session followed in each chamber, during which subcommittee members decided what changes, if any, were needed. A majority vote in the subcommittee reported the bill for implementing legislation out to the full committee. Upon passage by the full committee, the bill was voted on separately by each chamber. After both the House and Senate passed versions of the same bill, it was sent to a conference committee in order to resolve differences. Following agreement in the conference committee, both chambers approved the bill in separate votes, and it was sent to the President, who signed it into law.

Congressional approval of legislation implementing the Protocol for the United States took four years and played out in three overlapping realms. First, a protracted set of informal negotiations took place within and among agencies in the administration, including the US Department of State, the National Science Foundation (NSF), the Environmental Protection Agency (EPA), the National

---

[4] Senate Committee on Foreign Relations, *Protocol on Environmental Protection to the Antarctic Treaty*, Senate Executive Report No. 54, 102nd Cong., 2nd Sess. (1992), p. 1.

[5] *Hearings on Antarctic Scientific Research, Tourism, and Marine Resources Act of 1993* before the Senate Committee on Commerce, Science and Technology, 103rd Cong., 1st Sess. (1993) (testimony of R. Tucker Scully, Director, Office of Oceans, US Department of State), p. 3. For differences regarding self-executing nature of the Protocol in domestic legal systems of various countries, compare, e.g., the cases of Italy (Pineschi, Chapter 19 in this book), Norway (Njåstad, Chapter 20 in this book) and South Africa (Dodds, Chapter 21 in this book).

Oceans and Atmospheric Administration (NOAA) in the Department of Commerce, the Council on Environmental Quality (CEQ) and the Coast Guard. The second realm involved numerous ad hoc meetings that were convened by various NGOs to coordinate efforts and positions that would most strengthen protective measures in the US legislation. Chief among these groups were the Antarctica Project, Greenpeace and the Environmental Defense Fund, with supporting roles played by Friends of the Earth, the Humane Society of the United States, the National Wildlife Federation and the Sierra Club. Third, through a series of congressional hearings, these forces combined to set out their respective aims of and objections to various bills proposed during the 102nd, 103rd and 104th Congresses as drafts for implementing legislation. This legislative process culminated in 1996 in Public Law 104–227, entitled the Antarctic Science, Tourism and Conservation Act of 1996 (ASTCA).

KEY FEATURES OF PUBLIC LAW 104–227 (ASTCA)

The implementing legislation required the reconstruction of various US laws that regulated American activities in the Antarctic. Title I of the ASTCA amends the Antarctic Conservation Act of 1978[6] (ACA) to provide authority for implementing the Protocol's provisions on environmental impact assessment (EIA) (Article 8 and Annex I of the Protocol), the conservation of fauna and flora (Annex II), waste disposal and management (Annex III) and area protection and management (Annex V).[7] Critical here are various terms applied in the law and their relationship to provisions in the Protocol.

Definitions

The US implementing legislation avers as its purpose 'the conservation and protection of the fauna and flora of Antarctica, and of the ecosystem upon which such fauna and flora depend, consistent with the Antarctic Treaty and the Protocol'.[8] To this end, the ASTCA builds upon existing US laws to provide the authority necessary to give full effect to the Protocol's provisions. The legislation amends the definitions of terms in the ACA[9] so that they conform to designations in the Protocol.

---

[6] USC, Vol. 16, sections 2401ff.
[7] This fifth annex on area protection and management was adopted on 17 October 1991 at the XVI Consultative Meeting in Bonn; see *Final Report of the Sixteenth Antarctic Treaty Consultative Meeting, Bonn, Germany, 7–18 October 1991* (Bonn: German Federal Ministry of Foreign Affairs, 1991), pp. 116–125.
[8] USC, Vol. 16, section 2401(b).
[9] *Ibid.*, section 2402.

In the implementing legislation, the scope of 'Antarctica' conforms precisely to the definition in the Protocol and the Antarctic Treaty. 'Antarctica' thus is defined as 'the area south of 60 degrees south latitude'.[10] The Protocol asserts that activities be conducted in the Antarctic such that they avoid 'detrimental changes in the distribution, abundance or productivity of species or populations of species of fauna or flora'.[11] The ASTCA thus defines 'harmful interference' to include several activities, namely flying or landing helicopters; using vehicles, vessels, small boats, explosives or firearms that might disturb concentrations of wildlife, or disturb birds or seals that are breeding; damage terrestrial plants; or cause 'significant adverse modification of habitats of any species or population of native mammal, native bird, native plant, or native invertebrate'.[12]

Regarding wildlife, the law asserts that any act of 'harmful interference' or 'taking' is prohibited. The implementing legislation defines 'take' as 'to kill, injure, capture, handle, or molest a native mammal or bird, or to remove or damage such quantities of native plants that their local distribution or abundance would be significantly affected'.

Also notable is the concept of a 'person' under the ASTCA. As defined, 'person' refers to 'any person subject to the jurisdiction of the United States and any department, agency, or other instrumentality of the Federal government or any State or local government'.[13] Thus, not only are individual US citizens or US governmental agencies considered to be 'persons' under the law, but also to be non-US nationals working under US government auspices or situated within the area of a US Antarctic station.

Environmental Principles

Article 3 of the Environmental Protocol establishes a new set of environmental principles for Antarctic activities. These principles require that all activities be planned and conducted in order to limit adverse impacts on the Antarctic environment and dependent and associated ecosystems and that regular monitoring occur to allow assessments of ongoing activities. The Protocol obligates parties to follow EIA procedures for proposed activities, including both governmental and non-governmental activities, such as tourism, for which advance notice is required under the Antarctic Treaty. Parties must also provide prompt and effective action in response to environmental emergencies, including the development of joint contingency plans.

Article 3 directs each party to ensure that its nationals comply with certain 'environmental principles' when planning or conducting activity in Antarctica.

---

[10] *Ibid.*, section 2402(2).
[11] Protocol, Art. 3(2)(a)(iv).
[12] ASTCA, section 102(5)(F).
[13] *Ibid.*, section 102(14).

Those principles specifically mandate that all activities in the region be planned and conducted so as to limit 'adverse impacts on the Antarctic environment' and to 'avoid' effects on climate, weather patterns, air and water quality, flora and fauna, endangered species and specially managed and protected areas. Planning must be done by governments and operators on the basis of sufficient information and informed judgements about EIAs, verified by monitoring programmes.

Government officials and conservationists interpret the significance of Article 3 differently. In the view of the US administration, the principles of Article 3 reflect a statement of US policy rather than a set of integral legal obligations in the legislation. Although these environmental principles will be considered in the planning of US Antarctic activities, they will not be construed as legally binding obligations. Environmentalists, on the other hand, see the provisions in Article 3 as integral, legally binding elements of the Protocol. As these principles entail a binding set of obligations, they are indeed legal commands for governments and they should constitute an essential part of the regulatory framework that guides all actions in the Antarctic.[14]

The US implementing legislation does not directly address the legal status of Article 3. Instead, it integrates respect for the provision's principles into various dimensions of the law. Planning and conducting activities on the basis of sufficient information so as to prevent harmful effects on Antarctic ecosystems is met in the ASTCA by the requirement that EIAs be undertaken[15] and by the mandate that permits must be secured to conduct certain activities.[16] Also important in affirming the legal obligations of Article 3 are provisions that impose specific regulations on US activities for protecting and conserving native fauna and flora and for preserving specially managed areas in the Antarctic.[17] Furthermore, the Environmental Protection Agency is mandated to promulgate regulations for non-governmental activities, including tourism, that ensure that EIAs are performed for those activities.[18] Finally, while monitoring *per se* is not mentioned in the US implementing legislation, provision is made for the EPA to coordinate the review of information regarding EIAs received from other parties under the Protocol.[19] Promulgation of an EPA interim final rule in 1997 (and its extension in 1998 through 2001) makes monitoring essential in performing EIAs for activities having more than a minor or transitory impact on the Antarctic environment. The US legislation, moreover, designates certain prohibited acts that buttress protective measures in Article 3, particularly in application to Protocol Annex II (on the

---

[14] See 'Statement of Christopher C. Joyner', in *Hearings*, pp. 147–148.
[15] ASTCA, section 104.
[16] *Ibid.*, section 105.
[17] *Ibid.*, section 106.
[18] *Ibid.*, section 104.
[19] *Ibid.*, section 104(c)(1)(B).

conservation of fauna and flora) and Annex III (on waste disposal and waste management).

International Cooperation

Article 6 the Protocol calls for international cooperation in planning and conducting activities in the Antarctic Treaty area. As one of the 'Congressional findings' highlighted in the Antarctic, Tourism and Conservation Act of 1996, the Antarctic Treaty and the Protocol are aptly characterised as establishing 'a firm foundation for the conservation of Antarctic resources, for the continuation of international cooperation and the freedom of scientific investigation in Antarctica'.[20] The implementing legislation addresses the need for international cooperation under the rubric of 'federal activities carried out jointly with foreign governments', as well as through the circulation among the Antarctic Treaty Consultative Parties of documented proposals for activities that might have significant impacts upon the Antarctic environment.

Mineral Resources

The Environmental Protocol prohibits all activities relating to mineral resources in Antarctica, except scientific research, and provides that this ban may not be reviewed until at least fifty years have passed following entry into force of the Protocol.[21] The 1996 US implementing legislation tersely upholds this prohibition on mineral resource activities in Antarctica, albeit without elaboration on the definition of 'scientific research.' The legislation simply amends the Antarctic Protection Act of 1990[22] to make permanent the interim protection against drilling and mining activities provided for in that act. The ban on drilling or mining is clearly affirmed, as the law asserts that, 'It is unlawful for any person to engage in, finance, or otherwise knowingly provide assistance to any Antarctic mineral resource activity'.[23] This prohibition on minerals activities was incorporated into the legislation without serious controversy or political obstruction.[24] US legislation banning Antarctic mineral activities had been in force since 1990, and no commercial or corporate enterprise had expressed interest since then in pursuing any such activity.

---

[20] USC, Vol. 16, section 2401(a)(4).
[21] See Protocol, Arts. 7 and 25.
[22] USC, Vol. 16, sections 2461ff.
[23] ASTCA, Title II, section 202(a); USC, Vol. 16, section 2463.
[24] For possible law of the sea implications of domestic legislation in this respect, see Bastmeijer, Chapter 16, and Dodds, Chapter 21, in this book. See also D. Vidas, 'The Southern Ocean Seabed: Arena for Conflicting Regimes?', in D. Vidas and W. Østreng (eds.), *Order for the Oceans at the Turn of the Century* (The Hague: Kluwer Law International, 1999), pp. 291–314, especially at pp. 303–307.

The Annexes to the Protocol

Annex I to the Protocol, which is linked to qualifications set out in Article 8 of the Protocol, establishes a scheme for prior assessment of the environmental impact of proposed public and private activities in Antarctica. The ASTCA implements these provisions on environmental impact assessment through application of the 1969 National Environmental Policy Act (NEPA).[25] NEPA thus becomes the designated authority for implementing and enforcing EIAs for federal agency activities.[26] The US legislation, however, makes a notable exception for EIAs for joint activities in the Antarctic. When such joint activities are carried out in cooperation with foreign governments, EIAs will not be conducted by the United States unless the Secretary of State determines that another state, which is coordinating the environmental assessment and has acceded to the Protocol, is contributing to a major part of the activity another.[27] Concerns by environmentalists over this caveat were somewhat mollified, however, by an attendant stipulation that the Secretary of State is required to publish a notice in the *Federal Register* whenever a determination is made that a state other than the United States is responsible for coordinating the environmental assessment of such a joint activity, or whenever a draft comprehensive environmental evaluation is received.[28] Moreover, the Administrator of the Environmental Protection Agency is required to promulgate regulations to provide for the EIA of non-governmental activities, such as tourism.[29] It also articulates the federal agency's responsibilities with respect to the obligations of the Protocol, reaffirming the lead role of the National Science Foundation as manager of the US Antarctic research program.

Annex II to the Protocol (on the conservation of fauna and flora) strengthens and expands the protection system developed under the 1964 Agreed Measures. Similarly, Annex III to the Protocol (on waste disposal and waste management) sets forth detailed requirements regulating the generation, management and disposal of wastes within the Antarctic Treaty area. Much of Title I in the US implementing legislation addresses these dual concerns of environmental protection and waste disposal.

The ASTCA incorporates the obligations of Annex II to the Protocol by extending the 1978 ACA's scope of prohibited acts and system of permits. To manage and dispose of wastes in the Antarctic, the legislation stipulates thirteen

---

[25] See USC, Vol. 42, section 4332(2)(C), and section 102(2)(C).

[26] For purposes of this NEPA application, the term 'significantly affecting the quality of the human environment' was stipulated as being synonymous with the term 'more than a minor or transitory impact'; ASTCA, Title II, section 104(B), USC, Vol. 16, section 2463. For a discussion on different interpretations of the meaning of the term 'minor or transitory impact', see Bastmeijer, Chapter 16 in this book.

[27] ASTCA, Title I, section 104(4B).

[28] *Ibid.*, section 104(h); USC, Vol. 16, section 2403.

[29] *Ibid.*, section 104(c).

actions as prohibited, including the following: disposal of waste or pollutants onto land, ice shelves, or into circumpolar waters; open burning of wastes; damage or harm done to historic sites; failure to comply with regulations aimed at controlling tourism; interference with or resistance to authorised enforcement officials; and violation of regulations and permits.[30]

Annex IV to the Protocol concerns the prevention of marine pollution and places strict controls on the discharge of pollution from ships in Antarctic waters. Legislation initially favoured by the administration generally did not address the implementation of Annex IV; neither did it strongly oppose limits being placed on sovereign immunity for government vessels or did it prohibit government vessels from discharging wastes, including plastics and garbage, into Antarctic waters in contravention of the Act to Prevent Pollution from Ships (APPS).[31] In contrast, under the APPS, as of 31 December 1993 government vessels were prohibited from disposing of plastics and garbage at sea.[32]

In implementing the Protocol, US lawmakers were reluctant to go quite that far. The ASTCA amends section 2 of the Act to Prevent Pollution from Ships such that requirements in Annex IV to the Antarctic Protocol 'shall apply in Antarctica to all vessels over which the United States has jurisdiction'. Even so, that law retains provisions in the APPS that explicitly exclude application of the law to 'a warship, naval auxiliary, or other ship owned or operated by the United States when engaged in non-commercial service'.[33] While this caveat renders US naval and coast guard vessels in the Antarctic essentially immune from preventive measures that implement Annex IV, this situation does not give US government vessels the licence to pollute circumpolar southern seas. Provision is made in the APPS for the heads of federal departments and agencies to set standards for these vessels that 'ensure, so far as is reasonable and practicable without impairing the operations or operational capabilities of such ships, that such ships act in a manner that is consistent with the [Antarctic Environmental Protection] Protocol'.[34] Nonetheless, US implementing legislation retains provision that sovereign immunity applies to most US government vessels operating in the Antarctic. Despite increasing tourism in the Antarctic, no specific measures were proposed in US implementing legislation to prevent adverse environmental impacts from activities sponsored by American tour operators. In the ASTCA, the EPA is charged with the responsibility of promulgating regulations for tourism operators regarding making EIAs of their activities, for which United States is required to give advance notice under Article VII(5) of the Antarctic Treaty. Tourism conducted by US operators is thus regulated by the National Science

---

[30] *Ibid.*, Title I, section 103; *Statutes*, Vol. 110, section 3036.
[31] Public Law 96–478, of 21 October 1990, *Statutes*, Vol. 94, section 2297, as amended, USC, Vol. 33, section 1901.
[32] USC, Vol. 33, sections 1901–1912.
[33] *Ibid.*, section 1901.
[34] APPS, section 1902(g).

Foundation through a system of permits adopted since enactment of the implementing law.[35]

Compliance and Enforcement

The Protocol asserts in Article 13 that each party shall take appropriate enforcement measures to ensure compliance with the Protocol's provisions. Accordingly, the ASTCA provides ways and means to enforce compliance with the Protocol by constructing a system of permits and regulations, and by authorising which agencies should be competent for which actions. Title I amends the permits section of the Antarctic Conservation Act of 1978 to make it conform to requirements in the Protocol. Specified in Title 1 are the need for a permit for killing or capturing native birds or mammals in conjunction with scientific activities or the construction and operation of scientific support facilities.[36] The legislation also stipulates that, with two exceptions, NSF shall issue all regulations to implement the Protocol and Act, including Annex II, Annex V and Article 15 (Emergency Response Action) with respect to land areas and ice shelves.[37] Moreover, NSF is mandated, with the concurrence of EPA, to issue regulations to implement Annex III to the Protocol.[38] Finally, the Coast Guard is directed to issue regulations that implement Annex IV to Protocol and, with the concurrence of NSF, to issue regulations that implement Article 15 of the Protocol with respect to ships.[39]

Highly contentious during the debate over US legislation was which government agency should be given the lead responsibility for the US Antarctic Program. In previous US legislation pertaining to the Antarctic, different executive agencies were assigned various duties to ensure that US activities complied with US obligations under international law. The US Department of State coordinated Antarctic foreign policy and chaired the inter-agency Antarctic Policy Group. The National Science Foundation implemented the 1978 Antarctic Conservation Act (for fauna and flora), funded basic scientific research in the area, and contracted with Navy and private contractors for logistical support in Antarctica. The National Oceans and Atmospheric Administration implemented the 1990 Antarctic Protection Act (which asserted the mining ban), supported scientific research in Antarctica, and was charged in 1982 by the Antarctic Marine Living Resources Convention Act with

---

[35] As provided for under ASTCA, section 105.

[36] *Ibid.*, section 105.

[37] *Ibid.*, section 106. One exception is that the EPA Administrator must concur with NSF's issuance of regulations to implement waste disposal and management requirements of the Protocol under Annex III. USC, Vol. 16, section 2405(a)(2). Another exception is that the Coast Guard is mandated to issue regulations for implementing marine pollution prevention requirements under Annex IV of the Protocol and for conducting emergency response action for ships. ASTCA, section 106. See USC, Vol. 16, 2405(b).

[38] ASTCA, section 104, *Statutes*, Vol. 111, section 3038.

[39] ASTCA, Title I, section 106, *Statutes*, Vol. 111, section 3041.

implementing fisheries regulation in the Southern Ocean.[40] The Coast Guard was responsible for developing US policy on marine pollution from vessels and for coordinating this policy with the International Maritime Organisation. Finally, the Environmental Protection Agency and the Council on Environmental Quality participated in the Antarctic Policy Group and were responsible for reviewing agency proposals that affected the Antarctic environment.[41]

The issue over which US agency should assume the primary regulatory role became the greatest challenge to the adoption of implementing legislation. The National Science Foundation was assigned major responsibility for administering the US Antarctic Program by Congress in 1971[42] and was charged in 1982 by the Reagan administration with the responsibility for overseeing implementation of the Antarctic Conservation Act of 1978. By the early 1990s, however, impressions of NSF's role changed appreciably within the conservationist community. Environmentalists contended that while NSF should be permitted to support basic science and operate US bases and facilities in Antarctica, other agencies should have more prominent roles in enforcing the Protocol. Environmental groups took this attitude for three reasons: 1) because NSF had not issued pollutant regulations called for under the 1978 ACA until July 31, 1992; 2) NSF had failed, as highlighted by the Inspector General, to assert formal enforcement action against permittees and tour operators for violations of the ACA; and 3) NOAA was recognised as having had more experience with resource protection and management in the Antarctica than NSF.[43]

As adopted, the implementing legislation represents a quasi-victory for the administration in terms of agency enforcement. Throughout the legislative negotiations, the administration consistently advocated that NSF be allowed to self-permit and self-enforce activities affecting waste disposal, including incineration and sewage treatment. Environmentalists preferred that EPA be put in charge of activities concerning waste disposal and the review of EIAs and that NOAA to be given lead responsibility, or at the very least, mandated concurrence, for permitting logistics, takings and harmful interference, tourist activities and entry into protected

---

[40] USC, Vol. 16, sections 2431–2444 (1988), implementing the Convention on the Conservation of Antarctic Living Marine Resources (ILM, Vol. 19, 1980, pp. 837ff).

[41] For discussion of US decision-making in Antarctic matters, see C.C. Joyner and E. Theis, *Eagle Over the Ice: The U.S. in the Antarctic* (Hanover, NH: New England University Press, 1997), pp. 53–61. See also C.C. Joyner, 'The Role of Domestic Politics in Making United States Antarctic Policy', in O.S. Stokke and D. Vidas (eds.), *Governing the Antarctic: The Effectiveness and Legitimacy of the Antarctic Treaty System* (Cambridge University Press, 1996), pp. 412–427.

[42] National Science Foundation, *Facts about the United States Antarctic Program* (Washington, DC: National Science Foundation, 1988), p. 9.

[43] See 'Statement of Susan Sabella, Greenpeace', *Antarctic Environmental Protection Act: Hearings on H.R. 964 before the Subcommittee on the Science of the Senate Committee on Science, Space and Technology*, 103rd Cong., 1st Sess. (1993), pp. 34–35 and 'Statement of Bruce S. Manheim, Jr., Environmental Defense Fund', in *Hearings*, p. 204.

areas. In the end, though, NSF retains most of its oversight authority to issue permits and regulations and to supervise the process of environment impact assessments.

Regulations

The ASTCA provides for a *system of permits* issued under the authority of the Director of the NSF. In special circumstances, certain unlawful activities such as the disposal of wastes, incineration, introduction of non-native species, taking or harmful interference with Antarctic wildlife, or purchasing or transporting native wildlife from Antarctic land, ice shelves, or circumpolar seas, may be authorised as lawful through a system of permits. An exception is made to the unlawful character of certain prohibited acts if those acts were committed 'under emergency circumstances involving the safety of human life or of ships, aircraft, or equipment or facilities of high value, or the protection of the environment'.[44]

Permitting requirements under the ASTCA are relatively well defined and administered. The permitting system allows a permit holder to engage in any of the prohibited acts only as a permit-holder. The criteria for issuing a permit are strict, and applications require a detail description of and thorough justification for a proposed action.[45] Permits for taking specially protected species may be issued only if there is a compelling scientific purpose for doing so and if the actions allowed by the permit do not jeopardise the natural ecological system or survival of that species.[46]

Permits authorising entry into an Antarctic Specially Protected Area are issued only if that entry is consistent with an approved management plan, or if there is 'a compelling purpose for such entry which cannot be served elsewhere' and the resultant actions do not jeopardise the natural ecological system in that area.[47]

The Protocol also establishes a Committee for Environmental Protection (CEP) as an expert body to furnish advice and formulate recommendations to the Antarctic Treaty Consultative Meetings with respect to implementing the Environmental Protocol.[48] Since the CEP is only an advisory body, its establishment generated little controversy during the US legislative debate and no direct reference is made to it in US law.

---

[44] ASTCA, Title I, section 103, *Statutes*, Vol. 111, section 3036.

[45] In the ACA, permits authorising the taking or harmful interference with Antarctic mammals or birds are issued 'only for the purpose of providing specimens for scientific study or scientific information' [and] specimens for museums, zoological gardens, or other educational or cultural institutions or uses'. The implementing legislation adds a third possibility for obtaining a permit, to which environmentalists have expressed disappointment, namely 'for unavoidable consequences of scientific activities or the construction and operation of scientific support facilities'; USC, Vol. 16, section 2404 (e)(2)(A)(i).

[46] USC, Vol. 16, section 2404(e)(2)(B).

[47] USC, Vol. 16, section 2404(e)(2)(C).

[48] On the CEP see Orheim, Chapter 6 in this book.

No formal inspectorate is created by the ASTCA for enforcing the Protocol's provisions, notwithstanding assertions by some observers that such an agency is critical for ensuring compliance with the provisions.[49] Nevertheless, the new *inspection procedures* in the Protocol are intended to be important compliance and enforcement tools. The Environmental Protocol places the United States under obligation to ensure that inspections are carried out by observers to promote environmental protection and to ensure compliance with the Protocol.[50] Another innovation in the Protocol requires that a report be generated after each inspection. Each report will be sent to the party inspected for comment, circulated to all Consultative Parties and the CEP, and will be considered at the next Consultative Meeting.[51]

The ASTCA requires that the agency responsible for regulating Antarctic activities must comply fully with the Protocol's *reporting and notification* requirements.[52] This seems all the better, since the United States maintains the largest presence in Antarctica and possesses the most resources to use in reporting. That the United States made this commitment sends a strong signal to other states of the essential need for transparency and the need to make information available for compliance and planning purposes.

While inspections are not specifically inserted into the ASTCA, provision for them is available in other legislation modified by the new law. Particularly notable are inspections of reception facilities authorised by the Secretary of Commerce to determine whether conditions there meet standards and measures necessary for compliance with the Protocol.[53] Additionally, ship inspections may be performed to determine if violations have occurred of Annex IV or V of the Environmental Protocol. Discharge of a harmful substance in circumpolar waters that violates the 1973 International Convention for the Prevention of Pollution from Ships, as modified by the Protocol of 1978 (MARPOL 73/78),[54] is a violation of Annex IV to the Antarctic Protocol and is made subject to penalties.[55] Similarly, inspections may proceed to determine whether disposal of garbage by a US vessel in Antarctic waters has violated Article 5 of Annex IV to the Protocol.[56] US law in fact permits the Department of Commerce to inspect at any time a ship of 'United States registry or nationality or operating under the authority of the United States' to verify whether

---

[49] Antarctic and Southern Ocean Coalition, *ASOC Information Paper No. 2, Upon Closer Inspection* (1992), p. 1.
[50] Protocol, Art. 14(1).
[51] *Ibid.*, Art. 14(4).
[52] ASTCA, section 106.
[53] USC, Vol. 33, section 1905(f).
[54] ILM, Vol. 12, 1973, pp. 1,319ff (Convention); and ILM, Vol. 17, 1978, pp. 546ff (Protocol).
[55] USC, Vol. 33, section 1907(c).
[56] USC, Vol. 33, section 1907(d)(1).

that vessel has discharged either harmful substances or garbage into the Antarctic marine ecosystem.[57]

Regarding *emergency response action* (provided for in Article 15 of the Protocol), the unlawful nature of prohibited acts is excepted under US law if the 'person committing the act reasonably believed that the act was committed under emergency circumstances involving the safety of human life or of ships, aircraft, or equipment or facilities of high value, or the protection of the environment'.[58] These exceptions are not intended to function as loopholes in the law whereby activities impacting the Antarctic environment can be conducted without penalty, but rather they aim at providing room for more expeditious life-saving and other rescue operations in emergency circumstances.

Although *dispute settlement* procedures are stipulated in the Protocol, the ASTCA does not explicitly mention dispute settlement. The decision on how to proceed with resolving a dispute with a foreign government over issues in the Protocol is thus remanded to the US Department of State, rather than to the legislative prerogatives of Congress.

US PRACTICE SINCE 1996

US government agencies moved impressively to publicise and effect rules and regulations that enforce compliance with the ASTCA. This can be seen in three main developments: 1) the EPA's promulgation of certain rules for governing the conduct of non-governmental activities; 2) NSF's promulgation of certain procedures for regulating the conduct of US governmental activities; and 3) the proposal by both the EPA and NSF during 1998 of new regulations designed to strengthen existing rules governing the conduct of activities in the Antarctic.

The EPA's Interim Final Rule

The ASTCA directs the US government to implement federal regulations for governing the conduct of nationals in the Antarctic. This regulatory responsibility for activities by US non-governmental operators fell to the Environmental Protection Agency.

The EPA is required under the ASTCA to promulgate regulations that are consistent with Annex I to the Protocol and provide for: (a) the environmental impact assessment of non-governmental activities in Antarctica (including tourism)

---

[57] USC, Vol. 33, section 1907(e). Violation of Annex IV to the Protocol is designated a class D felony, and carries certain civil penalties. A person who violates Annex IV or the regulations issued thereunder is liable to a fine not to exceed USD 25,000 for each violation; moreover, false, fictitious or misleading statements given to Commerce Department officials are subject to additional penalties of up to USD 5000 each. USC, Vol. 33, section 1908(b).

[58] ASTCA, Title I, section 103(c), *Statutes*, Vol. 110, section 3036; USC, Vol. 16, section 2403(c).

and (b) coordination of review of information concerning environmental impact assessments received from other parties under the Protocol.[59] During 1996 and early 1997, the EPA investigated circumstances in the Antarctic that could affect or be affected by such regulations.[60] Then, on 30 April 1997, the EPA issued an interim final rule that establishes requirements for non-governmental operators in evaluating and assessing impacts from their activities, including tourism, on the Antarctic environment and its dependent and associated ecosystems.[61] The United States is required under Article VII(5) of the Antarctic Treaty to give notice of all expeditions organised in or proceeding from the United States. Although the rule applies to all non-governmental operators, the vast majority of non-governmental expeditions are carried out by organisations that sponsor *tourism* to the Antarctic.[62]

The interim final rule, upon entry into force of the Protocol, gave the United States legal capability to implement obligations in the Protocol. Even so, the interim final rule was intended to be limited in time and effect before adoption of a final rule. Thus, the interim rule only applied to non-governmental activities through the 1998/1999 austral summer. However, representatives from the affected industry and environmental non-governmental operators requested that more time be allowed so that greater experience could be obtained before a final rule was promulgated. As a result, on 14 July 1998, the EPA extended application of its interim final rule through the 2000–2001 austral summer.[63]

The Environmental Protection Agency's interim rule proffers clues to future US environmental policy in the Antarctic. The rule specifies procedures for performing EIAs of non-governmental activities, including tourism, and thus carries environmental significance. Not only does the rule serve as an inventory for US non-governmental activities bound for Antarctica, but it also sets out requirements for assessments and coordination of those activities and their impacts upon Antarctic ecosystems.

The requirements of the EPA's interim final rule apply to any person who organises or conducts a non-governmental expedition that proceeds from the United States (i.e., by an operator), or that is organised in the United States. The activities involved in this application of the rule are primarily related to ship-borne tourism. Promulgation of the interim rule was not expected to adversely affect the numbers of visitors to Antarctica or visits to sites frequently visited by US tour operators, nor

---

[59] ASTCA, Title I, section 104(c), *Statutes*, Vol. 110, section 3038; USC, Vol. 16, section 2403(c).

[60] See generally US Environmental Protection Agency, *Environmental Assessment of Proposed Interim Rules for Nongovernmental Activity in Antarctica*, 24 March 1997, available at <http://es.epa.gov/oeca/ofa/rcticea.html>.

[61] Environmental Protection Agency, CFR, Title 40, Part 8, 'Environmental Impact Assessment of Nongovernmental Activities in Antarctica', *Federal Register*, Vol. 62, 30 April 1997, pp. 23,538–23,549.

[62] On the issues involved in regulation of Antarctic tourism in the ATS, see Richardson, Chapter 4 in this book.

[63] US Environmental Protection Agency, 'Direct Amendment to Interim Final Rule', *Federal Register*, Vol. 63, April 15, 1998, pp. 18,323–18,326, at website <http://es.epa.gov/oeca/ofa/amend/html>.

was the rule expected to affect activities conducted at those sites; thus far, it appears that the promulgated rule has not had any negative effects on any of these circumstances.

Two other positive aspects of the promulgation of the rule are particularly notable. Firstly, the interim final rule allows for the documentation of planned migration and monitoring by tour operators. Secondly, the interim final rule was intended to facilitate the collection of data concerning the effects and intensity of activities by non-governmental visitors to the Antarctic. The clear intent here is to obtain a more regular, verifiable record of US activities in the Antarctic and to gauge more accurately their impacts upon Antarctic ecosystem.

The interim final rule stipulates requirements for documentation to ensure that all proposed activities are assessed so that the level of impact on the Antarctic environment can be determined. The assessment, documentation, disclosure and review facets of the interim final rule provide for the development of tour and expedition plans that must reflect knowledgeable analysis of the Antarctic environment. For all US non-governmental activities deemed to have less than minor or transitory impacts on the Antarctic environment, a US operator must provide a Preliminary Environmental Review Memorandum (PERM) at least 180 days before departure of an expedition to Antarctica. That PERM is reviewed by the Environmental Protection Agency to determine whether more detailed assessments are required, as authorised in Article 3 of Annex I to the Protocol.[64]

If a proposed activity is thought to have at least a minor or transitory impact on the Antarctic environment, an initial environmental evaluation (IEE) will be necessary (as provided for under Article 2 of Annex I to the Protocol), and the operator has 75 days to respond to the EPA. Documentation for an IEE should include projected impacts of activities associated with expeditions, such as air emissions, discharges into the ocean, noise from engines, landings for sightseeing and activities by visitors near wildlife.

For US activities thought to have greater than minor or transitory impacts on the environment, a comprehensive environmental evaluation (CEE) is required. In the EPA's view, a CEE could be triggered by either a proposed activity that entails 'a major departure from current non-governmental activities, resulting in a large increase in adverse environmental impact at a site' or by an activity that 'is likely to give rise to particularly complex, cumulative, large-scale or irreversible effects', such as perturbations in unique and very sensitive biological systems. For example, a CEE might be required to consider construction and operation of a new crushed-rock airstrip or runway.[65] The CEE is expected to provide a detailed analysis that comprehensively evaluates, *inter alia*, the activity, its impacts, any alternatives to

---

[64] *Federal Register*, Vol. 63, p. 23,541.

[65] Environmental Protection Agency, CFR, Title 40, part 8, 'Environmental Impact Assessment of Nongovernmental Activities in Antarctica', *Federal Register*, Vol. 42, 30 April 1997, p. 23,542.

and possible mitigation of the activity.⁶⁶ The final decision to proceed with a US activity for which a CEE has been prepared will not be taken until after the draft CEE has been considered by an Consultative Meeting, upon the advice of the CEP.⁶⁷

If a CEE is required, drafts of the CEE must be distributed to all Consultative Parties 120 days in advance of the subsequent Consultative Meeting at which the CEE will be addressed.⁶⁸ The interim rule mandates that the EPA, in consultation with other agencies (namely, the US Department of State and National Science Foundation) will review the CEE to determine whether it meets the requirements set out under Annex I to the Protocol.

The EPA interim final regulations are directed to apply to operators of non-governmental (tourist) expeditions organised in or proceeding from the United States to Antarctica. The regulations do not apply to individual US citizens who are on an expedition, but who are not the operator of that expedition. Nor do the regulations apply to US citizens on tours that are organised in or proceed from countries other than the United States or that are run in conjunction with governmental activities.⁶⁹

The interim final regulations require an operator to undertake procedures that assess and provide a verifiable record of actual impacts of any activity that proceeds on the basis of an IEE or a CEE. Collection of this data should be helpful in minimising any impacts on the environment. Operators must monitor key environmental indicators in order to assess and verify the impacts of activities for which an IEE or a CEE has been prepared. Violation of these interim regulations is unlawful, and US operators who violate the regulations are subject to criminal and civil penalties.⁷⁰

Upon receipt of a CEE from another Antarctic Treaty party, the US Department of State is to circulate copies to all interested federal agencies and coordinate responses of federal agencies back to the Treaty party. Copies of CEEs are to be made available to the public upon request. The US Department of State is also responsible for circulating information concerning IEEs and CEEs to all interested federal agencies as well as to the CEP and other Consultative Parties.

The EPA's interim final rule produces certain benefits, including improved awareness by US non-governmental operators of the goals, objectives and motives for Antarctic environmental protection. By promulgating such a rule, the US government standardises the approach to documenting environmental events and circumstances and stipulates requirements and procedures for undertaking EIAs. These conditions should reduce possible conflicts between US governmental and

---

⁶⁶ CFR, Title 40, Part 8, p. 23,540.
⁶⁷ *Ibid.*
⁶⁸ See Art. 3(4) of Annex I to the Protocol.
⁶⁹ *Federal Register*, Vol. 62, p. 23,540.
⁷⁰ Pursuant to sections 7–9 of ACA, as amended by the ASTCA. See also USC, Vol. 16, sections 2407–2409 and CFR, Title 45, Part 672.

non-governmental activities, as well as promote greater public trust and credibility in the industry regarding non-governmental activities in the Antarctic.

The NSF Environmental Assessment Procedures

Under the ASTCA, government activities are overseen and regulated by NSF.[71] As mandated in the implementing legislation, NSF in April 1997 issued its environmental assessment procedures for US governmental activities in the Antarctic.[72]

The assessment procedures aim to elicit and evaluate information so that NSF can be better informed about the possible environmental consequences from actions proposed by the United States Antarctic Program (USAP). Such information should more fully integrate environmental considerations into final decisions on whether to proceed with an activity.

NSF is supposed to ensure that the environmental effects of activities in Antarctica are appropriately identified and considered within the decision-making process.[73] To this end, a preliminary environmental review must be performed of each proposed action to consider its potential direct effects and reasonably foreseeable indirect effects on the Antarctic environment. If an action might have at least a minor or transitory impact on Antarctic environment, the NSF official responsible for preparing environmental documents must prepare either an IEE or a CEE, depending on the nature of the proposed action. The essential elements to be considered are whether and to what extent the proposed action might lead to any of the following impacts: adversely affect the Antarctic environment; adversely affect air and water quality; affect atmospheric, terrestrial, glacial and marine environments; detrimentally affect the distribution, abundance or productivity of species or populations of flora or fauna; further jeopardise endangered or threatened species; degrade or put at substantial risk areas of biological, scientific, historic, aesthetic or wilderness significance; or pose highly uncertain environmental effects or unique environmental risks.[74]

NSF has determined that some actions pose less than a minor or transitory impact on the Antarctic environment, and are therefore excluded from the NSF procedures for environmental impact assessment. The excluded activities include: 1) scientific research activities that involve the 'low volume collection of biological or geologic specimens, provided no more mammals or birds are taken than can normally be replaced by natural reproduction in the following season'; 2) small scale

---

[71] This regulatory function is carried out under the authority in the implementing legislation delegated to NSF to issue permits for the conduct of certain government activities in Antarctica. See ASTCA, Title I, section 105. See also USC, Vol. 16, section 2402.

[72] CFR, Title 45, Vol. 3, Part 641, sections 641(10)–641(22).

[73] *Ibid.*, section 641(11).

[74] *Ibid.*, section 641(16).

detonation of explosives related to seismic research in the continental interior of Antarctica where no potential exists for impact on native flora and fauna; 3) use of weather balloons, research rockets and retrievable automatic weather stations; and 4) use of radioisotopes, provided such use complies with applicable laws and regulations. Interior remodelling and renovation of facilities are excepted from NSF's list of excluded activities.[75]

The NSF procedures require that an IEE be undertaken if a proposed action may have more than a minor or transitory impact on the Antarctic environment. These procedures include information concerning the following: a description of the action, including its purpose, location, duration and intensity; consideration of alternatives to the action; and the potential impacts of the action on the Antarctic environment.[76]

The purpose of carrying out a CEE is to allow informed consideration about 'reasonably foreseeable potential environmental effects of a proposed action' and possible alternatives to such action. CEEs should contain, *inter alia*: 1) a description of the action, including its purpose, location, duration and intensity; 2) consideration of alternatives to the action; 3) a description of initial base-line environmental state to which predicted changes can be compared; 4) a description of methods and data for forecasting potential impacts; 5) an estimate of the 'nature, extent, duration, and intensity' of direct, indirect and cumulative potential impacts of proposed action; 6) identification of measures, to include monitoring, that might be used to minimise or prevent potential impacts of the proposed action; 7) consideration of potential impacts of proposed action on conduct of scientific research; and 8) identification of gaps in knowledge and uncertainties in compiling information.[77]

These NSF procedures do not apply to emergency situations concerning the safety of human life, or safety to ships, aircraft or equipment and facilities of high value; nor do the procedures apply when protection of the environment requires taking quick action without completion of an environmental review.[78] Thus, emergency and rescue operations can proceed immediately, unfettered by protracted regulatory or bureaucratic delays.

Recent Regulatory Developments

Recent regulatory developments indicate that the United States is broadening the regulation of its activities in Antarctica. In June 1998, the National Science Foundation announced its proposed rule for the conservation of Antarctic animals and plants,[79] which implements Annexes II and V to the Protocol. The proposed rule

---

[75] *Ibid.*, section 641(16)(c).
[76] *Ibid.*, section 641(17).
[77] *Ibid.*, section 641(18)(b).
[78] *Ibid.*, section 641(22).
[79] National Science Foundation, 'Conservation of Antarctic Animals and Plants', CFR, Title 45, part

would amend certain parts of the Antarctic Conservation Act of 1978 to bring it into conformity with the ASTCA.

Regarding Annex II, the proposed NSF rule makes the following activities unlawful for any 'person' under US jurisdiction: 1) engage in taking any native mammal, bird or plant; 2) engage in harmful interference; 3) enter Antarctic specially designated areas; 4) possess, sell, export or import native mammals, birds, or plants; 5) introduce non-indigenous animals or plants; or 6) violate permit conditions.[80] The rule stipulates procedures for permit applications, including the criteria for issuance, possession, denial, and evocation.[81] Also, the proposed rule enumerates designation of 23 species of native mammals, 44 species of birds, and 6 native plants.[82] The proposed NSF rule also stipulates conditions for issuing permits, including those necessary to take native mammals, birds and plants,[83] to enter Antarctic Specially Protected Areas,[84] and to import into and export from the United States native mammals, birds or plants.[85]

In June 1998, NSF proposed another rule that would require US tour operators who used non-US-flagged vessels for Antarctic expeditions to ensure that the vessel owner has an emergency response plan.[86] The need for this regulation is clear: Article 15 to the Protocol ('Emergency Response Action') requires that each party provide for prompt and effective response action to emergencies that might arise from activities in the Antarctic, including emergencies from tourism and other non-governmental activities. On 15 April 1998, the US Coast Guard issued regulations to implement Article 15 for US-flagged vessels operating in the Antarctic.[87] Because some US tour operators may charter non-US-flagged vessels for Antarctic expeditions, these regulations mandate that non-US flagged vessels used by American tour operators have on shipboard oil pollution emergency response plans that are consistent with Article 15. This proposed regulation also requires that US tour operators notify passengers of their obligations under the 1978 ACA.[88]

Finally, in June 1998, the National Science Foundation announced hikes in civil monetary penalties that can be imposed for inadvertent or deliberate violations of the Antarctic Conservation Act of 1978. Applied to the tourism guidelines, after 31 July

---

670; *Federal Register*, Vol. 63, 2 June 1998, pp. 29,963–29,970.

[80] *Federal Register*, Vol. 63, section 670(4). Exceptions are made in cases of 'extraordinary circumstances,' including under emergency conditions involving the safety of human life or equipment and aiding or salvaging native mammals or birds for scientific study. *Ibid.*, section 670(5).

[81] *Federal Register*, Vol. 63, sections 670(11)–(15).

[82] *Ibid.*, sections 670(19)–(21).

[83] *Ibid.*, section 670(23).

[84] *Ibid.*, section 670(26).

[85] *Ibid.*, sections 670(31)–(34).

[86] National Science Foundation, 'Antarctic Tourism', CFR, Title 45, parts 672 and 673; *Federal Register*, Vol. 63, 4 June 1998, pp. 30,438–30,440.

[87] See CFR, Title 33, Part 151.

[88] National Science Foundation, 'Antarctic Tourism', CFR, Title 45, parts 672 and 673; *Federal Register*, Vol. 63, 4 June 1998, pp. 30,438–30,440.

1999 the maximum civil penalties rose to USD 12,000 for unintentional violations and USD 25,000 for intentional violations.[89] These increases doubled the penalties for violations of the ACA and stand as a notable deterrent for future US operators. That these increases are concomitant with new rules and regulations by the EPA and NSF suggests that the US government intends to strengthen compliance with and enforcement of laws that aim at stronger environmental protection in the Antarctic.

CONCLUSION

US implementing legislation for the Antarctic Environmental Protection Protocol promotes US interests in protecting the value of Antarctica for scientific research. Enactment by Congress in 1996 of Public Law 104–227, ASTCA, brought the US government and persons under its jurisdiction into compliance with that instrument.

The new legislation amends the Antarctic Conservation Act of 1978 to make the regulation of US research activities in the Antarctic consistent with the requirements in the Protocol. This is accomplished by specifying definitions, prohibited acts, and stipulations for environmental impact assessments. The National Science Foundation remains the lead agency for managing the Antarctic science program and for issuing regulations and research permits. The law further amends the Antarctic Conservation Act to allow the use of established procedures of the 1969 National Environmental Policy Act to meet the Protocol's mandate for comprehensively assessing and monitoring effects of both governmental and non-governmental activities on the fragile Antarctic ecosystem. The ASTCA prohibits the introduction of prohibited products into Antarctica, the open burning of wastes, and the disposal of wastes onto ice-free land areas or into fresh water systems in Antarctica; a permit is required for any incineration, waste disposal, entry into special areas and takings or harmful interference. The ASTCA alters the Antarctic Protection Act to continue indefinitely a ban on Antarctic mineral resource activities and also amends the Act to Prevent Pollution from Ships to implement provisions of the Environmental Protocol that relate to the protection of marine resources, so that the Act conforms to MARPOL 73/78.

The Environmental Protocol obligates parties to certain conditions on the conduct of scientific and logistical support activities in the polar south. The ASTCA fulfils these obligations for the United States and enhances the prospects for more effectively protecting the environment of Antarctica and conserving the continent as a laboratory for the conduct of research that is vital to understanding the global environment. To this end, the recent adoption of more comprehensive rules and

---

[89] National Science Foundation, 'Antarctic Conservation Act of 1978, Civil Monetary Penalties', *Federal Register*, Vol. 63, 16 June 1998, pp. 32,761–32,762; CFR, Title 45, Part 672.

regulations intimates that environmental considerations have become increasingly salient in the formulation of US law and policy for the Antarctic. The greatest challenges for the US government in the future will be in monitoring compliance with these laws and ensuring enforcement of them as legal mandates.

Index

Adventure Network International, 86
African National Congress, 401
Agreed Measures for the Conservation of Antarctic Fauna and Flora, 34, 47, 48, 108, 234, 309, 347, 364
Amery Ice Shelf, 314
Antarctic and Southern Ocean Coalition, 24, 42, 79, 290
Antarctic Convergence, 7, 165, 246, 276, 411
Antarctic Environmental Officers Network, 116
Antarctic Management Committee (South Africa), 403
Antarctic Peninsula, 42, 72, 84, 97, 98, 166, 167, 263, 304, 318, 348
Antarctic Policy Group (USA), 425
Antarctic Site Inventory Project, 80
Antarctic Specially Managed Areas, 37, 81, 88, 280, 312
Antarctic Specially Protected Areas, 280, 281
Antarctic Treaty, 1, 5, 6, 7, 12, 13, 16, 17, 21, 33, 38, 46, 65, 94, 125, 140, 142, 143, 202, 207, 222, 263, 264, 265
Antarctic Treaty area, 7, 12, 16, 25, 31, 33, 53, 85, 164, 169, 174, 181, 182, 183, 184, 185, 189, 190, 193, 202, 207, 213, 233, 251, 254, 255, 264, 266, 279, 281, 297, 298, 299, 300, 302, 322

Antarctic cases, 263, 271
Aramco case, 59
Arctic, 8, 9, 81, 246, 247, 249, 250, 254, 255, 259
Arctowski station, 84
Areas of Special Tourist Interest, 74
Argentina, 12, 40, 45, 84, 97, 98, 101, 125, 130, 179, 181, 196, 200, 253, 263, 275, 292
Asylum case, 237, 238, 240–241
Aurora Australis, 319
Australia, 2, 12, 17, 56, 78, 87, 102, 125, 130, 137, 168, 173, 199, 253, 262, 263, 268, 275, 292, 297, 309–335
Australian Antarctic Division, 302, 321
Australian Antarctic Foundation, 262
Australian Antarctic Territory, 56, 63, 262, 313, 324
Australian Capital Territory, 322, 324
Australian Commonwealth, 319
Australian Fisheries Management Authority, 328
Arthur Harbour, 166

Bahia Paraiso, 166, 167, 175, 181, 182, 196, 199, 200
Balleny Islands, 281
baselines, 86, 249, 265, 269, 270
Belgium, 102, 125, 275
Bellinghausen Sea, 379
Bonaparte Point, 181

Bouvetøya, 379, 380, 388
Brazil, 179, 275
Bulgaria, 1, 275

Canada, 86, 176, 243, 247, 248, 249, 250, 251, 275
Casey base, 329
Chile, 12, 17, 45, 66, 75, 77, 97, 98, 101, 125, 171, 179, 193, 199, 200, 223, 253, 263, 271, 275, 292, 301, 337–353
Chilean Antarctic Institute, 346, 347, 348, 349, 352
Chilean SCAR Committee, 346
China, 130
Cold War, 274, 282, 283, 284
Commission on the Limits of the Continental Shelf, 16, 261, 262, 267, 268, 270, 271, 272, 314
Committee for Environmental Protection, 5, 8, 9, 13, 24, 30, 36, 54, 78, 94, 99, 100, 102–103, 104, 105, 106, 107–124, 129, 131, 135, 138, 140, 142, 152–158, 159, 222, 280, 281, 282, 290, 293, 301, 303, 305, 306, 307
comprehensive environmental evaluation, 30, 31, 32, 35, 78, 79, 119, 120–122, 131, 154, 203, 295, 299, 300
Consultative Meetings, 10, 13, 14, 15, 23, 24, 26, 29, 36, 46, 54, 65, 94, 104–106, 109, 126–127, 132, 134, 135, 136, 138, 150, 159, 164, 178, 180, 256, 273, 293, 300, 303, 305
 document distribution, 102, 112, 119, 127, 129, 141, 144, 145
 hosting, 113, 126, 128, 131, 137, 142–145, 150
 transparency, 102, 141, 147, 150, 156
 website, 105, 119, 121, 142–152, 155, 159
continental shelf, 16, 50, 248, 252, 261, 266, 270
 beyond 200 miles, 16, 248, 261–272, 313
contingency planning, 76, 86
Convention for the Conservation of Antarctic Seals, 16, 21, 48, 74, 100, 164, 169, 200, 208, 222, 251, 265, 274, 282, 311, 347
Convention on the Conservation of Antarctic Marine Living Resources, 2, 6, 7, 12, 16, 21, 22, 23, 24, 25, 28, 29, 33, 34, 35, 39, 41, 42, 43, 48, 49, 50, 51, 52, 53, 55, 56, 61, 67, 68, 74, 100, 101, 103, 104, 105, 107, 108, 111, 127, 136, 137, 139, 155, 165, 169, 200, 222, 236, 251–252, 265, 273–284
 Commission, 23, 28, 29, 35, 39, 42, 67, 100, 104–105, 128, 137, 139, 222, 175, 276, 279, 280, 281, 283
 conservation measures, 13, 23, 39, 42, 55, 67, 222, 275
 Scientific Committee, 24, 25, 35, 100, 104, 108, 109, 110, 111, 114, 116, 275, 280, 281
Convention on the Control of Transboundary Movements of Hazardous Wastes and Their Disposal, 126, 192, 193, 223, 224, 228–229, 233, 238, 239, 240, 369

Convention on the Law of Treaties, 39, 231, 232, 237, 238, 239
Convention on the Regulation of Antarctic Mineral Resource Activities, 2, 3, 4, 5, 6, 48, 50, 52, 53, 74, 100, 102, 108, 128, 165, 168–169, 170, 179, 180, 204, 213, 265, 274, 339
Council for Antarctic Policy (Chile), 346
Council on Environmental Quality (USA), 419
Council of Managers of National Antarctic Programmes, 13, 14, 24, 32, 36, 73, 100, 101, 103, 104, 105, 107, 110, 111, 116, 155, 164, 166, 167, 178, 180, 255
Covadonga Anchorage, 349

Davis base, 329
deep seabed, 193, 264, 268, 299
Denmark, 251
Department of Environmental Affairs and Tourism (South Africa), 399
departure state, *see under* jurisdiction
dependent and associated ecosystems, 7, 168, 182, 184, 185, 190, 195, 199. 202, 203, 206, 216, 234, 277, 278, 297, 298, 359, 363
Dronning Maud Land, 263, 379, 380, 298, 408

ecosystem approach, 25, 278, 411
Ecuador, 179
Eleventh Special Antarctic Treaty Consultative Meeting, 16, 53, 75, 102, 104, 105, 168, 169, 170, 274, 279, 284, 289, 296, 307

emergency response action, 200, 245, 429
enforcement, 11, 12, 17, 23, 30, 37, 48, 55, 76, 80, 114, 119, 175, 170, 216, 288, 296, 301, 303, 305, 307
environmental emergencies, 114, 175
environmental harm, 42, 128, 170
Environmental, Health and Safety Management System, 408
environmental impact assessment, 6, 16, 22, 23, 30, 31, 33, 34, 48, 55, 67, 76, 80, 81, 82, 86, 114, 115, 119, 120, 154, 175, 203, 304, 306
Environmental Protection Agency (USA), 418, 429
Esperanza station, 84
European Union, 235, 275
exclusive economic zone, 55, 56, 192, 244, 248, 252, 253, 259

Fildes Bay, 349
Finland, 64, 251, 275, 295
Fish Stocks Agreement, 41, 57, 58, 59, 60, 61, 67, 282
fishing vessels, 57, 59, 60, 68, 169
flag state, *see under* jurisdiction
flags of convenience, 65
Food and Agriculture Organisation, 24, 58, 61, 282
France, 2, 45, 75, 77, 80, 98, 102, 125, 168, 173, 263, 275

gateway ports, 12, 40, 65, 68, 89, 292, (*see also* jurisdiction, departure state)
General Agreement on Tasiffs and Trade, 60, 139

Germany, 40, 66, 75, 77, 82, 89, 171, 173, 174, 179, 275
Greenpeace, 73, 291
Greenwich Island, 349
Group of Specialists on Environmental Affairs and Conservation, 116, 117

Heard Island, 328
Historic Sites and Monuments, 84
Holland-America Line, 82, 85

ice shelves, 265, 269, 311
Iceland, 251
illegal, unregulated and unreported (IUU) fishing, 25, 29, 33, 43, 165, 264, 276
India, 275
informal meetings, 77
   on tourism, 77
information exchange, 142, 155, 304, 306
initial environmental evaluation, 203, 298, 299, 300, 408
inspection 49, 50, 53, 55, 62, 64, 67, 88
integrated environmental management, 407
Inter-Ministerial Polar Committee (Norway), 394
International Association of Antarctic Tour Operators, 26, 37, 73, 80, 82, 83, 84, 85, 86. 167, 259, 302, 306
International Centre for Antarctic Information and Research, 80, 88
International Civil Aviation Organisation, 222

International Convention for the Regulation of Whaling, 4, 16, 51, 53, 164, 169, 200, 233, 274
International Convention for the Safety of Life at Sea, 61, 66, 89, 244, 255
International Court of Justice, 186, 263, 271
International Database on Antarctic Tourism, 80, 88
International Geophysical Year, 96
International Hydrographic Organisation, 222, 259
International Law Commission, 186, 231
International Maritime Organisation, 15, 16, 24, 57, 61, 222, 243, 244, 247, 251, 254, 255, 256, 257, 258, 259, 349
   Special Area, 66
International Northern Sea Route Programme, 251
International Seabed Authority, 16, 268
Italy, 17, 61, 66, 77, 80, 82, 102, 173, 275, 355, 366
IUCN, *see* World Conservation Union

Japan, 247, 251, 275, 287, 297
Jervis Bay, 324
jurisdiction, 10-13, 17, 39, 45, 62, 138, 176, 178, 187, 188, 194, 202, 204, 206, 207, 212, 213, 214, 215, 216, 252, 292, 295, 300, 304, 305, 307
   aerial, 46, 53
   coastal state, 49, 55, 56, 60, 62, 67, 176, 252, 253, 265
   departure state, 12, 40, 68, 213, 253,

Index

flag state, 12, 39, 46, 47, 48, 49, 52, 55, 56, 64
maritime, 41, 46, 47, 53
port state, 40, 49, 56, 57, 61, 64, 76, 89, 176, 253, 292, 353

King George Island, 167
Korea, Democratic People's Republic of, 239
Korea, Republic of, 82, 275

liability for environmental damage, 105, 129, 163, 173, 180, 182–184, 194, 281, 294
  civil liability, 163, 183, 187–189, 190, 192, 199, 202, 204, 205, 207, 208, 210, 211, 214
  comprehensive approach, 46, 172, 194
  Eighth Offering, 170, 178, 179, 199, 202, 212
  environmental protection fund, 129, 172, 178, 203, 209, 210, 214, 215
  Group of Legal Experts on Liability, 14, 130, 163, 170, 171, 172, 173–178, 179, 180, 199, 208
  hazardous activities, 163, 194, 202
  insurance, 75, 172, 177, 179
  irreparable damage, 106, 172, 177, 178, 190, 192, 203, 209, 210, 214
  joint and several liability, 172, 176, 208
  liability *ex delicto*, 184–187, 204, 207, 213
  remedial measures, 170, 173, 192, 169, 203, 204, 205, 206, 210, 214, 215, 217
  residual state liability, 176, 207, 210, 212, 213–215
  response action, 173, 175, 177, 191, 196, 203–206, 209, 210, 214, 215, 217
  'source-oriented' liability regimes, 190, 194–195, 196
  standard of liability, 172, 176, 202, 208–209
  step-by-step approach, 172, 200, 201
  strict liability, 170, 176, 177, 192, 202, 208, 209, 210, 216
Lotus case, 61

Macquarie Island, 314, 315, 328
Malaysia, 294
Marine Expeditions, 85
Marion Island, 400, 411
Maritime Safety Committee (of the IMO), 16, 248, 254, 257, 258, 259
Maritime Territory and Merchant Marine Directorate (Chile), 349
Maritime Training Centre (Chile), 349
MARPOL 73/78, 40, 58, 61, 64, 65, 67, 88, 223, 233, 244, 255, 256, 257, 309, 371
Mawson base, 329
McDonald Islands, 328
measures under Article IX of the Antarctic Treaty
  decisions, 29, 88, 95, 133, 149, 159, 222
  measures, 26, 28, 62, 95, 127, 133, 139, 140, 222, 256

recommendations, 47, 48, 62, 73, 83, 86, 103, 112, 144, 222, 256
resolutions, 29, 88, 95, 133, 149, 159, 222
minor or transitory impact, 6, 8, 25, 30, 34, 204, 206, 299, 300
monitoring, 7, 30, 86, 99, 164, 166, 222
Mount Erebus, 72, 187–188
Multiple-use Planning Areas, 74
MV *Marco Polo*, 82
MV *Rotterdam*, 82, 85

National Committee for Environmental Impact Assessments (Chile), 346
National Environmental Commission (Chile), 346, 351
National Oceans and Atmospheric Administration (USA), 419
National Public Works Department (South Africa), 404
National Science Foundation (USA), 80, 433
Netherlands, 65, 80, 82, 85, 89, 171, 173, 174, 179, 199, 200, 275, 295, 297, 298
New Zealand, 12, 63, 64, 80, 88, 102, 125, 145, 168, 173, 179, 263, 277, 281, 292, 295, 297, 298, 299, 300, 307
non-governmental organisations, 3, 4, 54, 73, 74, 75, 77, 78, 85, 88, 101, 206, 210, 279, 290, 293, 294, 306, 407
Northern Sea Route, 247, 249, 250, 251
Northern Territory, 324
Norway, 15, 17, 45, 97, 110, 125, 141, 146, 149, 150, 152, 251, 258, 263, 275, 292, 293, 297, 298, 300, 301

Norwegian Antarctic Research Expedition, 379
Norwegian Polar Institute, 150, 155, 158, 301, 302, 387, 391

objective regime, 41
Orient Lines, 82, 84

Pakistan, 26, 33
Palmer station, 84, 181
Paradise Bay, 349
Paradise Harbour, 410
Patagonian toothfish, 12, 42, 165, 227, 264, 277, 279, 282, 284
Patriot Hills, 353
Permanent Commission of the South Pacific, 62
permit system, 32, 34, 35
Peru, 142, 152, 179, 275
Peter I Øy, 379, 380
Poland, 84, 275
Polar Class ships, 244, 245
Polar Code, 16, 243–259
Polar Congress, 338
Port Lockroy, 84, 410
Prat base, 342, 349
precautionary principle, 25, 244, 300
precautionary measures, 204, 205, 206, 214, 215, 217
preliminary assessment, 32, 298
Preliminary Environmental Review Memorandum, 431
Prince Edward Islands, 400, 410, 413
  Management Committee, 400, 403, 410, 412
  Management Plan, 400, 412
prior assessment, 25, 32, 87, 298

Protocol on Environmental Protection
to the Antarctic Treaty, 3, 53, 63,
69, 71, 77, 89, 94, 100, 128, 182,
199, 205, 222, 252, 273, 287, 289,
293, 294, 296
- Andersen draft, 103
- Annex I, 5, 30-32, 34, 81, 104, 115,
120, 122, 123, 131, 235, 279,
280, 296, 298, 303, 333, 346
- Annex II, 5, 34, 48, 104, 108, 115,
234, 235, 279, 280, 304, 333,
347, 357
- Annex III, 5, 104, 115, 211, 233,
280, 303, 348, 357, 368
- Annex IV, 5, 27, 40, 48, 62, 64, 67,
89, 104, 115, 211, 233, 256,
257, 280, 294, 309, 348
- Annex V, 5, 28, 34, 36, 78, 81, 88,
104, 115, 234, 235, 254, 280,
281, 349
- annex on liability, 14, 172, 177, 189,
200, 201, 202, 204, 205, 207,
209, 212, 281
- draft annex on tourism, 77, 78
- environmental principles, 7, 31, 234,
235, 278, 289, 297, 420
- mining ban, 6, 128, 164, 193, 194,
234, 267, 289, 299, 307, 333,
406

Puerto Soberania base, *see* Prat base

Quantas, 322

Reconstruction and Development
Programme (South Africa), 401
Region XII, of Magallanes (Chile), 342,
352
Ross Ice Shelf, 314
Ross Sea, 72, 84
Rules of Procedure, 101, 104, 109–113,
118, 123, 270, 271
Russia, 164, 247, 250, 251, 275, 283
SA *Agulhas*, 404, 409
Scientific Committee on Antarctic
Research, 14, 24, 36, 73, 100,
101, 103, 104, 105, 107, 109, 110,
111, 114, 116, 117, 155, 174, 175,
180, 221, 281
scientific research, 25, 27, 31, 42, 46,
49, 52, 71, 74, 166, 174, 208, 235,
295, 307
self-executing treaties, 17, 345
Shackleton Ice Shelf, 314
shipping, 16, 243, 247, 254, 255, 259
ships,
- *Bahia Paraiso*, 166, 167, 175, 181,
182, 196, 199, 200
- MV *Marco Polo*, 82
- MV *Rotterdam*, 82, 85
- SA *Agulhas*, 404, 409

South Africa, 12, 17, 125, 275, 292,
295, 399, 403
South African Committee for Antarctic
Research, 404
South African National Antarctic
Expedition , 410
South African National Antarctic
Programme, 400
South Atlantic Treaty Organisation, 402
South Georgia Islands, 55, 253
South Sandwich Islands, 55, 253
South Shetland Islands, 81, 167, 343,
348
Southern Ocean, 12, 16, 25, 43, 98, 165,
165, 222, 227, 229, 233, 236, 238,

246, 264, 275, 276, 278, 279, 261, 282
sovereignty, 10, 27, 41, 45, 97, 248, 251, 252, 253, 264, 267, 270, 300, 305
  claims, 39, 131, 164, 178, 264, 228, 251, 252, 263, 266, 267, 271, 272, 278, 287, 292, 298, 305
  disputes, 194, 251, 270, 275
Sovereignty Bay, 349
Soviet Union, 52, 101, 125, 249, 250, 283, (see also Russia)
Spain, 77, 200, 275
Svalbard, 81, 396
Sweden, 8, 83, 251, 275, 295, 297
Switzerland, 139, 221

Terra Australis, 337
third states, 67, 165, 188, 206, 257, 249, 293, 294
tourism, 10, 12, 26, 27, 32, 36, 40, 67, 71, 73, 81, 164, 167, 174, 215, 259, 300, 304, 307
tourist vessels, 42, 65, 66, 72, 82, 84, 88, 167, 181, 257, 324
Transitional Environmental Working Group, 104, 109, 280, 290
Turkey, 239

Ukraine, 275, 283
United Kingdom, 12, 40, 45, 55, 63, 64, 66, 74, 75, 78, 79, 80, 82, 98, 102, 125, 137, 179, 193, 253, 254, 255, 257, 263, 271, 283, 297, 298, 299, 300
United Nations, 24, 71, 93, 139, 221, 222, 223, 230, 240, 275

General Assembly, 2, 3, 4, 15, 221, 224
'Question of Antarctica', 15, 127
Secretary-General, 101, 127, 225, 268
United Nations Conference on Environment and Development, 229
United Nations Conference on the Human Environment, 224
United Nations Conference on the Law of the Sea, Third, 269
United Nations Convention on the Law of the Sea, 7, 16, 58, 60, 65, 66, 67, 89, 190, 193, 223, 252, 255, 256, 259, 261, 262, 263, 265, 266, 267, 268, 269, 270, 271, 272, 282, 288, 289, 301, 315, 406
United Nations Educational, Scientific and Cultural Organisation, 136
United Nations Environment Programme, 15, 192, 222, 223, 224–230, 238, 239, 241
United States, 17, 27, 40, 78, 79, 80, 82, 84, 85, 86, 87, 97, 98, 101, 102, 125, 130, 158, 164, 171, 173, 179, 187, 188, 200, 247, 250, 251, 254, 259, 275, 283, 292, 297, 298, 301
United States Antarctic Program, 433
Uruguay, 47, 179, 275

World Conservation Union, 24, 42, 225
World Meteorological Organisation, 222, 259
World Trade Organisation, 60

ENVIRONMENT & POLICY

1. Dutch Committee for Long-Term Environmental Policy: *The Environment: Towards a Sustainable Future.* 1994 ISBN 0-7923-2655-5; Pb 0-7923-2656-3
2. O. Kuik, P. Peters and N. Schrijver (eds.): *Joint Implementation to Curb Climate Change. Legal and Economic Aspects.* 1994 ISBN 0-7923-2825-6
3. C.J. Jepma (ed.): *The Feasibility of Joint Implementation.* 1995
ISBN 0-7923-3426-4
4. F.J. Dietz, H.R.J. Vollebergh and J.L. de Vries (eds.): *Environment, Incentives and the Common Market.* 1995 ISBN 0-7923-3602-X
5. J.F.Th. Schoute, P.A. Finke, F.R. Veeneklaas and H.P. Wolfert (eds.): *Scenario Studies for the Rural Environment.* 1995 ISBN 0-7923-3748-4
6. R.E. Munn, J.W.M. la Rivière and N. van Lookeren Campagne: *Policy Making in an Era of Global Environmental Change.* 1996 ISBN 0-7923-3872-3
7. F. Oosterhuis, F. Rubik and G. Scholl: *Product Policy in Europe: New Environmental Perspectives.* 1996 ISBN 0-7923-4078-7
8. J. Gupta: *The Climate Change Convention and Developing Countries: From Conflict to Consensus?* 1997 ISBN 0-7923-4577-0
9. M. Rolén, H. Sjöberg and U. Svedin (eds.): *International Governance on Environmental Issues.* 1997 ISBN 0-7923-4701-3
10. M.A. Ridley: *Lowering the Cost of Emission Reduction: Joint Implementation in the Framework Convention on Climate Change.* 1998 ISBN 0-7923-4914-8
11. G.J.I. Schrama (ed.): *Drinking Water Supply and Agricultural Pollution.* Preventive Action by the Water Supply Sector in the European Union and the United States. 1998 ISBN 0-7923-5104-5
12. P. Glasbergen: *Co-operative Environmental Governance: Public-Private Agreements as a Policy Strategy.* 1998 ISBN 0-7923-5148-7; Pb 0-7923-5149-5
13. P. Vellinga, F. Berkhout and J. Gupta (eds.): *Managing a Material World.* Perspectives in Industrial Ecology. 1998 ISBN 0-7923-5153-3; Pb 0-7923-5206-8
14. F.H.J.M. Coenen, D. Huitema and L.J. O'Toole, Jr. (eds.): *Participation and the Quality of Environmental Decision Making.* 1998 ISBN 0-7923-5264-5
15. D.M. Pugh and J.V. Tarazona (eds.): *Regulation for Chemical Safety in Europe: Analysis, Comment and Criticism.* 1998 ISBN 0-7923-5269-6
16. W. Østreng (ed.): *National Security and International Environmental Cooperation in the Arctic – the Case of the Northern Sea Route.* 1999 ISBN 0-7923-5528-8
17. S.V. Meijerink: *Conflict and Cooperation on the Scheldt River Basin.* A Case Study of Decision Making on International Scheldt Issues between 1967 and 1997. 1999
ISBN 0-7923-5650-0
18. M.A. Mohamed Salih: *Environmental Politics and Liberation in Contemporary Africa.* 1999 ISBN 0-7923-5650-0
19. C.J. Jepma and W. van der Gaast (eds.): *On the Compatibility of Flexible Instruments.* 1999 ISBN 0-7923-5728-0
20. M. Andersson: *Change and Continuity in Poland's Environmental Policy.* 1999
ISBN 0-7923-6051-6

ENVIRONMENT & POLICY

21. W. Kägi: *Economics of Climate Change: The Contribution of Forestry Projects*. 2000
 ISBN 0-7923-6103-2
22. E. van der Voet, J.B. Guinée and H.A.U. de Haes (eds.): *Heavy Metals: A Problem Solved?* Methods and Models to Evaluate Policy Strategies for Heavy Metals. 2000
 ISBN 0-7923-6192-X
23. G. Hønneland: *Coercive and Discursive Compliance Mechanisms in the Management of Natural Resourses*. A Case Study from the Barents Sea Fisheries. 2000
 ISBN 0-7923-6243-8
24. J. van Tatenhove, B. Arts and P. Leroy (eds.): *Political Modernisation and the Environments*. The Renewal of Environmental Policy Arrangements. 2000
 ISBN 0-7923-6312-4
25. G.K. Rosendal: *The Convention on Biological Diversity and Developing Countries*. 2000 ISBN 0-7923-6375-2
26. G.H. Vonkeman (ed.): *Sustainable Development of European Cities and Regions*. 2000 ISBN 0-7923-6423-6
27. J. Gupta and M. Grubb (eds.): *Climate Change and European Leadership*. A Sustainable Role for Europe? 2000 ISBN 0-7923-6466-X
28. D. Vidas (ed.): *Implementing the Environmental Protection Regime for the Antarctic*. 2000 ISBN 0-7923-6609-3; Pb 0-7923-6610-7

KLUWER ACADEMIC PUBLISHERS – DORDRECHT / BOSTON / LONDON